Aluminum Alloys: Properties, Processes and Applications

Aluminum Alloys: Properties, Processes and Applications

Edited by
Nathaniel Gordon

STATES
ACADEMIC PRESS
www.statesacademicpress.com

Published by States Academic Press,
109 South 5th Street,
Brooklyn, NY 11249, USA

ISBN: 978-1-63989-041-5

Cataloging-in-Publication Data

Aluminum alloys : properties, processes and applications / edited by Nathaniel Gordon.
p. cm.
Includes bibliographical references and index.
ISBN 978-1-63989-041-5
1. Aluminum alloys. 2. Aluminum alloys--Mechanical properties. 3. Light metal alloys. I. Gordon, Nathaniel.
TN775 .A48 2022
669.722--dc23

For information on all States Academic Press publications
visit our website at www.statesacademicpress.com

STATES
ACADEMIC PRESS

Contents

Preface

Aluminum alloys are the alloys in which the predominant metal is aluminum. Copper, magnesium, silicon, tin, zinc and manganese are the typical alloying elements. Such alloys are classified into two branches namely casting alloys and wrought alloys. These can be sub-divided into heat-treatable and non-heat-treatable alloys. Primarily aluminum is used for wrought products such as rolled plate, foils and extrusions. Aluminum alloys are majorly used in aerospace manufacturing and engineering structures due to their corrosion resistance and lightweight. This book consists of contributions made by international experts. It brings forth some of the most innovative concepts and properties of aluminum alloys. A number of latest researches have been included to keep the readers up-to-date with the global concepts in this area of study.

This book has been the outcome of endless efforts put in by authors and researchers on various issues and topics within the field. The book is a comprehensive collection of significant researches that are addressed in a variety of chapters. It will surely enhance the knowledge of the field among readers across the globe.

It gives us an immense pleasure to thank our researchers and authors for their efforts to submit their piece of writing before the deadlines. Finally in the end, I would like to thank my family and colleagues who have been a great source of inspiration and support.

Editor

1

Anodisation of Aluminium Alloys by Micro-Capillary Technique as a Tool for Reliable, Cost-Efficient, and Quick Process Parameter Determination

Daniela Nickel,[1,2] Dagmar Dietrich,[1] Roy Morgenstern,[1] Ingolf Scharf,[1] Harry Podlesak,[3] and Thomas Lampke[1]

[1]*Materials and Surface Engineering Group, Technische Universität Chemnitz, 09107 Chemnitz, Germany*
[2]*Center of Engineering Materials, State Materials Testing Institute Darmstadt (MPA), Chair and Institute for Materials Science (IfW), Technische Universität Darmstadt, 64283 Darmstadt, Germany*
[3]*Composite Materials Group, Technische Universität Chemnitz, 09107 Chemnitz, Germany*

Correspondence should be addressed to Dagmar Dietrich; dagmar.dietrich@mb.tu-chemnitz.de

Academic Editor: Gianluca Percoco

Anodisation is essential for improving surface properties of aluminium alloys and composites regarding wear and corrosion behaviour. Optimisation of the anodising process depends on microstructural constituents contained in aluminium alloys and represents a key task, consisting of the control of process parameters and electrolyte formulation. We applied the micro-capillary technique known from corrosion studies and modified it to form anodic aluminium oxide films on high-strength aluminium alloys in comparison to pure aluminium in sulphuric acid. A glass capillary with an opening of 800 μm in diameter was utilized. Corresponding electrochemical measurements during potentiodynamic and potentiostatic anodisation revealed anodic current responses similar to conventional anodisation. The measurement of film thickness was adapted to the thin anodised spots using ellipsometry and energy dispersive X-ray analysis. Cross sections prepared by focused ion beam milling confirm the thickness results and show the behaviour of intermetallic phases depending on the anodising potential. Consequently, micro-capillary anodising proved to be an effective tool for developing appropriate anodisation conditions for aluminium alloys and composites because it allows quick variation of electrolyte composition by applying low electrolyte volumes and rapid film formation due to short process durations at small areas and more flexible variation of process parameters due to the used set-up.

1. Introduction

The generation of microstructural constituents in tailor-made aluminium based materials resulted in substantial progress achieved in development and application of this material group. Kinds of microstructural constituents can be dislocation arrangements and grain boundaries (high- and low-angle boundaries), solid solution formation, and arrangements of intermetallic compound (IC) particles of varying sizes. As regards the latter, these second-phase particles in aluminium alloys affect the material behaviour depending on their size: constituent particles of fairly large size, dispersoids of mostly intermediate size or fine-scaled hardening precipitates. For example, the dominant constituent phase in AlZn5.5MgCu alloys is Al_7Cu_2Fe; additionally existent are (Al, Cu)$_6$(Fe, Cu); Al_6Fe; Mg_2Si, α-$Al_{12}Fe_3Si$, and SiO_2 [1]. Most of the second-phase particles possess complex chemistry, crystal structure, and morphology. Dispersoids in the AlZn5.5MgCu alloy show a variety of rod-like, triangular, or globular shapes [1]. The general properties of this alloy are mainly determined by equilibrium phases like $MgZn_2$ (M-phase, η phase), $Al_2Mg_3Zn_3$ (T-phase), and Al_2CuMg (S-phase). Responsible for high strength are the η phase, its precursors GP zones, and intermediate precipitates (η' phase). The transition phases are semi-coherent and mostly located inside the matrix grains on dislocations and low-angle boundaries. The incoherent equilibrium phases nucleate on high-angle boundaries and in the grain interior [2]. The dispersoids

TABLE 1: Chemical composition of the studied materials according to the specification of the producer.

Alloy	Si (wt.%)	Fe (wt.%)	Cu (wt.%)	Mn (wt.%)	Mg (wt.%)	Cr (wt.%)	Zn (wt.%)	Ti (wt.%)	Al
Al 5N	0.00079	0.00098	<0.0002	0.00004	0.00103	0.000059	0	0.000098	Bal.
1050	<0.25	<0.4	<0.05	<0.05	0–0.05	<0.03	<0.07	<0.05	Bal.
2024	0.50	0.50	3.5–4.9	0.3–0.9	1.2–1.8	0.10	0.25	0.15	Bal.
7075	0.40	0.50	1.2–2.0	0.30	2.1–2.9	0.18–0.28	5.1–6.1	0.20	Bal.

$Al_2Mg_3Zn_3$ and Al_2CuMg are known to affect hot workability and fracture properties. Compared to T561 temper, η' phase and η phase are coarsened by artificial overaging in temper T73, which is beneficial to increase the resistance to stress-corrosion cracking [2–4].

Automotive and aerospace applications require enhanced corrosion resistance, improved tribological properties, or a simply improved decorative appearance. This is usually accomplished by the formation of an anodised aluminium oxide (AAO) due to the conversion of the surface region of the aluminium alloy [5]. AAO films on pure aluminium are fairly homogeneous showing hexagonal arranged pores perpendicular to the surface which results from the dynamic equilibrium between film growth at the metal/oxide interface and field-assisted dissolution at the pore bases [6]. In contrast, the microstructure and the resulting properties of AAO films formed on aluminium alloys are severely influenced due to the conversion of solid solutions and the behaviour of second phases during anodisation [7, 8]. In particular, anodising of copper-containing alloys is remarkably limited due to inefficient film growth and enhanced porosity of the produced AAO film [9, 10]. Dissolution of second-phase particles results in voids and gas-pockets [11, 12]. Second-phase particles with low oxidation rates can be incorporated into the film because of the faster oxidation of the surrounding matrix [12–14]. This observation was similarly made for AAO films formed on aluminium matrix composites (AMC) with alumina or silicon carbide particles [15, 16]. Anodic oxidation rates of a variety of ICs have been examined in relation to that of the matrix [17]. Recent studies considered the detailed processes occurring at ICs by simultaneously recording the electrochemical data. Film formation was performed by either potentiostatic (constant potential), galvanostatic (constant current density), or potentiodynamic anodisation (increasing potential with defined rate). The current density response during potentiodynamic anodisation and the time-voltage response during galvanostatic anodisation were discussed with regard to several commercial alloys, different electrolytes, and occurring growth stages during AAO film formation [12–14]. Potentiodynamic anodising was proposed as an efficient tool for the development of appropriate anodising parameters since it provides a rapid characterisation of the complex electrolyte/alloy behaviour over a wide potential range [18]. Such studies comprise the application of newly designed electrolyte formulations for anodising of aluminium alloys and aluminium matrix composites. Nevertheless, using conventional anodising for this task necessitates the application of large electrolyte volumes in the scale of several litres with the disadvantage of high expenses for procurement and disposal.

The aim of this work is to transfer the micro-capillary technique known from local corrosion research [19–23] to local anodising studies. Recently, this technique was sophisticated by developing multichannel arrangements for improving scanning droplet cell microscopy [24]. We applied the micro-capillary technique on pure aluminium and three commercial aluminium alloys. Sulphuric acid was chosen as a common electrolyte to provide the comparison to conventional anodising and to demonstrate the feasibility of the technique. The simultaneous electrochemical characterisation during initiation and progress of the film growth has been studied and correlated with microstructural features, in particular with the behaviour of second-phase particles and the development of the film thickness on the example of the high-strength and stress-corrosion resistant alloy EN AW-7075 in T73 condition that is artificially aged for providing stress-corrosion resistance. Film thickness measurement was adapted to thin and small AAO areas. Finally, a discussion on potentials and limitations of the presented method is given.

2. Materials and Methods

Pure aluminium (Al 5 N, Hydro Aluminium Rolled Products GmbH) and three commercial aluminium alloys in extruded condition and the given temper (EN AW-1050, EN AW 2024 T4, EN AW-7075 T651, and T73; for chemical composition see Table 1) were used. Samples were etched in 3 wt.% sodium hydroxide (NaOH) for 60 s at 50°C and desmutted in 30 vol.% nitric acid (HNO_3) for 30 s at room temperature. After anodising, the surface was rinsed carefully with deionized water and ethanol.

The samples were anodised in 0.4 M sulphuric acid (H_2SO_4) at room temperature (22 ± 0.5) °C using a micro corrosion cell (Swiss Microcell System) based on a glass capillary technique. Parts of the three-electrode set-up are a glass capillary with a ground tip of 800 μm in diameter and a seal of a silicone rubber layer to prevent leakage acting as anodising electrode, a 0.5 mm platinum wire acting as counter electrode, and a silver/silver chloride (3 M KCl) electrode acting as reference electrode. An exact positioning of the capillary was permitted by mounting the electrode set-up at the nosepiece of an optical microscope (OM) and fixing the samples at the stage of the OM. A detailed description of the microcell system has been published elsewhere [23] [BÖH95]. Each test was performed at least twice. Anodising was accompanied by micro-electrochemical measurements using a high precision voltage source (DIGISTANT Model 4462, Burster Präzisionsmesstechnik), a potentiostat (IMP83 PCT-BC, Jaissle Elektronik), and a digital multimeter (Model

2000, Keithley Instruments) providing a noiseless data acquisition with a rate of 9.1 Hz matching one point every 1.85 mV during the used sweep rate. Potentiodynamic anodising was done from the open circuit potential to 10 V at a sweep rate of 17 mV/s. Potentiostatic anodising was performed for 600 s using potentials between 0.3 V and 10 V. For comparison with conventional anodising, a three-electrode corrosion cell with the working electrode (exposed area 10 mm in diameter), Pt counter electrode, and Ag/AgCl (3 M KCl) reference electrode was used. Macro-scaled anodising was performed in stirred electrolyte at controlled temperature (22 ± 0.5) °C at an anodising potential of 10 V for 1200 s and 3600 s. The ratio of electrolyte volume to sample surface area was around 3 mL/mm^2 in the micro-capillary set-up and 50 mL/mm^2 in the macroscopic set-up. All potentials designated in this paper refer to the Ag/AgCl (3 M KCl) electrode if not indicated differently.

Microstructural studies were done using a scanning electron microscope (SEM, Zeiss NEON 40EsB crossbeam) equipped with a focused ion beam mill (FIB, Orsay) and an energy dispersive X-ray spectrometer (EDS, EDAX). Alloys were examined by low-voltage scanning transmission electron microscopy (LV-STEM) at 30 kV in combination with EDS. Electron transparent samples were prepared by grinding, electropolishing, and finally ion-polishing on both sides using 3 kV Ar ions with 6° incidence angle. AAO films were studied by scanning electron microscopy. Surface imaging with secondary electrons (SE) was done without carbon coating at 2 kV using Everhart-Thornley detector. EDS spectra were acquired at 5 kV. Cross sections of the 800 μm circular AAO film areas were prepared by focused ion beam (FIB) milling after deposition of a tungsten protective coating. 30 kV Ga ions with a current of 10 pA for the last polishing step were used. Imaging of the cross sections was accomplished at 3 kV using the InLens detector of the SEM.

Ellipsometry data for film thickness and porosity evaluation were collected with a rotating analyser ellipsometer (M-2000, J.A. Woollam Co.) in the spectral range from 1690 to 245 nm with incidence angles of 45, 50, 55, and 60°. The diameter of the polarised light beam was 1 mm resulting in an elliptical measuring area comparable to the anodised area. Modelling, regression analysis and parameter fit was done using the software CompleteEASE (J.A. Woollam Co). B-spline was used as polynomial function for the substrate and an effective medium approximation for the films. Modelling of the pores in the layer was done by a Cauchy layer of (100x) vol% alumina and x vol% air.

3. Results and Discussion

Potentiodynamic linear polarisation curves were recorded during anodisation in sulphuric acid in a micro-capillary cell arrangement (Figure 1). The curves of the alloys consist of four stages: (i) starting with a current peak around 0 V which is most pronounced for EN AW-2024; (ii) followed by monotone rising up to 3 V, (iii) a shallow peak between 3 and 6 V for EN AW-2024 and correspondingly between 4 and 9.5 V for EN AW-7075; (iv) further rising with different slopes

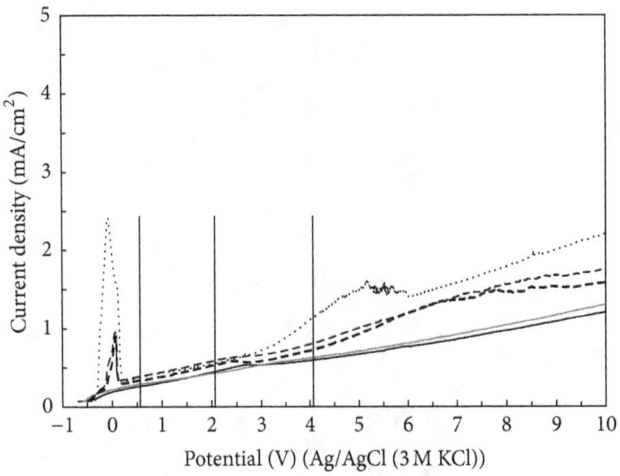

FIGURE 1: I/V behaviour of Al 5 N, EN AW-1050, EN AW-2024 T4, EN AW-7075 T73, and EN AW-7075 T651 during potentiodynamic anodising in 0.4 M sulphuric acid using the micro-capillary technique; vertical lines mark the potentials selected for potentiostatic anodising (as shown in Figures 4 and 5).

depending on the material. Different temper states of the alloy EN AW-7075 slightly affect the overall current density level and the peak width in (i).

As discussed in the Introduction, the alloy EN AW-7075 was selected for further experiments since it includes a large variety of Zn, Mg, Cu, Fe, and Si containing IC particles with high influence on the microstructure of the AAO film. The film after potentiodynamic anodisation is characterised by numerous voids due to the oxidation and dissolution of second-phase particles. Using secondary electrons, the arrangement of the hardening IC particles in the alloy is shown on the alloy surface before anodisation (Figure 2(a)). A cross section of the alloy with the AAO film on top is shown in Figure 2(c) after. Using backscattered electrons, grain boundaries of partially elongated grains of the alloy parallel to the extrusion direction are visible due to the orientation contrast. The coarsened hardening precipitates are located on the grain boundaries; finer intermetallic particles are located inside the grains. A rather corresponding arrangement of voids in the AAO film can be observed in Figure 2(b) in planar view. Larger voids seem to mirror former grain boundaries and finer voids are located inside the thus bordered areas. A large constituent particle appears mostly unchanged embedded in the film. A detailed image of the film morphology at higher magnification (Figure 2(d)) clearly reveals shape and size of the voids originating from the IC particles and the pore structure in between, respectively. The pores originally straight perpendicular directed to the alloy surface show a frequent change of the direction as a result of the wavy interface which evolves between the alloy and the anodised film. However, the pore structure is not as complex as it develops on EN AW-2024. The branching of pores is associated with copper enrichment during anodisation of the alloys which is delayed

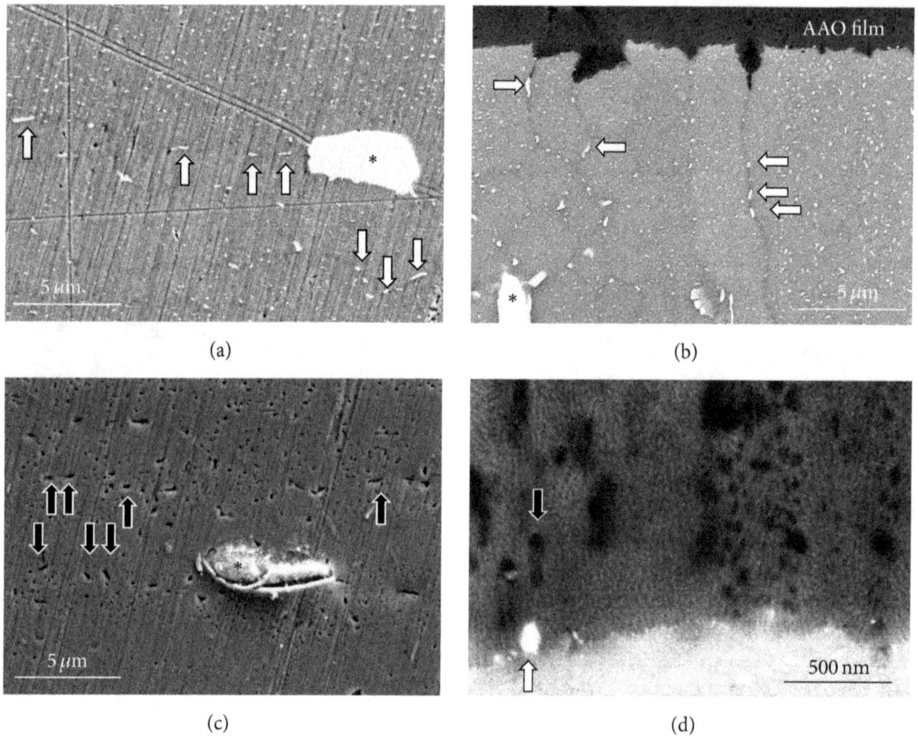

FIGURE 2: The microstructure of EN AW-7075 T73 in (a) planar view and (b) cross section (with the AAO film on top), and the AAO film formed by potentiodynamic anodising of this alloy in 0.4 M sulphuric acid in (c) planar view and (d) cross section showing the behaviour of coarse constituent particles resistant to anodising (marked by asterisks) and IC particles along grain boundaries (some ones are marked by white arrows) and in the grain interior resulting in voids in the AAO film (some ones are marked by black arrows).

due to the lower concentration of copper in EN AW-7075 and thus permits the development of straight pores to a larger extent.

TEM imaging of the grain interior of the alloy (Figure 3(a)) reveals a variety of dispersoids. Additionally, uniformly arranged hardening precipitates are shown. The grain boundary decorating particles were confirmed as η phase ($MgZn_2$). EDS analyses verified also the existence of a range of other ICs: Mg_2Si, $MgZn_2$, $Mg_{32}Zn_{49}$, Mg(AlCu), Al_2CuMg, and Al_7Cu_2Fe. However, it should be noted that it was not possible to identify all particles exactly by EDS analysis due to the excitation volume which mostly comprises the matrix too. Moreover, it was not possible to make a certain correlation between shape and stoichiometry. The AAO film is shown in the same magnification in Figure 3(b) giving evidence of a multitude of pore mouths typical for AAO films, larger voids as a result of complete dissolution of IC particles and embedded particles probably originating from chemically transformed ICs.

Potentiostatic anodisation experiments using the capillary technique were conducted to get a deeper insight into the film formation and the behaviour of second-phase particles in EN AW-7075. Four anodisation potentials were selected as indicated in Figure 1. The corresponding current density response as a function of the duration of anodisation is shown for anodising small spots under the micro-capillary (Figure 4) and for the first 600 s of the larger areas 10 mm in diameter (Figure 5). Both experiments show a fairly similar behaviour. Differences in the absolute current density are noticed only for both lower potentials which is attributed to the influence of electrolyte alteration in the micro-capillary set-up without a continuous electrolyte flow and stirring. The overall behaviour shows increasing current density with increasing anodising potential. Generally, an initial current density drop is followed by a plateau. This behaviour is associated with an uneven film growth in the initiation phase followed by a rather uniform thickening in the steady current density region. Curioni et al. [13] found a similar but more pronounced transient region between the initial current density drop and the plateau region for the alloy 2024-T3 above 3 V (SCE).

The AAO films formed by potentiostatic anodisation are shown in Figure 6. Their circular areas correspond to the capillary opening. The spots show distinct interference colours with regard to thickness of the formed films as expected from the variation of the applied charge transfer. Naturally, this is the case for all studied alloys as well as for pure aluminium. According to Faraday's law, the theoretically achievable thickness of alumina films on pure aluminium was calculated first. Several methods were used to determine the actual thickness on the alloy because the measurement of submicrometre films at fairly small areas with inhomogeneous composition and varying porosity is a nontrivial task. An estimation based on the interference colours was detailed by ellipsometry. The results were proved by SEM imaging on FIB cross sections and confirmed by EDS analysis in planar view.

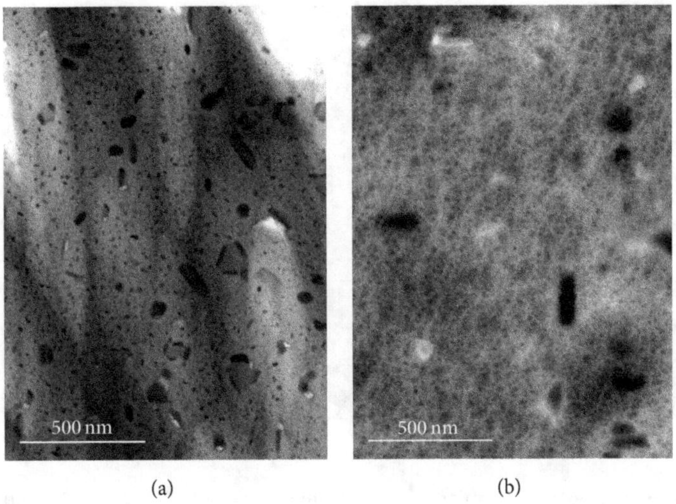

FIGURE 3: (a) LV-STEM image of IC particles in the grain interior of EN AW-7075 T73 and (b) SE image of a corresponding AAO film formed by potentiodynamic anodising of this alloy in 0.4 M sulphuric acid showing typical pore mouths, voids caused by IC particle dissolution, and (transformed) IC particles embedded in the film.

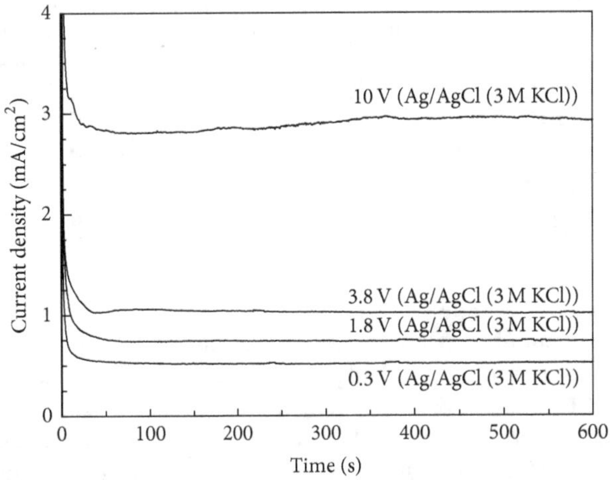

FIGURE 4: Current density-time behaviour during potentiostatic anodising in 0.4 M sulphuric acid on EN AW-7075 T73 using the micro-capillary technique to form circular oxide areas 800 μm in diameter.

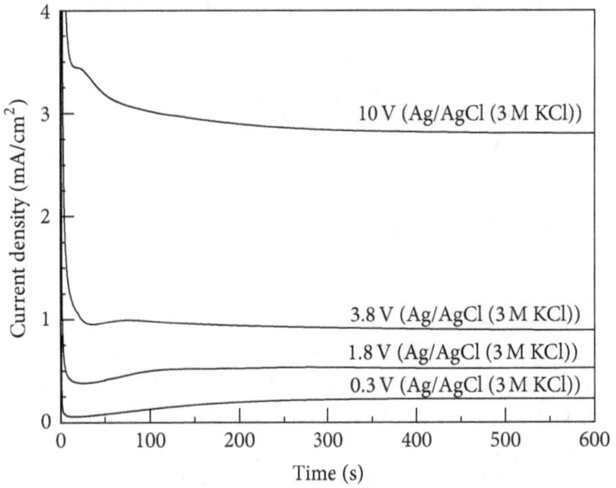

FIGURE 5: Current density-time behaviour during potentiostatic anodising in 0.4 M sulphuric acid on EN AW-7075 T73 using the conventional technique to form circular areas 10 mm in diameter.

The thickness of an anodic oxide film (d) formed on pure aluminium can be estimated according to Faraday's law of electrolysis using the quantity of charge transfer at the electrode (Q), the molar mass (M) of the formed film, the number of electrons transferred per ion (n), the Faraday constant (F), and the film density (ρ). Consider

$$d = \frac{QM}{nF\rho}. \quad (1)$$

Using the charge transfer according to the averaged current densities (Figure 4) and the anodisation time of 600 s, M = 0.102 kg/mol, n = 6 transferred electrons per reaction ($2Al^{3+} + 3O^{2-} \rightarrow Al_2O_3$), and ρ = 3940 kg/m^3 for bulk alumina, a film thickness d_F is calculated (Table 2, column 3). The figures obtained are a rough approximation. Calculated for pure aluminium, they represent the upper limit of alumina formation on one side. Otherwise, the density of the alumina film is overestimated, in particular due to the inherent pore formation in AAO films.

The interference colours of the AAO films vary from tan to blue and green under broadband incident light (Figure 6). The refractive index of alumina (corundum) is around 1.8 in the visible wavelength range which is between the corresponding refractive indices of silicon dioxide (~1.5) and silicon nitride (~2.0). Hence, it is reasonable to refer to corresponding colour/thickness charts for SiO_2 and Si_3N_4 related to microfabrication processes [26]. Accordingly, the tan colour indicates a thickness d_C of around 50 nm, the blue to bluish colour corresponds to values between 100 and 200 nm,

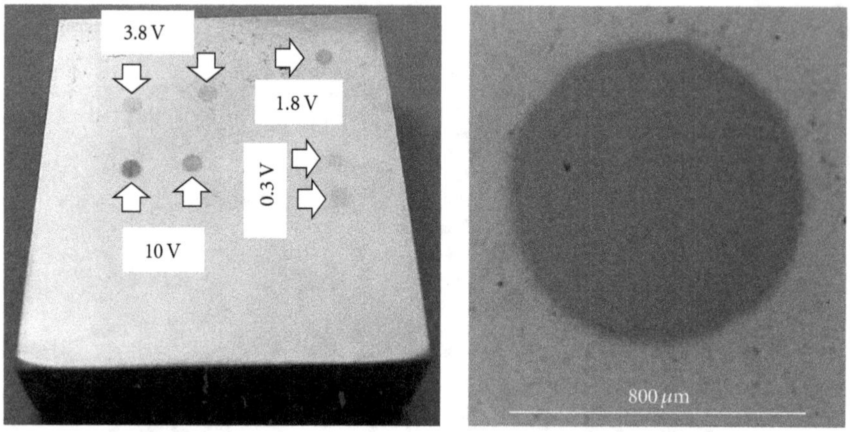

(a) (b)

FIGURE 6: Photograph of the circular areas formed by micro-capillary anodising of EN AW-7075 T73 in the potentiostatic regime in 0.4 M sulphuric acid at 0.3 V, 1.8 V, 3.8 V, and 10 V for 600 s showing interference colours from tan to blue and green under broadband incident light due to their film thickness (a) and an example in SEM imaging (b).

TABLE 2: Film thickness d_F calculated by Faraday's law (on pure Al), film thickness d_c according to colour, film thickness d_E estimated by ellipsometry, film thickness d_{SEM} measured on cross sections (all on EN AW-7075 T73), O/Al ratio according to the film/substrate contribution calculated using the EDS spectra, and O/Al ratio derived from EDS analysis as a function of the anodisation potential U (Ag/AgCl (3 M KCl)).

Film	U (V)	d_F (nm)	d_c (nm)	d_E (nm)	d_{SEM} (nm)	O/Al ratio calculated	O/Al ratio measured
1 (Figure 7(a))	0.3	150	~50	49 ± 15	25 ± 15	0.05	0.08 ± 0.05
2 (Figure 7(c))	1.8	205	~100	92 ± 15	85 ± 20	0.46	0.45 ± 0.05
3 (Figure 7(e))	3.8	275	~200	165 ± 15	145 ± 40	1.12	1.02 ± 0.05
4 (Figure 7(g))	10.0	745	~500	599 ± 15	590 ± 50	—	1.73 ± 0.05

and the green colour approximates 500 nm. Undoubtedly, the assumption of a uniform refractive index for all AAO films regardless of the anodising potential is not reliable because of differences in composition and porosity influencing their optical properties. This became apparent during the procedure of ellipsometry data analysis which iteratively converts the complex reflectance ratio as the primary data into the wavelength dependent optical constant and thickness. In the visible wave length region (i.e., 600 nm), the refractive index rises from 1.65 for the thinnest film formed at 0.3 V to 1.9 for the other films. Correspondingly, the mean film porosity rises from 43% for films formed between 0.3 V and 3.8 V to 49% for the film formed at 10 V; porosity gradients were not considered. The corresponding thickness d_E derived from the iterative procedure is shown in Table 2, column 5.

For verification, cross sections located in the centre of the anodised spots were prepared by FIB. The achieved cross sections are shown in Figures 7(a), 7(c), 7(e), and 7(g). It should be noted that, prior to ion milling, the AAO films were covered by a protection layer on top. This layer is a tungsten coating which was deposited to avoid the influence of the high-energy 30 kV Ga ions. Additionally and for further characterisation of the film microstructure, the surface of the AAO films is shown in Figures 7(b), 7(d), 7(f), and 7(h). The SEM images of the cross sections reveal the thickness d_{SEM} (Table 2, column 6). Differences between ellipsometry and SEM values are low and suspected to originate mainly from optical anisotropy. Nevertheless, variations in the film thickness cannot be excluded because the cross sections prepared by FIB show only a limited part of the centre of the film.

EDS analysis of a $500 \times 500\ \mu m^2$ large area in planar view was used to evaluate the homogeneity of the film thickness (Figure 8). An increasing oxygen signal in the EDS spectra with increasing film thickness is expected because the excitation volume as the source of the characteristic of O Kα and Al Kα X-rays includes the thin AAO films and the aluminium alloy below as well. Figure 8 shows to which extent film and substrate have to be considered in the case of the electron energy of 5 keV which is for certain high enough to excite Al Kα X-rays. Considering the paths of 500 electrons using the Monte Carlo (MC) simulation method, the mean excitation depth is around 300 nm for aluminium oxide and aluminium, as well. The calculated depth is nearly equal for both the film and the substrate because of the opposing influences of the mean atomic number and the density of the materials giving a spherical excitation volume of $8\ \mu m^3$. This means that excitation occurs completely in the film in the case of the thickest film formed at 10 V. The nominal O/Al ratio of stoichiometric alumina is 1.5, but the measured O/Al ratio of this film is 1.73 ± 0.05. This is not unexpected since AAO films are known to contain additional OH-groups at the pore walls. The O/Al ratio is reduced due to the rising contribution of the base material to the excitation volume in the case of the thinner films. Basing on the film thickness derived from

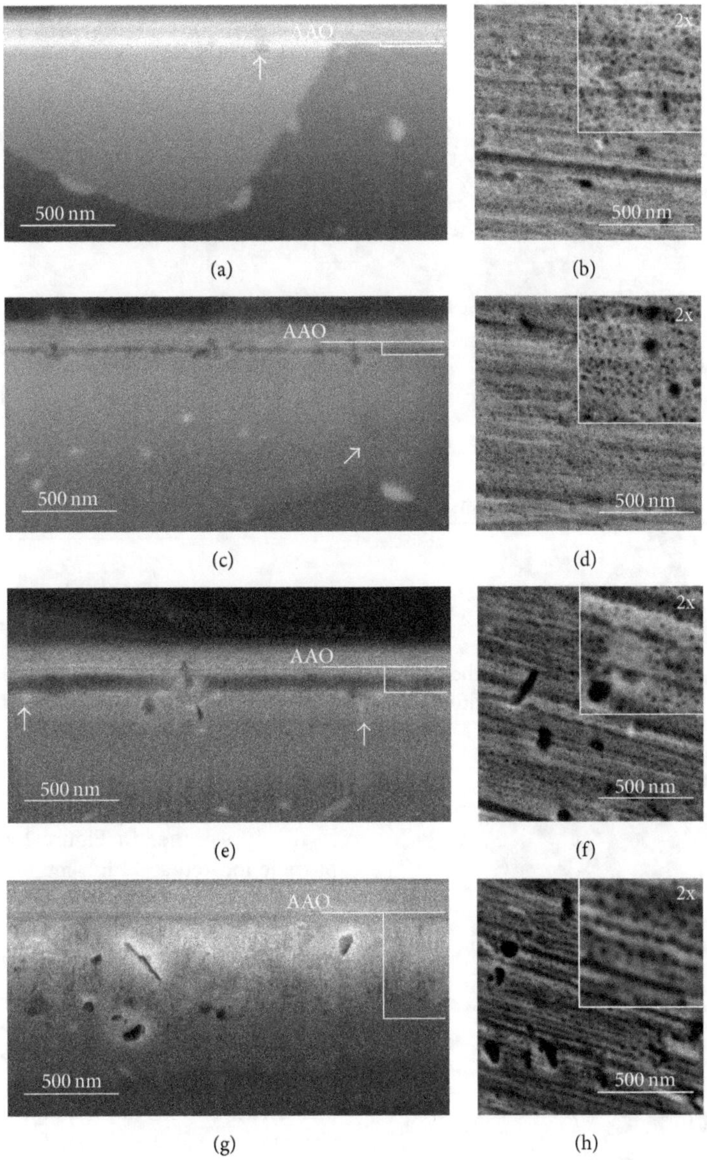

Figure 7: Cross section (left) and surface (right) of AAO films on EN AW-7075 T73 after potentiostatic anodising in 0.4 M sulphuric acid at (a, b) 0.3 V, (c, d) 1.8 V, (e, f) 3.8 V, and (g, h) 10 V; the AAO films are marked in the cross sections to clearly distinguish the protective coating on top (bars indicate 500 nm referring to identical magnification except for the twice enlarged inserts).

SEM imaging, the expected O/Al ratios are calculated and compared to the measured values (Table 2). Apart from the thinnest film whose thickness estimation remains uncertain, a satisfactory agreement in the limits of the accuracy of the method is stated. This in turn verifies that the films of localised and conventional anodisation have a fairly uniform thickness and the values obtained from cross sections by SEM are representative for the entire film.

The correlation between the transferred charge per unit area and the film thickness on potentiostatic anodised EN AW-7075 T73 is shown in Figure 9. The results of anodising a macroscopic area at 10 V for durations of 1200 s and 3600 s are added to demonstrate that the results achieved by the capillary technique are comparable to anodising in a thermostatically controlled conventional process at room temperature. Macroscopic anodising at lower anodisation potentials resulted in thin films comparable to the corresponding micro-capillary AAO films. However, because of the long duration times needed for macroscopic anodisation at low anodisation potentials, preponderance of film redissolution over film formation compromised the thickness evaluation.

The main purpose of the cross section preparation was the thickness evaluation of the potentiostatic anodised films. An additional benefit is provided by revealing the behaviour of the IC particles in dependence of the applied potential for film formation which is shown exemplarily for the AAO films

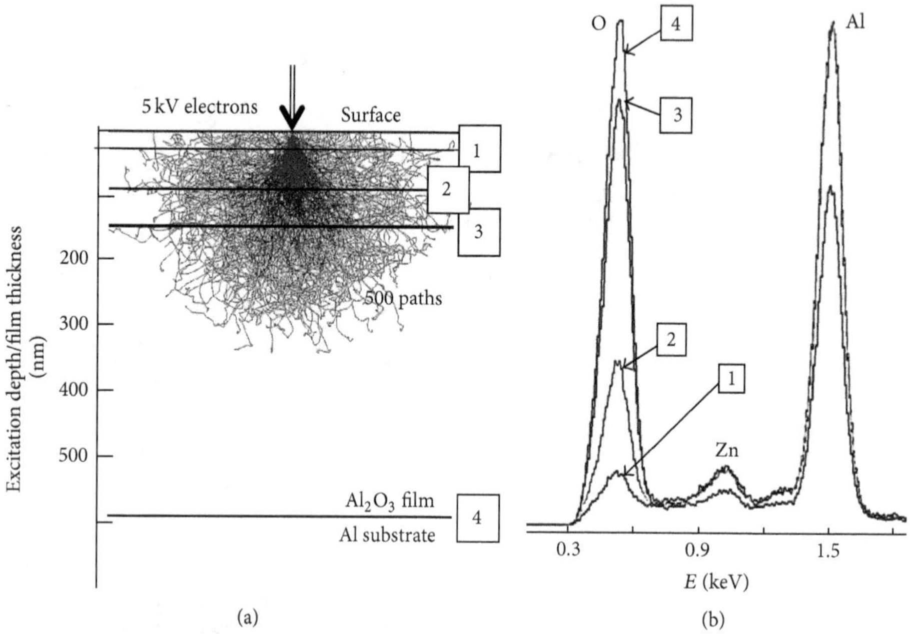

FIGURE 8: Monte Carlo simulated electron paths showing the excitation volume in the alumina films on aluminium (a) and the corresponding EDS spectra (referred to as numbered in Table 2) representing the dependence of O Kα and Al Kα intensities on film thickness (b).

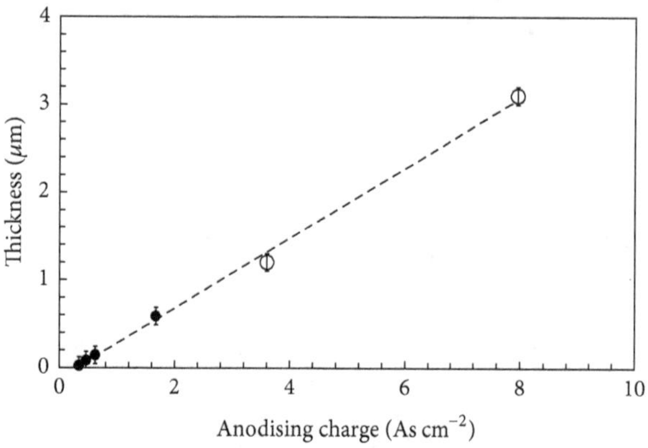

FIGURE 9: Thickness of the AAO films on EN AW-7075 T73 after potentiostatic anodising in 0.4 M sulphuric acid formed by the micro-capillary technique at 0.3 V, 1.8 V, 3.8 V, and 10 V (solid circles) for 600 s and by conventional anodising of circular oxide areas 10 mm in diameter at 10 V for 1200 s and 3600 s (open circles) as a function of the transferred charge per area unit.

on EN AW-7075 T73. For further characterisation of the film microstructure, the surface of the AAO films is shown in Figures 7(b), 7(d), 7(f), and 7(h).

The AAO film formed at 0.3 V (Figure 7(a)) is difficult to notice because of its low thickness. The dominating feature is a large iron-rich IC particle (most likely Al_3Fe or Al_7Cu_2Fe) in the alloy which shows a bright layer with a thickness similar to the AAO film thickness. The SE image of the surface shows shallow pores (Figure 7(b)).

As expected, the AAO film formed at 1.8 V (Figure 7(c)) is thicker and shows some voids. $MgZn_2$ particles in the alloy are supposed to decorate the grain boundaries corresponding to the appearance in Figure 2(a). Clearly visible is a larger particle located at a high-angle grain boundary near a triple junction in the right of the figure. At the site (marked by an arrow) where the corresponding vertical grain boundary cuts the surface, the AAO film formation proceeded concurrently with the chemical dissolution of the $MgZn_2$ phase. Similar particle dissolution that occurred simultaneously with the progress of AAO formation is observed in two other positions in the figure. At such locations, the film starts to become uneven around the developed voids. Similarly, those voids are also observed at the film surface together with a multitude of pore mouths (Figure 7(d)).

The AAO film formed at 3.8 V (Figure 7(e)) shows progressing film thickening and dissolution of a former large rod-like particle in the centre of the image. Interestingly, not only enclosed the segment of the former particle now by the formed AAO film, but also the other half is already completely dissolved. Varying chemical behaviour of diverse IC particles can be observed. Notably, some globular second-phase particles located at the film formation front and near the rod-like void appear unaltered. They will become embedded in the AAO layer, supposedly in unchanged or transformed state if the film formation would proceed. The surface of the film reveals evidence of such embedded particles additionally to pore mouths and voids (Figure 7(f)). The pore number is slightly reduced.

The AAO film formed at 10 V (Figure 7(g)) shows all morphological features known from conversion layers of the studied alloy. Apart from voids and transformed second-phase particles, irregularities of the pore morphology can be noticed. A pronounced uneven front of film growth has been developed due to the fast dissolution of second-phase

TABLE 3: IC particles in Al alloys and their behaviour during sulphuric acid anodisation.

Phase	Evidence (this work)	Behaviour during sulphuric acid anodisation (published)
Mg_2Si (β)	Verified	Dissolution, cavity formation [17]
$MgZn_2$ (η, M, σ)	Verified	Passivation above 5 V (SCE) [25]
$Mg_{32}Zn_{49}$ (T′)	Verified	Partial dissolution, partial oxidation [17]
Al_2CuMg (S)	Verified	Preferential oxidation above 0 V (SCE) [12]
Al_6Mn		Partial dissolution, embedment [17]
Al-2%Cu (α)		Yellow coating, enhanced porosity [17], copper enrichment at interface below 3 V (SCE) [9, 13]
Al_2Cu (θ)		Fast oxidation [17], preferential oxidation above 1.8 V (SCE) [12]
Al_7Cu_2Fe	Verified	Formation of Cu sponge [25], preferential oxidation above 1.8 V (SCE) [12]
Al_3Fe (β)		Partial dissolution, embedment [17], preferential oxidation above 2.2 V [12]
AlFeSi phases (α)		Partial dissolution, partial oxidation [17]

particles at the first contact with the electrolyte at the wavy interface. Branching of pores has proceeded with growing film thickness. Pore mouths at the surface are remarkably reduced in number and enhanced in diameter (Figure 7(h)).

4. Discussion

The pronounced peaks in the polarisation curve during potentiodynamic anodising (Figure 1) are associated with the oxidation and dissolution of IC particles in the aluminium alloys. Conventional "macro-area" anodising studies by Saenz de Miera et al. [12] and Curioni et al. [13] reveal similar potentiodynamic behaviour. This can be stated for all studied alloys but is particularly obvious in the case of EN AW-2024. According to [13], the striking peak around 0 V (SCE) is expected for EN AW-2024 and EN AW-7075 attributed to the S-phase (Al_2CuMg) with the peak intensity indicating the higher concentration in EN AW-2024.

Similarly shown for high purity aluminium in conventional experiments [12], no peaks appear in the curves of the micro-capillary cell experiment of Al 5 N and EN AW-1050 due to the absence of copper-containing second-phase particles. The thickening of the natural oxide on aluminium starts around 0.35 V (SCE) [12] and the following region of monotone current rise is connected with further oxide formation and film thickening.

According to the results of Keller et al. [6] and O'Sullivan and Wood [27], the thickness of the barrier layer and the diameter of the pores theoretically extend from 0.5 to 12 nm with rising anodising potential from 0.3 to 10 V. Correspondingly, the cell diameter enhances from 0.6 to 22 nm. Although these findings relate to pure aluminium, a qualitatively similar behaviour is observed for the pore and cell dimensions of the alloy EN AW-7075 T73 (Figures 7(b), 7(d), 7(f), and 7(h)). Corresponding to the pore and cell sizes in pure aluminium, the calculated pore volume decreases from 55% to 23% [27]. In practice, pore volume is generally greater than calculated because dissolution occurs at inner and outer surfaces of the coating during the time of anodisation. Additionally, the porosity in alloys is enhanced by voids formed due to the IC particle dissolution. Therefore considering the rising quantity of IC particles which underwent dissolution and leave additional voids in the film, porosity values between 43% and 49% derived from the ellipsometry study are sound and support the thickness results of ellipsometry.

The correlation between the transferred charge and the film thickness is similar comparing the results of micro-capillary anodising in a stagnant electrolyte and of macroscopic anodising in a thermostatically controlled and agitated electrolyte (Figure 9). A rise in temperature during the micro-capillary process might be expected due to the heat released above the spots inside the capillary. However, the transferred charge remains very low because of the small area to be anodised and the small duration (600 s) consequently needed. Nevertheless, parameter control by temperature stabilisation and composition balance of the electrolyte has to be kept in view for further development.

The anodising behaviour of second-phase particles was reviewed by Jelinek [17]. Recent detailed studies were published by Paez et al. [9], Saenz de Miera et al. [12, 14], Curioni et al. [13], and Tardelli and Rocca [25]. The phases with relevance to the studied alloys are listed in Table 3. Their verification by EDS on electron transparent samples of the alloy we focused on in this work is also indicated in the table. Although the micro-capillary cell technique has been applied for corrosion studies either on synthesised analogues of intermetallic compounds by Birbilis and Buchheit [22] modelling the second phases present in aluminium alloys or on a 7075 alloy in solution heat-treated condition and in T6 temper by Andreatta et al. [21], the applicability for local anodising was missing even after several modifications in the technical equipment. As presented for corrosion studies [21, 22], single precipitates could not be examined, but their galvanic coupling revealed changes from anodic to cathodic behaviour with respect to the temper state of the alloy matrix. Comparing our local results of anodising with the dissolution behaviour in chloride containing medium by Cavanaugh et al. [28] and Wloka and Virtanen [29], for example, similar behaviour can be stated. Cavanaugh et al. found anodic behaviour of the phases $MgZn_2$ and Mg_2Si and cathodic behaviour of the phases Al_7Cu_2Fe, Al_3Fe, and Al_2CuMg with respect to the matrix of AA7075 T6.

Wloka and Virtanen [29] confirmed anodic behaviour of $MgZn_2$ and cathodic behaviour of α-Al_7Cu_2Fe with respect to the matrix and evidenced that $MgZn_2$ is first dissolved due the high dissolution rate. Mg_2Si undergoes a selective dissolution

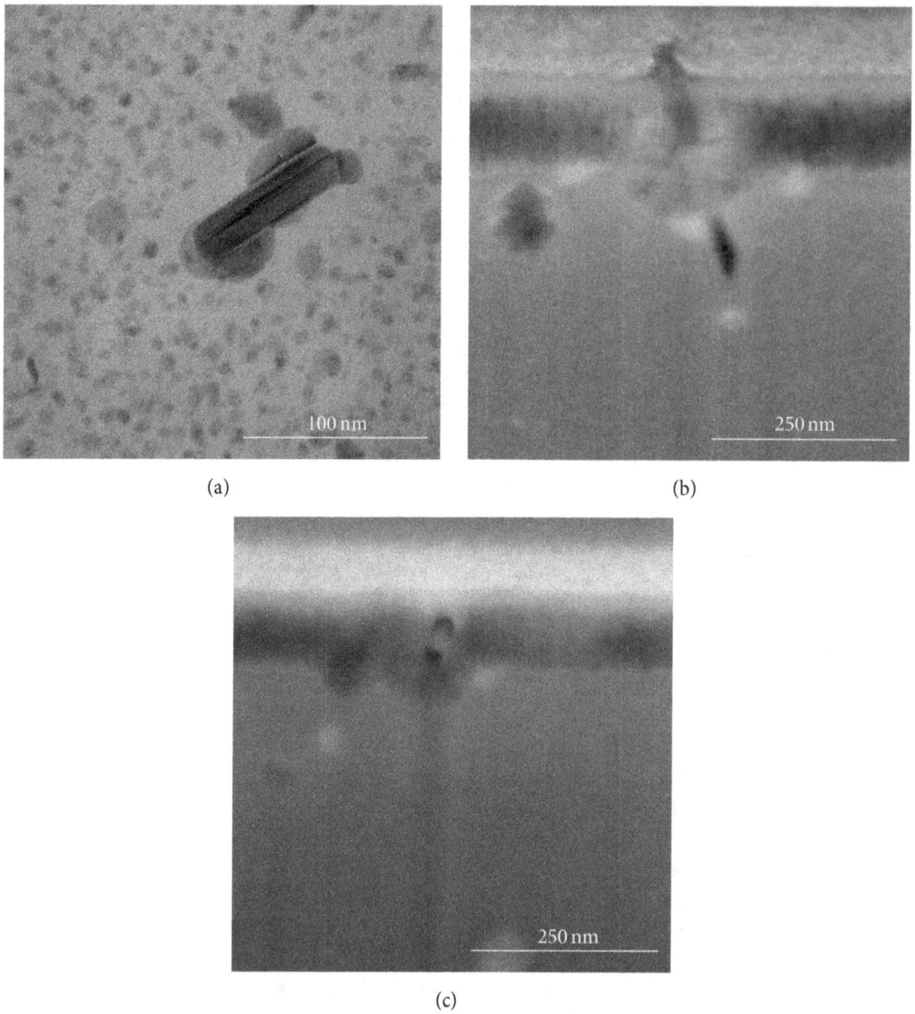

FIGURE 10: Comparison of a (a) coarsened $MgZn_2$ particle (η-phase) surrounded by Mg_2Si or $Mg_{32}Zn_{49}$ particles in the alloy EN AW-7075 with the corresponding appearance in the AAO film on that alloy after anodisation at 3.8 V showing (b) dissolution of $MgZn_2$ and (c) transformation of Mg_2Si to SiO_x.

of Mg leaving SiO_x particles behind; oxygen reduction takes place at the cathodic Al_7Cu_2Fe phase with pitting corrosion of the matrix. The behaviour of the second phases during anodisation in sulphuric acid shows similarities. Tardelli and Rocca [25] anodised model phases in sulphuric acid and evidenced selective dissolution of Al_7Cu_2Fe accompanied by formation of a nanometric sponge of copper at the surface. Bulk $MgZn_2$ formed an oxidic passivation layer above 5 V (SCE) with a high redissolution rate. Saenz de Miera et al. [12, 14] prepared intermetallic compound layers as macroscopic model specimens. The oxidation of the S-phase (Al_2CuMg) was found to start around 0 V (SCE) with the magnesium hindering the formation of a stable oxide. Oxidation of copper and iron containing particles like Al_7Cu_2Fe started above 2 V (SCE), but at reduced rate. Depending on their initial size, corresponding particles were suspected to be partially or completely oxidised supporting a relatively stable oxide [12]. We found a thin metallic film formation on the large constituent phase (apparently Al_7Cu_2Fe) after anodisation at 0.3 V which is correlated to the selective dissolution found by Tardelli and Rocca [25]. We observed fast and complete dissolution of $MgZn_2$ particles resulting in voids of corresponding shape in agreement with Tardelli and Rocca [25] and Cavanaugh et al. [28]. Transformed particles embedded in the AAO film are supposed to be mainly SiO_x as a result of selective Mg dissolution of Mg_2Si particles [17, 22, 28].

To illustrate this, $MgZn_2$ particles coarsened by overaging and surrounded by Mg_2Si or $Mg_{32}Zn_{49}$ particles were studied before and after anodisation by TEM and SEM imaging, respectively (Figure 10). The behaviour of the IC particles and their influence on the morphology of the anodised film is shown on the cross section after FIB milling (Figures 10(b) and 10(c)). Because of the combination of milling and imaging, the SE image quality is influenced by the compromises concerning the sample position with respect to the objective lens and moreover limited due to the beam sensitivity of the AAO film and the residual curtaining left after preparation. However, in contrast to metallographic preparation of cross sections, ion beam milling provides more detail concerning embedded particles inside of voids. Such particles regularly disappear during the grinding and polishing processes for sample preparation as can be seen in Figure 2(d). The TEM image in Figure 10(a) shows a coarse $MgZn_2$ particle with surrounding Mg_2Si or $Mg_{32}Zn_{49}$ particles in the alloy. The shape of a corresponding void in the oxide film after particle dissolution is evidenced in Figure 10(b), a magnified detail of Figure 7(e). The void is surrounded by relatively stable particles which seem to get anodised or transformed and embedded in the film (Figure 10(c)). This appears to be fairly similar to the behaviour of the IC particles observed under corrosive attack [28].

The microscopic study of the cross sections combined the study of the oxide morphology with the microstructure of the alloy below the formed AAO film. As a result of the presented work, above 1.8 V the dissolution of the hardening phase in EN AW-7075 T73 during anodisation produces a coating with numerous voids of the size of the IC particles and moreover with local thickness variations due to the dissolution of these microstructural constituents at the anodisation frontier.

One advantage of the used method is that electrochemistry and microstructure of the IC particles contained in the alloy and transformed by anodising can be studied correspondingly. Unfortunately, the micro-capillary method cannot offer the possibility to size down the anodised area to the size of single secondary particles. Using a potential minimal capillary diameter around 15 μm would at least enable the study of small areas with different phases side by side. The smaller the diameter of the capillary the more necessary a continuous exchange of the electrolyte to avoid alteration and to control temperature. This requires a continuous electrolyte flow in the micro-capillary cell as already presented for a multi-scanning droplet cell presented in [30] used as example for anodising titanium between 11 and 18 V.

5. Conclusions

The successful development of electrolyte formulations for anodising of aluminium alloys and aluminium matrix composites requires measurement of electrochemical parameters, thickness monitoring, and visualising the formation of the film and the transformation of the containing IC particles. Using conventional anodising for this task necessitates the application of large electrolyte volumes in the scale of several litres with the drawback of high expenses for procurement and disposal.

The micro-capillary technique provides an advantageous approach for a more efficient investigation of the anodising process. The influence of the electrolyte composition and anodisation regime on the electrochemical response of alloys with different composition and condition can be investigated by applying low electrolyte volumes and short anodisation durations at small areas. Methods of film evaluation (thickness, microstructure, and morphology) were adapted to the anodised spots of limited thickness (i.e., below 1 μm) and area (i.e., 0.5 mm^2). This was accomplished by using ellipsometry, energy dispersive X-ray analysis, and preparation of cross sections by focused ion beam milling and exemplarily shown for AlZn5.5MgCu. Thus, the method provides an efficient tool to study different electrolyte formulations. Additionally, insights into film formation and transformation of IC particles are provided by the applied techniques and can be extended to study particle strengthened aluminium matrix composites.

A comparison of the behaviour of the studied alloy during conventional anodising with a large circulated electrolyte volume at controlled temperature demonstrated the general utility and the limitations of the micro-capillary method. Further work will include the improvement of the microcell setup. In particular, a continuous electrolyte flow should be provided to avoid electrolyte alteration and heating.

Conflict of Interests

The authors declare that there is no conflict of interests regarding the publication of this paper.

Acknowledgments

Financial support was granted by Deutsche Forschungsgemeinschaft (German Research Foundation) in the collaborative research centre “High-Strength Aluminium-Based Lightweight Materials for Safety Components” (DFG SFB 692) and in the cluster of excellence “MERGE Technologies for Multifunctional Lightweight Structures” (DFG EXC 1075). The authors wish to thank Peter Richter and Professor D. R. T. Zahn, Semiconductor Physics, TU Chemnitz, for providing ellipsometry measurements and analyses. TEM preparation and study of the alloy were performed by Anne Schulze, Materials Engineering Group, TU Chemnitz. Thanks to Dagobert Spieler, Materials and Surface Engineering Group, TU Chemnitz, comprehensive electrochemical measurements, careful anodisation, and competent service of the microcell equipment have been accomplished.

References

[1] R. Ayer, J. Y. Koo, J. W. Steeds, and B. K. Park, “Microanalytical study of the heterogeneous phases in commercial Al-Zn-Mg-Cu alloys,” *Metallurgical Transactions A*, vol. 16, no. 11, pp. 1925–1936, 1985.

[2] S. T. Lim, I. S. Eun, and S. W. Nam, “Control of equilibrium phases (M,T,S) in the modified aluminum alloy 7175 for thick forging applications,” *Materials Transactions*, vol. 44, no. 1, pp. 181–187, 2003.

[3] D. J. Lloyd and M. C. Chaturvedi, “A calorimetric study of aluminium alloy AA-7075,” *Journal of Materials Science*, vol. 17, no. 6, pp. 1819–1824, 1982.

[4] J. K. Park and A. J. Ardell, “Microstructures of the commercial 7075 Al alloy in the T651 and T7 tempers,” *Metallurgical Transactions A*, vol. 14, no. 10, pp. 1957–1965, 1983.

[5] P. G. Sheasby and R. Pinner, *The Surface Treatment and Finishing of Aluminum and Its Alloys*, vol. 2, ASM International, Materials Park, Ohio, USA, 2001.

[6] F. Keller, M. S. Hunter, and D. L. Robinson, “Structural features of oxide coatings on aluminum,” *Journal of the Electrochemical Society*, vol. 100, no. 9, pp. 411–419, 1953.

[7] G. E. Thompson, H. Habazaki, K. Shimizu et al., "Anodizing of aluminium alloys," *Aircraft Engineering and Aerospace Technology*, vol. 71, no. 3, pp. 228–238, 1999.

[8] D. Nickel, T. Lampke, G. Alisch, and S. Steinhäuser, "Microstructural effect on the wear behaviour of the hard-anodised aluminium alloys EN AW-6082 and EN AW-7075," *Materialwissenschaft und Werkstofftechnik*, vol. 40, no. 7, pp. 523–531, 2009.

[9] M. A. Paez, O. Bustos, G. E. Thompson, P. Skeldon, K. Shimizu, and G. C. Wood, "Porous anodic film formation on an Al-3.5 wt % Cu alloy," *Journal of the Electrochemical Society*, vol. 147, no. 3, pp. 1015–1020, 2000.

[10] S. Feliu Jr., M. J. Bartolomé, J. A. González, V. López, and S. Feliu, "Passivating oxide film and growing characteristics of anodic coatings on aluminium alloys," *Applied Surface Science*, vol. 254, no. 9, pp. 2755–2762, 2008.

[11] Y.-S. Huang, T.-S. Shih, and J.-H. Chou, "Electrochemical behavior of anodized AA7075-T73 alloys as affected by the matrix structure," *Applied Surface Science*, vol. 283, pp. 249–257, 2013.

[12] M. Saenz de Miera, M. Curioni, P. Skeldon, and G. E. Thompson, "Modelling the anodizing behaviour of aluminium alloys in sulphuric acid through alloy analogues," *Corrosion Science*, vol. 50, no. 12, pp. 3410–3415, 2008.

[13] M. Curioni, M. Saenz de Miera, P. Skeldon, G. E. Thompson, and J. Ferguson, "Macroscopic and local filming behavior of AA2024 T3 aluminum alloy during anodizing in sulfuric acid electrolyte," *Journal of the Electrochemical Society*, vol. 155, no. 8, pp. C387–C395, 2008.

[14] M. Saenz de Miera, M. Curioni, P. Skeldon, and G. E. Thompson, "The behaviour of second phase particles during anodizing of aluminium alloys," *Corrosion Science*, vol. 52, no. 7, pp. 2489–2497, 2010.

[15] D. Dietrich, S. Nehrkorn, M. Händel et al., "A hardness–microstructure correlation study of anodised powder-metallurgical Al–Cu alloy composites," *Surface and Coatings Technology*, vol. 242, pp. 118–124, 2014.

[16] R. Morgenstern, D. Nickel, D. Dietrich, I. Scharf, and T. Lampke, "Anodic oxidation of AMCs: influence of process parameters on coating formation," *Materials Science Forum*, vol. 825-826, pp. 636–644, 2015.

[17] T. W. Jelinek, *Oberflächenbehandlung von Aluminium*, Eugen G. Leuze, Saulgau, Germany, 1997.

[18] M. Curioni, T. Gionfini, A. Vicenzo, P. Skeldon, and G. E. Thompson, "Optimization of anodizing cycles for enhanced performance," *Surface and Interface Analysis*, vol. 45, no. 10, pp. 1485–1489, 2013.

[19] H. Böhni, T. Suter, and A. Schreyer, "Micro- and nanotechniques to study localized corrosion," *Electrochimica Acta*, vol. 40, no. 10, pp. 1361–1368, 1995.

[20] M. M. Lohrengel, "Electrochemical capillary cells," *Corrosion Engineering Science and Technology*, vol. 39, no. 1, pp. 53–58, 2004.

[21] F. Andreatta, M. M. Lohrengel, H. Terryn, and J. H. W. de Wit, "Electrochemical characterisation of aluminium AA7075-T6 and solution heat treated AA7075 using a micro-capillary cell," *Electrochimica Acta*, vol. 48, no. 20-22, pp. 3239–3247, 2003.

[22] N. Birbilis and R. G. Buchheit, "Electrochemical characteristics of intermetallic phases in aluminum alloys—an experimental survey and discussion," *Journal of the Electrochemical Society*, vol. 152, no. 4, pp. B140–B151, 2005.

[23] T. Suter and H. Böhni, "A new microelectrochemical method to study pit initiation on stainless steels," *Electrochimica Acta*, vol. 42, no. 20–22, pp. 3275–3280, 1997.

[24] J. P. Kollender, M. Voith, S. Schneiderbauer, A. I. Mardare, and A. W. Hassel, "Highly customisable scanning droplet cell microscopes using 3D-printing," *Journal of Electroanalytical Chemistry*, vol. 740, pp. 53–60, 2015.

[25] J. Tardelli and E. Rocca, "Electrochemical behaviour of intermetallic phases in sulfuric acid at high voltage—anodization of Al_7Cu_2Fe and $MgZn_2$ phase," http://sites.poli.usp.br/org/emcr2012/CD/PDF/Oral%2026_Rocca.pdf.

[26] J. Henrie, S. Kellis, S. M. Schultz, and A. Hawkins, "Electronic color charts for dielectric films on silicon," *Optics Express*, vol. 12, no. 7, pp. 1464–1469, 2004.

[27] J. P. O'Sullivan and G. C. Wood, "The morphology and mechanism of formation of porous anodic films on aluminium," *Proceedings of the Royal Society of London Series A: Mathematical and Physical Sciences*, vol. 317, no. 1, pp. 511–543, 1970.

[28] M. K. Cavanaugh, R. G. Buchheit, and N. Birbilis, "Evaluation of a simple microstructural-electrochemical model for corrosion damage accumulation in microstructurally complex aluminum alloys," *Engineering Fracture Mechanics*, vol. 76, no. 5, pp. 641–650, 2009.

[29] J. Wloka and S. Virtanen, "Detection of nanoscale η-MgZn2 phase dissolution from an Al-Zn-Mg-Cu alloy by electrochemical microtransients," *Surface and Interface Analysis*, vol. 40, no. 8, pp. 1219–1225, 2008.

[30] J. P. Kollender, A. I. Mardare, and A. W. Hassel, "Multi-Scanning Droplet Cell Microscopy (multi-SDCM) for truly parallel high throughput electrochemical experimentation," *Electrochimica Acta*, vol. 179, pp. 32–37, 2015.

2

A Review on Superplastic Formation Behavior of Al Alloys

Xiao-guo Wang,[1,2] Qiu-shu Li,[1] Rui-rui Wu,[1] Xiao-yuan Zhang,[2] and Liyun Ma[3]

[1]*School of Materials Science and Engineering, Taiyuan University of Science and Technology, Taiyuan 030024, China*
[2]*Department of Mechanical and Electrical Engineering, College of Information, Shanxi Agricultural University, Taigu 030800, China*
[3]*Department of Mining Engineering, Luliang University, Lishi 033000, China*

Correspondence should be addressed to Qiu-shu Li; liqiushu@tyust.edu.cn

Academic Editor: Guru P. Dinda

Superplastic formation technology is considered to be an effective and promising method to conquer formation difficulties of Al alloys, especially thin-walled or complex structure. In this paper, fundamentals of superplastic deformation of Al alloys are summarized and the superplastic deformation behaviors of main kinds of Al alloys are summarized and compared, including Al-Mg alloys, Al-Li alloys, and Al-Zn-Mg alloys as well as aluminum matrix composites. Then, the superplastic deformation mechanism for different Al alloys is thoroughly discussed. Last but not the least, the factors affecting the superplastic deformation process are fully analyzed.

1. Introduction

1.1. Superplasticity of Materials. Generally speaking, superplasticity [1–6] is a characteristic of some materials with equiaxed and fine grains, when they are deformed in tension at relatively high temperature ($(0.5\text{–}0.7)T_m$) and low strain rate [7, 8], which has been discovered in various materials. Remarkable elongation is a notable symbol of superplasticity, for example, Sn–Bi eutectic alloys [9] under superplastic tensile tests exhibiting outstanding elongation, as much as 1950%. The superplastic elongations of main Al alloys can vary from 200% to 2000%, and the elongation of some superplastic materials can be up to 8000%. The superplastic characteristics of materials tremendously improve their formability with reduced deformation resistance and viscous/semiviscous flow during plastic deformation, particularly for complicated structure components.

It is well acknowledged that strain hardening occurs during plastic deformation, of which the constitutive equation [10] is affected by strain-hardening exponent n, as shown in Equation (1). The constitutive equation of superplastic deformation at relatively low temperature abides to viscoplastic Equation (2) proposed by Rossard, which indicates that the process combines strain hardening with strain rate hardening. The constitutive equation of superplastic deformation at relatively high temperature acts up to flow stress Equation (3) proposed by Backofen et al., implying that the process is mainly manipulated by strain rate sensitivity exponent m:

$$\sigma = K\varepsilon^n, \quad (1)$$

$$\sigma = K\varepsilon^n\dot{\varepsilon}^m, \quad (2)$$

$$\sigma = K\dot{\varepsilon}^m, \quad (3)$$

where σ is the true stress, ε is the true strain, $\dot{\varepsilon}$ is the strain rate, n is the strain-hardening exponent, m is the strain rate sensitivity, and k is a parameter for material.

1.2. Superplastic Formability of Al Alloys. Because of their low density and high specific strength combined with exceptional corrosion resistance, Al [11–14] are the most widely utilized alloys in aerospace, automobiles industry, mechanical manufacturing, marine business, and chemical industry. Superplastic forming technology (SFT), which can dramatically decrease flow/residual stress and improve formation quality, has been commonly used to manufacture complex shapes in sheet or tube. The Al alloys used are capable of achieving equivalent thickness strains in excess of 300% at low strain rates. Companies like Superform and Alcoa have successfully applied SFT to produce Al alloy automobile parts (decklid, fenders, door inners, and brake disk) with 50% weight saving [15]. In 1991, 8090 Al alloys were used to produce military aircraft doors by SFT with 23% weight reduction and 68% cost reduction.

To enhance the superplastic formability of Al alloys, it is necessary to make the alloys to go through several advanced forming technologies, such as flake powder metallurgy [16] and cryo-deformation [17–20] as well as severe plastic deformation (SPD) [21], including friction stir processing (FSP) [22–28], equal-channel angular pressing (ECAP) [29–35], high-pressure torsion (HTP) [36], accumulative roll bonding (ARB) [37], multiaxial alternative forging (MAF) [38], and repetitive corrugation and straightening (RCS) [39, 40]. After SPD process, ultrafine grain, high superplasticity, and low flow stress are available for Al alloys. Yet, due to high cost, additionally required technologies may hold back the direct and full commercial application of SPD technology. Meanwhile, more and more energy and resources are pouring into developing new and convenient hot/cold mechanical processing for superplasticity of various Al alloys. Therefore, thermal-mechanical processing (TMP) [41–43], which is mainly based on the principles of static recrystallisation [44–47] and dynamic recrystallisation [48, 49], is also an important way to develop superplasticity of Al alloys.

2. Fundamentals of Superplastic Deformation of Al Alloys

2.1. Stacking Fault Energy. It is well known that Al and Al alloys are with high stacking fault energy (SFE) which is a significant intrinsic material parameter that would change mechanisms and grain refinement during superplastic deformation. In Al and Al alloys, recovery is not totally suppressed during superplastic deformation. Additionally, lowering the SFE would promote deformation twinning over dislocation slip and increase dislocation slip-twinning transition temperature [50, 51].

2.2. Strain-Hardening Ability. It is clear that strain hardening is a typical phenomenon in plastic deformation. In the primary stage of superplastic deformation, flow stress rapidly rises with flow strain due to appearance of massive defects (vacancies/dislocations/twinning). As the process goes on, strain softening, due to recovery/recrystallisation/cross slip, starts to offset or partially offset the increase of flow stress until there is a dynamic balance between the hardening and softening processes, namely, steady stage of superplastic deformation. The strain-hardening ability is strongly dependant on temperature and strain rate of superplastic deformation, as shown in Figure 1. The true stress increases with increase in strain rate and decrease in temperature, which is explained in Section 4.2.

2.3. Superplastic Deformation Behavior. After rolling or SPD, Al alloys consist of banded microstructure with subgrain boundaries (SGBs) and low-angle boundaries (LABs). Following homogenization annealing, Al alloys are recrystallised and the microstructure combined with equiaxed recrystallised grains with weakened banded grains is formed before superplastic tensile tests. The early stage of superplastic deformation of Al alloys is usually accompanied with static/dynamic grain growth and anisotropic grain elongation. It should be noticed that grains are more elongated in the tensile direction than in the transverse direction [53, 54]. Yet, the static/dynamic grain growth is closely related to the distribution of intermetallic precipitates, which is fully discussed in Section 4.2. With the increase in strain, the previous microstructure is replaced by a more equiaxed and uniform grain structure and the banded grains are vanished.

2.4. Crystallographic Texture. Crystallographic texture is a preferred orientation structure of grains in ploycrystalline materials.

Being subjected to the formation process (casting, rolling, recrystallisation, and heat treatment), Al and Al alloys, typical FCC ploycrystalline materials, would produce corresponding texture. The main crystallographic texture components in FCC metals are listed in Table 1.

Prior to superplastic deformation, the microtexture of rolled/SPD Al alloys is of a strong brass texture [17, 55], which is a dominant plane strain hot deformation texture for Al alloys. After homogenization annealing for proper period, Al alloys exhibit cube, near cube, rotated cube, or decreased brass component. As the strain of superplastic deformation increases, a randomization of the microtexture happens along with the increase in misoriented boundaries; therefore, the alloys exhibit a very weak texture component as cube, near cube, rotated cube, or decreased brass component. Figure 2 shows ODFs representative sections of Al-Zn-Mg-0.25Sc-0.10Zr during superplastic deformation. In summary, the overall texture intensity of Al alloys is substantially reduced following superplastic deformation and no new orientation is developed.

2.5. Strain Rate Sensitivity (m) and Deformation Activation Energy (Q). Usually, superplastic deformation can be described by an equation for power-law creep in the following form:

$$\dot{\varepsilon} = A\frac{D_0 G b}{RT}\left(\frac{b}{d}\right)^p\left(\frac{\sigma}{G}\right)^{1/m}\exp\left(-\frac{Q}{RT}\right), \tag{4}$$

where $\dot{\varepsilon}$ the is strain rate, Q is the activation energy of an appropriate diffusion process, A and p are the empirical material constants, G is the shear modulus, b is the Burgers vector, R is the gas constant, T is the absolute temperature, D_0 is the frequency factor, and m is the strain rate sensitivity exponent ($1/m$ is the stress exponent n, $n = 1/m$).

Strain rate sensitivity (m) is a critical parameter for superplastic deformation, as its value characterizes the ability of an alloy to resist necking spread during

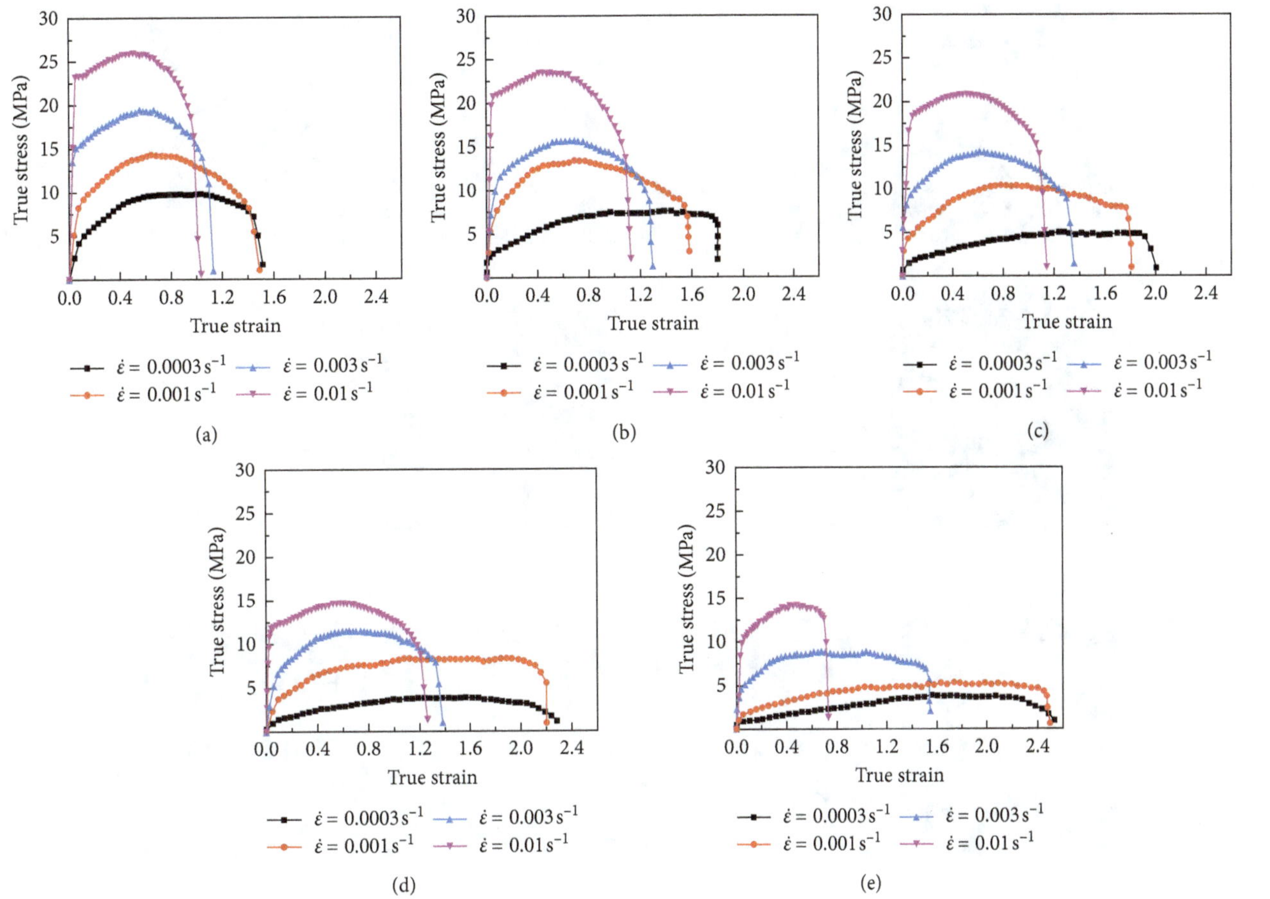

Figure 1: True stress-true strain curves of 7B04 Al alloy after superplastic deformation under different parameters: (a) 470°C; (b) 485°C; (c) 500°C; (d) 515°C; (e) 530°C [52].

Table 1: Main texture components in FCC metals.

Texture	{*hkl*}	<*uvw*>	Φ_1 (°)	Φ (°)	Φ_2 (°)
Cube	{0 0 1}	<1 0 0>	00	0	0
Rotated cube	{0 0 1}	<1 1 0>	45	0	0
Copper	{1 1 2}	<1 1 1>	90	35	45
Brass	{0 1 1}	<2 1 1>	35	45	0
Goss	{0 1 1}	<1 0 0>	0	45	0
S	{1 2 3}	<6 3 4>	59	37	63
R	{1 2 4}	<2 1 1>	57	29	63

deformation. The larger the value of strain rate sensitivity is, the greater the superplastic elongation will be. Additionally, it should be aware that m constantly varies during superplastic deformation. According to Equation (3), after logarithmic transformation, strain rate sensitivity can be stated as

$$m = \frac{\partial \ln \sigma}{\partial \log \dot{\varepsilon}}\bigg|_T . \quad (5)$$

Normally, the varying value of the strain rate sensitivity indicates different mechanisms of superplastic deformation. When the value of the strain rate sensitivity is around 0.3, the main mechanism of superplastic deformation of Al alloys is GBS accompanied with other accommodation mechanisms (Section 4.1). When the value of the strain rate sensitivity is around 0.5, the dominant mechanism of superplastic deformation of Al alloys is GBS. With increasing value of strain rate sensitivity, GBS gradually becomes the main mechanism of superplastic deformation of Al alloys and is responsible for the majority part of superplastic deformation of Al alloys.

During plastic deformation of metals, deformation activation energy Q represents the energy required for the transition atoms in each mole to jump across the barrier. Essentially, superplastic deformation is a thermal activation process. Equation (6) is applied to calculate the activation energy (Q), which is a critical dynamic parameter to indicate the level of difficulty of plastic deformation of materials at relatively high temperature. The values of the activation energy (Q) for different deformation mechanisms are different. During superplastic deformation, the value of Q for the dislocation pipe diffusion in aluminum is 82 kJ/mol, the value of the grain-boundary sliding is 84 kJ/mol, and the value of lattice self-diffusion of pure aluminum is 142 kJ/mol:

$$Q = \frac{R}{m} \frac{\partial \ln \sigma}{\partial (1/T)}\bigg|_{\dot{\varepsilon}}, \quad (6)$$

where R is the gas constant, m is the strain rate sensitivity, σ is the flow stress, and T is the absolute temperature.

Average values of m and Q during superplastic deformation of different aluminum alloys are listed in Table 2.

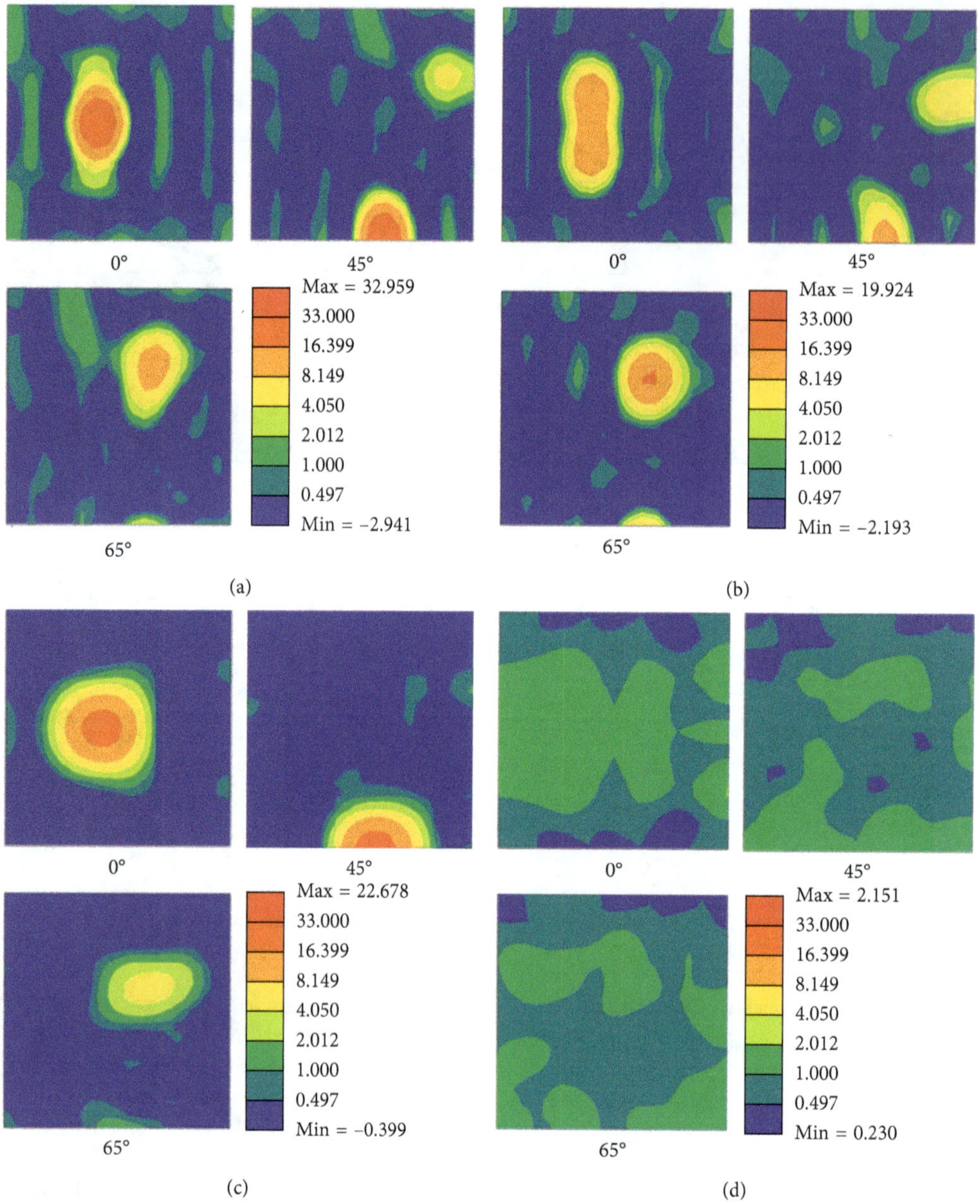

Figure 2: ODFs representative sections of the Al-Zn-Mg-0.25Sc-0.10Zr alloy deformed at 500°C and $1 \times 10^{-2} \cdot s^{-1}$ after interrupting the tensile test at different strains: (a) $\varepsilon = 0$; (b) $\varepsilon = 0.69$; (c) $\varepsilon = 1.10$; (d) $\varepsilon = 2.40$ [42].

Table 2: Reference data to illustrate range and average values of m and Q during superplastic deformation.

Alloy type	T (°C)	Strain rate (s^{-1})	M	Q (kJ/mol)	Reference
Al-Mg-Sc-Zr	450–525	1.67×10^{-3}–1×10^{-1}	0.36–0.42 0.39	90–126 107	[43]
Al-Zn-Mg-Sc-Zr	450–500	1×10^{-3}–1×10^{-1}	0.35–0.40 0.37	83.8–85.0 84.5	[42]
Al-Zn-Mg-Sc	450–500	1×10^{-3}–1×10^{-2}	0.46–0.51 0.50	83.8–85.6 85	[56]
Al-Mg-Sc-Zr	425–500	1×10^{-3}–1×10^{-1}	0.35–0.53 0.43	93.269–177.517 143.762	[57]
Al-Mg-Li	430–570	5×10^{-4}–1×10^{-3}	0.41–0.85 0.61	—	[58]
Al_3Ti/2024Al	450–540	0.15–1.5	—	206	[59]

3. The Superplastic Behaviors of Al Alloys

3.1. Al-Mg Alloys. Ye et al. [58, 60–67] developed a thermal-mechanical process to produce fine-grained Al-Mg-Si alloy plates with the chemical composition of Al-5.2%Mg-2.1%Li-0.12%Zr(wt.%) for superplasticity research. The uniaxial tensile tests were conducted at initial strain rates ranging from 0.01 to 0.0005/s in the temperature interval of 450°C–570°C. It turned out that great elongations of 580–915% were obtained at 510°C–540°C and initial strain rate between 0.001/s and 0.0005/s, in which the optimal elongation of about 915% occurred at a temperature of 525°C and an initial strain rate of 0.001/s.

Mikhaylovskaya et al. [68] put emphasis on the different superplastic deformation behaviors of two fine-grained AA5083 alloys, with and without chromium content, at elevated temperature from 500°C to 560°C and constant strain rates of 0.0005/s, 0.001/s, and 0.002/s. The sigmoidal shape of the stress-strain rate curves obtained for both alloys imply that they were subjected to superplastic deformation.

Duan et al. [41] studied the superplasticity of Al-Mg-0.15Sc-0.10Zr alloy, using a simple thermal-mechanical processing. Their work revealed that the cold rolled alloy sheet exhibited excellent superplasticity (elongation of ≥800%) at a temperature range of 450°C–500°C and a high strain rate range of 0.01–0.1/s. A maximum elongation of 1579% was achieved at 475°C and a high strain rate of 0.05/s.

Xu et al. [69] performed a research on high strain rate superplasticity of an Al-6.1Mg-0.25Sc-0.1Zr (wt.%) alloy by original asymmetrical rolling technology. The alloy sheets were subjected to both plane strain deformation and additional shear component along the rolling direction, simultaneously, owing to the different circumferential velocities of the two rotary rolls. The studied alloy exhibited fantastic superplasticity (elongation >1000%) at the temperature interval of 450°C–500°C and relatively high strain rates ranging from 0.01/s to 0.25/s. The highest ductility of 3200% is achieved at 500°C and 0.05/s. Compared with the same materials produced by traditional rolling, asymmetrical rolling technology can increase the optimum strain rate up to 10 times.

Li et al. [43] dealt with the superplastic behavior of an Al-5.8Mg-0.4Mn-0.25Sc-0.1Zr (wt.%) alloy at the temperature interval of 450°C–525°C and strain rates ranging from 0.00167/s to 0.1/s. Experimental results showed that the alloys exhibited nice elongation, ranging from 230% to 740%, and the maximum elongation of 740% was achieved at 500°C and the initial strain rate of 0.00667/s.

3.2. Al-Li Alloys. Xinming Zhang et al. [70–76] did superplastic investigation on 2A97 Al-Li alloy at the temperature range of 480°C–490°C and strain rates of 0.001–0.0025/s. The results indicated that the alloy exhibited promising elongation, ranging from 600% to 850%, and the highest elongation of 850% was achieved at 490°C and the initial strain rate of 0.002/s.

Zhang et al. [55] conducted superplastic investigation on 5A90 Al-Li alloy at the temperature range of 450°C–500°C and strain rate of 0.0003–0.0018/s. The results indicated that the alloy exhibited promising elongation, ranging from 310% to 1050%, and the greatest elongation of 1050% was achieved at 475°C and the initial strain rate of 0.0008/s.

3.3. Al-Zn-Mg Alloys. Wang et al. [77–86] performed superplastic investigation on the friction stir processed (FSP) Al-Zn-Mg-Cu alloy at the temperature of 500°C–535°C and a high strain rate of 0.01/s. The results proved that the FSP Al-Zn-Mg-Cu alloy possesses excellent thermal stability even up to incipient melting temperature; furthermore, a highest elongation of 3250% was obtained at 535°C and 0.01/s.

Mikhaylovskaya et al. [87] compared the superplastic deformation behavior of two aluminum AA7XXX alloys with Sc and Zr additions distinguished by the presence and absence of coarse eutectic Al_9FeNi particles. The experimental results were that the alloy with Sc(0.1%) and Zr (0.2%) additions and eutectic Al_9FeNi particles (a size of 1.8 μm and a volume fraction of 0.085) exhibits high superplasticity (elongation of 300%–915%) at the temperature interval of 400°C–500°C and strain rates ranging from 0.005/s to 0.01/s. The highest ductility of 915% is achieved at 480°C and 0.01/s. On the contrary, the alloy with same Sc (0.1%) and Zr(0.2%) additions and without Al_9FeNi particles did not elongate more than 340%, and there is no superplasticity at high strain rates (more than 0.01/s).

The superplastic performances of the abovementioned aluminum alloys are listed in Table 3.

3.4. Aluminum Matrix Composites. A great number of researchers [90, 91] have been constantly focusing on developing high strain rate and low temperature superplastic formation technology for aluminum matrix composites. The most recent works about superplasticity of aluminum matrix composites are listed in Table 4.

4. Discussion

4.1. Mechanisms of Superplastic Deformation. Enormous efforts and researches have been done on what is the exact mechanism of superplastic deformation, yet there is not a unified conclusion.

Generally speaking, superplastic deformation is mainly controlled by three mechanisms: grain-boundary sliding (GBS), dislocation slip/creep, and diffusion creep or directional diffusional flow [53]. Meanwhile, superplastic deformation is accompanied by various processes, for instance, grain-boundary migrations and static/dynamic grain growth, grain rotation [80, 99], recovery, recrystallisation, and polygonisation [87, 100, 101]. It is commonly accepted that GBS at low strain rates and creep by glide plus climb, also named slip creep, at high strain rates are the dominant controlling mechanisms that operate during high-temperature deformation of fine-grained metallic materials [102–104]. The contribution of GBS in the total strain varies with different Al alloys.

The work of Duan [105], Liu [106], and Cao [107] showed that GBS is a predominant mechanism of superplastic deformation of Al-Zn-Mg-Sc(Zr) alloys. Bate [108]

Table 3: Deformation conditions of optimal superplastic elongation of Al alloys.

Al alloys (wt.%)	Temperature (°C)	Strain rate (s^{-1})	Elongation (%)	Reference
Al-5.2Mg-2.1Li-0.12Zr	525	0.001	915	[58]
Al-Mg-0.15Sc-0.10Zr	475	0.05	1579	[41]
Al-6.1Mg-0.25Sc-0.1Zr	500	0.05	3200	[69]
Al-5.8Mg-0.4Mn-0.25Sc-0.1Zr	500	0.00667	740	[43]
Al-5.3Mg-2.1Li-0.11Zr	475	0.0008	1050	[55]
Al-5Mg-0.2Sc	520	0.056	2300	[88]
Al-3Mg-0.2Sc	400	0.033	2280	[89]
Al-3.8Cu-1.4Li(2A97)	490	0.002	850	[76]
Al-5.85Zn-2.56Mg-1.89Cu	535	0.01	3250	[86]
Al-6Zn-2.3Mg-1.7Cu(7B04)	530	0.0003	1105	[52]
Al-5.6Zn-2.2Mg-1.5Cu(7475)	515	0.0005	590	[46]
Al-5.41Zn-1.9Mg-0.25Sc-0.1Zr	500	0.01	1523	[76]
Al-5.36Zn-1.9Mg-0.1Sc-0.1Zr	500	0.005	1050	[56]
Al-5.4Zn-2Mg-0.25Sc-0.1Zr	200	0.01	539	[57]

Table 4: Deformation conditions of optimal superplastic elongation of aluminum matrix composites.

Composites	Temperature (°C)	Strain rate (s^{-1})	Elongation (%)	Reference
20wt.%Si_3N_{4P}/5052	545	0.001	700	[92]
10wt.%SiC_P/2024	480	0.001	345	[93]
20wt.%Si_3N_{4P}/2124	510	0.04	840	[94]
15wt.%SiC_P/IN9021	550	0.005	610	[95]
17.8wt.%SiC_P/2124	490	0.083	425	[96]
15wt.%SiC_P/6A02	560	0.00089	250	[97]
12wt.%SiC_P/5052	550	0.0011	405	[98]
5wt.%Al_3Ti/2024	510	0.15	640	[59]

concluded that the major superplastic deformation mechanism of Al-6Cu-0.4Zr alloy may be an intragranular dislocation slip. Katsas [109] considered that dynamic recrystallisation process results in the superplastic deformation of Al-1Zr alloy. Sotoudeh [110] and Bricknell [111] believed that diffusion creep is the main superplasticity mechanism of "supral type" alloys with banded microstructure. del Valle et al. [112] concluded that Al alloys that were composed of fine-equiaxed grains at the start of the superplastic test deformed mainly by GBS at low strain rate with only small evidence of dislocation activity and slip creep was the controlling mechanism at high strain rate superplastic deformation. The research of Wang et al. [86] indicated that the presence of a liquid phase because high experimental temperature plays a vital role in the extraordinary elongation of the alloy by relaxing the stress concentration and suppressing the the appearance of cavities during deformation. Jiao [59] also supported that a certain amount of liquid phase during superplastic deformation can properly enhance the superplasticity of aluminum alloys. Meanwhile, Johannes and Mishra [113] concluded that fiber (whisker) formation (Figure 3) was the evidence of the existence of liquid phases along grain boundaries. Furthermore, Duan et al. [56] thought the existence of fiber as an indirect evidence for GBS.

Commonly, fine and equiaxed grains are typical features of superplastic materials. It is fully established by a great number of theories and researches that grain-boundary sliding is the dominant reason that fine-grained materials are able to develop superplastic deformation. During deformation, materials migration and grain-boundary sliding happen simultaneously in order to assure intergranular accommodation. Moreover, it is necessary for another mechanism to accommodate stress concentration resulting from GBS. There are two main accommodation mechanisms for GBS, diffusion creep and dislocation sliding.

After a thorough research, Ashby and Verrall proposed the classic mechanism theory of GBS with diffusion creep for superplastic deformation, the theoretical model of which is shown in Figure 4. The Ashby–Verrall model (Figure 3) interprets how the equiaxed grains move by means of GBS. Though the shape of grains stay the same after sliding,the overall morphology changes, which is in agreement with macroelongation. Moreover, Ashby and Verrall explained that the strain difference between the initial and final circumstances is accommodated by grain-boundary diffusion and volume diffusion, as well as the ability of grain-boundary diffusion is much stronger than that of volume diffusion.

Currently, there are three models for theories of grain-boundary sliding accommodated by dislocation sliding, which are introduced in Table 5.

Additionally, it is considered that the mechanism of superplastic deformation for Al alloys is related to the microstructure status of the alloys prior to the deformation. After TMP, there are two ways to obtain fine grains for the alloys, which would result in different mechanisms of superplastic deformation. On one hand, if the thermal-mechanically treated alloys are subjected to recrystallisation annealing treatment, the main deformation mechanism of superplastically deformed alloys is concurrently controlled by

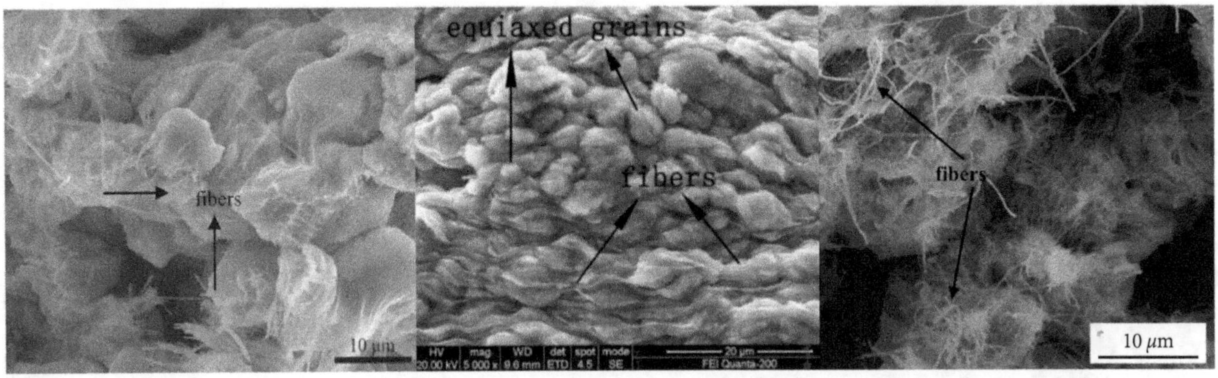

FIGURE 3: SEM image of fracture morphology of superplastically deformed Al alloys [41, 42, 57].

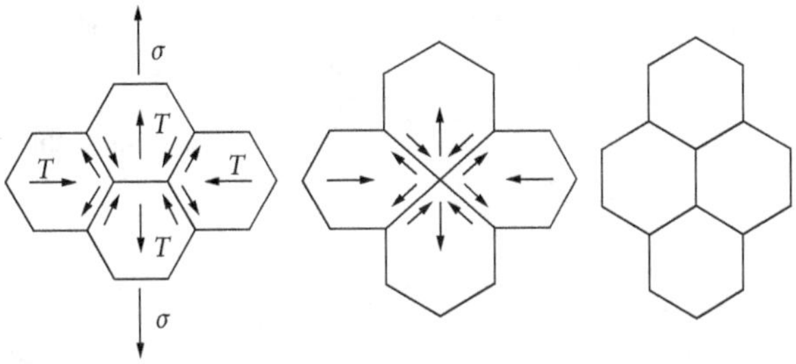

FIGURE 4: The Ashby–Verrall theoretical model for superplastic deformation. (a) Initial. (b) Intermediate. (c) Final.

both grain-boundary sliding (GBS) and accompanied accommodation mechanism which include dislocation motion [116–118], creep diffusion [119], and liquid phase [86], aiming to relax the stress concentration. On the other hand, when the thermal-mechanically treated alloys directly experience superplastic deformation, during the primary stage of deformation, the mechanism is controlled by deformation-induced continuous recystallisation (DICR) due to the occurrence of dynamic recrystallisation. Subsequently, the superplastic deformation mechanism is similar to static recrystallisation, after dynamic recrystallisation is completed.

4.2. Factors Affecting Superplastic Deformation Behavior

4.2.1. Temperature of Superplastic Deformation. Fundamentally speaking, superplasticity is a thermally activated phenomenon; therefore, the deformation temperature is a crucial factor. It is commonly accepted that there is a basic requirement for the Al alloys to exhibit superplasticity, which is that the deformation temperature shall be more than $0.5T_{\mathrm{m}}$ (melting temperature). As the temperature keeps rising, the viscous forces within grain boundaries gradually reduce and grain boundaries become more and more unstable. As a result, the grain boundary begins to slide with relative ease, which makes the tensile test sample to show better elongation. However, if the temperature is too high, grain boundaries are excessively softened and grain-boundary binding force dramatically decreases, which lead to less elongations. It can be seen from Figure 5 that the superplastic elongations are intensively changed with different temperatures of superplastic deformation and there is an optimal temperature for each kind of aluminum alloy.

Zhang et al. [52] found out that superplastic property of Al alloys started to decline and became insensitive to deformation temperature when the alloys were superplastically deformed at relatively high strain rates.

4.2.2. Strain Rates of Superplastic Deformation. Besides deformation temperature, stain rates also could notably influence the superplasticity of Al alloys. As shown in Equation (3), the true stress σ is highly dependent on the strain rate $\dot{\varepsilon}$, so it also means that the superplastic deformation process of Al alloys is significantly affected by the strain rate.

Researches [120, 121] indicate that one precondition of superplasticity is that the stain rates should be in the range of 10^{-4}–10^{-1}/s. Figure 6 shows the correlation between superplastic elongation and varying initial strain rates at different temperatures of different Al alloys. It can be seen that relatively low strain rates are in favor of better superplasticity. Deformed at lower initial strain rates, the Al alloys exhibit an extended strain-hardening phenomenon and decreased strain-hardening rates due to dislocation relaxation triggered by accommodation mechanisms of GBS. At the same time, the microstructure of Al alloys after superplastic deformation is partially determined by strain rates. Low strain rates could result in coarseness of the microstructure ascribed to recovery and recrystallisation processes. Better superplasticity resulting from relatively low strain rates comes at a price of great amount of time, which strongly restricts further application of superplastic deformation of Al alloys. Consequently, high strain rate superplasticity (HSRSP) [122–127], which can not only dramatically reduce the consuming time but also develop high superplasticity, is attracting more and more attention.

4.2.3. Precipitates Particles. Prior to superplastic deformation, particles with different size, locations, and shapes are distributed in the grains and at the boundaries, which will play a very important role during superplasticity.

By adding a few amount of trace element, high density of coherent fine precipitates ($Al_3(Zr,Sc)$, ~30 nm [128–131]; $Al_6(Mn,Cr)$, 38 ± 7 nm [68]) are formed and dispersed both at grain boundaries and within the grain interiors, which will strongly retard the grain growth during

Table 5: Three models for theories of grain-boundary sliding accommodated by dislocation sliding.

Models	Schematic illustration of the mechanism of superplasticity	Introduction of the models
Ball–Hutchison model [114]	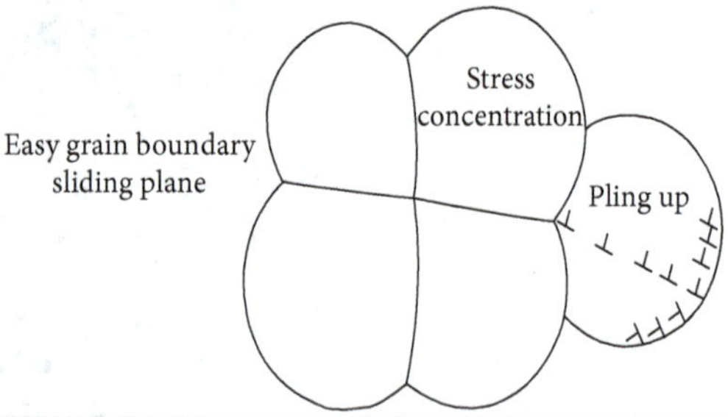	During deformation, grain boundaries are properly aligned and slide as groups. When the sliding is blocked by other grains, the local stresses would result in dislocation in the blocking grain, piling up at the opposite grain boundary until the back stress stops further generation of dislocation. The piled up dislocations could climb into and along the grain boundary. Thus, grain-boundary sliding, which is governed by the kinetics of climb along grain boundary to annihilation areas, is feasible due to the constant replacement of the dislocation.
Mukherjee model [115]	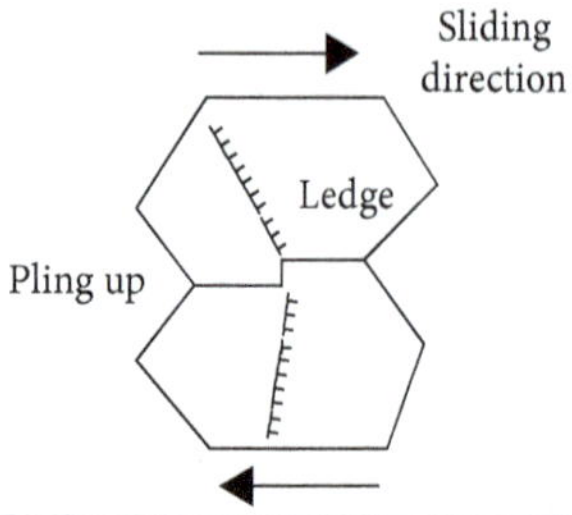	The model elaborates dislocation accommodated sliding mechanism of single grain. Dislocations are generated at scraggy surfaces of grain boundaries. The alignment mechanism of the generated dislocations is the same as that of the Ball–Hutchison model.
Gifkins model [116]	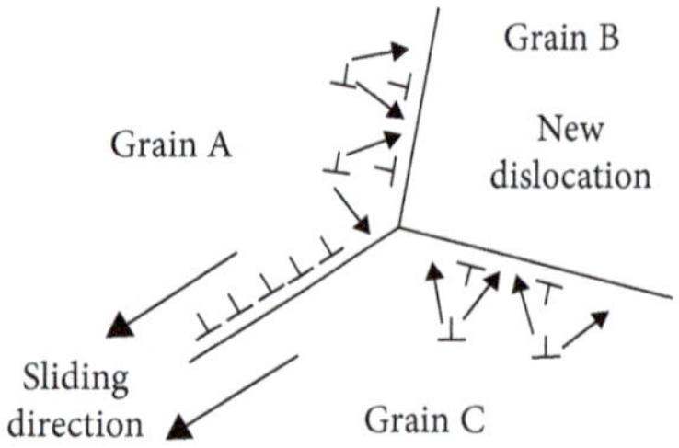	Gifkins described grain-boundary sliding and accommodation process as grain-boundary dislocation motion and proposed the core-mantle model. Grain-boundary sliding around grain triple junctions is accommodated by generation of new dislocation and their climbing along the boundaries. Dislocation motion is constrained to occur at grain boundary and mantle area rather than in the core of grain. According to the model, grains are able to slide by switching places in three-dimensional space.

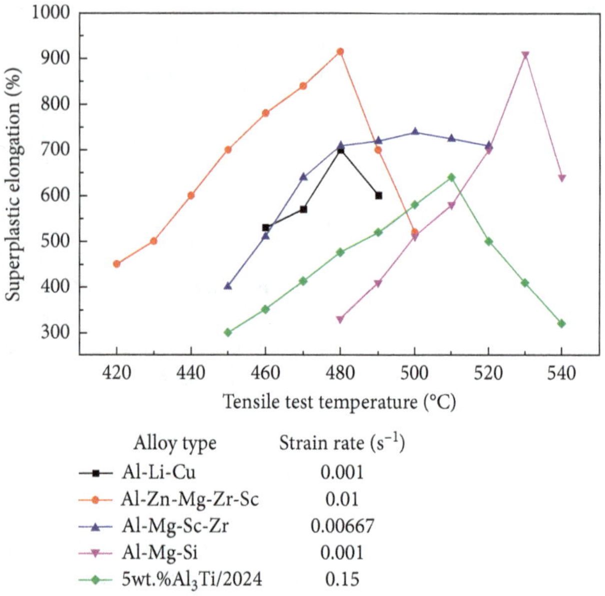

Figure 5: Variation of superplastic elongations of different Al alloys with initial strain rates at various temperatures.

recrystallisation and inhibit the movement of grain boundaries, subgrain boundaries, and dislocation by the Zener pinning effect (Figure 7). Furthermore, the stability of cavity can be partially attributed to additional interfacial energy of $Al_3(Zr,Sc)$ particles.

Besides nanoscale particles, there are also existence of other coarse particles (Al_9FeNi, ~1.8 μm, Figure 8 [87]) that affect superplastic deformation. These coarse particles can form extensive deformation zones during rolling or SPD process, which can accelerate nucleation at preheating and at superplastic deformation ascribed to particles stimulated nucleation (PSN) and lead to grain refinement.

In addition, the presence of particle-depleted zones (PDZs) associated with diffusion creep is often detected in the Al alloys with uniform distribution of nanosized particles during superplastic deformation. The PDZ is often considered to facilitate the total elongation of Al alloys during SPF.

4.2.4. Thermal-Mechanical Processing. As stated in Section 1.2, thermal-mechanical processing plays a vital pole for Al alloys to obtain superplasticity. Even the alloys with experimentally approximate composition, such as Al-5.8Mg-0.4Mn-0.25Sc-0.10Zr [43] and Al-6.10Mg-0.3Mn-0.25Sc-0.10Zr [69], may exhibit superplasticity with huge difference, simply owing to distinguished thermal-mechanical processes. Both the elongations of the Al-5.8Mg-0.4Mn-0.25Sc-0.10Zr and Al-6.10Mg-0.3Mn-0.25Sc-0.10Zr alloys sample were tested at the temperatures ranging from 450°C to 475°C and at strain rates of 0.005/s to 0.1/s, however,

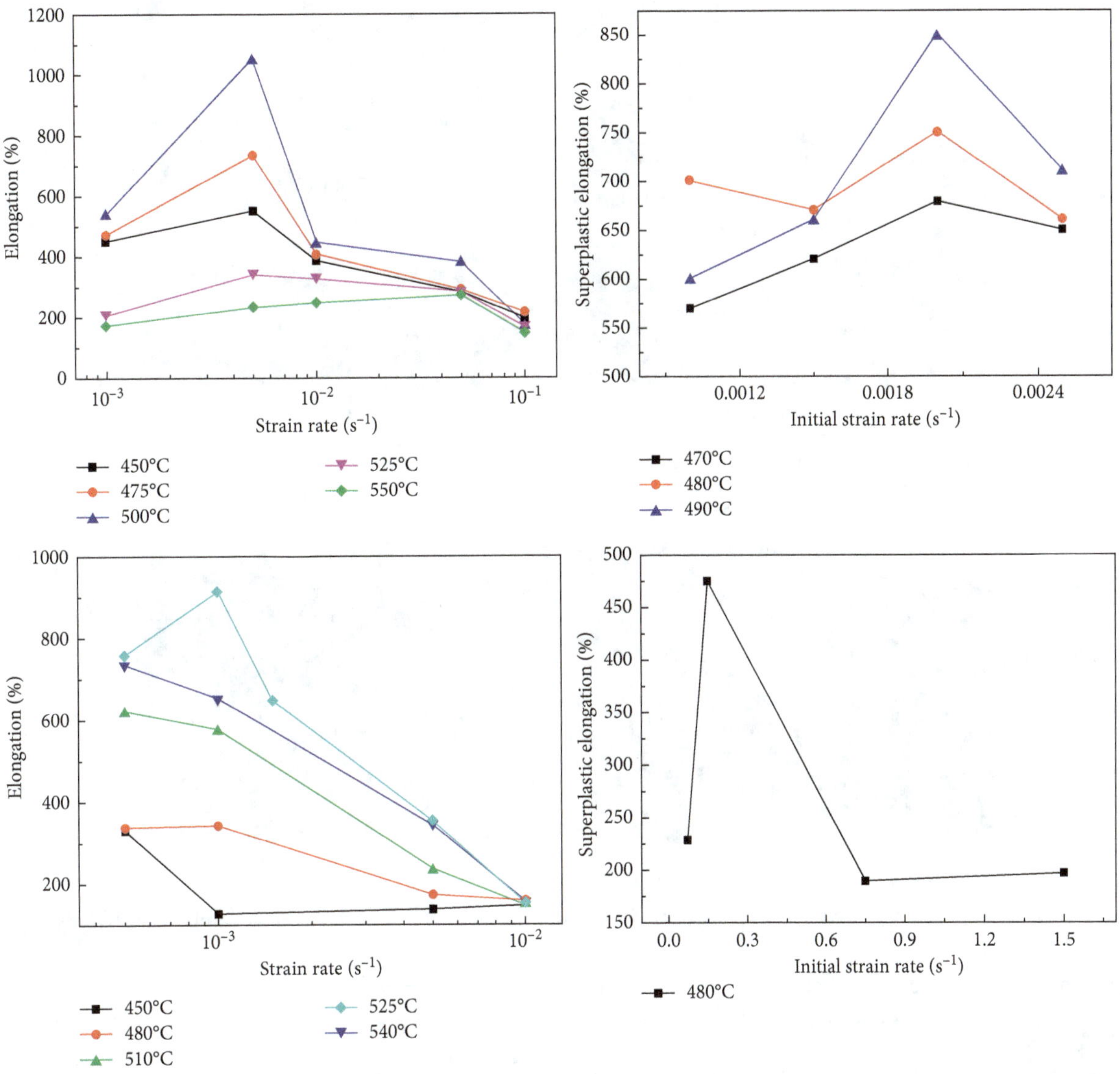

FIGURE 6: Variation of superplastic elongations of different Al alloys with varying initial strain rates at various temperatures. (a) Al-Zn-Mg-Sc-Zr [56]; (b) Al-Li-Cu; (c) Al-Mg-Si [47]; (d) 5wt.%Al_3Ti/2024.

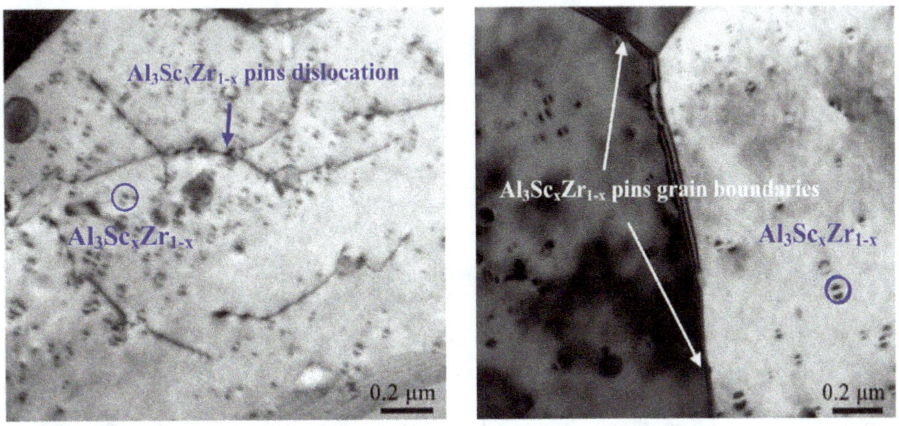

(a) (b)

FIGURE 7: Continued.

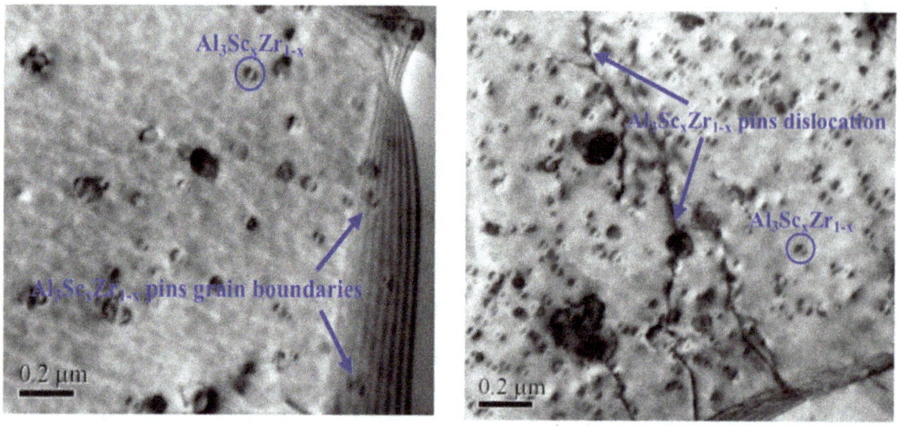

(c) (d)

FIGURE 7: TEM microstructures of superplastically deformed Al-Zn-Mg-Sc-Zr alloy [42, 56].

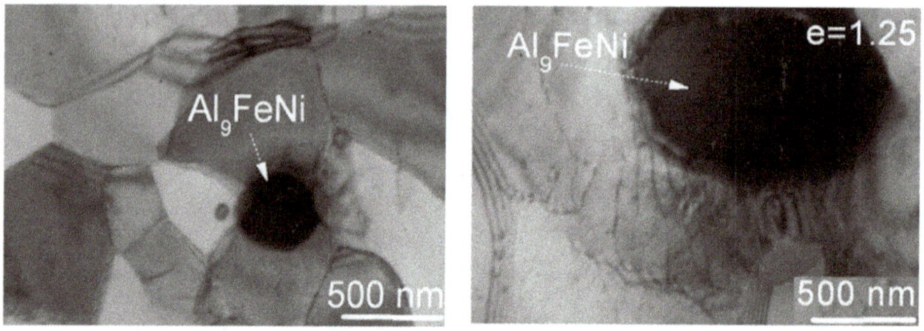

(a) (b)

FIGURE 8: TEM microstructures of Al-Zn-Mg-Sc-Zr alloy: (a) 20 minutes annealed at 480°C; (b) strain of 1.25 [87].

subjected to different thermal-mechanical processes, simple rolling and asymmetrical rolling, respectively (please check [43, 69] for details).

Figure 9 shows the comparison of maximum elongations of the two Al-Mg-Mn-Sc-Zr alloys under 4 tested temperatures of all tested stain rates. It is well established that the asymmetrical rolled alloy. The maximum elongations of asymmetrical rolled Al-Mg-Mn-Sc-Zr alloy is at least 3.1 times that of simple rolled Al-Mg-Mn-Sc-Zr alloy.

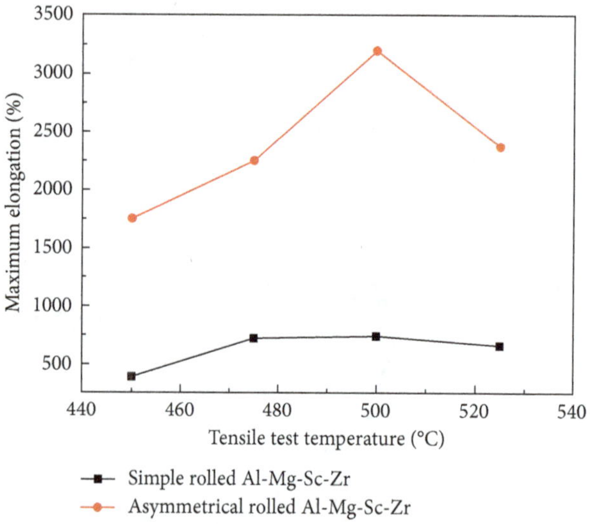

FIGURE 9: Comparison of maximum elongations of the two Al-Mg-Mn-Sc-Zr alloys under 4 tested temperatures of all tested stain rates.

4.3. Recovery and Recrystallisation. As described in Section 2.2, recovery and recrystallisation are of great importance in structural softening mechanism of Al alloys. Owing to high SFE of Al alloys, dynamic recovery is even more important to the softening process. Therefore, the following sections illustrate the softening process caused by recovery and recrystallisation in detail.

4.3.1. Dynamic Recovery. High density of defects (vacancies and dislocations) is generated and accumulated in the microstructure of plastic deformed Al alloys. High SFE and solute strengthening effect work together to make Al alloys undergo dynamic recovery mechanisms, which include (a) annihilation resulting in neutralization of dislocation dipoles and reduction in dislocation density, (b) polygonisation which involves systemic organization of randomly oriented dislocation into more stable microstructural elements called cells, and (c) proliferation of cells into fully developed subgrains [18]. Because

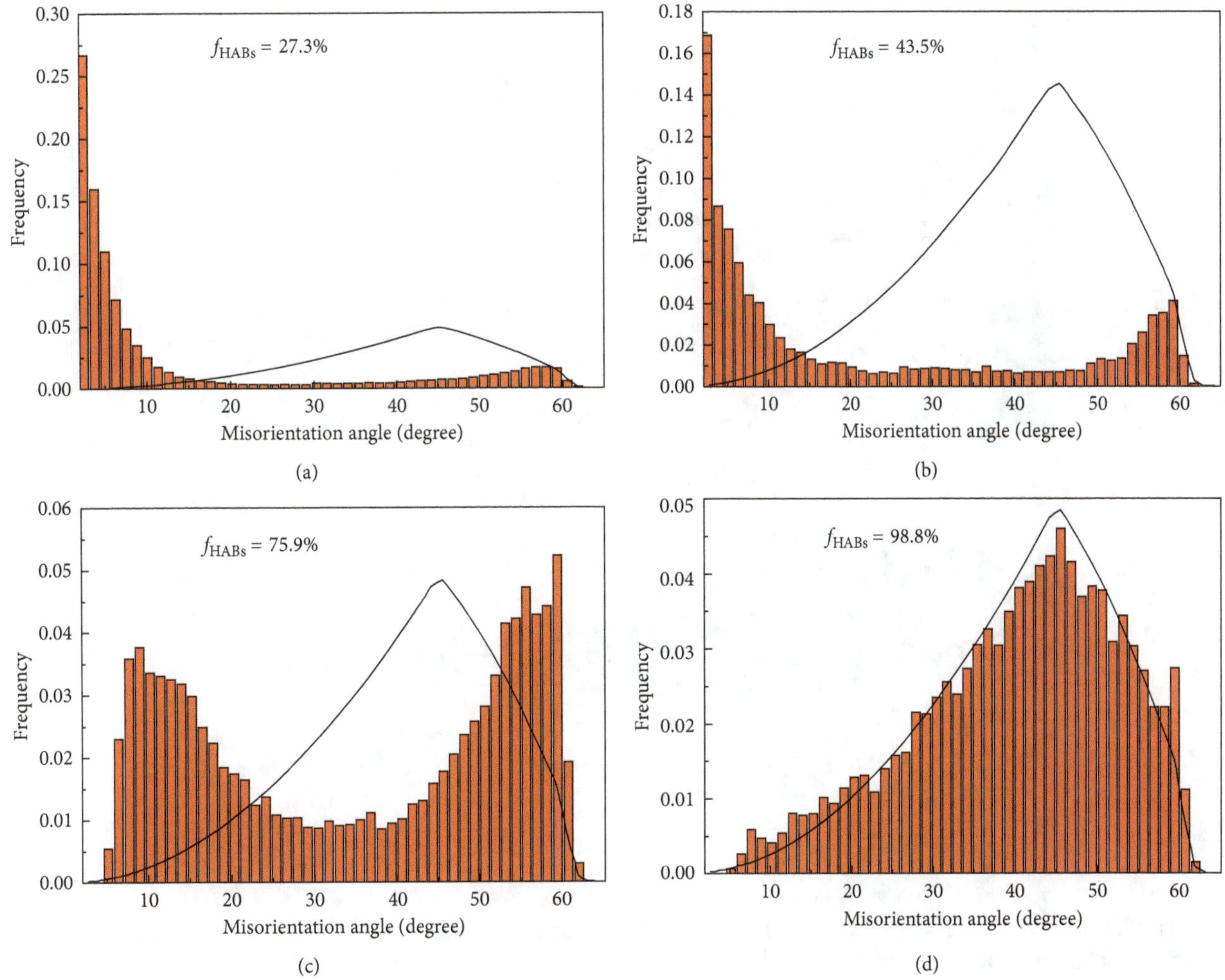

FIGURE 10: Misorientation angle distribution of the Al-Zn-Mg-0.25Sc-0.1Zr alloy deformed at 500°C and $1 \times 10^{-2} \cdot s^{-1}$ after interrupting the tensile test at different strains: (a) $\varepsilon = 0$; (b) $\varepsilon = 0.69$; (c) $\varepsilon = 1.10$; (d) $\varepsilon = 2.40$ [42].

of work hardening, tangled microstructure is formed and transformed into an ordered one by rearrangement of dislocations into micronscale entities called cells. Densly tangled dislocations are located along the boundaries of the cells, while less dislocations are found within the cell interiors resulting from annihilation of the internal dislocations. Subsequently, the cell boundaries turn into structured subgrain boundaries with higher misorientation.

4.3.2. Deformation-Induced Continuous Recrystallisation. If the alloy sheets subjected to severe rolling/SPD process are directly superplastically deformed without static recrystallisaion annealing, it is the fibrous structure rather than equiaxed grain structure that participate in the superplastic deformation. Thus, there are two stages existing during superplastic deformation, according to structure transformation. The initial microstructure of the deformed Al alloys consists of a large amount of LABs, which are not favorable for GBS because of their low misorientations, while it is well proved that HABs are desired microstructure for superplastic flow by GBS [132, 133]. In the first stage, when the superplastic deformation induced continuous recrystallisation, initial fibrous grain structures, typical LABs, along the rolling direction gradually shift into equiaxed grains, HABs keep increasing until the transformation of grains is completed (Figure 10), grain orientation is gradually randomized (Figure 11), and texture is weakened. After dynamic recrystallisation is finished, the second stage of superplastic deformation is mainly controlled by grain-boundary sliding.

5. Existing Challenges and Limitations

In spite of the inspiring and exciting research outcomes about superplastic formation behavior of Al alloys, there are still enormous unknown yet critical territories waiting to be found out and developed.

(i) There are still long way to go to develop a synthetical theory to utterly reveal the essence of superplasticity of Al alloys.

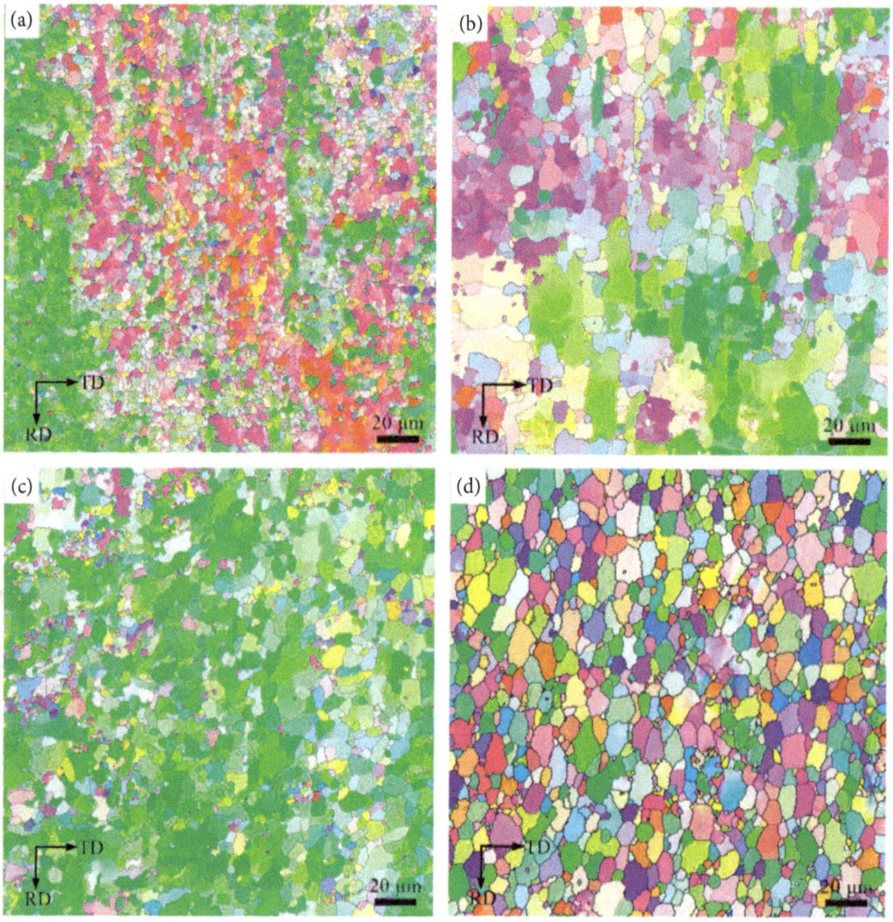

Figure 11: EBSD analyses of the Al-Zn-Mg-0.25Sc-0.10Zr alloy deformed at 500°C and $1 \times 10^{-2} \cdot s^{-1}$ after interrupting the tensile test at different strains: (a) $\varepsilon = 0$; (b) $\varepsilon = 0.69$; (c) $\varepsilon = 1.10$; (d) $\varepsilon = 2.40$ [42].

(ii) To exhibit superplasticity, the Al alloys need to meet the strict microstructure requirements, which are for the alloys to hold equiaxed and fine grains as well as the ability to recrystallisation.

(iii) Superplasticity of Al alloys also holds specific deformation requests, including high temperatures and low strain rates, which obviously limit the further industrialized and commercial applications of superplastic deformation.

(iv) For now, the applications of advanced severe plastic deformation (SPD) technologies, which are often used to activate superplasticity of Al alloys, are basically limited to scientific research and study. There is no sign to make SPD technologies fully utilize in commercial process.

(v) There are few researches about combining superplastic deformation with other forming technologies, such as welding and casting, to form complex-shaped components.

(vi) It is necessary to keep developing Al alloys with exceptional and comprehensive properties, including strength, ductility, superplastic formation ability, and stress corrosion resistance, so that the potentials of Al alloys can be fully released.

6. Conclusions

In this paper, the fundamentals and superplastic deformation behaviors of several series of Al alloys are described and it can be seen that currently the main different kinds of Al alloys all have exhibited outstanding superplasticity. Furthermore, the mechanisms of superplastic deformation are discussed and various theories regarding superplastic mechanism are summarized and analyzed, including theory of grain-boundary sliding with accommodation mechanism and deformation-induced continuous recrystallisation. Meanwhile, factors that affect superplastic deformation process of Al alloys are explained in detail, including temperatures, stain rates, and thermal-mechanical process. Subsequently, two significant superplastic deformation parameters of different Al alloys, strain rate sensitivity (m) and deformation activation energy (Q), are compared to help understand the relationship within these two parameters and superplastic deformation mechanisms. Finally, the existing challenges and limitations of current superplastic deformation of Al alloys are properly summarized.

Conflicts of Interest

The authors declare that they have no conflicts of interest.

References

[1] G. Giuliano, *Superplastic Forming of Advanced Metallic Materials: Methods and Applications*, Elsevier, 2011.

[2] T. Y. M. Al-Naib and J. L. Duncan, "Superplastic metal forming," *International Journal of Mechanical Sciences*, vol. 12, no. 6, pp. 463–470, 1970.

[3] JIS H 7007, *Glossary of Terms used in Metallic Superplastic Materials*, Japanese Standards Association, Tokyo, Japan, 1995.

[4] V. N. Perevezentsev, V. V. Rybin, and V. N. Chuvil'deev, "The theory of structural superplasticity—III. Boundary migration and grain growth," *Acta Metallurgica et Materialia*, vol. 40, pp. 907–914, 1992.

[5] A. H. Chokshi, A. K. Mukherjee, and T. G. Langdon, "Superplasticity in advanced materials," *Materials Science and Engineering: R: Reports*, vol. 10, no. 6, pp. 237–274, 1993.

[6] C. H. Hamilton, "Simulation of static and deformation-enhanced grain growth effects on superplastic ductility," *Metallurgical Transactions A*, vol. 20, no. 12, pp. 2783–2792, 1989.

[7] P. S. Bate, N. Ridley, and B. Zhang, "Mechanical behaviour and microstructural evolution in superplastic Al–Li–Mg–Cu–Zr AA8090," *Acta Materialia*, vol. 55, no. 15, pp. 4995–5006, 2007.

[8] G. Sakai, Z. Horita, and T. G. Langdon, "Grain refinement and superplasticity in an aluminum alloy processed by high-pressure torsion," *Materials Science and Engineering A*, vol. 393, no. 1-2, pp. 344–351, 2005.

[9] C. E. Pearson, "Viscous properties of extruded eutectic alloys of Pb-Sn and Bi-Sn," *Journal of Institute Metals*, vol. 54, no. 1, pp. 111–123, 1934.

[10] O. Ruano and O. D. Sherby, "On constitutive equations for various diffusion-controlled creep mechanisms," *Revue de Physique Appliquee*, vol. 23, pp. 625–637, 1988.

[11] B. M. Watts, M. J. Stowell, B. L. Baikie, and D. G. E. Owen, "Superplasticity in Al-Cu-Zr alloys. Part 1: material preparation and properties," *Metal Science*, vol. 10, pp. 189–197, 1976.

[12] S. Mehta, P. K. Sengupta, K. J. L. Iyer, and K. Nair, "Studies of the superplasticity of high strength aluminium alloy," *Aluminium*, vol. 68, pp. 234–237, 1992.

[13] J. Xinggang, W. Qingliing, C. Jianzhong, and M. Longxiang, "A study of the improvement," *Metallurgical Transactions A*, vol. 24, no. 11, pp. 2596-2597, 1993.

[14] J. Xinggang, C. Janzhong, and M. Longxiang, "A study of grain refinement and superplasticity of high strength 7475 Al alloy," *Z.Metallkd*, vol. 84, pp. 216–219, 1993.

[15] M. Mabuchia and K. Higashib, "On accommodation helper mechanism for superplasticity in metal matrix composites," *Acta Materialia*, vol. 47, no. 6, pp. 1915–1922, 1999.

[16] H. Huang, G. Fan, and Z. Tan, "Superplastic behavior of carbon nanotube reinforced aluminum composites fabricated by flake powder metallurgy," *Materials Science and Engineering: A*, vol. 699, pp. 55–61, 2017.

[17] K. Arun Babu, V. Subramanya Sarma, and C. N. Athreya, "Experimental verification of grain boundary-sliding controlled steady state superplastic flow in both continually and statically recrystallizing Al alloys," *Materials Science and Engineering: A*, vol. 657, pp. 185–196, 2016.

[18] A. Dhal, S. K. Panigrahi, and M. S. Shunmugam, "Insight into the microstructural evolution during cryo-severe plastic deformation and post-deformation annealing of aluminum and its alloys," *Journal of Alloys and Compounds*, vol. 726, pp. 1205–1219, 2017.

[19] H. P. Pu, F. C. Liu, and J. C. Huang, "Characterization and analysis of low-temperature superplasticity in 8090 Al-Li alloys," *Metallurgical and Materials Transactions A*, vol. 26, pp. 1153–1166, 1995.

[20] F. C. Liu and Z. Y. Ma, "Contribution of grain boundary sliding in low-temperature superplasticity of ultrafine-grained aluminum alloys," *Scripta Materialia*, vol. 62, no. 3, pp. 125–128, 2010.

[21] R. Valiev, "Nanostructuring of metals by severe plastic deformation for advanced properties," *Nature Materials*, vol. 3, no. 8, pp. 511–516, 2004.

[22] L. B. Johannes, I. Charit, R. S. Mishra, and R. Verma, "Enhanced superplasticity through friction stir processing in continuous cast AA5083 aluminum," *Materials Science and Engineering: A*, vol. 464, pp. 351–357, 2007.

[23] I. Charit and R. S. Mishra, "Evaluation of microstructure and superplasticity in friction stir processed 5083 Al alloy," *Materials Research*, vol. 19, no. 11, pp. 3329–3342, 2004.

[24] M. A. García-Bernal, R. S. Mishra, R. Verma, and D. Hernández-Silva, "High strain rate superplasticity in continuous cast Al–Mg alloys prepared via friction stir processing," *Scripta Materialia*, vol. 60, no. 10, pp. 850–853, 2009.

[25] Z. Y. Ma and R. S. Mishra, "Development of ultrafine-grained microstructure and low temperature (0.48 Tm) superplasticity in friction stir processed Al–Mg–Zr," *Scripta Materialia*, vol. 53, no. 1, pp. 75–80, 2005.

[26] A. Orozco-Caballero, M. Álvarez-Leal, P. Hidalgo-Manrique, C. MaríaCepeda-Jiménez, O. Antonio Ruano, and F. Carreño, "Grain size versus microstructural stability in the high strain rate superplastic response of a severely friction stir processed Al-Zn-Mg-Cu alloy," *Materials Science and Engineering A*, vol. 680, pp. 329–337, 2017.

[27] J. A. del Valle, P. Rey, D. Gesto, D. Verdera, and J. A. Jiménez, "Mechanical properties of ultra-fine grained AZ91 magnesium alloy processed by friction stir processing," *Materials Science and Engineering: A*, vol. 628, pp. 198–206, 2015.

[28] A. Orozco-Caballero, P. Hidalgo-Manrique, and C. M. Cepeda-Jiménez, "Strategy for severe friction stir processing to obtain acute grain refinement of an Al–Zn–Mg–Cu alloy in three initial precipitation states," *Materials Characterization*, vol. 112, pp. 197–205, 2016.

[29] H. Akamatsu, T. Fujinami, Z. Horita, and T. G. Langdon, "Influence of rolling on the superplastic behavior of an Al-Mg-Sc alloy after ECAP," *Scripta Materialia*, vol. 44, no. 5, pp. 759–764, 2001.

[30] Z. Horita, M. Furukawa, M. Nemoto, A. J. Barnes, and T. G. Langdon, "Superplastic forming at high strain rates after severe plastic deformation," *Acta Materialia*, vol. 48, no. 14, pp. 3633–3640, 2000.

[31] K. T. Park, H. J. Lee, C. S. Lee, W. J. Nam, and D. H. Shin, "Enhancement of high strain rate superplastic elongation of a modified 5154 Al by subsequent rolling after equal channel angular pressing," *Scripta Materialia*, vol. 51, no. 6, pp. 479–483, 2004.

[32] R. Z. Valiev and T. G. Langdon, "Principles of equal-channel angular pressing as a processing tool for grain refinement," *Progress in Materials Science*, vol. 51, no. 7, pp. 881–981, 2006.

[33] R. B. Figueiredo and T. G. Langdon, "Evaluating the Superplastic Flow of a Magnesium AZ31 Alloy Processed by

Equal-Channel Angular Pressing," *Metallurgical and Materials Transactions A*, vol. 45, no. 8, pp. 3197–3204, 2014.

[34] C. M. Cepeda-Jiménez, J. M. García-Infanta, O. A. Ruano, and F. Carreño, "Mechanical properties at room temperature of an Al–Zn–Mg–Cu alloy processed by equal channel angular pressing," *Journal of Alloys and Compounds*, vol. 509, pp. 8649–8656, 2011.

[35] C. M. Cepeda-Jiménez and J. M. García-Infanta, "Influence of Processing Severity During Equal-Channel Angular Pressing on the Microstructure of an Al-Zn-Mg-Cu Alloy," *Metallurgical and Materials Transactions A*, vol. 43, no. 11, pp. 4224–4236, 2012.

[36] L. Jiao, Y. T. Zhao, and W. Yue, "The development of high strain rate and low temperature superplasticity in aluminum matrix composites," *Materials Review*, vol. 27, pp. 119–123, 2013.

[37] N. Tsuji, K. Shiotsuki, and Y. Saito, "Superplasticity of ultra-fine grained al–mg alloy produced by accumulative roll-bonding," *Materials Transactions, JIM*, vol. 40, no. 8, pp. 765–771, 1999.

[38] M. Node, M. Hirohashi, and K. Funami, "Low temperature superplasticity and its deformation mechanism in grain refinement of Al-Mg Alloy by multi-axial alternative forging," *Materials Transactions*, vol. 44, no. 11, pp. 2288–2297, 2003.

[39] Y. T. Zhu, T. C. Lowe, H. Jiang et al., *United States Patent Application*, USP6197129, 2001.

[40] Y. T. Zhu, H. Jiang, J. Y. Huang et al., "A new route to bulk nanostructured metals," *Metallurgical and Materials Transactions A*, vol. 32, no. 6, pp. 1559–1562, 2001.

[41] Y. L. Duan, L. Tang, Y. Deng, X. W. Cao, G. F. Xu, and Z. M. Yin, "Superplastic behavior and microstructure evolution of a new Al-Mg-Sc-Zr alloy subjected to a simple thermomechanical processing," *Materials Science and Engineering A*, vol. 669, pp. 205–217, 2016.

[42] Y. L. Duan, G. F. Xu, X. Y. Peng, Y. Deng, Z. Li, and Z. M. Yin, "Effect of Sc and Zr additions on grain stability and superplasticity of the simple thermal–mechanical processed Al–Zn–Mg alloy sheet," *Materials Science and Engineering A*, vol. 648, pp. 80–91, 2015.

[43] M. Li, Q. Pan, Y. Shi, X. Sun, and H. Xiang, "High strain rate superplasticity in an Al–Mg–Sc–Zr alloy processed via simple rolling," *Materials Science and Engineering A*, vol. 687, pp. 298–305, 2017.

[44] X. G. Jiang, Q. L. Wu, J. Z. Cui et al., "A new grain-refinement process for superplasticity of high-strength 7075 aluminum alloy," *Journal of Materials Science*, vol. 28, pp. 6038-6039, 1993.

[45] D. H. Shin, C. S. Lee, and W. Kim, "Superplasticity of fine-grained 7475 Al alloy and a proposed new deformation mechanism," *Acta Materialia*, vol. 45, no. 12, pp. 5195–5202, 1997.

[46] A. Smolej, M. Gnamus, and E. Slacek, "The influence of the thermomechanical processing and forming parameters on superplastic behaviour of the 7475 aluminium alloy," *Journal of Materials Processing Technology*, vol. 118, no. 1-3, pp. 397–402, 2001.

[47] L. Y. Ye, X. M. Zhang, D. W. Zheng et al., "Superplastic behavior of an Al–Mg–Li alloy," *Journal of Alloys and Compounds*, vol. 487, no. 1-2, pp. 109–115, 2009.

[48] H. P. Pu and J. C. Huang, "Processing routes for inter-transformation between low- and high-temperature 8090 Al–Li superplastic sheets," *Scripta Metallurgica et Materialia*, vol. 33, no. 3, pp. 383–389.

[49] J. Wadsworth and A. R. Pelton, "Superplastic behavior of a powder-source aluminum-lithium based alloy," *Scripta Metallurgica*, vol. 18, no. 4, pp. 387–392, 1984.

[50] E. El-Danaf, S. R. Kalidindi, and R. G. Doherty, "Influence of grain size and stacking-fault energy on deformation twinning in fcc metals," *Metallurgical and Materials Transactions A*, vol. 30, no. 5, pp. 1223–1233, 1999.

[51] A. Rohatgi, K. S. Vecchio, and G. T. Gray III, "A metallographic and quantitative analysis of the influence of stacking fault energy on shock hardening in Cu and Cu–Al alloys," *Acta Materialia*, vol. 49, pp. 427–438, 2001.

[52] N. Zhang, Y. Wang, and H. Hou, "Superplastic deformation behavior of 7B04 Al alloy," *Journal of Materials Engineering*, vol. 45, pp. 27–33, 2017.

[53] A. V. Mikhaylovsaya, O. A. Yakovtseva, and M. N. Sitkina, "Comparison between superplastic deformation mechanisms at primary and steady stages of the fine grain AA7475 aluminium alloy," *Materials Science and Engineering A*, vol. 718, pp. 277–286, 2018.

[54] M. Kh Rabinovich and V. G. Trifonof, "Dynamic grain growth during superplastic deformation," *Acta Materialia*, vol. 44, no. 5, pp. 2073–2078, 1996.

[55] P. Zhang, L. Ye, X. Zhang, G. Gu, H. Jiang, and Y. Wu, "Grain structure and microtexture evolution during superplastic deformation of 5A90 Al-Li alloy," *Transactions of Nonferrous Metals Society of China*, vol. 24, no. 7, pp. 2088–2093, 2014.

[56] Y. Duan, G. Xu, L. Zhou, and D. Xiao, "Achieving high superplasticity of a traditional thermal-mechanical processed non-superplastic Al–Zn–Mg alloy sheet by low Sc additions," *Journal of alloys and compounds*, vol. 638, pp. 364–373, 2015.

[57] H. Xiang, Q. L. Pan, X. H. Yu, X. Huang, X. Sun, and X. D. Wang, "Superplasticity behaviors of Al-Zn-Mg-Zr cold-rolled alloy sheet with minor Sc addition," *Materials Science Engineering A*, vol. 676, pp. 128–137, 2016.

[58] L. Ye, X. Zhang, and D. Zheng, "Superplastic behavior of an Al–Mg–Li alloy," *Journal of alloys and compounds*, vol. 487, pp. 109–115, 2009.

[59] L. Jiao, *Research on Plastic Deformation Behavior and Properties of In Situ Particulate Reinforced Aluminum Matrix Composites*, Jiangsu University, Suzhou, China, 2014.

[60] O. Roder, T. Gysler, and G. Lütjering, "Fatigue properties of Al-Mg alloys with and without scandium," *Materials Science and Engineering A*, vol. 234–236, pp. 181–184, 1997.

[61] R. J. Dashwood, R. Grimes, A. W. Harrison, and H. M. Flower, "The Development of a High Strain Rate Superplastic Al-Mg-Zr Alloy," *Materials Science Forum*, vol. 357–359, pp. 339–344, 2001.

[62] R. Grimes, R. J. Dashwood, A. W. Harrison, and H. M. Flower, "Development of a high strain rate superplastic Al–Mg–Zr alloy," *Materials Science and Technology*, vol. 16, no. 11-12, pp. 1334–1339, 2000.

[63] D. H. Bae and A. K. Ghosh, "Grain size and temperature dependence of superplastic deformation in an Al–Mg alloy under isostructural condition," *Acta Materialia*, vol. 48, no. 6, pp. 1207–1224, 2000.

[64] A. Smolej, D. Klobčar, B. Skaza et al., "Superplasticity of the rolled and friction stir processed Al–4.5 Mg–0.35Sc–0.15Zr alloy," *Materials Science and Engineering A*, vol. 590, pp. 239–245, 2014.

[65] M. Furukawa, P Berbon, and T. G. Langdon, "Age hardening and the potential for superplasticity in a fine-grained Al-Mg-

Li-Zr alloy," *Metallurgical and Materials Transactions A*, vol. 29, no. 1, pp. 169–177, 1998.

[66] M. Furukawa, Y. Iwahashi, Z. Horita, and M. Nemoto, "Structural evolution and the Hall-Petch relationship in an Al-Mg-Li-Zr alloy with ultra-fine grain size," *Acta Materialia*, vol. 45, no. 11, pp. 4751–4757, 1997.

[67] L. Man-ping, S. Shao-chun, H. J. Roven et al., "Deformation defects and electron irradiation effect in nanostructured Al–Mg alloy processed by severe plastic deformation," *Transactions of Nonferrous Metals Society of China*, vol. 22, no. 8, pp. 1810–1816, 2012.

[68] A. V. Mikhaylovskaya, O. A. Yakovtseva, I. S. Glolovin, and A. V. Pozdniakov, "Superplastic deformation mechanisms in fine-grained Al–Mg based alloys," *Materials Science and Engineering A*, vol. 627, pp. 31–41, 2015.

[69] G. Xu, X. Cao, T. Zhang, Y. Duan, X. Peng, and Y. Deng, "Achieving high strain rate superplasticity of an Al-Mg-Sc-Zr alloy by a new asymmetrical rolling technology," *Materials Science and Engineering A*, vol. 672, pp. 98–107, 2016.

[70] R. J. Rioja and J. Liu, "The Evolution of Al-Li base products for aerospace and space applications," *Metallurgical and Materials Transactions A*, vol. 43, no. 9, pp. 3325–3337, 2012.

[71] H. Yin, H. Li, X. Su, and D. Huang, "Processing maps and microstructural evolution of isothermal compressed Al–Cu–Li alloy," *Materials Science and Engineering A*, vol. 586, pp. 115–122, 2013.

[72] Y. Lin and Z. Zheng, "Microstructural evolution of 2099 Al Li alloy during friction stir welding process," *Materials Characterization*, vol. 123, pp. 307–314, 2017.

[73] Y. Lin, Z. Zheng, S. Li, X. Kong, and Y. Han, "Microstructures and properties of 2099 Al-Li alloy," *Materials Characterization*, vol. 84, pp. 88–99, 2013.

[74] A. A. Mazilkin and M. M. Myshlyaev, "Microstructure and thermal stabilityof superplastic aluminium—lithium alloy after severe plastic deformation," *Journal of Materials Science*, vol. 41, no. 12, pp. 3767–3772, 2006.

[75] R. Kaibyshev, K. Shipilova, and F. Musin, "Achieving high strain rate superplasticity in an Al–Li–Mg alloy through equal channel angular extrusion," *Materials Science and Technology*, vol. 21, no. 4, pp. 408–418, 2005.

[76] X. Zhang, L. Xie, L. Ye, and Panzhang, "Effect of aging treatment on grain refinement and superplasticity of 2A97 aluminum-lithium alloy," *Heat Treatment of Metals*, vol. 39, pp. 88–92, 2014.

[77] M. Taheri-Mandarjani, A. Zarei-Hanzaki, and H. R. Abedi, "Hot ductility behavior of an extruded 7075 aluminum alloy," *Materials Science and Engineering A*, vol. 637, pp. 107–122, 2015.

[78] K. Ma, T. Hu, H. Yang, T. Topping, A. Yousefiani, and E. J. Lavernia, "Coupling of dislocations and precipitates: Impact on the mechanical behavior of ultrafine grained Al–Zn–Mg alloys," *Acta Materialia*, vol. 103, pp. 153–164, 2016.

[79] P. K. Rout, M. M. Ghosh, and K. S. Ghosh, "Microstructural, mechanical and electrochemical behaviour of a 7017 Al–Zn–Mg alloy of different tempers," *Materials Characterization*, vol. 104, pp. 49–60, 2015.

[80] K. Sotoudeh and P. S. Bate, "Diffusion creep and superplasticity in aluminium alloys," *Acta Materialia*, vol. 58, no. 6, pp. 1909–1920, 2010.

[81] J. Liu and D. J. Chakrabarti, "Grain structure and microtexture evolution during superplastic forming of a high strength Al–Zn–Mg–Cu alloy," *Acta Materialia*, vol. 44, no. 12, pp. 4647–4661, 1996.

[82] X. Cao, G. Xu, Y. Duan, Z. Yin, and L. Lu, "Achieving high superplasticity of a new Al–Mg–Sc–Zr alloy sheet prepared by a simple thermal–mechanical process," *Materials Science and Engineering A*, vol. 647, pp. 333–343, 2015.

[83] X. Xu, J. Zheng, Z. Li, R. Luo, and B. Chen, "Precipitation in an Al-Zn-Mg-Cu alloy during isothermal aging: Atomic-scale HAADF-STEM investigation," *Materials Science and Engineering A*, vol. 691, pp. 60–70, 2017.

[84] K. Wen, Y. Fan, G. Wang, L. Jin, X. Li, and Z. Li, "Aging behavior and fatigue crack propagation of high Zn-containing Al-Zn-Mg-Cu alloys with zinc variation," *Materials Science and Engineering A*, vol. 27, pp. 217–227, 2017.

[85] F. Jiang, H. S. Zurob, G. R. Purdy, and H. Zhang, "Characterizing precipitate evolution of an Al-Zn-Mg-Cu-based commercial alloy during artificial aging and non-isothermal heat treatments by in situ electrical resistivity monitoring," *Materials Science and Engineering: A*, vol. 117, pp. 47–56, 2016.

[86] K. Wang, F. C. Liu, Z. Y. Ma, and F. C. Zhang, "Realization of exceptionally high elongation at high strain rate in a friction stir processed Al–Zn–Mg–Cu alloy with the presence of liquid phase," *Scripta Materialia*, vol. 64, no. 6, pp. 572–575, 2011.

[87] A. V. Mikhaylovskaya, O. A. Yakovtseva, V. V. Cheverikin, and A. D. kotov, "Superplastic behaviour of Al-Mg-Zn-Zr-Sc-based alloys at high strain rates," *Materials Science and Engineering A*, vol. 659, pp. 225–233, 2016.

[88] R. Kaibyshev, E. Avtokratova, A. Apollonov, and R. Davies, "High strain rate superplasticity in an Al–Mg–Sc–Zr alloy subjected to simple thermomechanical processing," *Scripta Materialia*, vol. 54, no. 12, pp. 2119–2124, 2006.

[89] I. Charit and R. S. Mishra, "Low temperature superplasticity in a friction-stir-processed ultrafine grained Al–Zn–Mg–Sc alloy," *Acta Materialia*, vol. 53, no. 15, pp. 4211–4223, 2005.

[90] R. Raj, "A mechanistic basis for high strain rate superplasticity of aluminum based metal matrix composites," *Materials Science and Engineering*, vol. 215, no. 1-2, pp. 1–8, 1996.

[91] V. N. Perevezentsev and K. Higashi, "Theoretical Investigation of High Strain Rate Superplasticity," *Materials Science Forum*, vol. 304–306, pp. 217–224, 1999.

[92] M. Mabuchi, K. Higashi, K. Inoue et al., "Experimental investigation of superplastic behavior in a 20 vol% aluminum composite," *Scripta Metallurgica et Materialia*, vol. 26, no. 12, pp. 1839–1844, 1992.

[93] K. Higashi, T. Okada, T. Mukai et al., "Superplastic behavior in a mechanically alloyed aluminum composite reinforced with SiC particulates," *Scripta Metallurgica et Materialia*, vol. 26, no. 2, pp. 185–190, 1992.

[94] M. Mabuchi, K. Higashi, and T. G. Langdon, "An investigation of the role of a liquid phase in Al–Cu–Mg metal matrix composites exhibiting high strain rate superplasticity," *Acta Metallurgica et Materialia*, vol. 42, no. 5, pp. 1739–1745, 1994.

[95] C. Tang, M. Li, Z. Li et al., "Superplasticity of Al composite fabricated by spray deposition processing," *Journal of Materials Engineering*, vol. 3, pp. 17–19, 1996.

[96] G. H. Zahid, R. I. Todd, and P. B. Prangnell, "Superplasticity in an alulllinium alloy 2124/SiCp composite," *Materials Science and Technology*, vol. 14, no. 9, pp. 901–905, 1998.

[97] X. Xu, W. Wang, and L. Cai, "Superplasticity of A SiCP/6A02 Al composite," *Materials for Mechanical Engineering*, vol. 26, no. 9, pp. 20-21, 2002.

[98] X. Xu, K. Chen, F. Dai, and L. Cai, "Investigation on superplasticity of SiC particulate reinforced aluminum matrix composite," *Acta Metallurgica Sinica*, vol. 38, no. 5, pp. 544–548, 2000.

[99] F. R. Cao, H. Ding, Y. L. Li, G. Zhou, and J. Z. Cui, "Superplasticity, dynamic grain growth and deformation mechanism in ultra-light two-phase magnesium–lithium alloys," *Materials Science and Engineering A*, vol. 527, no. 9, 2010.

[100] I. I. Novikov, V. K. Portnoy, V. S. Levchenko, and A. O. Nikiforov, "Subscolidus superplasticity of aluminium alloys," *Materials Science Forum*, vol. 243–245, pp. 463–468, 1997.

[101] J. C. Tan and M. J. Tan, "Superplasticity and grain boundary sliding characteristics in two stage deformation of Mg–3Al–1Zn alloy sheet," *Materials Science and Engineering A*, vol. 339, no. 1-2, pp. 81–89, 2003.

[102] O. A. Ruano and O. D. Sherby, "On constitutive equations for various diffusion-controlled creep mechanisms," *Revue de Physique Appliquee*, vol. 23, no. 4, pp. 625–637, 1988.

[103] T. G. Nieh, J. Wadsworth, and O. D. Sherby, *Superplasticity in Metals and Ceramics*, Cambridge University Press, Cambridge, UK, 1997.

[104] M. E. Kassner and M. T. Pérez-Prado, "Five-power-law creep in single phase metals and alloys," *Progress in Materials Science*, vol. 45, no. 1, pp. 1–102, 2000.

[105] Y. L. Duan, G. F. Xu, D. Xiao, L. Q. Zhou, Y. Deng, and Z. M. Yin, "Excellent superplasticity and deformation mechanism of Al–Mg–Sc–Zr alloy processed via simple free forging," *Materials Science and Engineering A*, vol. 624, pp. 124–131, 2015.

[106] J. Liu and D. J. Chakrabarti, "Grain structure and microtexture evolution during superplastic forming of a high strength Al–Zn–Mg–Cu alloy," *Acta Materialia*, vol. 44, no. 12, pp. 4641–4661, 1996.

[107] X. Cao, G. Xu, Y. Duan, Z. Yin, and Y. Wang, "Achieving high superplasticity of a new Al–Mg–Sc–Zr alloy sheet prepared by a simple thermal–mechanical process," *Materials Science and Engineering A*, vol. 647, pp. 333–343, 2015.

[108] P. S. Bate, F. J. Humpherys, N. Ridley, and B. Zhang, "Microstructure and texture evolution in the tension of superplastic Al–6Cu–0.4Zr," *Acta Materialia*, vol. 53, no. 10, pp. 3059–3069, 2005.

[109] S. Katsas, R. Dashwood, R. Grimes, M. Jackson, G. Todd, and H. Henein, "Dynamic recrystallisation and superplasticity in pure aluminium with zirconium addition," *Materials Science and Engineering A*, vol. 444, no. 1-2, pp. 291–297, 2007.

[110] K. Sotoudeh and P. S. Bate, "Diffusion creep and superplasticity in aluminium alloys," *Acta Materialia*, vol. 58, no. 6, pp. 1909–1920, 2010.

[111] R. H. Bricknell and J. W. Ddington, "Superplasticity in the commercial Al-Zn-Mg alloy BA 708," *Metallurgical Transactions A*, vol. 7, no. 1, pp. 153-154, 1976.

[112] J. A. del Valle, M. Teresa Pérez–Prado, and O. A. Ruano, "Symbiosis between grain boundary sliding and slip creep to obtain high-strain-rate superplasticity in aluminum alloys," *Journal of the European Ceramic Society*, vol. 27, no. 11, pp. 3385–3390, 2007.

[113] L. B. Johannes and R. S. Mishra, "Multiple passes of friction stir processing for the creation of superplastic 7075 aluminum," *Materials Science and Engineering A*, vol. 464, no. 1-2, pp. 255–260, 2007.

[114] A. Ball, "Superplasticity in the aluminium-zinc eutectoid—an early model revisited," *Materials Science and Engineering A*, vol. 234–236, pp. 365–369, 1997.

[115] A. K. Mukherjee, "The rate controlling mechanism in superplasticity," *Materials Science and Engineering*, vol. 8, no. 2, pp. 83–89, 1971.

[116] R. G. Gifkins, "Grain-boundary sliding and its accommodation during creep and superplasticity," *Metallurgical Transactions A*, vol. 7, no. 8, pp. 1225–1232, 1976.

[117] A. Ball and M. M. Hutchison, "Superplasticity in the Aluminium–Zinc Eutectoid," *Journal of Metal Science*, vol. 3, pp. 1–7, 1969.

[118] K. Mukherjee, "The rate controlling mechanism in superplasticity," *Materials Science and Engineering*, vol. 8, no. 2, pp. 83–89, 1971.

[119] M. F. Ashby and R. A. Verrall, "Diffusion-accommodated flow and superplasticity," *Acta Metallurgica*, vol. 21, no. 2, pp. 149–163, 1973.

[120] W. A. Backofen, I. R. Turner, and D. H. Avery, "Superplasticity in an Al-Zn alloy," *Transactions of ASM*, vol. 57, pp. 980–990, 1964.

[121] K. I. Sugimoto, U. Noboru, M. Kobayashi et al., "Effects of volume fraction and stability of retained austenite on ductility of TRIP-aided dual-phase steels," *ISIJ International*, vol. 32, no. 12, pp. 1311–1318, 1992.

[122] A. Orozco-Caballero and A. Orozco-Caballero, "Grain size versus microstructural stability in the high strain rate superplastic response of a severely friction stir processed Al-Zn-Mg-Cu alloy," *Materials Science and Engineering A*, vol. 680, pp. 329–337, 2017.

[123] N. Fakhar, F. Fereshteh-Saniee, and R. Mahmudi, "High strain-rate superplasticity of fine- and ultrafine-grained AA5083 aluminum alloy at intermediate temperatures," *Materials and Design*, vol. 85, pp. 342–348, 2015.

[124] N. Fakhar, F. Fereshteh-Saniee, and R. Mahmudi, "Grain refinement and high strain rate superplasticity in alumunium 2024 alloy processed by high-pressure torsion," *Materials Science and Engineering A*, vol. 622, pp. 139–145, 2015.

[125] J. Lei, W. Xiaolu, and L. Hui, "High strain rate superplasticity of in situ Al 3 Zr/6063Al composites," *Rare Metal Materials and Engineering*, vol. 45, no. 11, pp. 2798–2803, 2016.

[126] B. Kim, J. C. Kim, S. Lee, and K.-S. Lee, "High-strain-rate superplasticity of fine-grained Mg–6Zn–0.5Zr alloy subjected to low-temperature indirect extrusion," *Scripta Materialia*, vol. 141, pp. 138–142, 2017.

[127] A. Orozco-Caballero, C. M. Cepeda-Jiménez, and P. Hidalgo-Manrique, "Lowering the temperature for high strain rate superplasticity in an Al–Mg–Zn–Cu alloy via cooled friction stir processing," *Materials Chemistry and Physics*, vol. 142, no. 1, pp. 182–185, 2013.

[128] Z. Y. Ma, R. S. Mishra, M. W. Mahoney, and R. Grimes, "High strain rate superplasticity in friction stir processed Al–Mg–Zr alloy," *Materials Science and Engineering A*, vol. 351, no. 1-2, pp. 148–153, 2003.

[129] Z. Y. Ma and R. S. Mishra, "Development of ultrafine-grained microstructure and low temperature (0.48 Tm) superplasticity in friction stir processed Al–Mg–Zr," *Scripta Materialia*, vol. 53, pp. 75–80, 2005.

[130] T. G. Nieh, L. M. Hsiung, J. Wadsworth, and R. Kaibyshev, "High strain rate superplasticity in a continuously recrystallized Al–6%Mg–0.3%Sc alloy," *Acta Materialia*, vol. 46, no. 8, pp. 2789–2880, 1998.

[131] K. L. Kending and D. B. Miracle, "Strengthening mechanisms of an Al-Mg-Sc-Zr alloy," *Acta Materialia*, vol. 50, no. 16, pp. 4165–4175, 2002.
[132] Y. N. Wang and J. C. Huang, "Comparison of grain boundary sliding in fine grained Mg and Al alloys during superplastic deformation," *Scripta Materialia*, vol. 48, no. 8, pp. 1117–1122, 2003.
[133] K. T. Park, S. H. Myung, D. H. Shin, and C. S. Lee, "Size and distribution of particles and voids pre-existing in equal channel angular pressed 5083 Al alloy: their effect on cavitation during low-temperature superplastic deformation," *Materials Science and Engineering*, vol. 371, no. 1-2, pp. 178–186, 2004.

3

Modeling of Flow Stress of 2026 Al Alloy under Hot Compression

Zheng-bing Xiao,[1,2] Yuan-chun Huang,[1,3] and Yu Liu[3]

[1]*School of Mechanical and Electrical Engineering, Central South University, Changsha 410083, China*
[2]*Collaborative Innovation Center of Advanced Nonferrous Structural Materials and Manufacturing, Central South University, Changsha 410083, China*
[3]*Light Alloy Research Institute, Central South University, Changsha 410012, China*

Correspondence should be addressed to Zheng-bing Xiao; xiaozb@csu.edu.cn

Academic Editor: Sutasn Thipprakmas

In order to investigate the workability and to optimize the hot forming parameters for 2026 Al alloy, hot compression tests were performed in the temperature range of 350~450°C with strain rates of 0.01~10 s^{-1} and 60% deformation degree on a Gleeble-1500 thermosimulation machine. The true stress-strain curves obtained exhibit that the stress increases dramatically at small strains and then moves forward to a steady state, showing dynamic flow softening. Meanwhile, on the basis of Arrhenius equation, a constitutive equation on the flow stress, temperature, and strain rate was proposed. Yet, the values of the predicated stress from the equation and the true stress differ by as much as 50.10%. Given the intricate impact of precipitation of the second phases on the strength of 2026 Al alloy, the introduction of a revised equation with the reinforcement of temperature was carried out, fitting well with the experiment data at peak stresses. What is more, both pictures obtained by scanning electron microscopy (SEM) and transmission electron microscopy (TEM) were compatible with all the inferences.

1. Introduction

Understanding the materials flow behavior during hot deformation is important for materials processing designers, especially for those dealing with metal processes like hot forging, rolling, and extrusion [1–3]. Usually, constitutive equations describing relationships on the strain rate, temperature, and flow stress, which can be used in the finite element simulation, are helpful in selecting the appropriate working parameters for a given metal to process. A great deal of investigation had attempted to set up constitutive equations to predict the flow stress, especially for the frequently used alloys like aluminum alloys, steels, and nickel based super alloys [4–10]. Generally, constitutive models now used can be classified into phenomenological type, physical-based type, and artificial neural network (ANN) type [11]. The Arrhenius equation of phenomenological type was the main one for its simplicity and accuracy [6, 12–16]. Numerous papers have been published based on the equations proposed by Sellars and Zener [17–19], either the constitutive equations or the modified ones, especially in the hot working of steels, all agreeing well with experiment results, assisting in selecting the hot deformation parameters [12, 20–27].

But, unlike steels, microstructures of aluminum alloys will significantly change due to precipitation in the hot deformation processing, not to mention that those can be strengthened by heat treatment, 2xxx and 7xxx alloys, for example, which usually serve as skins and frames of airplanes. Second phases, like Al_2Cu in 2xxx alloys and $MgZn_2$ in 7xxx alloys, usually appear in the hot working and heat treatment, which need substantial increase of the stress to continue the hot deformation [15, 28–32]. Prediction of the peak stress for those alloys is hard for their temperature-dependent nature, but the stress-strain relationship, especially the peak stress, is crucial for processing, since it determines the selection of the machines to fulfill the upper compressive stress needed.

As an improved version of 2024 Al alloy, 2026 Al alloy with lower Fe and Si content but minor addition of Zr to inhibit recrystallization during hot working was established by Alcoa in 1999, satisfying demand for better safety and durability without sacrifice of strength and toughness [33]. As a new alloy of high strength and high damage tolerance, 2026 Al alloy is used as the upholstery skin material for lower aero foil in airplanes like A380 from hot rolling.

Unlike the traditional materials, works about 2026 Al alloy are rare to find, and the most existent investigations on 2026 Al alloy mainly focus on the microstructure evolution during hot processing or the precipitation of the second phases [33, 34]. To the authors' knowledge, there is little information available in the literature about flow behavior of 2026 Al alloy; further study is needed to perform numerical simulation and to select proper process parameters. This paper will put forward and verify a new constitutive equation about the flow behavior of 2026 Al alloy under hot deformation.

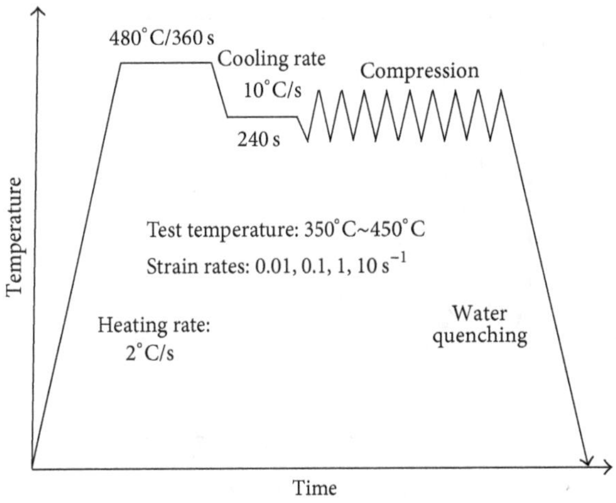

FIGURE 1: Experimental procedure for hot compression tests.

2. Experiments

A commercial 2026 Al alloy of chemical composition (wt.%) 1.33Mg-4.20Cu-0.55Mn-0.15Zr-(bal.)Al, with minor elements like Si and Fe below 0.05 and 0.07 (wt.%), respectively, was used in the investigation. Specimens were prepared with diameter of 10 mm and height of 15 mm from the direct cast ingot homogenized for 490°C × 12 h + 520°C × 12 h. Friction between the specimens and upsetting dies was minimized by entrapping the lubricant with composition of 75% graphite + 20% machine oil + 5% nitric acid trimethyl benzene grease in a depth of 0.1 mm recessed into both ends of the specimens.

As depicted in Figure 1, the specimens were heated to 480°C at a heating rate of 2°C/s and held for 6 min and then cooled to the deformation temperature at 10°C/s and held for 4 min before test. The hot compression tests were carried out on a Gleeble-1500 thermosimulation machine at four different strain rates ($0.01\,s^{-1}$, $0.1\,s^{-1}$, $1\,s^{-1}$, and $10\,s^{-1}$) and four different temperatures (300°C, 350°C, 400°C, and 450°C), with 60% final reduction in height. After the hot compression, specimens were quenched in water immediately for micrographic observation.

3. Results and Discussion

3.1. True Stress and Strain. The true stress-strain curves obtained from hot compression tests for 2026 Al alloy are shown in Figure 2. For aluminum alloys of high stacking fault energy, dynamic recovery is the main reason for the softening phenomenon [35], which is confirmed by the occurrence of dynamic softening on the curves in this work, for the flow stresses increase monotonically up to an almost steady state (e.g., samples deformed at strain rate of $0.01\,s^{-1}$). As can be seen from the figures, the strain rate and temperature significantly affect the flow stresses. The true stress level will rise once deformation temperature decreases and strain rate increases.

Under low temperatures, restraint of the number of slip systems reduces the mobility at grain boundaries, thus increasing the flow stress. This was approved further by the dislocation walls in Figure 3, for the TEM observation of the sample at 300°C/10 s^{-1}, which shows the highest flow stress during compression.

By raising the deformation temperature, dislocation slip, climb, and annihilation correspondingly increase, together with vacancy diffusion rate, resulting in easier activation of dynamic recovery and even dynamic recrystallization to overcome the hardening. Figure 4 shows, for the sample deformed at $0.01\,s^{-1}$/450°C, subgrains formed at grain boundaries, evidence of dynamic recrystallization, which can greatly contribute to the decrease of the flow stress. However, with the increase of strain rate, the time for energy accumulation, dynamic recovery, and recrystallization is shortened, and the true stress would increase.

3.2. Deformation Constitutive Equation. Usually, the relationship regarding flow stress, deformation temperature, and strain rate can be expressed by Arrhenius equation. Zener-Hollomon parameter Z in an exponent-type equation represents the effects of strain rate and temperature on the deformation behaviors. Consequently, Arrhenius equation can be used to connect Z parameter with stress [18, 19]. Hence,

$$Z = \dot{\varepsilon} \exp\left(\frac{Q}{RT}\right), \tag{1}$$

$$\dot{\varepsilon} = AF(\sigma) \exp\left(-\frac{Q}{RT}\right), \tag{2}$$

where

$$F(\sigma) = \begin{cases} \sigma^{n'} & \alpha\sigma < 0.8 \\ \exp(\beta\sigma) & \alpha\sigma > 1.2 \\ \sinh(\alpha\sigma)^{n} & \forall\sigma, \end{cases} \tag{3}$$

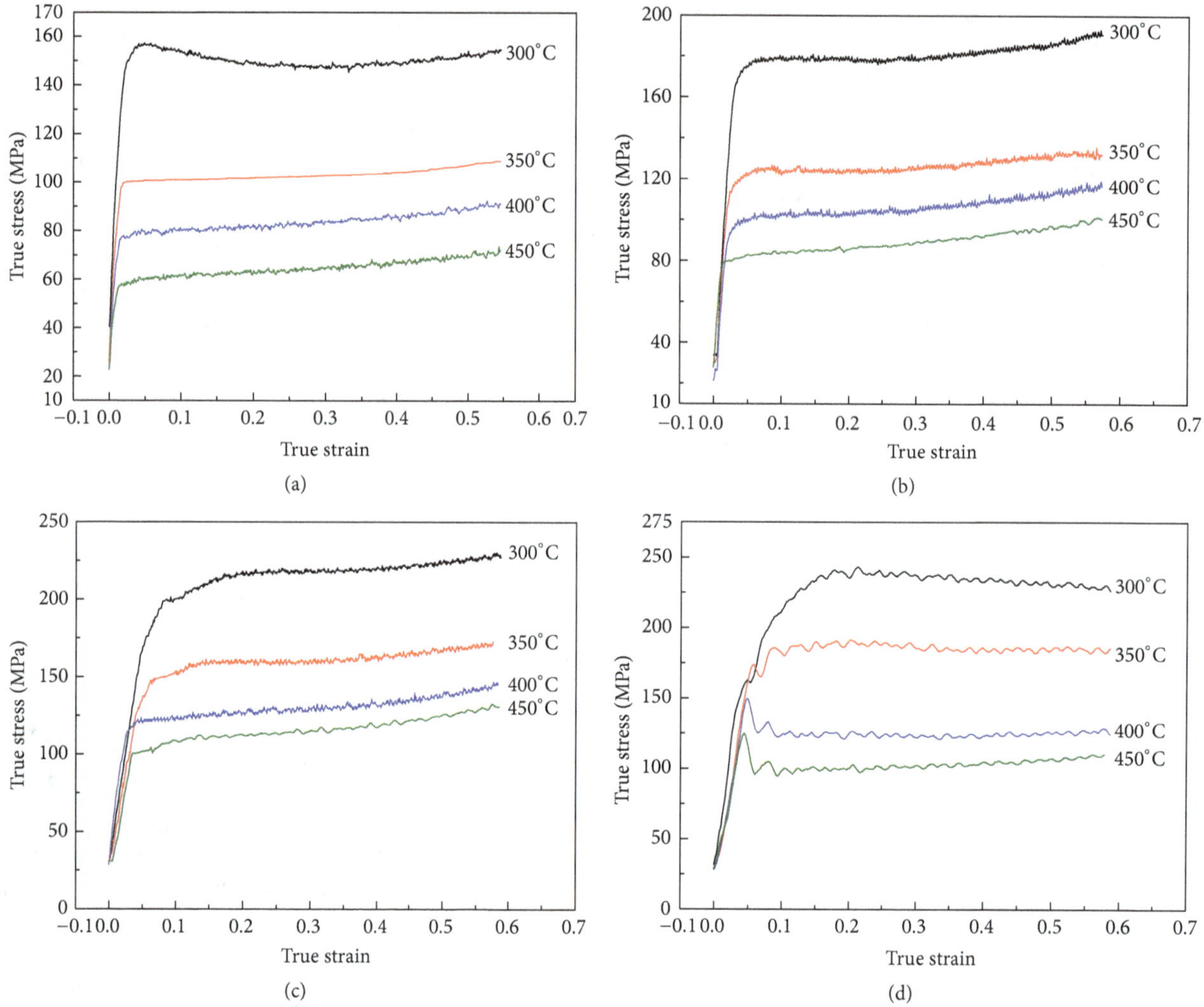

FIGURE 2: Typical true stress-strain curves for 2026 Al alloy under different deformation temperatures with strain rates of (a) $0.01\,s^{-1}$; (b) $0.1\,s^{-1}$; (c) $1\,s^{-1}$; and (d) $10\,s^{-1}$.

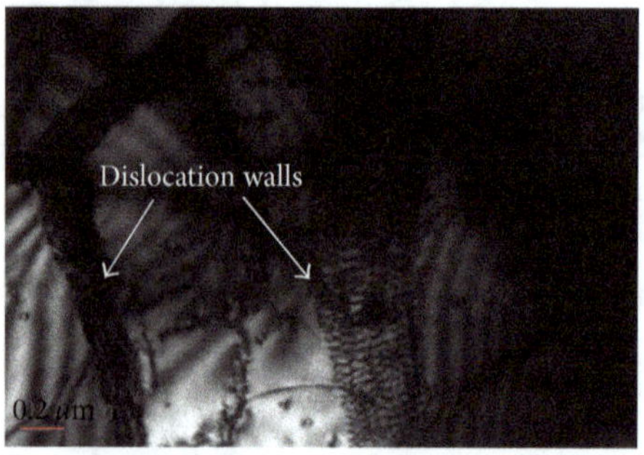

FIGURE 3: TEM micrograph of the sample deformed at temperature of 300°C with the strain rate of $10\,s^{-1}$ with dislocation walls.

in which $\dot{\varepsilon}$ is the stain rate (s^{-1}), Q is the activation energy of hot deformation ($kJmol^{-1}$), R is the universal gas constant ($8.31\,Jmol^{-1}K^{-1}$), T is the absolute temperature (K), and σ is the flow stress (MPa) for a given strain, while A, α, and n, n' are the material constants, $\alpha = \beta/n'$.

Equations (4) and (4) are the substitution of the power law and exponential law of $F(\sigma)$ into (1) for the low-stress ($\alpha\sigma < 0.8$) and high-stress level ($\alpha\sigma > 1.2$), respectively:

$$\dot{\varepsilon} = B\sigma^{n'}, \tag{4}$$

$$\dot{\varepsilon} = B' \exp\left(\beta\sigma\right)\frac{1}{2}, \tag{5}$$

where B and B' are the material constants. To find out the material constants in the above equations, deformation strain of peak stress is used here as an example.

Equations (6) and (6) are the logarithm form of (4) and (4), respectively:

$$\ln\left(\sigma\right) = \frac{1}{n'}\ln\left(\dot{\varepsilon}\right) - \frac{1}{n'}\ln\left(B\right), \tag{6}$$

$$\sigma = \frac{1}{\beta}\ln\left(\dot{\varepsilon}\right) - \frac{1}{\beta}\ln\left(B'\right). \tag{7}$$

Substituting the values of the peak stress and corresponding strain rate at different deformation temperatures into (6)

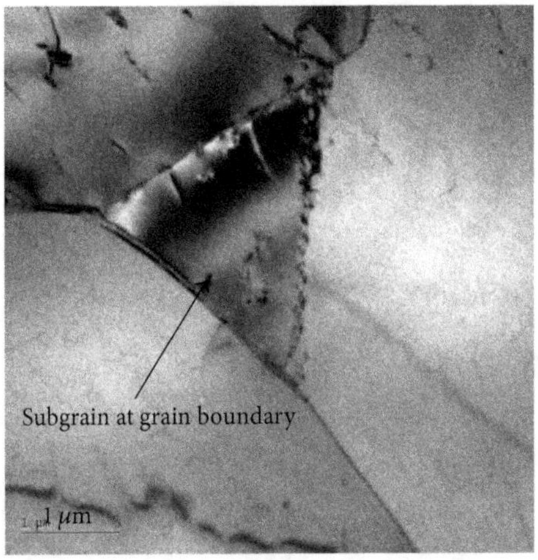

FIGURE 4: TEM micrograph of the sample deformed at temperature of 450°C with the strain rate of 0.01 s^{-1}.

and (6) can obtain the relationship regarding the stress and strain rate (in logarithm form) shown in Figure 5.

As can be seen in Figure 5, a group of straight lines approximated is close to parallel. The slopes of these lines have similar values and values of n' and β at different deformation temperatures can be derived by linear fitting method. As $\alpha = \beta/n'$, each corresponding value of α can be calculated. By average, the mean value of α is 8.0128×10^{-3}.

Equation (1) can be expressed as follows, fit for all the stress level σ:

$$\dot{\varepsilon} = A\left[\sinh(\alpha\sigma)\right]^{n} \exp\left(-\frac{Q}{RT}\right). \tag{8}$$

Taking the logarithm of (8) for both sides gives

$$\ln\left[\sinh(\alpha\sigma)\right] = \frac{\ln\dot{\varepsilon}}{n} + \frac{Q}{nRT} - \frac{\ln A}{n}. \tag{9}$$

Substituting the peak stress and relevant strain rate for all the tested temperatures into (9) gives the relationship shown in Figure 6. It is not hard to obtain the value of n, 9.343.

For a fixed strain rate, derivative of (8) gives

$$Q = Rn\frac{d\left\{\ln\left[\sinh(\alpha\sigma)\right]\right\}}{d(1/T)}. \tag{10}$$

According to (10), $\ln[\sinh(\alpha\sigma)]$ can be regarded as a function of $1/T$, shown in Figure 7, and the activation energy can be computed as 229.546 kJ/mol by averaging the values of Q at four different strain rates.

On the basis of the value of n and Q known, it is easy to get the value of A, $5.17 \times 10^{22}\ s^{-1}$, $9.53 \times 10^{22}\ s^{-1}$, $8.3 \times 10^{22}\ s^{-1}$, and $3.63 \times 10^{22}\ s^{-1}$ for 300°C, 350°C, 400°C, and 450°C, respectively.

Then, (8) can be expressed below by substituting the values obtained above:

$$\dot{\varepsilon} = A\left[\sinh\left(8.0128 \times 10^{-3}\sigma\right)\right]^{9.343} \cdot \exp\left(\frac{-2.29546 \times 10^{5}}{RT}\right). \tag{11}$$

Finally, Zener-Hollomon parameter Z in (1) can be represented as follows:

$$Z = \dot{\varepsilon}\exp\left(\frac{2.29546 \times 10^{5}}{RT}\right) = A\left[\sinh\left(8.0128 \times 10^{-3}\sigma\right)\right]^{9.343}. \tag{12}$$

According to (8) and (1), relationships between the flow stress σ and Zener-Hollomon parameter Z can be described below:

$$\sigma = \frac{1}{\alpha}\ln\left\{\left(\frac{Z}{A}\right)^{1/n} + \left[\left(\frac{Z}{A}\right)^{2/n} + 1\right]^{1/2}\right\}. \tag{13}$$

Or

$$\sigma = 124.8\ln\left\{\left(\frac{Z}{A}\right)^{1/9.343} + \left[\left(\frac{Z}{A}\right)^{2/9.343} + 1\right]^{1/2}\right\}. \tag{14}$$

Depending on the deformation temperature, the constant A in (11) to (14) is $5.17 \times 10^{22}\ s^{-1}$, $9.53 \times 10^{22}\ s^{-1}$, $8.3 \times 10^{22}\ s^{-1}$, and $3.63 \times 10^{22}\ s^{-1}$ for 300°C, 350°C, 400°C, and 450°C, respectively.

3.3. Modification of Zener-Hollomon Parameter. The above-developed constitutive equations for 2026 Al alloy at peak stress state were verified by comparison of the values from experiments and constitutive equation; differences between them are shown in Table 1. In the table, error is defined as

$$\text{Error} = \frac{\left|\sigma_c - \sigma_m\right|}{\sigma_m} \times 100\%. \tag{15}$$

σ_c is the stress calculated by the constitutive equation, and σ_m is the peak stress measured by experiments.

Apparently, errors between the experiment and predicated results range from 9.86% to 50.10%, indicating the need to modify the constitutive equation.

Usually, four different stages, namely, work hardening stage (stage 1), stable stage (stage 2), softening stage (stage 3), and steady stage (stage 4), show up on the true strain-true stress curves of typical steels, which can be explained well from the viewpoint of dynamic recovery and dynamic recrystallization under hot deformation [36, 37]. However, for almost all of the Al alloys, precipitation of the second phases inevitably occurs in the hot deformation procedure; 2026 Al alloy is no exception. From Figure 8, the SEM and TEM microstructures of the sample deformed at 350°C with the strain rate of 1 s^{-1} and the final strain of 0.6; the matrix was occupied by numerous tiny white particles (most of them are Al_2Cu), a combining effect of temperature and strain whose influence on flow stress cannot be neglected.

Theoretically, by hindering recrystallization or being coherent with α-Al, precipitation can enhance the material strength to some degree, reflected by the rise of the true stress level on the true stress-true strain curve [33, 34]. Typically speaking, precipitation itself is a complicated factor whose impact on the true stress is relevant to temperature. Therefore,

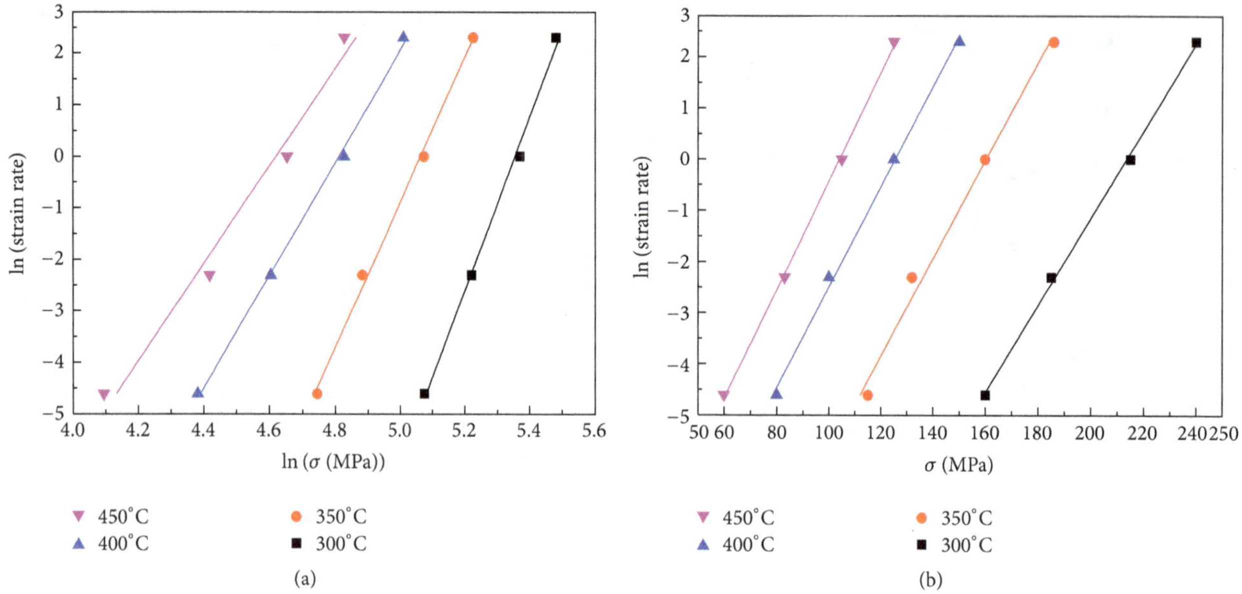

FIGURE 5: Relationships between strain rate and peak stresses: (a) $\ln \dot{\varepsilon}$ and $\ln \sigma$; (b) $\ln \dot{\varepsilon}$ and σ.

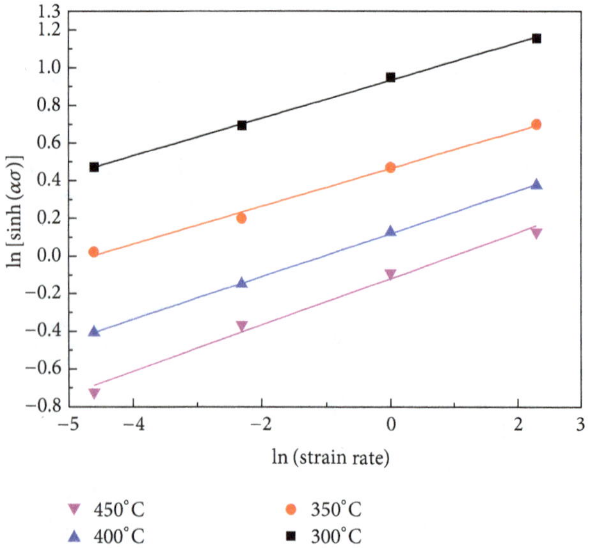

FIGURE 6: Relationship between $\ln [\sinh(\alpha\sigma)]$ and $\ln \dot{\varepsilon}$.

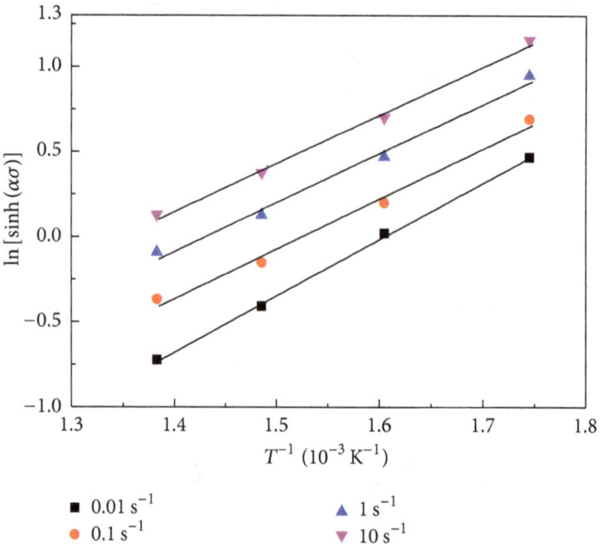

FIGURE 7: Relationship between $\ln [\sinh(\alpha\sigma)]$ and temperature.

the influence of temperature on the flow stresses should be enforced here for the nature of 2026 Al alloy.

Through numerous repeated calculations and validation, it is found that, to make the Z parameter better conform to the real, (1) should multiply with $\exp(22.93975X - 135.508)$. So, the revised Z' parameter can be expressed as

$$Z' = \exp(22.93975X - 135.508)\,\dot{\varepsilon}\exp\left(\frac{Q}{RT}\right), \tag{16}$$

where factor X represents the deformation temperature in °C; it represents the influences of the temperature on the precipitation behavior of the alloy.

Therefore, the flow stress constitutive equation expressed as (13) can be renewed as

$$\sigma = \frac{1}{\alpha}\ln\left\{\left(\frac{Z'}{A}\right)^{1/n} + \left[\left(\frac{Z'}{A}\right)^{2/n} + 1\right]^{1/2}\right\}. \tag{17}$$

Parameters of α, n, and A in the equation are not changed as before, while Z is the modified one from (16).

3.4. Verification of the Revised Constitutive Equation. Table 2 shows the comparisons between the peak stresses calculated from the revised constitutive equation and the experiment data.

Table 1: Comparisons of the predicated (σ_P) and measured (σ_E) peak stress (unit: MPa) before the modification of Z parameter.

Strain rate (s^{-1})	Temperature (°C)	σ_E	σ_P	Error analysis		
				Error (%)	Mean (%)	Standard deviation (%)
0.01	300	160	239.18	49.49	29.20	19.42
	350	115	126.34	9.86		
	400	80	67.53	15.58		
	450	72	41.85	41.87		
0.1	300	191	277.79	45.44	27.49	15.24
	350	132	150.90	14.32		
	400	100	84.64	15.36		
	450	83	54.09	34.83		
1	300	228	318.32	39.61	25.35	13.74
	350	160	177.83	11.14		
	400	125	104.57	16.34		
	450	105	68.99	34.30		
10	300	240	360.24	50.10	26.83	17.67
	350	186	206.67	11.11		
	400	150	127.03	15.31		
	450	125	86.51	30.79		

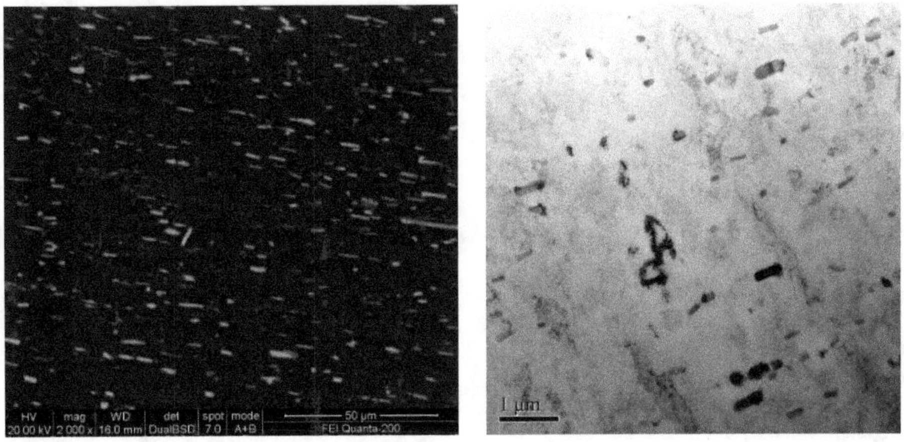

(a) (b)

Figure 8: Microstructure after hot deformation at true strain of 0.6 with the deformation temperature of 350°C and strain rate of 1 s^{-1}: (a) backscattered electron image and (b) TEM micrograph.

As observed from the values, agreement between the calculated and measured values is good, and the maximum relative error is only 5.87%. The results indicated that the modified Z parameter can give a good estimate of the peak stresses for 2026 Al alloy under different hot compression temperature and strain rate and can be used in guidance of the hot processing of 2026 Al alloy.

4. Conclusions

In this study, the effects of deformation parameters on the true stress-strain behavior of 2026 Al alloy were investigated. According to the experiment data, a constitutive equation incorporating the effects of strain rates and temperatures is obtained, while the errors between the experiment and predicated results range from 9.86% to 50.10%. So a revised constitutive equation is derived by compensation of temperature considering the impact of the second phase on the true stress; a good agreement between the predicated and the experiment results indicated that the revised constitutive equation can give an accurate prediction of the peak stress for 2026 Al alloy and can be used in the hot processing of the 2026 Al alloy. In addition, pictures about the microstructures were obtained using scanning electron microscopy (SEM) and transmission electron microscopy (TEM) for verification and assistance.

Competing Interests

The authors declare that they have no competing interests.

Table 2: Comparisons of the predicated (σ_P) and measured (σ_E) peak stress (unit: MPa) after the modification of Z parameter.

Strain rate (s^{-1})	Temperature (°C)	σ_E	σ_P	Error (%)	Error analysis Mean (%)	Standard deviation (%)
0.01	300	160	169.37	5.86	3.03	2.4
	350	115	115.28	0.24		
	400	80	81.72	2.15		
	450	72	69.22	3.87		
0.1	300	191	202.22	5.87	4.14	2.03
	350	132	138.54	4.96		
	400	100	101.21	1.21		
	450	83	86.76	4.53		
1	300	228	238.21	4.48	2.53	1.44
	350	160	164.37	2.73		
	400	125	123.29	1.37		
	450	105	106.60	1.52		
10	300	240	246.77	2.82	2.62	0.72
	350	186	192.33	3.40		
	400	150	147.50	1.66		
	450	125	128.22	2.58		

Acknowledgments

This work was supported by the National Program on Key Basic Research Project of China (no. 2012CB619504 and no. 2014CB046702). The authors thank Lin Y. C. and his coworkers in the School of Mechanical and Electrical Engineering, Central South University, Changsha, for their kind and helpful theoretical guidance and discussion.

References

[1] A. Łukaszek-Sołek and J. Krawczyk, "The analysis of the hot deformation behaviour of the Ti-3Al-8V-6Cr-4Zr-4Mo alloy, using processing maps, a map of microstructure and of hardness," *Materials & Design*, vol. 65, pp. 165–173, 2015.

[2] T.-D. Kil, J.-M. Lee, and Y.-H. Moon, "Quantitative formability estimation of ring rolling process by using deformation processing map," *Journal of Materials Processing Technology*, vol. 220, pp. 224–230, 2015.

[3] S. Saadatkia, H. Mirzadeh, and J.-M. Cabrera, "Hot deformation behavior, dynamic recrystallization, and physically-based constitutive modeling of plain carbon steels," *Materials Science and Engineering: A*, vol. 636, pp. 196–202, 2015.

[4] H. Mirzadeh, J. M. Cabrera, and A. Najafizadeh, "Constitutive relationships for hot deformation of austenite," *Acta Materialia*, vol. 59, no. 16, pp. 6441–6448, 2011.

[5] M. El Mehtedi, F. Gabrielli, and S. Spigarelli, "Hot workability in process modeling of a bearing steel by using combined constitutive equations and dynamic material model," *Materials & Design*, vol. 53, pp. 398–404, 2014.

[6] M. El Mehtedi, F. Musharavati, and S. Spigarelli, "Modelling of the flow behaviour of wrought aluminium alloys at elevated temperatures by a new constitutive equation," *Materials and Design*, vol. 54, pp. 869–873, 2014.

[7] T.-H. Pham, J. J. Kim, and S.-E. Kim, "Estimating constitutive equation of structural steel using indentation," *International Journal of Mechanical Sciences*, vol. 90, pp. 151–161, 2015.

[8] C.-L. Gan, K.-H. Zheng, W.-J. Qi, and M.-J. Wang, "Constitutive equations for high temperature flow stress prediction of 6063 Al alloy considering compensation of strain," *Transactions of Nonferrous Metals Society of China (English Edition)*, vol. 24, no. 11, Article ID 63492, pp. 3486–3491, 2014.

[9] Y. Liu, Z. Yao, Y. Ning et al., "The flow behavior and constitutive equation in isothermal compression of FGH4096-GH4133B dual alloy," *Materials and Design*, vol. 63, pp. 829–837, 2014.

[10] X. Peng, H. Guo, Z. Shi, C. Qin, and Z. Zhao, "Constitutive equations for high temperature flow stress of TC4-DT alloy incorporating strain, strain rate and temperature," *Materials and Design*, vol. 50, pp. 198–206, 2013.

[11] Y. C. Lin and X.-M. Chen, "A critical review of experimental results and constitutive descriptions for metals and alloys in hot working," *Materials & Design*, vol. 32, no. 4, pp. 1733–1759, 2011.

[12] D. Samantaray, C. Phaniraj, S. Mandal, and A. K. Bhaduri, "Strain dependent rate equation to predict elevated temperature flow behavior of modified 9Cr-1Mo (P91) steel," *Materials Science and Engineering: A*, vol. 528, no. 3, pp. 1071–1077, 2011.

[13] N. Haghdadi, A. Zarei-Hanzaki, and H. R. Abedi, "The flow behavior modeling of cast A356 aluminum alloy at elevated temperatures considering the effect of strain," *Materials Science and Engineering A*, vol. 535, pp. 252–257, 2012.

[14] B.-J. Lv, J. Peng, D.-W. Shi, A.-T. Tang, and F.-S. Pan, "Constitutive modeling of dynamic recrystallization kinetics and processing maps of Mg-2.0Zn-0.3Zr alloy based on true stress-strain curves," *Materials Science and Engineering A*, vol. 560, pp. 727–733, 2013.

[15] M. Zhou, Y. C. Lin, J. Deng, and Y.-Q. Jiang, "Hot tensile deformation behaviors and constitutive model of an Al–Zn–Mg–Cu alloy," *Materials & Design*, vol. 59, pp. 141–150, 2014.

[16] H. Mirzadeh, "Constitutive modeling and prediction of hot deformation flow stress under dynamic recrystallization conditions," *Mechanics of Materials*, vol. 85, pp. 66–79, 2015.

[17] C. M. Sellars, "The kinetics of softening processes during hot working of austenite," *Czechoslovak Journal of Physics B*, vol. 35, no. 3, pp. 239–248, 1985.

[18] C. M. Sellars and W. J. McTegart, "On the mechanism of hot deformation," *Acta Metallurgica*, vol. 14, no. 9, pp. 1136–1138, 1966.

[19] C. Zener and J. H. Hollomon, "Effect of strain rate upon plastic flow of steel," *Journal of Applied Physics*, vol. 15, no. 1, pp. 22–32, 1944.

[20] G. Z. Quan, Y. Tong, G. Luo, and J. Zhou, "A characterization for the flow behavior of 42CrMo steel," *Computational Materials Science*, vol. 50, no. 1, pp. 167–171, 2010.

[21] L.-Z. Wu, X.-S. Li, J. Chen, H.-B. Zhang, and Z.-S. Cui, "Predicting critical conditions and stress-strain curves for dynamic recrystallization in SPHC steel," *Journal of Iron and Steel Research International*, vol. 17, no. 7, pp. 51–57, 2010.

[22] I. Mejía, A. Bedolla-Jacuinde, C. Maldonado, and J. M. Cabrera, "Determination of the critical conditions for the initiation of dynamic recrystallization in boron microalloyed steels," *Materials Science and Engineering: A*, vol. 528, no. 12, pp. 4133–4140, 2011.

[23] N. Park, A. Shibata, D. Terada, and N. Tsuji, "Flow stress analysis for determining the critical condition of dynamic ferrite transformation in 6Ni–0.1C steel," *Acta Materialia*, vol. 61, no. 1, pp. 163–173, 2013.

[24] Y. C. Lin, M.-S. Chen, and J. Zhong, "Constitutive modeling for elevated temperature flow behavior of 42CrMo steel," *Computational Materials Science*, vol. 42, no. 3, pp. 470–477, 2008.

[25] Y. C. Lin, M.-S. Chen, and J. Zhong, "Microstructural evolution in 42CrMo steel during compression at elevated temperatures," *Materials Letters*, vol. 62, no. 14, pp. 2136–2139, 2008.

[26] Y.-C. Lin, M.-S. Chen, and J. Zhang, "Modeling of flow stress of 42CrMo steel under hot compression," *Materials Science and Engineering: A*, vol. 499, no. 1-2, pp. 88–92, 2009.

[27] Y. C. Lin, M.-S. Chen, and J. Zhong, "Numerical simulation for stress/strain distribution and microstructural evolution in 42CrMo steel during hot upsetting process," *Computational Materials Science*, vol. 43, no. 4, pp. 1117–1122, 2008.

[28] M. I. A. E. Aal, "Influence of the pre-homogenization treatment on the microstructure evolution and the mechanical properties of Al–Cu alloys processed by ECAP," *Materials Science and Engineering: A*, vol. 528, no. 22-23, pp. 6946–6957, 2011.

[29] J.-H. Jang, D.-G. Nam, Y.-H. Park, and I.-M. Park, "Effect of solution treatment and artificial aging on microstructure and mechanical properties of Al–Cu alloy," *Transactions of Nonferrous Metals Society of China*, vol. 23, no. 3, pp. 631–635, 2013.

[30] C. Giummarra, G. H. Bray, and D. J. Duquette, "Fretting fatigue in 2XXX series aerospace aluminium alloys," *Tribology International*, vol. 39, no. 10, pp. 1206–1212, 2006.

[31] Y. Chen, M. Weyland, and C. R. Hutchinson, "The effect of interrupted aging on the yield strength and uniform elongation of precipitation-hardened Al alloys," *Acta Materialia*, vol. 61, no. 15, pp. 5877–5894, 2013.

[32] H. C. Fang, H. Chao, and K. H. Chen, "Effect of Zr, Er and Cr additions on microstructures and properties of Al-Zn-Mg-Cu alloys," *Materials Science and Engineering A*, vol. 610, pp. 10–16, 2014.

[33] D. F. Lam, C. C. Menzemer, and T. S. Srivatsan, "A study to evaluate and understand the response of aluminum alloy 2026 subjected to tensile deformation," *Materials & Design*, vol. 31, no. 1, pp. 166–175, 2010.

[34] X. Huang, H. Zhang, Y. Han, W. Wu, and J. Chen, "Hot deformation behavior of 2026 aluminum alloy during compression at elevated temperature," *Materials Science and Engineering A*, vol. 527, no. 3, pp. 485–490, 2010.

[35] F. J. Humphreys and M. Hatherly, "The deformed state," in *Recrystallization and Related Annealing Phenomena*, F. J. H. Hatherly, Ed., chapter 2, p. 11, Elsevier, New York, NY, USA, 2004.

[36] Z. Q. Sheng and R. Shivpuri, "Modeling flow stress of magnesium alloys at elevated temperature," *Materials Science and Engineering A*, vol. 419, no. 1-2, pp. 202–208, 2006.

[37] R. E. Smallman and R. J. BiShop, "Chapter 7—mechanical behaviour of materials," in *Modern Physical Metallurgy and Materials Engineering*, pp. 197–258, Butterworth-Heinemann, Boston, Mass, USA, 6th edition, 1999.

4

Nanoindentation Hardness Distribution and Strain Field and Fracture Evolution in Dissimilar Friction Stir-Welded AA 6061-AA 5A06 Aluminum Alloy Joints

Guangjian Peng,[1,2] Yi Ma,[1] Jiangjiang Hu,[1] Weifeng Jiang,[1] Yong Huan,[3] Zhitong Chen,[4] and Taihua Zhang[1]

[1]*College of Mechanical Engineering, Zhejiang University of Technology, Hangzhou 310014, China*
[2]*Department of Mechanical Engineering, Johns Hopkins University, Baltimore, MD 21218, USA*
[3]*State Key Laboratory of Nonlinear Mechanics (LNM), Institute of Mechanics, Chinese Academy of Sciences, Beijing 100190, China*
[4]*Department of Mechanical and Aerospace Engineering, The George Washington University, Washington, DC 20052, USA*

Correspondence should be addressed to Zhitong Chen; zhitongchen@gwu.edu and Taihua Zhang; zhangth@zjut.edu.cn

Academic Editor: Stanislaw Dymek

Aluminum alloy AA 6061-T651 and 5A06-H112 rolled plates were successfully welded by friction stir welding (FSW) at three rotation speeds of 600, 900, and 1200 rpm with two transverse speeds of 100 and 150 mm/min. Mechanical properties and strain field evolution of FSW AA 6061-AA 5A06 were characterized by the uniaxial tension and digital image correlation (DIC) tests. Furthermore, the hardness distribution map of whole cross section was obtained via the nanoindentation method with 700 indents. Both DIC and nanoindentation results reveal that the heat-affected zone (HAZ) of AA 6061 alloy is the softest area in the weldment, and the fracture happens in this region. The microstructure evolution characterized by electron-backscatter diffraction (EBSD) indicates that the continuous dynamic recrystallization is the primary grain structure evolution in the stirring zone, and the grain refinement helps improve the mechanical properties. Analyses of the micro- and macrofeatures of the fracture surfaces via scanning electron microscopy (SEM) and optical microscope suggest that the increasing of heat input could enlarge the size of HAZ and reduce the slant angle of HAZ and thus lead the fracture angle to decrease and cause the dimples change from inclined ones to normal ones.

1. Introduction

In the past decades, aluminum alloys have been used as an alternative to steel due to their high strength-to-weight ratio, superior formability and machinability, and excellent corrosion resistance [1–3]. However, conventional fusion welding methods are not suitable for aluminum alloys, particularly for dissimilar aluminum alloys. It has been clearly confirmed that conventional fusion welding methods have a series of severe issues, including the formation of secondary brittle phases, porosity and cracking during solidification, large welding deformation, and high residual stresses [4–6]. In order to solve problems of conventional fusion welding, it is of great importance to introduce a low-cost, environmentally friendly manufacturing process, with high productivity. Friction stir welding (FSW) is a solid-state welding method, which is invented by The Welding Institute (TWI) and has been demonstrated particularly suitable for joining aluminum alloys [7, 8].

During the FSW, a rotating tool with pin and shoulder is inserted into the workpiece and then moves along the welding direction. Frictional heat causes the alloys to soften and allows the tool to traverse along the joint line. There is no melting in the FSW process, which is beneficial for the FSW to avoid the welding issues occurring in the conventional fusion welding mentioned earlier. Therefore, the FSW should be considered as an effective and reliable process for the welding of aluminum alloys [7]. The FSW process generates three distinct microstructural zones: the nugget zone (NZ), the thermomechanically affected zone (TMAZ), and the heat-affected zone (HAZ) [9–11]. The NZ is the region that experiences dynamic recrystallization process because of the effect of strong mechanical stirring and high heat input, which generally consists of fine and equiaxed grains. The TMAZ adjacent to the NZ exhibits plastic deformation under the thermomechanical effects. The HAZ experiences only heating effect without any mechanical action. There have been a number of papers highlighting the microstructural and mechanical changes from the parent material to the welding nugget zone [12–17]. The microstructure and mechanical properties depend on the relative plastic deformation and heat distribution during FSW process. Coarsening of strengthening precipitates and dissolution as well as the formation of wide precipitate-free zones have been found. The local nucleation, growth, and coarsening processes for strengthening precipitates are a function of temperature, which in turn is now a function of distance from the weld nugget [18–20]. Mechanical failure of the welds can occur in the NZ, HAZ, or TMAZ depending on the amount of heat input, which is controlled by welding parameters such as pin profile, tool rotation, and travel speeds [21–23]. In FSW, the process parameters exerted significant impact on the microstructural characteristics, mechanical properties, and residual stress distributions of welds; therefore, it is necessary to optimize the process parameters to improve the welding quality and decrease welding defects [24, 25].

Nonheat treatable 5A06 aluminum alloys (Al-Mg alloy) have a better anticorrosion performance and excellent mechanical properties, whereas heat treatable 6061 aluminum alloys (Al-Mg-Si alloy) are versatility in strength and toughness [26]. Combination of these aluminum alloys could improve their performance and broaden applications in aerospace, automotive, and marine. In this paper, AA 6061-T651 and AA 5A06-H112 were joined together via friction stir weld at different welding parameters. Digital image correlation (DIC) and uniaxial tension were employed to investigate the variations of mechanical properties including yield strength, ultimate tensile strength, strain field evolution, and fracture features. A large number of nanoindentation tests were also carried out to obtain the hardness distribution to reveal the softest region in the weldment [27, 28]. A comprehensive analysis of mechanical properties and microstructure in FSW AA 6061 and AA 5A06 was proposed. Although the FSW process is now well established in joining of Al-alloys in similar combinations, there is still a need to fully understand the joint characteristics in dissimilar FSWed Al-alloy joints.

2. Experiments

2.1. Materials and FSW. Friction stir welds were carried out on 5 mm-thick AA 6061-T651 and AA 5A06-H112 rolled plates whose chemical compositions are listed in Table 1. Both aluminum plates were cut into rectangular shape measuring 200 mm in length and 100 mm in width. A stir tool with a shoulder of 16 mm diameter and a conical probe of 4.6 mm length was used. The conical probe has three grooves, and the specific dimensions are shown in Figure 1(b). The principal directions of FSW geometry were denoted as follows: welding direction (WD), rolling direction (RD), and normal direction (ND). Single-pass butt friction stir welds were performed perpendicular to the rolling direction of the plates by placing AA 6061 on the advancing side as shown in Figure 1(a). During the FSW process, the tool was put on a 2° tilt angle and 4.7 mm plunge depth. Three tool rotation speeds (RS) and two welding transverse speeds (TS) as listed in Table 2 were considered.

2.2. Uniaxial Tension and Digital Image Correlation Tests. Three tensile specimens were obtained from each welded plate according to ASTM E8M, as shown in Figure 1(c). For each specimen, 1 mm thickness of the top surface and 0.5 mm thickness of the bottom surface were machined away, to make the specimen flat with uniform cross section in the gauge. The uniaxial tensile tests were carried out in a material testing system MTS 810 with a constant extension speed of 1.5 mm/s. The yield strength and ultimate tensile strength were obtained, and the average value was used for analysis. During tensile tests, a two-dimensional digital image correlation (DIC) system was employed to monitor the stain field near the joint.

2.3. Nanoindentation Tests. Nanoindentation specimens with the length of 29 mm and thickness of 3 mm were cut from the joint area as shown in Figure 1(c). Then, the side surface of the specimen was polished to mirror finish via an automatic polishing machine with 0.3 μm alumina powder. An Agilent Nano Indenter G200 system with a modified Berkovich indenter was used to conduct nanoindentation tests. To quantitatively evaluate the distribution of indentation hardness in the joint area, an array that consists of 700 indents was carried out on the side surface of the specimen, with 400 μm space between very two indents. For each nanoindentation test, the indentation load was increased linearly to a maximum load of 120 mN in 30 seconds, then held for 10 seconds, and finally linearly decreased to zero in 30 seconds. During indentation tests, the thermal drift rate was within ±0.05 nm/s. The possible maximum thermal drift was ±3.5 nm for a test lasting 70 seconds, which was much smaller than the indentation depth (about 2 μm).

2.4. Microstructure Analysis. Microstructural characterization of the weld region was performed by electron-backscatter diffraction (EBSD). An SEM equipped with an EBSD system was used to conduct the high-resolution EBSD analysis. Scanning electron microscopy (SEM) was employed to analyze the microstructure of the fracture surface of tensile specimens. And energy dispersive spectroscopy (EDS) analysis was also performed to assess alloy elements of fracture surface.

TABLE 1: Chemical composition of AA 5A06, AA 6061, and the fracture surface of the welded specimen (wt.%).

	Si	Fe	Cu	Mn	Mg	Cr	Zn	Ti	Al
AA 5A06	0.30~0.40	≤0.40	0.05~0.10	0.50~0.80	5.80~6.80	—	0.10~0.20	0.02~0.10	Bal.
AA 6061	0.40~0.80	≤0.70	0.15~0.40	≤0.15	0.80~1.20	0.04~0.35	≤0.25	≤0.15	Bal.
Fracture surface	0.79	0.70	0.35	0.08	1.46	0.17	0.08	0.21	96.16

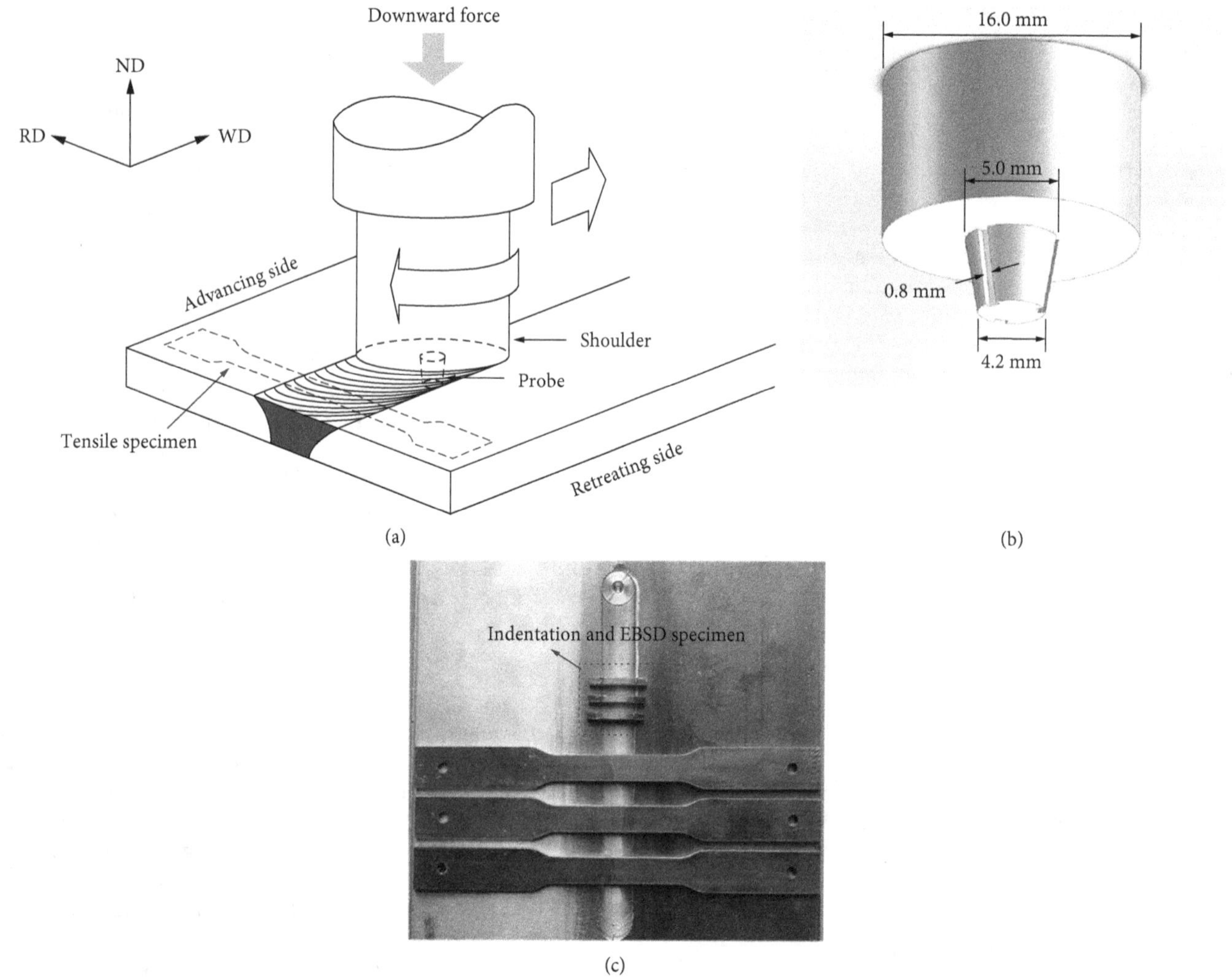

FIGURE 1: (a) Schematic drawing of the FSW joining process, (b) specific dimensions of the tool, and (c) the locations where tensile, indentation, and EBSD specimens were obtained.

TABLE 2: Investigated welding parameters in experimental scale-up processing.

Samples	RS (rpm)	TS (mm/min)	RS/TS (r/mm)
Sample 1	600	150	4
Sample 2	900	150	6
Sample 3	1200	150	8
Sample 4	600	100	6
Sample 5	900	100	9
Sample 6	1200	100	12

3. Results and Discussion

3.1. Tensile Properties. The representative engineering stress-strain curves of FSW joints and original AA 6061 and AA 5A06 are shown in Figure 2. For AA 6061 plates, the average yield strength is 236.17 MPa, the average ultimate tensile strength is 277.67 MPa, and the average elongation at break is 13.08%; for AA 5A06 plates, the average yield strength, ultimate tensile strength, and elongation at break are 125.78 MPa, 307.87 MPa, and 22.13%, respectively. It should be pointed out that the engineering stress-strain curves of FSW joints overlapping with the curve of AA 5A06 at the beginning is a coincidence (Figure 2). Comparison of the alloy elements as listed in Table 1 reveals that the yielding, necking, and fracture all occur at the AA 6061 side. This means the yield strength drops 50% to ~115 MPa and the ultimate tensile strength reduces from 277.67 MPa to ~190 MPa for AA 6061 in the FSW joints. Generally, the heat input introduced by FSW could affect the mechanical properties of joints a lot. For the present FSW joints, the

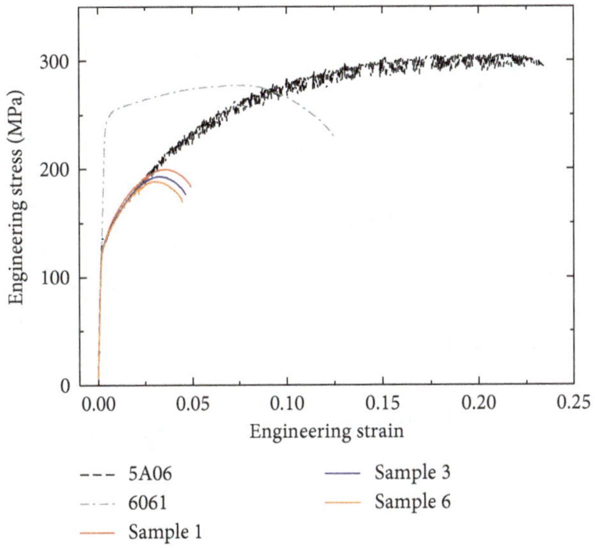

FIGURE 2: Engineering stress-strain curves of the FSW AA 6061-AA 5A06 specimens as compared to those of base metals AA 6061 and AA 5A06.

heat input definitely affects the mechanical properties of the AA 6061. As Peel et al. pointed out, this is mainly because the rolled AA 6061 stays in an extremely work hardened state and has a highly unstable microstructure, which will recrystallize readily under the elevated temperature generated during FSW [29]. The recrystallization will destroy the work hardened state and cause the mechanical properties to be weakened. It is commonly regarded that the higher R/T ratio (the ratio of rotational speed to transverse speed) leads to higher temperature and heat input [30–32]. Figure 3 shows both the yield strength and the ultimate tensile strength of FSW joints reduce slightly (within 10%) even while the R/T ratio increases from 4 r/mm to 12 r/mm. This means the increase of heat input does not affect the yield strength and ultimate tensile strength much. It could be inferred that the temperature generated by FSW at the R/T ratio of 4 is high enough to trigger the recrystallization of rolled AA 6061 and destroy the work hardened state.

3.2. Strain Field Evolution. The evolution of strain field near the joints was monitored by DIC during uniaxial tensile tests. Due to the limitation of space, the strain field evolution of Sample 1 is illustrated in Figure 4 as a representative one. Figure 4(a) shows that the strain distributes uniformly before yielding. With the extension of the specimen, yielding first occurs in the HAZ of the AA 6061 side as shown in Figure 4(b). It is clear from Figures 4(c)–4(e) that another weak area (red area) appears as the extension increases. The newly appeared weak area indicates the joint interface which could be clearly seen in Figure 5(a). Though the ratio of rotational speed to transverse speed increases from 4 r/mm to 12 r/mm, a joint interface exists in the NZ for all FSW samples. This means AA 6061 and AA 5A06 are difficult to merge to form "onion ring." In the current FSW samples, the HAZ of AA 6061 and the joint interface are two weak areas. During the tensile tests, there is a competition between the HAZ of AA 6061 and the joint interface. The evolution of strain field shown in Figures 4(e)–4(j) demonstrates that the joint interface won the competition and the HAZ of AA 6061 is the weakest area.

3.3. Cross Section and Nanoindentation Results. Figure 5(a) shows an optical micrograph of the cross section of FSW AA 6061-AA 5A06, after etching in Keller's reagent at 0°C for 15 seconds. Differences in the absolute shear strain rate between the advancing and the retreating sides of the weld result in an asymmetric shape. The advancing side of the weld is where the rotational velocity of the tool has the same direction as its travel velocity, whereas on the retreating side of the weld, the two velocity components having opposite directions. The NZ, TMAZ, HAZ, and based material (BM) were labelled in Figure 5(a). Generally, the order of average grain size for each zone is HAZ > TMZ > NZ [33]. The size and shape of grains in the extruded section are highly dependent on the extrusion ratio. The nanoindentation hardness distribution is shown in Figure 5(b), and it is evident that the lowest hardness locates in the HAZ of AA 6061. Since the yield strength being proportional to the hardness for metals, this could intuitively explain why the yielding first occurs in the HAZ of AA 6061 during tensile tests. It could be noted that the shape and orientation of the yielding band shown in Figure 4(b) agree well with that of the lowest hardness area as shown in Figure 5(b). Figure 6 shows the representative nanoindentation load-depth curves for different areas of the welds (the position of selected indents were marked in Figure 5(a)). It is clear from Figure 6(a) that the load-depth curves for BM, HAZ, and NZ of the AA 6061 side deviate from each other. Typically, a steeper load-depth curve indicates higher nanoindentation hardness. Thus, the hardness for each zone is BM > NZ > HAZ for the AA 6061 side. This demonstrates that the mechanical properties of AA 6061 were weakened after FSW. In Figure 6(b), the good superposition of the load-depth curves for different areas of the AA 5A06 side indicates that the change of mechanical properties for each zone is small after FSW. The hardness in NZ and TMAZ is slightly higher than that in BM (the reason is discussed in Section 3.4). It could be summarized from Figures 5(b) and 6 that the mechanical properties of AA 6061 are more sensitive to the heat input than those of AA 5A06.

3.4. Grain Structure of the Weld Region. The EBSD grain images in Figure 7 show that the grain size in the stirring zones is much smaller than that in the base materials. This can be related to the effect of dynamic recovery and recrystallization. Dynamic recovery and recrystallization are the main mechanisms for grain refinement during FSW [34, 35]. With increase of strain, dislocation is generated and a rapid multiplication of dislocation occurs. A progressive increase in dislocation density with increasing strain leads to three-dimensional arrays of low-angle grain boundaries transforming to high-angle grain boundaries

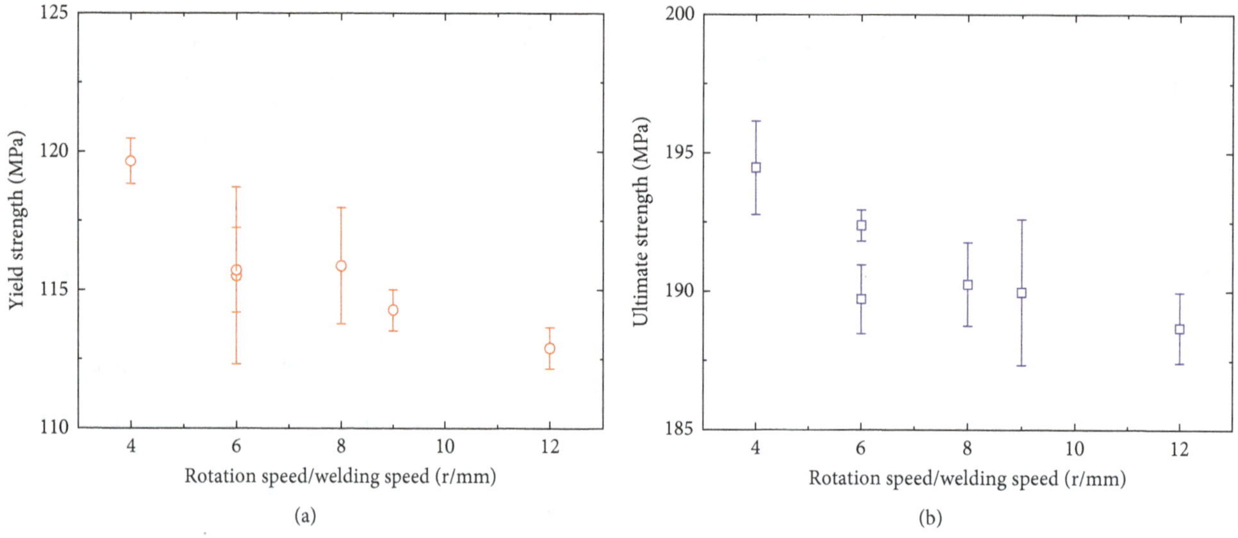

FIGURE 3: The variation of (a) yield strength and (b) ultimate tensile strength with the ratio of rotation speed to welding speed.

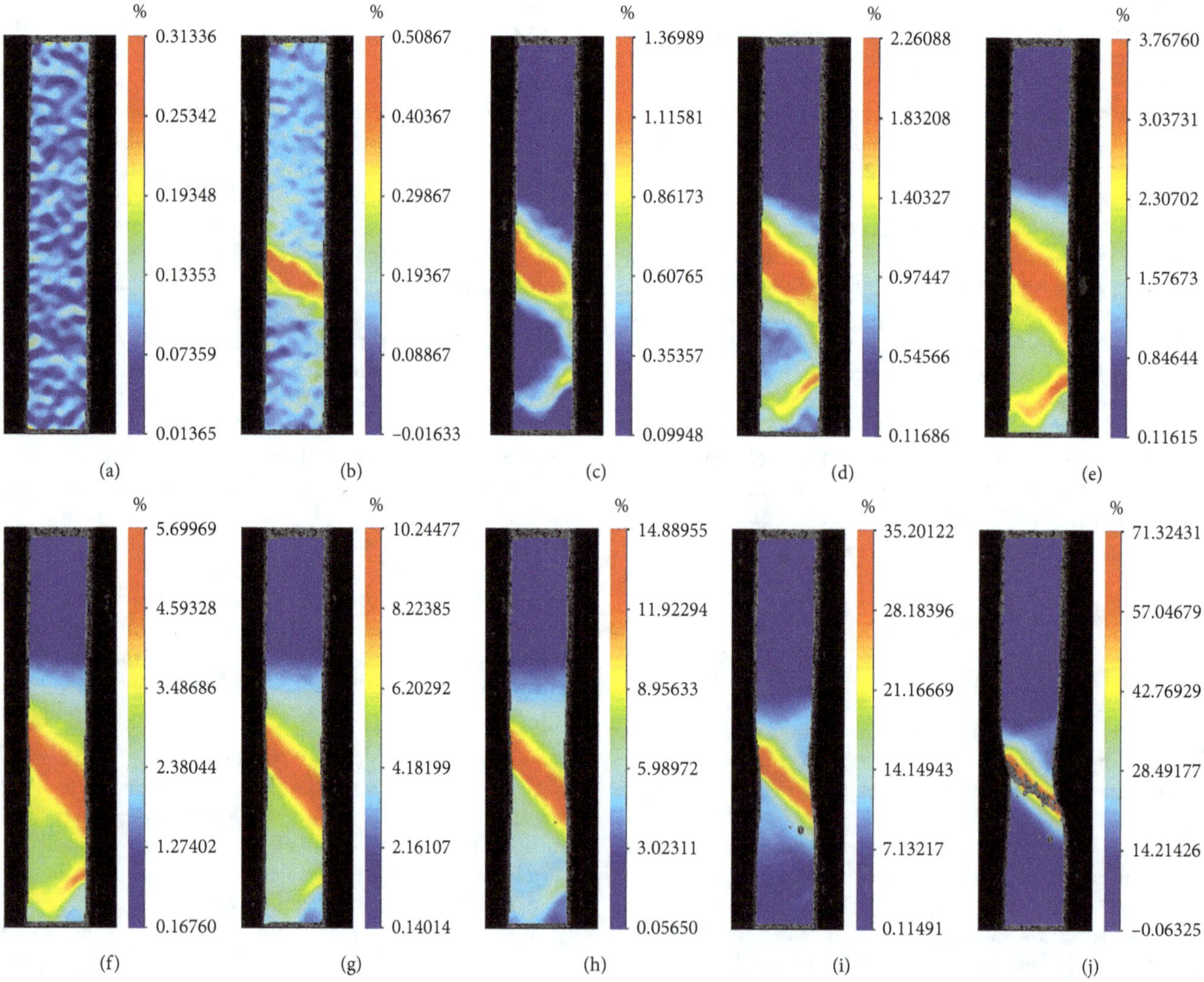

FIGURE 4: The strain field evolution of the weld region of Sample 1 monitored by DIC.

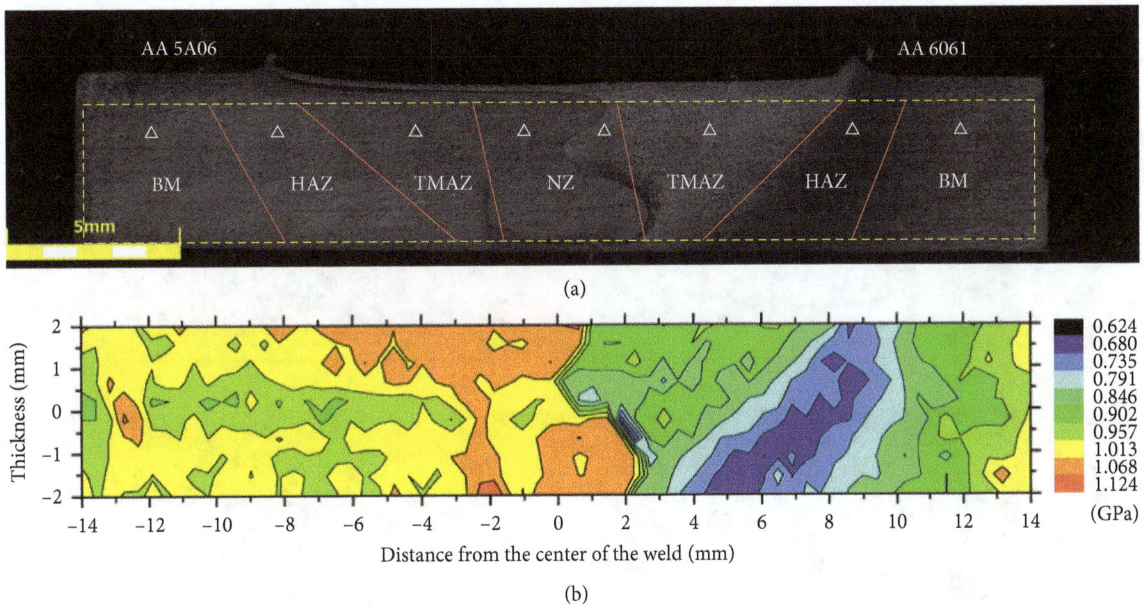

FIGURE 5: (a) The cross-sectional view of FSW AA 6061-AA 5A06, and (b) nanoindentation hardness distribution on the cross section.

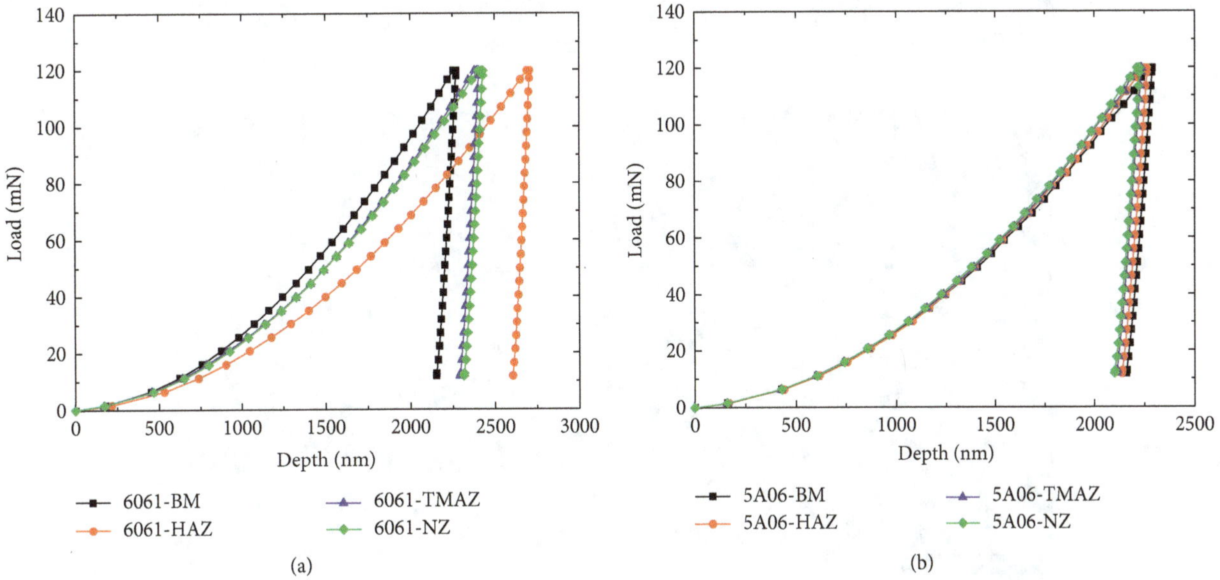

FIGURE 6: Typical nanoindentation load-depth curves for different areas of the FSW joint at (a) the AA 6061 side and (b) the 5A06 side.

during FSW. Individual segments of high-angle grain boundaries replace subgrains evolved at small strains, and grain size refinement occurs. Presence of vibration partially increases the temperature in the weld region, and additionally enhances the work hardening and leads to more strain in the zone [36]. Both grain size refinement and work hardening in the NZ and TMAZ could contribute to the improvement of mechanical properties in such areas. This helps us to explain the reason for each area having different mechanical properties, as shown in Figure 5(b). For AA 6061, the high temperature during FSW destroys the original work hardened state in the NZ, TMAZ, and HAZ and leads the mechanical properties to weaken greatly. However, the grain size refinement and work hardening generated by stirring in the NZ and TMAZ improve the mechanical properties with some extent. As a result, the HAZ is the weakest area in FSW AA 6061. Since AA 5A06 is nonheat treatable, the high temperature during FSW does not affect the mechanical properties. With the strengthening of grain size refinement and work hardening, the hardness in the NZ and TMAZ of AA 5A06 becomes the strongest (Figure 5(b)).

3.5. Macro- and Microstructures of Fracture. Figures 8(a)–8(c) show the optical macrograph of the side view of fracture surface for samples 1, 2, and 3, and the fracture angle of the surface decreases from 35° to 10°. The decrease of the fracture angle

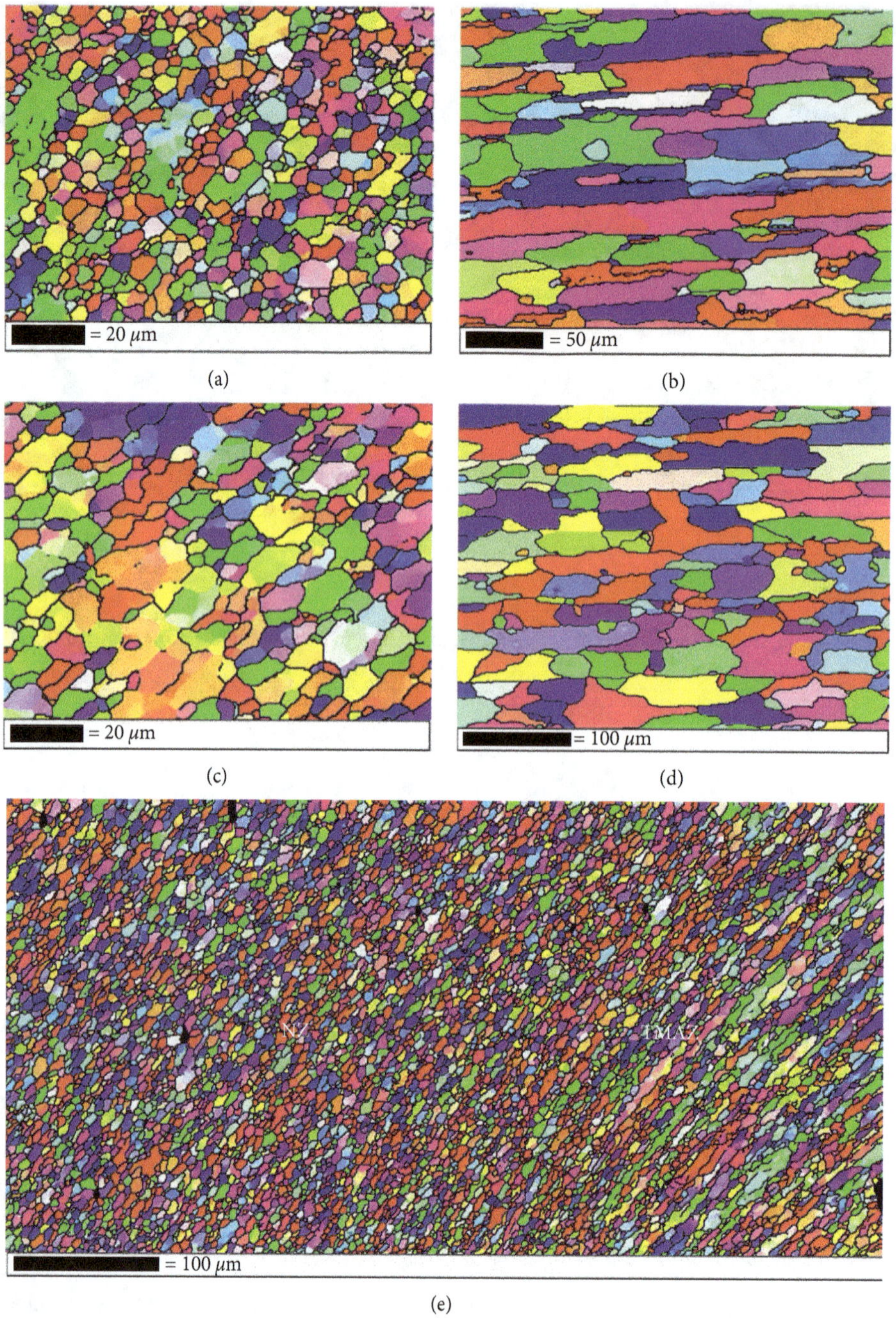

FIGURE 7: EBSD grain maps of FSW AA 6061-AA 5A06: (a) 5A06 NZ; (b) 5A06 BM; (c) 6061 NZ; (d) 6061 BM; (e) 6061 NZ-TMAZ.

could attribute to the change of the size and orientation of the HAZ. For samples 1, 2, and 3, the welding transverse speed was fixed at 150 mm/min and the rotation speeds were 600 rpm, 900 rpm, and 1200 rpm, respectively. With the increase of rotation speed, more heat was generated during FSW. The increasing heat input could enlarge the size of HAZ and change the orientation of HAZ. As discussed above, the HAZ of AA 6061 is the weakest area, and yielding firstly occurs in this region. Thus the initial yielding band shown in the DIC strain field could reflect the size and orientation of HAZ. From the strain field of samples 1, 2, and 3 at the beginning of yielding shown in Figures 8(a)–8(c), it could be found that the yielding band becomes thicker and the slant angle decreases gradually with increasing heat input. The change of size and orientation of HAZ was then intuitively revealed. The fracture angle agrees well with the slant angle of the HAZ. It could be concluded that the fracture angle of the welded specimen is determined by the size and orientation of the HAZ. Similar conclusion was obtained from Figures 8(d)–8(f) for samples 4, 5, and 6, where the rotation speed increases from 600 rpm to 1200 rpm with a fixed welding transverse speed of 100 mm/min.

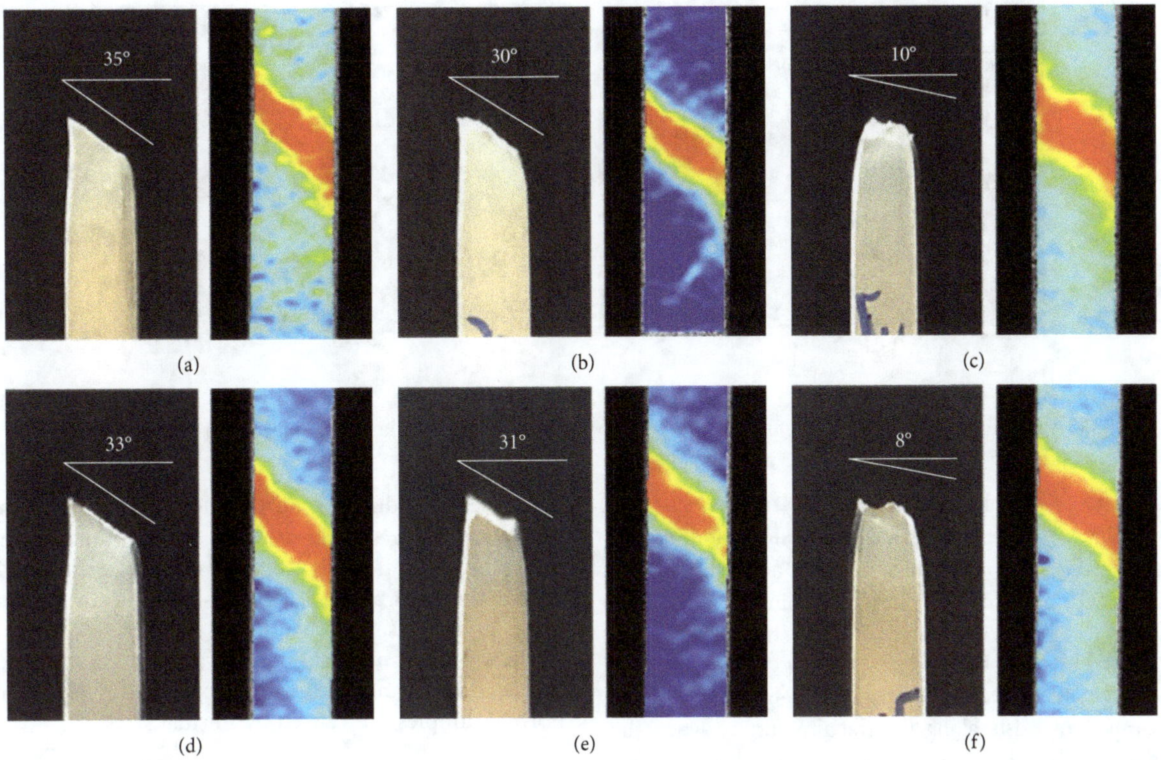

Figure 8: Optical macrograph of the side view of fracture surface and the corresponding strain field at the beginning of yielding for (a) Sample 1; (b) Sample 2; (c) Sample 3; (d) Sample 4; (e) Sample 5; (f) Sample 6.

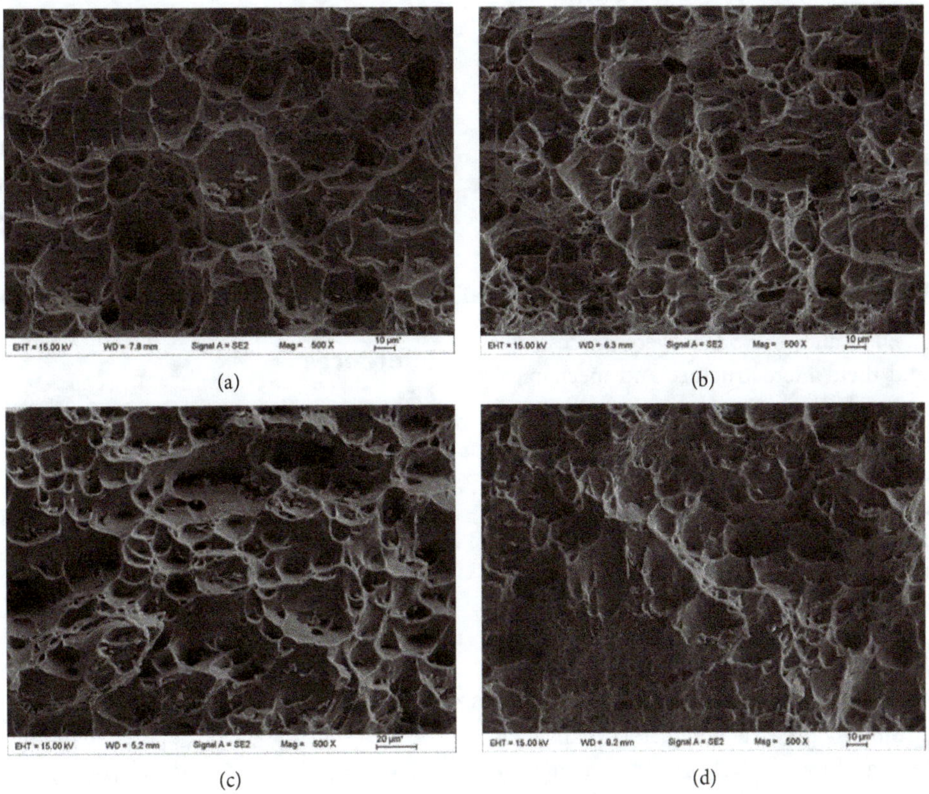

Figure 9: Continued.

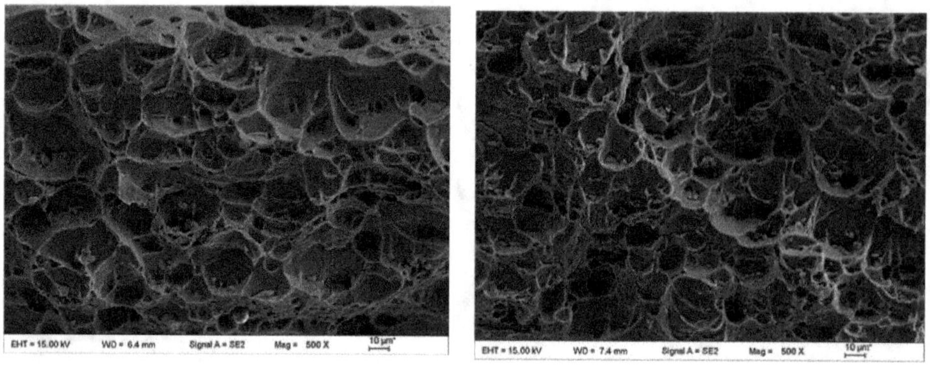

(e) (f)

Figure 9: SEM fractographs of FSW specimens: (a) Sample 1; (b) Sample 2; (c) Sample 3; (d) Sample 4; (e) Sample 5; (f) Sample 6.

From the SEM micrographs of the fracture surfaces shown in Figure 9, river patterns, cleavage steps, and tearing with small irregularly distributed dimples are observed. In Figures 9(a), 9(b), and 9(d)–9(f), most of the dimples are shallow and incline towards one direction. These characteristics are mainly caused by shear deformation. Due to the large fracture angle as shown in Figures 8(a), 8(b), and 8(d)–8(f), a large shear stress component exists along the fracture surface when the specimens are stretched. The shear stress could drive the materials to move relatively along the fracture surface to form shallow and inclined dimples. Thus, shear fracture should be the primary failure mode for samples 1, 2, 4, and 5. However, the equiaxed and normal dimples shown in Figures 8(c) and 8(f) indicate that the fracture of samples 3 and 6 is caused by normal stress. Since the macrofracture angles as shown in Figures 8(c) and 8(f) are small, the shear stress component along the fracture surface is ignorable, whilst the normal stress component is dominant. Under normal stress, the materials on both sides of the fracture surfaces eventually were pulled out from each other to form equiaxed and normal dimples.

4. Conclusions

In this paper, AA 6061 and AA 5A06 were successfully welded by FSW, and their microstructure and mechanical properties were analyzed. The grain structure evolution in stir zone is dominated by continuous dynamic recrystallization. The grain size refinement helps improve the mechanical properties in the HAZ and TMAZ of the weldments. Both the strain field evolution obtained by DIC and the nanoindentation hardness distribution map intuitively reveal that the welded HAZ of AA 6061 is the weakest region. Thus, the yielding, necking, and fracture occur in this zone. After welding, the yield strength of AA 6061 drops 50% to about 115 MPa and the ultimate tensile strength reduces from 277 MPa to about 190 MPa. Such a significant weakening of mechanical properties is mainly because the unstable work hardened state of rolled AA 6061 was destroyed by elevated temperature generated in FSW. The SEM analysis of fracture surface demonstrates that fractures in all tensile specimens are ductile fracture due to the presence of dimples. Furthermore, analyzing the micro- and macrofeatures of the fracture surfaces indicates that the increasing of heat input could enlarge the size of HAZ and reduce the slant angle of HAZ and thus leads the fracture angle to decrease and cause the dimples to change from inclined ones to normal ones. It was found that shear stress forms shallow and inclined dimples, whilst equiaxed and normal dimples are caused by normal stress.

Conflicts of Interest

The authors declare that they have no conflicts of interest.

Acknowledgments

The authors would like to gratefully acknowledge the support from the National Natural Science Foundation of China (Grant Nos. 11727803, 11772302, 11672356, and 11402233) and Public Welfare Project of Zhejiang Province (2015C31074).

References

[1] T. Dursun and C. Soutis, "Recent developments in advanced aircraft aluminium alloys," *Materials & Design*, vol. 56, pp. 862–871, 2014.

[2] C. Menzemer and T. Srivatsan, "The effect of environment on fatigue crack growth behavior of aluminum alloy 5456," *Materials Science and Engineering: A*, vol. 271, no. 1-2, pp. 188–195, 1999.

[3] N. Guo, Y. Fu, Y. Wang, Q. Meng, and Y. Zhu, "Microstructure and mechanical properties in friction stir welded 5A06 aluminum alloy thick plate," *Materials & Design*, vol. 113, pp. 273–283, 2017.

[4] B. Gibson, D. Lammlein, T. Prater et al., "Friction stir welding: process, automation, and control," *Journal of Manufacturing Processes*, vol. 16, no. 1, pp. 56–73, 2014.

[5] M. Ericsson and R. Sandström, "Influence of welding speed on the fatigue of friction stir welds, and comparison with MIG and TIG," *International Journal of Fatigue*, vol. 25, no. 12, pp. 1379–1387, 2003.

[6] A. von Strombeck, G. Cam, J. F. dos Santos, V. Venzke, and M. Kocak, "A comparison between microstructure, properties, and toughness behavior of power beam and friction stir welds in Al-alloys," in *Proceedings of the TMS 2001 Annual Meeting Aluminum, Automotive and Joining*, pp. 12–14, New Orleans, LA, USA, February 2001.

[7] R. S. Mishra and Z. Ma, "Friction stir welding and processing," *Materials Science and Engineering: R: Reports*, vol. 50, no. 1-2, pp. 1–78, 2005.

[8] G. Çam and G. İpekoğlu, "Recent developments in joining of aluminum alloys," *International Journal of Advanced Manufacturing Technology*, vol. 91, no. 5–8, pp. 1851–1866, 2017.

[9] C. He, K. Kitamura, K. Yang, Y. J. Liu, Q. Y. Wang, and Q. Chen, "Very high cycle fatigue crack initiation mechanism in nugget zone of AA 7075 friction stir welded joint," *Advances in Materials Science and Engineering*, vol. 2017, Article ID 7189369, 10 pages, 2017.

[10] Z. Chen, S. Li, and L. H. Hihara, "Microstructure, mechanical properties and corrosion of friction stir welded 6061 aluminum alloy," 2015, http://arxiv.org/abs/1511.05507.

[11] Z. Chen, S. Li, K. Liu, and L. H. Hihara, "A study on the mechanical property and corrosion sensitivity of an AA5086 friction stir welded joint," 2015, http://arxiv.org/abs/1511.04990.

[12] J.-Q. Su, T. Nelson, R. Mishra, and M. Mahoney, "Microstructural investigation of friction stir welded 7050-T651 aluminium," *Acta Materialia*, vol. 51, no. 3, pp. 713–729, 2003.

[13] B. Yang, J. Yan, M. A. Sutton, and A. P. Reynolds, "Banded microstructure in AA2024-T351 and AA2524-T351 aluminum friction stir welds: part I," *Metallurgical Studies, Materials Science and Engineering: A*, vol. 364, no. 1-2, pp. 55–65, 2004.

[14] P. Cavaliere, R. Nobile, F. Panella, and A. Squillace, "Mechanical and microstructural behaviour of 2024–7075 aluminium alloy sheets joined by friction stir welding," *International Journal of Machine Tools and Manufacture*, vol. 46, no. 6, pp. 588–594, 2006.

[15] B. B. Wang, F. F. Chen, F. Liu, W. G. Wang, P. Xue, and Z. Y. Ma, "Enhanced mechanical properties of friction stir welded 5083Al-H19 joints with additional water cooling," *Journal of Materials Science & Technology*, vol. 33, no. 9, pp. 1009–1014, 2017.

[16] Y. Mao, L. Ke, Y. Chen, F. Liu, and L. Xing, "Inhomogeneity of microstructure and mechanical properties in the nugget of friction stir welded thick 7075 aluminum alloy joints," *Journal of Materials Science & Technology*, vol. 34, no. 1, pp. 228–236, 2018.

[17] Z. Chen, S. Li, and L. H. Hihara, "Electrochemical and mechanical behaviors of dissimilar friction stir welding between 5086 and 6061 aluminum alloy," 2018, http://arxiv.org/abs/1802.03460.

[18] M. Mahoney, C. Rhodes, J. Flintoff, W. Bingel, and R. Spurling, "Properties of friction-stir-welded 7075 T651 aluminum," *Metallurgical and Materials Transactions A*, vol. 29, no. 7, pp. 1955–1964, 1998.

[19] K. A. Hassan, A. Norman, D. Price, and P. Prangnell, "Stability of nugget zone grain structures in high strength Al-alloy friction stir welds during solution treatment," *Acta Materialia*, vol. 51, no. 7, pp. 1923–1936, 2003.

[20] G. Cam, G. Ipekoglu, and H. T. Serindag, "Effects of use of higher strength interlayer and external cooling on properties of friction stir welded AA6061-T6 joints," *Science and Technology of Welding and Joining*, vol. 19, no. 8, pp. 715–720, 2014.

[21] H. Jamshidi Aval, S. Serajzadeh, N. A. Sakharova, A. H. Kokabi, and A. Loureiro, "A study on microstructures and residual stress distributions in dissimilar friction-stir welding of AA5086–AA6061," *Journal of Materials Science*, vol. 47, no. 14, pp. 5428–5437, 2012.

[22] S. Mironov, T. Onuma, Y. S. Sato, and H. Kokawa, "Microstructure evolution during friction-stir welding of AZ31 magnesium alloy," *Acta Materialia*, vol. 100, pp. 301–312, 2015.

[23] J. C. V. Juarez, G. M. D. Almaraz, R. G. Hernandez, and J. J. V. Lopez, "Effect of modified pin profile and process parameters on the friction stir welding of aluminum alloy 6061-T6," *Advances in Materials Science and Engineering*, vol. 2016, Article ID 4567940, 9 pages, 2016.

[24] H. Zhang, M. Wang, W. Zhou et al., "Microstructure–property characteristics of a novel non-weld-thinning friction stir welding process of aluminum alloys," *Materials & Design*, vol. 86, pp. 379–387, 2015.

[25] H. Liu, Y. Zhao, Y. Hu, S. Chen, and Z. Lin, "Microstructural characteristics and mechanical properties of friction stir lap welding joint of alclad 7B04-T74 aluminum alloy," *International Journal of Advanced Manufacturing Technology*, vol. 78, no. 9–12, pp. 1415–1425, 2015.

[26] J.-R. Kim, E.-Y. Ahn, H. Das et al., "Effect of tool geometry and process parameters on mechanical properties of friction stir spot welded dissimilar aluminum alloys," *International Journal of Precision Engineering and Manufacturing*, vol. 18, no. 3, pp. 445–452, 2017.

[27] Y. Ma, J. H. Ye, G. J. Peng, D. H. Wen, and T. H. Zhang, "Nanoindentation study of size effect on shear transformation zone size in a Ni-Nb metallic glass," *Materials Science and Engineering: A*, vol. 627, pp. 153–160, 2015.

[28] Y. Ma, G. J. Peng, Y. H. Feng, and T. H. Zhang, "Nanoindentation investigation on the creep mechanism in metallic glassy films," *Materials Science and Engineering: A*, vol. 651, pp. 548–555, 2016.

[29] M. Peel, A. Steuwer, M. Preuss, and P. Withers, "Microstructure, mechanical properties and residual stresses as a function of welding speed in aluminium AA5083 friction stir welds," *Acta Materialia*, vol. 51, no. 16, pp. 4791–4801, 2003.

[30] K. Elangovan and V. Balasubramanian, "Influences of pin profile and rotational speed of the tool on the formation of friction stir processing zone in AA2219 aluminium alloy," *Materials Science and Engineering: A*, vol. 459, no. 1-2, pp. 7–18, 2007.

[31] M. A. Gharacheh, A. Kokabi, G. Daneshi, B. Shalchi, and R. Sarrafi, "The influence of the ratio of "rotational speed/ traverse speed"(ω/v) on mechanical properties of AZ31 friction stir welds," *International Journal of Machine Tools and Manufacture*, vol. 46, no. 15, pp. 1983–1987, 2006.

[32] Z. Zhang, J. Bie, Y. Liu, and H. Zhang, "Effect of traverse/ rotational speed on material deformations and temperature distributions in friction stir welding," *Journal of Materials Science & Technology*, vol. 24, no. 6, pp. 907–914, 2008.

[33] D. Wadeson, X. Zhou, G. Thompson, P. Skeldon, L. D. Oosterkamp, and G. Scamans, "Corrosion behaviour of friction stir welded AA7108 T79 aluminium alloy," *Corrosion Science*, vol. 48, no. 4, pp. 887–897, 2006.

[34] R. Kaibyshev, K. Shipilova, F. Musin, and Y. Motohashi, "Continuous dynamic recrystallization in an Al-Li-Mg-Sc alloy during equal-channel angular extrusion," *Materials*

Science and Engineering: A, vol. 396, no. 1-2, pp. 341–351, 2005.

[35] M. Rahmi and M. Abbasi, "Friction stir vibration welding process: modified version of friction stir welding process," *International Journal of Advanced Manufacturing Technology*, vol. 90, no. 1–4, pp. 141–151, 2017.

[36] O. Barooni, M. Abbasi, M. Givi, and B. Bagheri, "New method to improve the microstructure and mechanical properties of joint obtained using FSW," *International Journal of Advanced Manufacturing Technology*, vol. 93, no. 9–12, pp. 4371–4378, 2017.

5

Evaluation of Parameters Affecting Magnetic Abrasive Finishing on Concave Freeform Surface of Al Alloy via RSM Method

Mehrdad Vahdati and SeyedAlireza Rasouli
Department of Mechanical Engineering, K. N. Toosi University of Technology, Tehran 19395-1999, Iran

Correspondence should be addressed to SeyedAlireza Rasouli; sarasouli@mail.kntu.ac.ir

Academic Editor: Donato Sorgente

The attempts of researchers in industries to obtain accurate and high quality surfaces led to the invention of new methods of finishing. Magnetic abrasive finishing (MAF) is a relatively new type of finishing in which the magnetic field is used to control the abrasive tools. Applications such as the surface of molds are ones of the parts which require very high surface smoothness. Usually this type of parts has freeform surface. In this study, the effect of magnetic abrasive process parameters on freeform surfaces of parts made of aluminum is examined. This method is obtained through combination of magnetic abrasive process and Control Numerical Computer (CNC). The use of simple hemisphere for installation on the flat area of the magnets as well as magnets' spark in curve form is a measure done during testing the experiments. The design of experiments is based on response surface methodology. The gap, the rotational speed of the spindle, and the feed rate are found influential and regression equations governing the process are also determined. The impact of intensity of the magnetic field is obtained using the finite element software of Maxwell. Results show that in concave areas of the surface, generally speaking, the surface roughness decreases to 0.2 μm from its initial 1.3 μm roughness. However, in some points the lowest surface roughness of 0.08 μm was measured.

1. Introduction

Finishing is among latest processes which are performed on the workpiece. This process considerably influences the performance and lifetime of the product. There is a direct relation between the machining accuracy, surface roughness, and lifespan of the produced parts. The increasing growth of the automation in industries and increase in numerical control machines and robots lead to the achievement of high accurate surfaces with very fine surface roughness [1]. In some branches of products, like molds, surface roughness determines the lifespan. Furthermore, surface quality of molds remarkably influences the quality of the produced parts. Usually, molds possess three directional intricate freeform surfaces. Generally finishing freeform surfaces faces limitations. Currently, this process is done mostly manually which relies on the expertise of the worker which is far behind the development of the new modern manufacturing process. This hand-performed task is not practically feasible and it is very sensitive to human factors. Thus, material removal and surface roughness are hardly similar in molds or similar components with intricate curves and surfaces [2].

On the other hand, one of finishing processes developed in recent years is magnetic abrasive finishing (MAF). Flexible magnetic abrasive brush can easily deform due to magnetic field strength and match itself to the shape of the surface. The advantage is that the tool is not rigid and there is no dressing; otherwise, it would be necessary to manufacture different tools for different surfaces. Size of abrasive particles is some microns and this lets the abrasive brush reach any complicated surface and nonachievable region [3]. Self-sharpening, high adoptability, negligible increase in temperature during finishing in comparison with other finishing processes, suitable control on imposing force on particles, causing very little residual stress on the surface of the workpiece, and avoiding initiation of the microcracks are some of MAF advantages which increase its application [4]. In Figure 1 a schematic of MAF process can be seen [5]. Moreover, in case of its application in mold manufacturing, recast layers caused by electro-discharge machining process provide suitable space

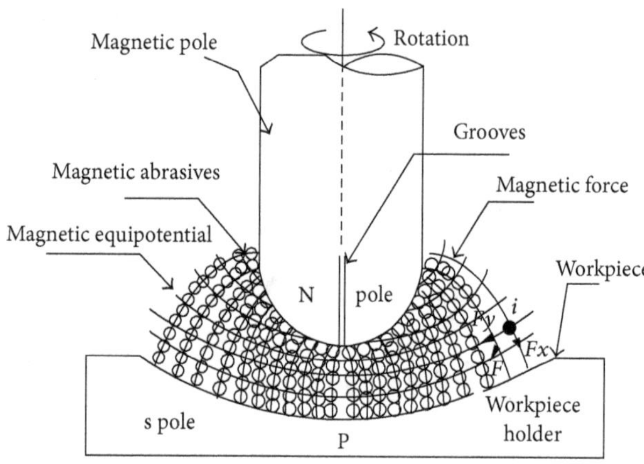

FIGURE 1: Schematic of MAF process.

for microcrack initiation; this method can simply remove these layers from the surface [6]. Also, considering lower normal force on abrasive particles (around or less than some μNs) penetration depth of the abrasive particles is about 1 μm and less and therefore this finishing process does not change geometrical and dimensional tolerances of the workpiece.

Combining Computer Numerical Control, CNC, system and MAF, the shape of the brush can match with the intricate 3D curves and finishing of the complicated freeform surfaces of the molds becomes possible. Therefore, with an automatic control of the process parameters a very smooth surface with fine uniform surface roughness can be achieved.

In most reported researches and in industrial practices MAF has been performed on parts with plane surfaces and internal/external cylinders [7]. Lin et al., using Taguchi method, studied parameters like rotational speed of spindle, feed speed, amount of lubricant, gap, and amount of abrasive powder on the surface of a stainless steel 304 workpiece with low curvature. They could decrease surface roughness value from 0.158 μm to 0.102 μm [8]. They found out that amount of abrasive particles, feed speed, and finishing gap are significant parameters on surface roughness. Gap was the most effective one and its optimum amount was in the range of 1 to 2 mm. Yin and Shinmura investigated the effect of vibration on the finishing process of the surface with low curvature [9]. Workpiece in horizontal and vertical and combined orientations was vibrated in high amplitudes and low frequencies and its influence was investigated on magnetic field, finishing pressure on abrasive particles, and process performance. Vibration in vertical and horizontal directions had the highest influence in reducing of the surface roughness. Y. J. Kim et al. introduced a novel method in finishing of the molds surface. In this new technique there is no need for CNC and instead, using guiding wheels passing the curved path with constant gap, magnetic abrasive finishing is performed. It should be mentioned that applied workpieces have concave and convex curvature with low slope. Zhang et al. in 2009 analyzed the theory of NURB curves on freeform surfaces and investigated their effects on MAF process [10]. Ding et al. considered time and its effect on surface roughness reduction of the concave and convex surfaces [11]. Anzai et al. studied the MAF on spark machined mold surfaces and they could considerably reduce the thickness of the recast layer [12].

Reviewing the reported researches one can say that MAF on mold freeform surfaces with high slope has never been modeled. To investigate significant factors on reduction average surface roughness, modeling and optimization of MAF are performed using Response Surface Method. In order to estimate the force on abrasive particles by magnetic field strength, PHVWE Teslameter in the range of 0–2 Tesla is used to measure and also Maxwell finite element software is employed to simulate magnetic flux density.

In this study design and manufacturing of tools, measuring and simulation, design and manufacture of workpiece, design of experiments, selection of suitable parameters, modeling, and simulation of MAF are discussed.

2. Experimental Setup

2.1. Tool. MAF required magnetic field is produced using an electrical or permanent magnet. Producing a variable magnetic field and possibility of easy separation of magnetic particles from the magnet surface are among this electrical magnet's advantages. To create an electromagnet generally many turns of wire are closely spaced and consequently they form a magnetic field. The central part of an electromagnet is a magnetic core created from a ferromagnetic material such as iron and this core is covered with many turns of wire which is usually made of copper. The magnetic core is the main part making a magnetic flux and with higher concentrations there will be a more powerful magnet [13].

Equation (1) can be used to calculate the density of magnetic flux in the center of solenoid:

$$B = K\mu_0 nI, \tag{1}$$

where I is the electric current and n is the number of solenoids per unit of length which is calculated using (2) and μ_0 is relative magnetic permeability in vacuum (Wb/(A-m)):

$$n = \frac{N}{L}, \tag{2}$$

where N is the total number of solenoids and L is the length of the wire. Moreover, K is the coefficient of relative magnetic permeability of the core.

The use of materials with high relative magnetic permeability in the core of solenoid can have a significant impact in creating higher magnetic flux density. Usually, materials such as pure iron, nickel, and cobalt are used for this purpose.

However, when the radius of the solenoid is small in comparison with its length and its rings are very close together, the magnetic field inside the solenoid will be uniform at the points far from the edges. In such a condition, the impact of radius of the core could be neglected.

To obtain the desired magnetic flux density it is possible to use the following ways: increasing the number of wire turns per unit of length of wire and increasing the electric current. Usually copper wires with small diameter (1 mm) are used to create a solenoid. As a result, because of the high length and small diameter of the wires, the wire resistance will be increased. Therefore, when using a fixed voltage, it is not possible to increase the flow rate too much. On the other hand, with increasing the diameter of copper wire used, the volume of the magnet will increase. The increase in the number of turns of wire not only increases the amount of heat due to the loss of electricity but also increases the size of electric magnets.

The following section presents information and details about some electrical magnets which have been used in the MAF process.

Singh et al. used an electrical magnet for MAF process. They used a magnet with a diameter of 77 mm and height of 55 mm. A copper wire turn was used to create magnetic field. In addition, to achieve a higher concentration of field lines and to create magnets, an iron core with a diameter of 38 was used. Using a power supply of 0–25 volts and an electric current of 0-1 amperes, a magnetic field of 0–0.44 Tesla was created [14]. Moreover, the magnet used in the research laboratory at Baghdad University had the following characteristics [15]:

The material of the iron core: C15 low carbon steel.

The cross section of the iron core: $A = 14$ cm.

The length of the iron core: $L = 75$ mm.

The diameter of the copper wire of the magnetic coil: $\varnothing = 1$ mm.

The number of turns: $N = 2400$.

A round flat faced electromagnet with a diameter of 100 mm and height of 57 mm was used for experimentation in Chandigarh Engineering College by Gill [16]. Taking into consideration the dimensions of the magnets used, it is clear that electrical magnets occupy too much space. Moreover, as in MAF on straight and freeform surfaces, the magnet rotates and wire cannot be used to send electrical energy to the coil. To overcome this problem slip ring with graphite is used. To cool down the coils during operation a cooling system is required. Considering all aforementioned problems regarding application of electrical magnets, for its lower volume, easy mounting, and lower price a permanent magnet is applied in this study.

In MAF process on straight surfaces due to rotation of the tool, cylindrical magnets are used to produce magnetic fields. It is clear that in finishing of freeform surfaces magnets with surface straight cannot be applied due to the curvature of the surfaces because finishing path's required constant gap between magnet and workpiece surface will lead to collision between magnet and the surface. In this process, similar to finishing of freeform surfaces, ball nose tools should be used. Spherical heads of these tools move on the intricate surfaces without collision. Based on slip curvature, diameter of these tools commonly is up to 20 mm. In MAF the head of the magnet should be formed in the shape of a hemisphere. Two ways can be followed to do this; first hemispherical iron head can be connected to the magnet. Figure 2 shows hemispherical iron heads and magnet fixture.

In order to facilitate the movement of the magnetic field, materials with higher relative magnetic permeability should be used. In materials with high relative magnetic permeability, pure iron (99.8%) is more easily approachable. To identify the passing magnetic flux density both measurements and simulations are followed.

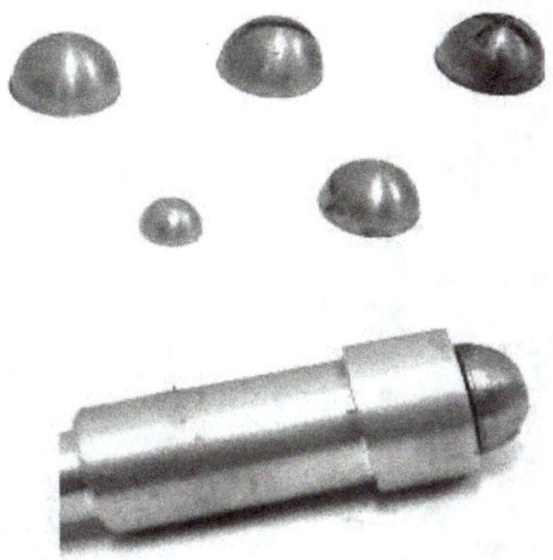

FIGURE 2: Hemispheres made of 99.8% pure iron.

FIGURE 3: Spark machined magnet with its fixture.

In simulations, iron hemisphere with radius of 8 mm is connected to a cylindrical magnet with diameter of 16 and height of 20 mm. It should be noted that the relative magnetic permeability of the iron changes with variation of the magnetic field strength. Here, as intensity of the magnetic field produced by the magnet is much higher than 400 (A per M), the relative magnetic permeability is close to 2400. Creating curvature on the surface of the magnet is the second possible way which is followed here. Machining and electro-discharge machining can be used to manufacture a magnet with hemispherical shape. Due to brittle nature of sintered NdFeB commercial magnet with trade code N35, mechanical machining faces problems and does not provide a uniform surface. EDM can produce a better surface. Figure 3 shows spark machined magnet with its fixture. 2D and 3D simulation models are illustrated in Figure 4.

Figures 5 and 6 can be used to compare the effect of either iron hemisphere or hemispherical magnet on magnetic flux

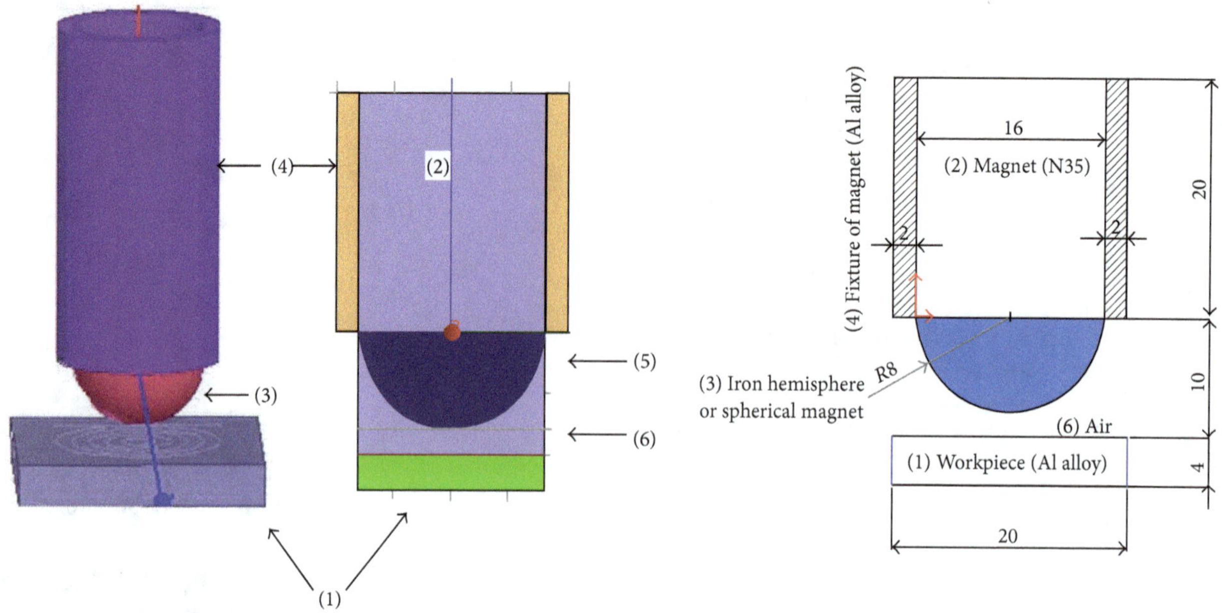

FIGURE 4: 2D and 3D simulation models: (1) workpiece (20 × 20 × 5 mm Al alloy), (2) magnet (Ø16 × 20 mm N35), (3) spherical magnet (Ø16 mm), (4) magnet fixture (Ø22 × Ø16 × 20 mm Al alloy), (5) iron hemisphere (Ø16 mm), and (6) air (thickness between iron hemisphere and workpiece = 2 mm).

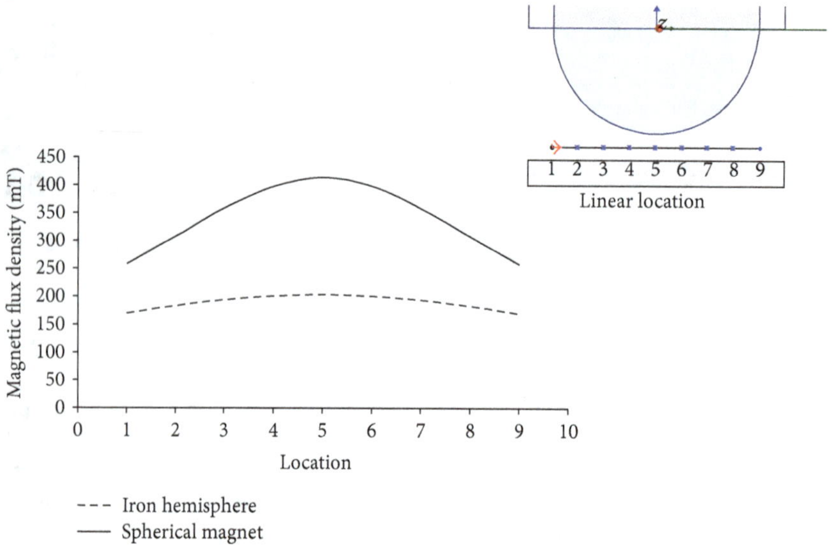

FIGURE 5: Magnetic flux density distribution at 2 mm distance from the surface in linear direction based on simulation results.

density based on simulation results. Magnetic flux density is measured at 2 mm distance from the external surface of the hemisphere circular and linear directions (Figure 5). It should be mentioned that workpiece which is located in front of hemisphere is made of non-ferromagnetic aluminum.

To verify simulation results, measurements are performed using calibrated Teslameter at 2 mm distance from the surface. Measuring method, equipment, and results are illustrated in Figures 7 and 8. All equipment applied for measuring magnetic flux density is made of non-ferromagnetic aluminum. Measurement path as it was for simulation was both circular and linear. Comparing Figures 6 and 8 it can be derived that measurements correlate well with simulation results. Average density of magnetic flux is presented in Table 1 for both simulation and measurements. Figure 9 depicts how that abrasive powder locates around the hemispherical magnet. Considering both simulation and experiment it can be concluded that the magnetic flux density around the hemispherical magnet is higher than that around the iron hemisphere. This fact causes better and more uniform distribution of the abrasive powder around the hemispherical magnet. In high velocities, iron hemisphere

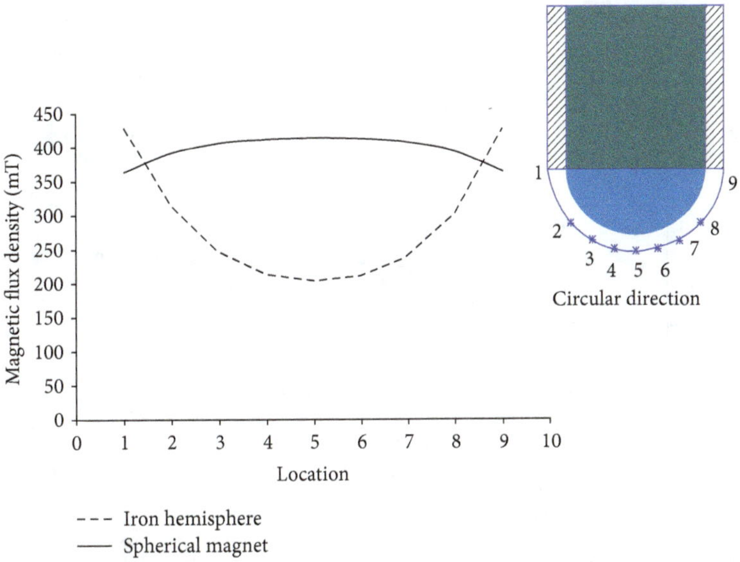

FIGURE 6: Magnetic flux density distribution at 2 mm distance from the surface in circular direction based on simulation results.

TABLE 1: Average measured and simulated magnetic flux density (mT).

	Linear distance 2 mm (simulation)	Linear distance 2 mm (measurement)	Difference (%)	Circular distance 2 mm (simulation)	Circular distance 2 mm (measurement)	Difference (%)
Iron hemisphere	0.185	0.148	25%	0.266	0.217	**22%**
Spherical magnet	0.32	0.26	23%	0.42	0.363	**15%**

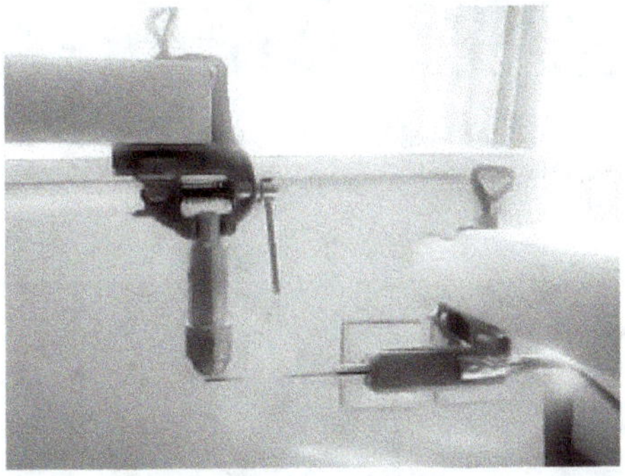

FIGURE 7: Measuring magnetic flux density at 2 mm from the surface of iron hemisphere.

cannot hold the abrasive power on its surface. Due to centrifugal forces and higher magnetic flux density in corners, much of abrasive powder goes to the corners and lower part of the hemisphere stays bare with almost no abrasive powder. Nevertheless, if the part in front of either hemispherical magnet or iron hemisphere is ferromagnetic, the behavior of circular curves will alter with their distance from the magnet.

2.2. Workpiece. Workpiece is modeled using freeform feature in Solidworks software package. On each specimen, some experiments are performed. Therefore, in order to make a uniform process, path does not have a curvature in the third direction. It should be noted that in the presented model, Figure 10, convex and concave areas possess smooth and high slopes.

For machining this workpiece a 3-axis CNC milling machine is used. G codes for rough and fine machining are extracted from Powermill software package. Workpiece is made of aluminum 7075. Machining parameters for finishing are kept constant but average surface roughness varies with curvature type. In this study investigation was performed only on the concave area of workpiece. Figure 11 shows measurements performed on this area of the model.

2.3. Condition of Experiment. In order to perform MAF process similar to finish machining, G codes are extracted from Powermill software package. Finishing strategy is used and in strategy remaining thickness is assumed equal to the gap. Rotational speed of the machining head and feed rate are easily set in the software. Moreover, amount of step-over can be set in each step. Step-over is defined as displacement of tool after passing the same path. In experiments step-over is assumed to be 0.1 mm. Other parameters like rotational speed, feed rate, and gap are variables which will be discussed later. It should be noted that by defining thickness of 1 mm on the path the minimum distance between tool and workpiece surface remains equal to 1 mm. Therefore it is possible that the distance in some areas becomes more. In Figures 12 and

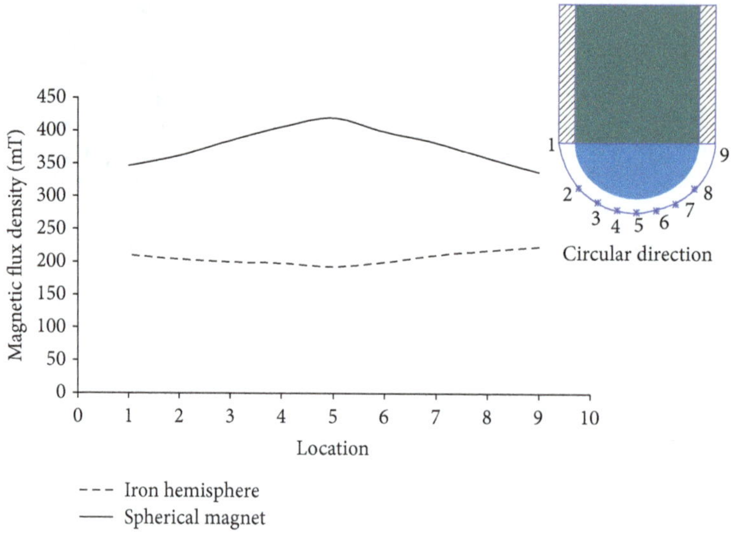

FIGURE 8: Measuring magnetic flux density at 2 mm from the surface of hemisphere in circular direction.

FIGURE 9: Positioning of the magnetic abrasive powder on hemispherical magnet.

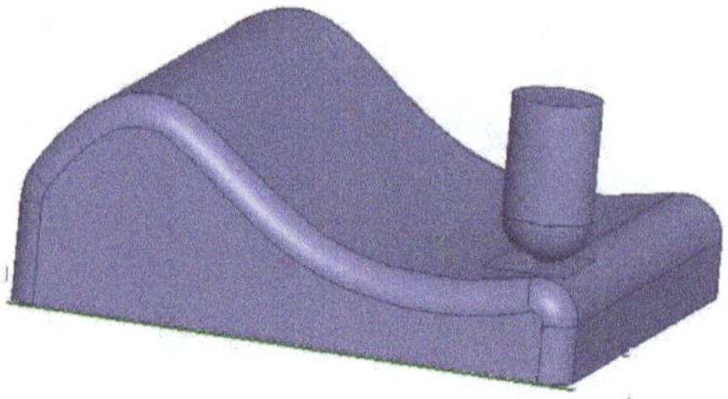

FIGURE 10: Model of workpiece with magnet with hemispherical head.

13, path of the process and the distance which is governed by G codes derived from Powermill software package are illustrated.

The most response of MAF process is the change in average surface roughness of the finished surface and the higher this change, the better and the more effective. Before doing the experimentation, finished surfaces are kept in acetone for 20 minutes. Then Surtronic 3+ surface roughness measuring set with cutoff of 0.8 based on DIN EN IS 0274:1998 is

FIGURE 11: Measuring magnetic flux density between magnet and workpiece surface.

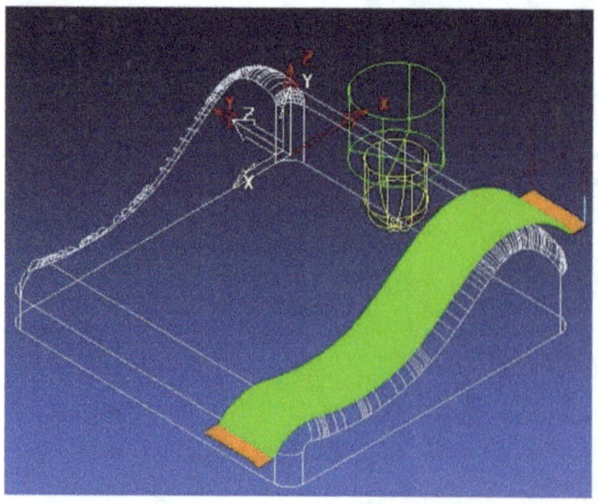

FIGURE 12: Path of finishing process.

applied for measurement [17]. In each area measurements are repeated for some times and average values are considered. It is clear that surface does not have a uniform roughness and different areas yield different roughness values. Figures

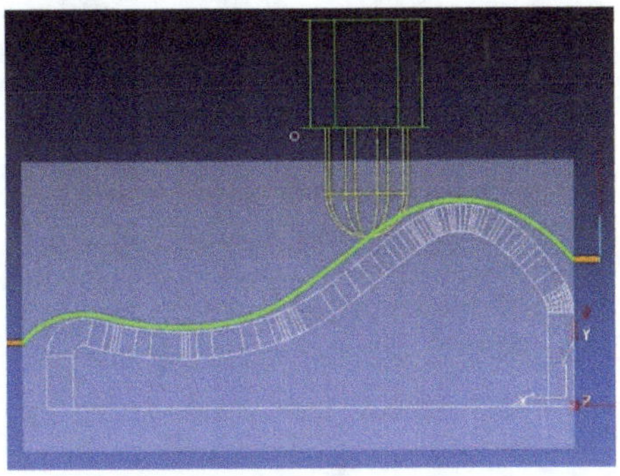

Figure 13: Distance between hemispherical head tool and workpiece surface.

Figure 14: Workpiece surface after machining.

14 and 15 show workpiece surface and measuring procedure, respectively. It should be said that during measuring measuring probe should be kept perpendicular to the path and, considering freeform surfaces, the workpiece should be rotated in different directions for making the measurement possible.

To do MAF process unbonded magnetic abrasive powder was used. Iron powder with mesh size of 400# and alumina abrasive powder with the mesh size of 1200# with weight ratio of 3/1 are stirred using a mechanical stirring machine for 20 minutes and become almost homogenous. To make a better connection of the abrasive particles to magnetic ones and avoiding separation in high speeds some amount of SAE30 lubricant is used in the mixture (weight ratio of 5%). Lubricant acts as a cooling agent and also it reduces frictional forces. Moreover it makes joint between iron and abrasive particles similar to joint between iron particles due to magnetic forces. It should be mentioned that absence of lubricant makes penetration of iron particles into the workpiece surface. Furthermore, rotational movement of the magnet makes abrasive particles separate from magnetic particles in short time. Effect of lack of lubricant usage is illustrated in Figure 16. Regarding distance of 2 mm from spherical head magnet surface to workpiece surface, the available volume is computed and 2 grams of powder is considered. Experimental constants are listed in Table 2.

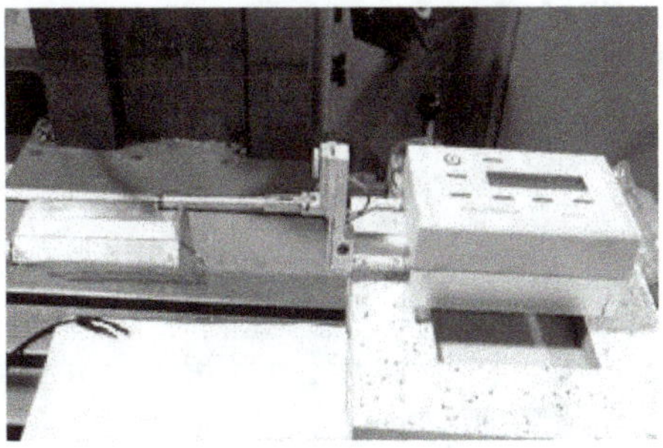

Figure 15: Measuring procedure with rotating workpiece.

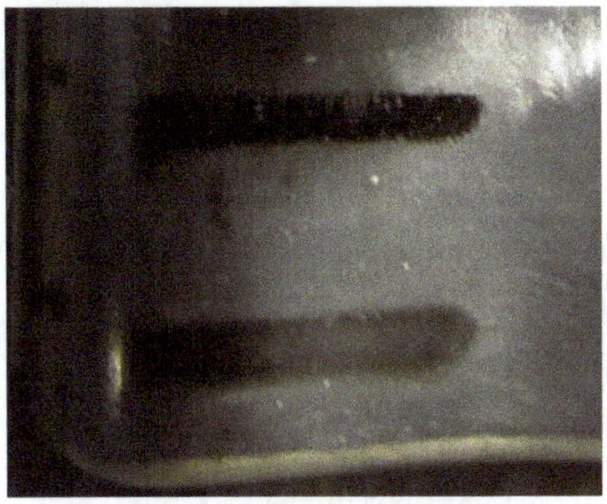

Figure 16: Process without applying lubricant.

Table 2: Constants in experiments.

Item	Condition
Workpiece	Al alloy
Abrasive size	Mesh #1200
Fe size	Mesh #400
Abrasive	Al2o3
Weight ratio (magnetic/abrasive)	3/1
Lubricant	30SAE
Lubricant ratio in powder	5%
Permanent magnet	N35

2.4. Response Surface Method. Suitable design of experiments can decrease the number of experiments. In some conventional methods for design of experiments like factorial design, the number of required experiments increases with the number of parameters. A high number of experiments are both time consuming and expensive. Response surface method is a mathematical-statistical method which is used to model and analyze problems which are affected by bunch of variables [18]. This method also can make a model between inputs and outputs of a problem and reveals it as a mathematical equation. In this method corresponding to the levels and factors constant number of experiments is required. In a complete factorial (general condition) if there are 4 factors varying in 3 levels, total required experiments equal 81. Using response surface method this can decrease to 30. In these 30 experiments excluding that for central point, no experiment is repetitive. RSM consists of a group of mathematical and

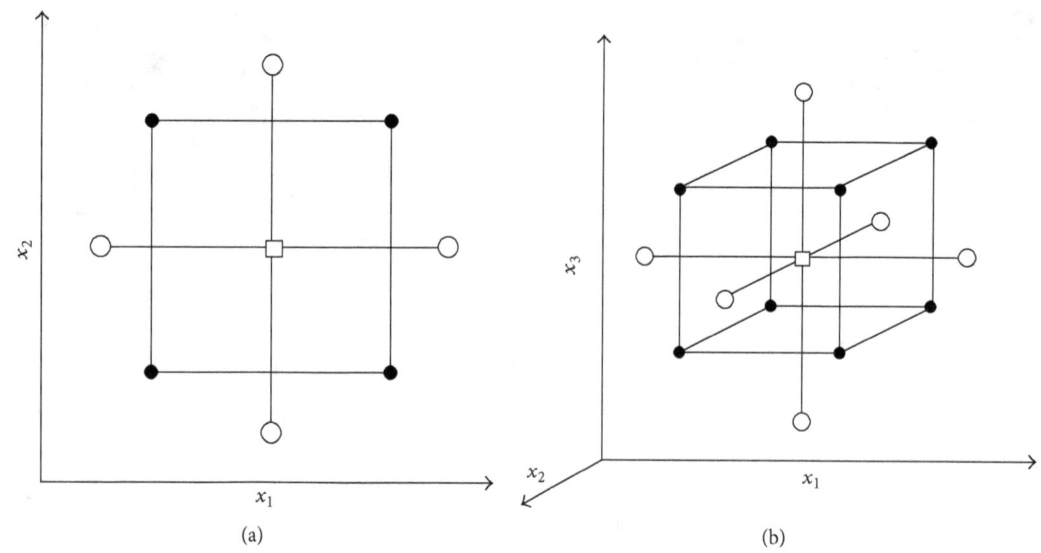

FIGURE 17: Central composite designs for optimization: (a) two variables ($\alpha = 1.41$) and (b) three variables ($\alpha = 1.68$); (●): points of factorial design, (○): axial points, and (□): central point.

statistical techniques that are based on the fit of empirical models to the experimental data obtained in relation to experimental design. Toward this objective, linear or square polynomial functions are employed to describe the system studied and, consequently, to explore (modeling and displacing) experimental conditions until its optimization [19].

Some stages in the application of RSM as an optimization technique are as follows: (1) the selection of independent variables of major effects on the system through screening studies and the delimitation of the experimental region, according to the objective of the study and the experience of the researcher; (2) the choice of the experimental design and carrying out the experiments according to the selected experimental matrix; (3) the mathematic-statistical treatment of the obtained experimental data through the fit of a polynomial function; (4) the evaluation of the model's fitness; (5) the verification of the necessity and possibility of performing a displacement in direction to the optimal region; and (6) obtaining the optimum values for each studied variable.

The simplest model which can be used in RSM is based on a linear function. For its application, it is necessary that the responses obtained are well fitted to the following equation:

$$y = \beta_0 + \sum_{i=1}^{k} \beta_i x_i + \varepsilon, \tag{3}$$

where k is the number of variables, β_0 is the constant term, i represents the coefficients of the linear parameters, x_i represents the variables, and ε is the residual associated to the experiments. Therefore, the responses should not present any curvature. To evaluate curvature, a second-order model must be used. Two-level factorial designs are used in the estimation of first-order effects, but they fail when additional effects, such as second-order effects, are significant. So, a central point in two-level factorial designs can be used for evaluating curvature. The next level of the polynomial model should contain additional terms, which describe the interaction between the different experimental variables. This way, a model for a second-order interaction presents the following terms:

$$y = \beta_0 + \sum_{i=1}^{k} \beta_i x_i + \sum_{1\le i\le j}^{k} \beta_{ij} x_i x_j + \varepsilon, \tag{4}$$

where ij represents the coefficients of the interaction parameters. In order to determine a critical point (maximum, minimum, or saddle), it is necessary for the polynomial function to contain quadratic terms according to the equation presented as follows:

$$y = \beta_0 + \sum_{i=1}^{k} \beta_i x_i + \sum_{i=1}^{k} \beta_{ii} x_i^2 + \sum_{1\le i\le j}^{k} \beta_{ij} x_i x_j + \varepsilon, \tag{5}$$

where ii represents the coefficients of the quadratic parameter. To estimate the parameters in (5), the experimental design has to assure that all studied variables are carried out in at least three factor levels. Thus, two modeling, symmetrical response surface designs are available. Among the more known second-order symmetrical designs are the three-level factorial design, Box-Behnken design, central composite design, and Doehlert design. These symmetrical designs differ from one another with respect to their selection of experimental points, number of levels for variables, and number of runs and blocks.

In this study central composite design (CCD) is used. This design consists of the following parts: (1) a full factorial or fractional factorial design; (2) an additional design, often a star design in which experimental points are at a distance α, from its center; and (3) a central point. Figures 17(a) and 17(b) illustrate the full central composite design for optimization of two and three variables. Full uniformly routable central composite designs present the following characteristics:

FIGURE 18: Performing MAF process.

(1) They require an experiment number according to $N = k^2 + 2k + C_P$, where k is the factor number and (C_P) is the replicate number of the central point.

(2) α-values depend on the number of variables and P value can be calculated by $\alpha = 2^{(K-P)/4}$. But P values less than 0.05 are near to zero. For two, three, and four variables, they are, respectively, 1.41, 1.68, and 2.00.

(3) All factors are studied in five levels $(-\alpha, -1, 0, +1, +\alpha)$.

Factorial points which perform the approximation of the interaction terms of the second-order model can be valued in the range of −1 and 1. Axial points approximate second-order terms in the second-order regression model. Not only do central points approximate the internal errors but also they contribute to approximation of the second-order terms. The desired response of the design of experiments is the variation percentage of surface roughness which can be computed using

$$\Delta\mathrm{Ra}\,(\%) = \frac{\text{initial roughness surface} - \text{finish roughness surface}}{\text{initial roughness surface}} \times 100. \quad (6)$$

In this paper effect of rotational speed of spindle (cutting speed), feed rate, gap, and amount of powder is investigated. Workpiece is divided into 8 equal sections. In each section, an experiment with its settings is performed. As the path of all experiments is constant, related to step-over and feed rate, time in which each experiment is completed varies. After performing 8 experiments using a ball nose tool, the whole workpiece is machined and measuring is performed again. In each new experiment new and unused powder is applied. MAF is performed on the whole path on the workpiece surface but here only data for concave part is reported. Figure 18 depicts the procedure of MAF on workpiece.

2.4.1. Mathematical Model. First step in usage of response surface method is to find mathematical approximate relations between outputs of the experiments and input variables. In this study experiments are designed based on central composite design. Considered variables are listed in Table 3 in five levels with their values. Mathematical model for the experiments is formed as a linear second-order regression equation which passes through the points listed in the table or it occupies the closest values to them.

TABLE 3: Coded values of experiment input factors.

Factors	$-\alpha$	−1	0	1	α
Amount of powder, w (gr)	0.35	0.7	1.05	1.4	**1.75**
Gap, G (mm)	0.5	1	1.5	2	**2.5**
Feed rate, F (mm/min)	10	20	30	40	**50**
Rotational speed, V (rpm)	100	600	1100	1600	**2100**

MiniTab v16 is used to analyze the response and also to derive governing mathematical equation. For statistical analysis variance analysis is used.

Considering designed experiments based on response surface method for 4 factors and specifying $\alpha = 2$, thirty experiments are required and together with the values of change in percentage of surface roughness are listed in Table 4. In order to avoid possible errors, experiments were done randomly and not based on order presented in the list.

3. Data Analysis and Results

Considering results of surface roughness change, analysis and modification of the model to omit insignificant factors, variance analysis table, and coded governing regression equation's coefficients are presented in Table 5. It should be noticed that those variables are significant in that their P value is less than 0.05 when the reliability is 95.

Taking into account the outputs from variance analysis and modified model, corresponding explicit equation for the output, based on machining parameters, can be computed using

$$\begin{aligned}\Delta\mathrm{Ra}\,(\%) = {} & 41.06 - 3.37G - 3.19F + 2.3533V \\ & - 1.2529V^2 - 2.23W \times F - 2.03G \times V \\ & - 2.78F \times V,\end{aligned} \quad (7)$$

where model R-Sq is 88.63% and R-Sq (adj) is 84.5% which shows acceptable accuracy of the presented model based on RSM method. Diagrams for scatter of the residuals are presented in Figure 19 which are suitable and matched. Lack of fit of the model becomes insignificant.

3.1. The Effect of Significant Factor. To accurately analyze the effect of parameters, the effect of main parameters and their interactions should be carefully analyzed. In this section the effect of each parameter is discussed. In analysis of the interaction of parameters, other variables are considered at experiment's equilibrium (central point).

3.1.1. The Effect of Gap between the Magnet and Workpiece. Magnetic flux density decreases with distance from the magnet surface. As can be seen in Figure 20, decreasing gap increases the change in surface roughness. The less the distance is, the higher the magnetic force acts on magnetic particles and therefore stiffness of the magnetic abrasive

Table 4: Performed experiments and their values.

Number	Rotational speed, V (rpm)	Feed rate, F (mm/min)	Gap, G (mm)	Amount of powder, w (g)	ΔRa (%)
1	−1	−1	−1	−1	35.61
2	−1	−1	−1	1	40.59
3	−1	−1	1	−1	35.39
4	−1	−1	1	1	37.32
5	−1	1	−1	−1	42.11
6	−1	1	−1	1	32.96
7	−1	1	1	−1	37.30
8	−1	1	1	1	30.87
9	1	−1	−1	−1	52.76
10	1	−1	−1	1	57.02
11	1	−1	1	−1	42.19
12	1	−1	1	1	45.58
13	1	1	−1	−1	43.62
14	1	1	−1	1	40.50
15	1	1	1	−1	32.84
16	1	1	1	1	30.34
17	0	0	0	−2	39.46
18	0	0	0	2	40.58
19	0	0	−2	0	45.87
20	0	0	2	0	31.99
21	0	−2	0	0	47.62
22	0	2	0	0	37.30
23	−2	0	0	0	35.11
24	2	0	0	0	45.55
25	0	0	0	0	40.56
26	0	0	0	0	46.06
27	0	0	0	0	43.48
28	0	0	0	0	41.03
29	0	0	0	0	40.27
30	0	0	0	0	41.57
31	0	0	0	0	38.07

brush increases. On the other hand, with an increase in gap, magnetic force cannot withstand centrifugal forces and therefore much of the powder is thrown away and separated from the magnet surface and process efficiency decreases remarkably. In particular, this case is true for mixed powders in comparison with sintered powder. Considering the type and size of the magnet and amount of magnetic flux, even within the minimum possible distance the imposing force is in order of some tenth or hundredth of micro Newton. Micro forces imposing on abrasive particles make micro chipping occur well and because of the fine size of the particles the penetration depth in workpiece surface is in the order of some tenth or hundredth of microns. Therefore they can easily delete hills and valleys on the surface without producing new ones. Based on previous reports, in MAF on straight surfaces, high reduction in gap introduces roughness on the surface. For curved surfaces, since the selected path can hold the minimum gap of 1 mm, many areas on the spherical surface of the magnet experience higher distances to the workpiece surface and thus in concave areas decreasing gap led to less surface roughness and better finishing.

3.1.2. Effect of Feed Rate. Generally in machining processes the less the feed rate is, the better the surface quality becomes. MAF is not excluded from this general law. High feed rates increase shear force and vibration of the abrasive particles and consequently the quality of the surface decreases. Furthermore, high feed rates change the mechanism of deformation and separation of the abrasive particles and increase surface roughness. Moreover, in lower feed rates abrasive powder in the contact interface with workpiece surface has more time for finishing. It should be noted that lowering feed rate increases machining time and production costs. Figure 21 depicts the influence of feed rate.

3.1.3. Effect of Cutting Speed. Regression coefficient of rotational speed almost equals that of feed rate with this difference that higher values of cutting speed positively affect surface

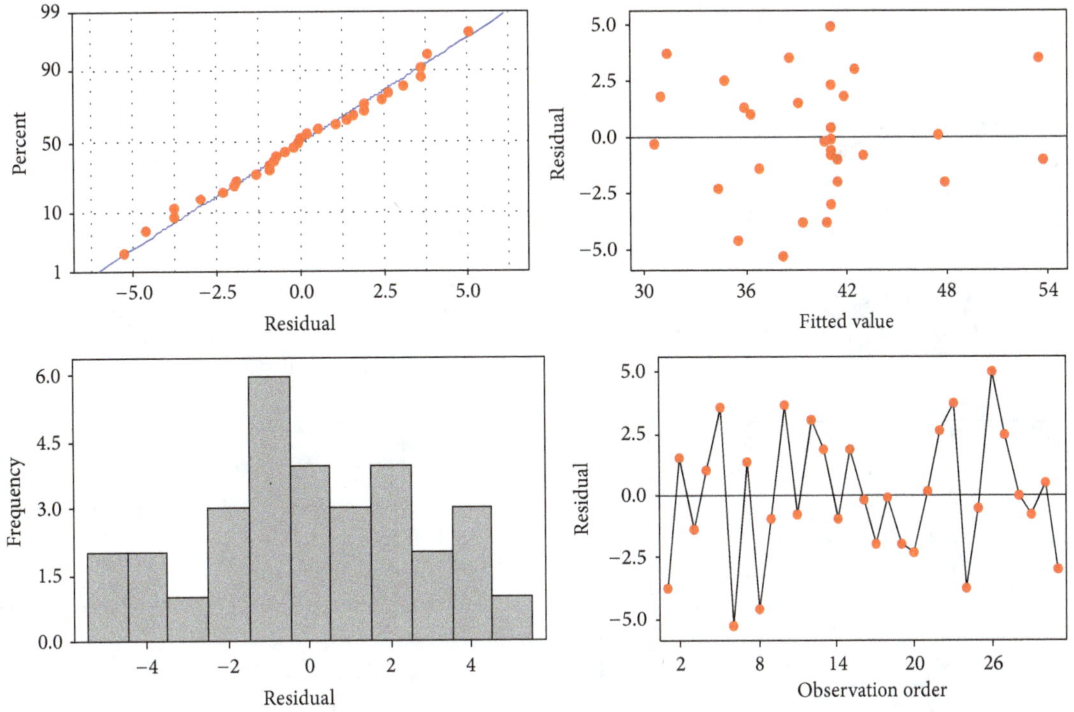

FIGURE 19: Diagrams for residual distribution.

TABLE 5: Regression model coefficients and lack of fit in primary and modified models.

Modified model		Initial model		Terms
Coefficient regression	*P* value	Coefficient regression	*P* value	
41.06	0.000	41.57	0.000	Constant
−0.18	0.709	−0.18	0.725	w
−3.37	0.000	−3.37	0.000	G
−3.19	0.000	−3.19	0.000	F
2.35	0.000	2.35	0.000	V
—	—	−0.31	0.512	$w \times w$
—	—	−0.58	0.229	$G \times G$
—	—	0.29	0.53	$F \times F$
−1.25	0.004	−1.30	0.013	$V \times V$
—	—	−0.03	0.95	$G \times w$
−2.23	0.001	−2.23	0.003	$F \times w$
—	—	0.66	0.30	$V \times w$
—	—	−0.14	0.81	$F \times G$
−2.03	0.002	−2.03	0.005	$V \times G$
−2.78	0.000	−2.78	0.000	$V \times F$
—	0.655	—	0.553	Lack of fit

quality. Due to changes in material plastic behavior in higher cutting speeds, shear forces decrease and consequently surface quality increases. Figure 22 shows the effect of cutting speed (rotational speed of machining head).

As can be seen with increasing cutting speed to 1600 rpm, the change in surface roughness increases but higher increase in cutting speed decreases surface quality. In high cutting speeds centrifugal forces surpass magnetic forces and abrasive particles are thrown away. Reduced amount of powder causes higher surface roughness. It should be noted that interaction of rotational speed and gap and also rotational speed and feed rate are also significant on the process. Figures 23 and 24 are the result of analysis of experiments showing the effect of interaction between factors. It can be seen from Figure 23 that with a low feed rate increasing cutting speed reduces surface roughness and aforementioned limit for increasing rotational speed does not exist. Figure 24 shows that in the minimum gap suitable results can be achieved.

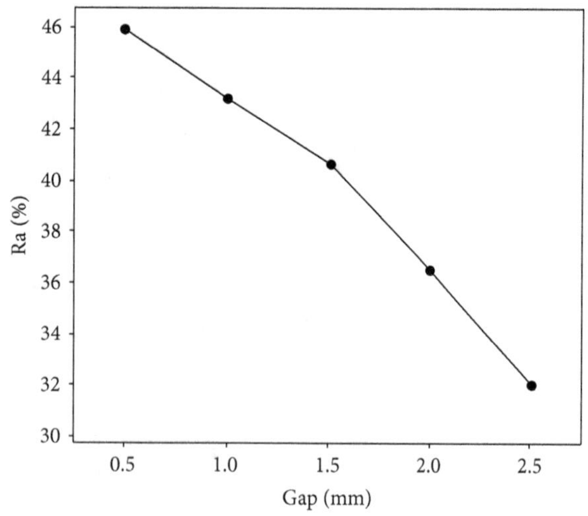

FIGURE 20: The effect of magnet-workpiece gap.

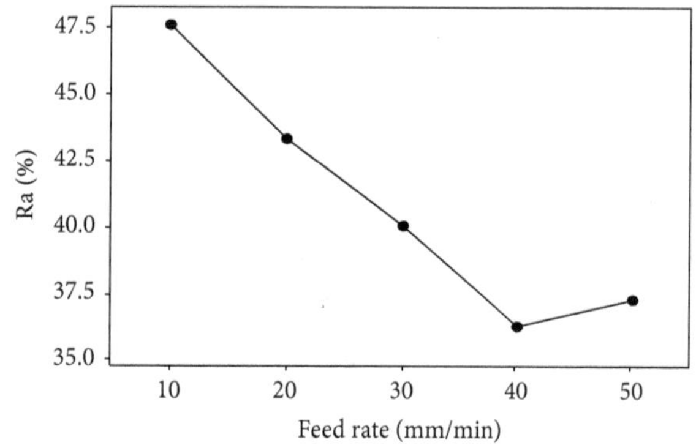

FIGURE 21: The effect of feed rate.

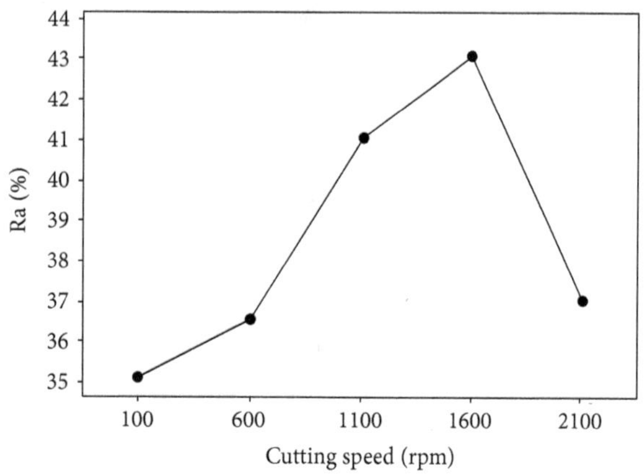

FIGURE 22: The effect of cutting speed.

3.1.4. Effect of Powder Amount. Considering variance analysis and Figure 25, this factor is not effective on the process and increasing powder amount does not affect surface roughness. However, the magnet in the defined gap can carry only limited amount of powder. Otherwise, due to sticking character of the combined magnetic abrasive powder caused by the

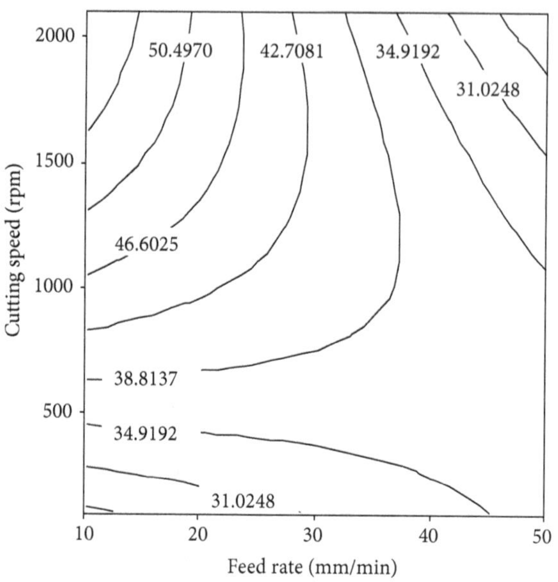

FIGURE 23: Interaction effect of cutting speed and feed rate.

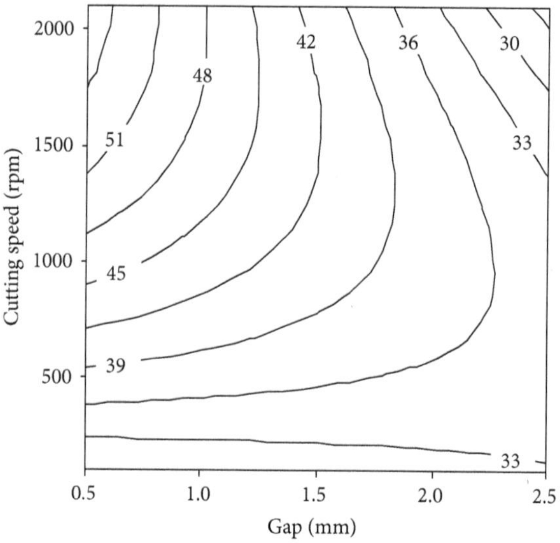

FIGURE 24: Interaction effect of cutting speed and gap.

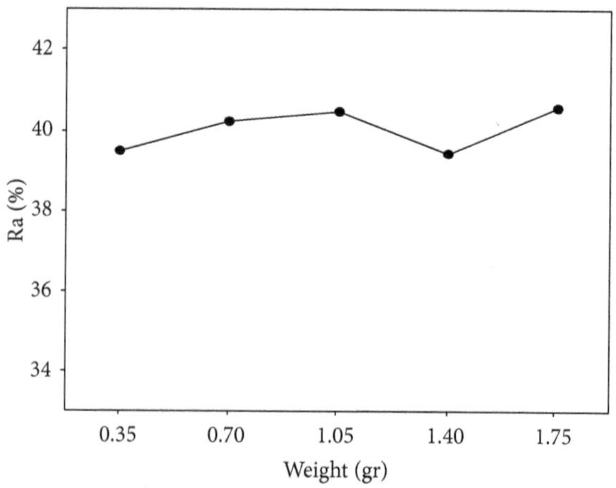

FIGURE 25: Effect of amount of powder.

Table 6: Optimum results.

Optimization	Cutting speed	Feed rate	Gap	Amount of powder	ΔRa (%)
Simulation	2100	10	0.5	1.75	%75.2
Experiment	2100	10	0.5	1.75	%62.1

Figure 26: Remaining powder on the workpiece surface.

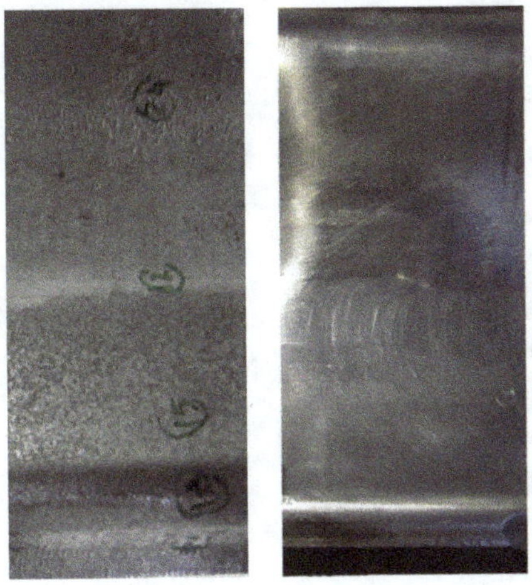

(a) (b)

Figure 27: Workpiece surface (a) before MAF and (b) after MAF.

lubricant, part of it will remain on the surface. Figure 26 shows this phenomenon.

3.2. Finding Optimum Condition for MAF Operation. Considering results from analysis of diagrams and governing mathematical model, software suggests an optimum solution and predicts the change in surface roughness. It can be seen that results taken from analysis of the experiments are confirmed with high accuracy and change in surface roughness can be achieved up to 65%. These results are presented in Table 6. Figure 27 shows process effect on the workpiece surface.

4. Conclusion

In this study in addition to simulation and measuring the magnetic flux density of iron hemisphere and hemispherical magnet, appropriate tool was manufactured for MAF process. Due to better distribution and availability of magnetic flux density on hemispherical magnet surface, it is used as MAF tool. The behavior of hemispherical magnet passing concave surfaces and its magnetic flux density is also investigated. The discrepancy between simulated and measured magnetic flux density varies between 15% and 26%. Simulation overestimates magnetic flux density. Average magnetic flux density imposing on aluminum workpiece surface in 2 mm distance from the magnet surface is 0.36 Tesla.

Based on results, gap possesses the highest influence on the change of surface roughness. In lower gaps this change is higher.

According to regression coefficients the effects of feed rate and cutting speed are equal. However, lower feed rate and higher cutting speed produce better results. Based on the interaction effect of cutting speed and feed rate and also cutting speed and gap, the limitation in increasing cutting speed is vanished because with decreasing gap and feed rate high cutting speed can be selected for the process. Powder amount has a negligible effect on the process even though higher amount of powder has a slightly positive influence. The suggested regression model correlates well with the experiment outputs. Optimum parameters are gap size of 0.5 mm, feed rate of 10 mm/min, rotational speed of 2100 rpm, and powder amount of 1.75 gr. This condition can improve surface roughness up to 75% based on simulation results. In performed experiments, based on the ranges of the variables, improvement was reported up to 62%. Difference was 13% percent, which is acceptable.

On concave surfaces, surface roughness value changes from 1.3 μm to 0.2 μm and in some points measurements show surface roughness of .08 μm.

Competing Interests

The authors declare that they have no competing interests.

References

[1] V. K. Jain, "Magnetic field assisted abrasive based micro-/nano-finishing," *Journal of Materials Processing Technology*, vol. 209, no. 20, pp. 6022–6038, 2009.

[2] S. M. Ji, Y. M. Xu, G. D. Chen, and M. S. Jin, "Comparative study of magnetic abrasive finishing in free-form surface based on different polishing head," *Materials Science Forum*, vol. 675, pp. 593–596, 2011.

[3] S. Jha and V. K. Jain, "Nanofinishing techniques," in *Micromanufacturing and Nanotechnology*, N. P. Mahalik, Ed., pp. 171–195, Springer, Berlin, Germany, 2006.

[4] D. K. Singh, V. K. Jain, and V. Raghuram, "Parametric study of magnetic abrasive finishing process," *Journal of Materials Processing Technology*, vol. 149, no. 1–3, pp. 22–29, 2004.

[5] B. Girma, S. S. Joshi, M. V. G. S. Raghuram, and R. Balasubramaniam, "An experimental analysis of magnetic abrasives finishing of plane surfaces," *Machining Science and Technology*, vol. 10, no. 3, pp. 323–340, 2006.

[6] B.-H. Yan, G.-W. Chang, J.-H. Chang, and R.-T. Hsu, "Improving electrical discharge machined surfaces using magnetic abrasive finishing," *Machining Science and Technology*, vol. 8, no. 1, pp. 103–118, 2004.

[7] R. S. Mulik and P. M. Pandey, "Magnetic abrasive finishing of hardened AISI 52100 steel," *The International Journal of Advanced Manufacturing Technology*, vol. 55, no. 5–8, pp. 501–515, 2011.

[8] C.-T. Lin, L.-D. Yang, and H.-M. Chow, "Study of magnetic abrasive finishing in free-form surface operations using the Taguchi method," *The International Journal of Advanced Manufacturing Technology*, vol. 34, no. 1-2, pp. 122–130, 2007.

[9] S. Yin and T. Shinmura, "A comparative study: polishing characteristics and its mechanisms of three vibration modes in vibration-assisted magnetic abrasive polishing," *International Journal of Machine Tools and Manufacture*, vol. 44, no. 4, pp. 383–390, 2004.

[10] M. D. Zhang, M. Lv, and H. L. Chen, "Theoretical research on polishing free-form surface with magnetic abrasive finishing," *Key Engineering Materials*, vol. 392–394, pp. 404–408, 2009.

[11] Y. H. Ding, X. G. Yao, X. X. Wang, and S. C. Yang, "Study on the performances of the ferromagnetic poles based on the curved surface magnetic abrasive finishing," *Key Engineering Materials*, vol. 359-360, pp. 365–368, 2008.

[12] M. Anzai, T. Yoshida, and T. Nakagawa, "Magnetic abrasive automatic polishing of curved surface, focused on experimental equipments and characterization," *RIKEN Review*, no. 12, pp. 15–16, 1996.

[13] L.-D. Yang, C.-T. Lin, and H.-M. Chow, "Optimization in MAF operations using Taguchi parameter design for AISI304 stainless steel," *The International Journal of Advanced Manufacturing Technology*, vol. 42, no. 5-6, pp. 595–605, 2009.

[14] D. K. Singh, V. K. Jain, and V. Raghuram, "Experimental investigations into forces acting during a magnetic abrasive finishing process," *The International Journal of Advanced Manufacturing Technology*, vol. 30, no. 7-8, pp. 652–662, 2006.

[15] N. K. M. Naif, "Study on the parameter optimization inmagnetic abrasive polishing forbrass cuzn33plate using Taguchi method," *The Iraqi Journal for Mechanical and Material Engineering*, vol. 12, no. 3, pp. 596–615, 2012.

[16] R. S. Gill, "Effect of the process parameters on the surface roughness during magnetic abrasive finishing process on ferromagnetic stainless steel workpieces," *International Journal of Mechanical Engineering and Technology*, vol. 4, pp. 310–316, 2013.

[17] http://www.taylor-hobson.com/.

[18] D. C. Montgomery, *Design and Analysis of Experiments*, John Wiley & Sons, 2008.

[19] M. A. Bezerra, R. E. Santelli, E. P. Oliveira, L. S. Villar, and L. A. Escaleira, "Response surface methodology (RSM) as a tool for optimization in analytical chemistry," *Talanta*, vol. 76, no. 5, pp. 965–977, 2008.

6

Strain Rate-Dependent Constitutive and Low Stress Triaxiality Fracture Behavior Investigation of 6005 Al Alloy

Yong Peng[1,2,3] **Xuanzhen Chen,**[1,2,3] **Shan Peng,**[1,4] **Chao Chen**[5] **Jiahao Li,**[1,6] **and Guowei Liu**[1,2,3]

[1]*School of Traffic and Transportation Engineering, Central South University, Changsha 410075, China*
[2]*Key Laboratory of Traffic Safety on Track, Central South University, Ministry of Education, Changsha 410075, China*
[3]*National and Local Joint Engineering Research Center of Safety Technology for Rail Vehicle, Changsha 410075, China*
[4]*Smart Transport Key Laboratory of Hunan Province, Changsha 410075, China*
[5]*Hunan Industry Polytechnic, Changsha 410208, China*
[6]*Joint International Research Laboratory of Key Technology for Rail Traffic Safety, Changsha 410075, China*

Correspondence should be addressed to Chao Chen; hngyzycc@163.com

Academic Editor: Luigi Nicolais

In order to study the dynamic and fracture behavior of 6005 aluminum alloy at different strain rates and stress states, various tests (tensile tests at different strain rates and tensile shearing tests at five stress states) are conducted by Mechanical Testing and Simulation (MTS) and split-Hopkinson tension bar (SHTB). Numerical simulations based on the finite element method (FEM) are performed with ABAQUS/Standard to obtain the actual stress triaxialities and equivalent plastic strain to fracture. The results of tensile tests for 6005 Al show obvious rate dependence on strain rates. The results obtained from simulations indicate the feature of nonmonotonicity between the strain to fracture and stress triaxiality. The equivalent plastic strain reduces to a minimum value and then increases in the stress triaxiality range from 0.04 to 0.30. A simplified Johnson-Cook (JC) constitutive model is proposed to depict the relationship between the flow stress and strain rate. What is more, the strain-rate factor is modified using a quadratic polynomial regression model, in which it is considered to vary with the strain and strain rates. A fracture criterion is also proposed in a low stress triaxiality range from 0.04 to 0.369. Error analysis for the modified JC model indicates that the model exhibits higher accuracy than the original one in predicting the flow stress at different strain rates. The fractography analysis indicates that the material has a typical ductile fracture mechanism including the shear fracture under pure shear and the dimple fracture under uniaxial tensile.

1. Introduction

Al-Mg-Si-Cu alloy, due to its excellent processing property and mechanical behavior with medium strength, superior corrosion resistance, and advanced welding performance, has been playing an important role in various industrial applications and scientific research [1]. It is crucial to ensure impact safety, while lightweight materials are extensively used to improve train speed [2–4]. A large number of studies [5–8] on the energy absorption components of the train show that the dynamic mechanical response of the material has a great influence on the performance of crash energy absorption. Previous studies [9–11] have investigated the dynamic response of different aluminum alloys. 6005 Al alloy, as a common manufacturing material for high-speed train, attracted extensive attention. The effect of temperature on the micromechanics of ductile fracture and the continuous cooling precipitation, including temperature- and time-dependent precipitation behavior, were investigated in [12, 13]. The influence of the age-hardening behavior and microstructural characterization of precipitates are also studied by electron microscopy [14].

To study the dynamic mechanical properties of materials, experimental studies, constitutive model, and numerical simulations have been developed under various loading conditions. Electronic testing machines and the split-Hopkinson bar (SHPB) system have been used to investigate mechanical behavior under tension, compression, or other loading conditions [15–17]. Constitutive models including the physical model [18] and phenomenological model [19] have been developed to describe the flow stress and mechanical behavior of materials under various loading conditions. Numerical simulation, based on the finite element method (FEM), is always used to calculate the flow behavior of variety of materials under specified loading conditions [20, 21]. The constitutive behavior model plays an important role in numerical simulation. The Johnson-Cook model [22], one of the most widely used phenomenological models, is used to describe the coupling factors of strain hardening, strain-rate hardening, and thermal. This model is also widely employed in numerical simulation because the five material constants can be easily acquired from the experimental data.

Ductility, considered as an important mechanical property, has been investigated on many materials [10, 23, 24]. Stress triaxiality, defined as the ratio of the hydrostatic stress and von Mises stress to depict the stress state of materials, is found to be an important factor that influences fracture strain greatly according to previous studies [23, 24]. Several fracture models [25, 26] have been proposed to describe the dependency of fracture on the stress triaxiality. The JC fracture model [25] is a common model widely used in the commercial finite element code. However, the JC fracture model is not very accurate especially for the low triaxiality region. Bao and Wierzbicki (BW) [27] developed a more flexible fracture locus to reflect the fracture behavior in the fracture strain-triaxiality space. Wierzbicki et al. [28] compared seven representative criteria and found that the X-W criterion was performing well in a wide triaxiality range. Even if the BW fracture locus is more practical than the JC one, it still does not consider the Lode angle effect. Recently, the importance of the two parameters, triaxiality and Lode angle, has been stressed for the ductile fracture of metals. The MMC is a modification of the original criterion whose description and derivation can be found in [29] and has been successfully used by Gilioli et al. [30] to study the ballistic impact failure of aluminum 6061-T6. Previous studies [10, 11, 24] mostly focus on high triaxialities, which means that the material is under axial tension loading. The fracture behavior of 6005 Al alloy under the tension and shearing loading would be studied.

In this paper, various loading conditions are set to obtain the dynamic behavior of 6005 Al alloy. Not only the strain-rate effect would be investigated through a series of experiments at different strain rates using the MTS and SHTB testing systems, but also the mechanical behavior under the combined force with tension and shearing was studied. There are several strain rates including quasi-static and dynamic ($10^{-3}\,s^{-1}$, $10^{-2}\,s^{-1}$, $10^{-1}\,s^{-1}$, $700\,s^{-1}$, $1100\,s^{-1}$, and $3300\,s^{-1}$). To investigate the fracture strain and ultimate strength of the material under the combined force with tension and shearing, tensile shearing tests under the different angles (0°, 30°, 45°, 60°, and 90°, resp.) are performed. Furthermore, a modified Johnson-Cook constitutive model would be obtained to describe the dynamic behavior of 6005 Al alloy. Model predictions would be compared with the testing data. Finally, a fracture criterion would be proposed to describe the relationship between fracture strain and stress triaxiality at the stress state of low triaxialities.

2. Methods

2.1. Materials and Experiments. The present study has been carried out on 6005 Al alloy. The chemical composition of the material is listed in Table 1. Two groups of experiments are set to study the mechanical property. They are tensile tests at various strain rates and tensile shearing tests of double-notch specimens. Three repeated tests under the same condition are carried out to ensure the accuracy and reliability of results, especially for dynamic loading tests.

2.1.1. Low Strain Rate Tensile Tests. To more reasonably study the mechanical behavior of 6005 Al alloy under the quasi-static and low strain-rate loading condition, tension tests based on four strain rates (from $10^{-4}\,s^{-1}$ to $10^{-1}\,s^{-1}$) are adopted by the Mechanical Testing and Simulation (MTS) 647 hydraulic wedge grip. The maximum axial load ability of the machine is 200 kN. The tension specimens are used with the dimensions of 100 mm in the gauge length L_0 and 20 mm in the width of the cross section and 8 mm in the thickness of the cross section, as shown in Figure 1(a). The engineering stress σ_E and engineering strain ε_E can be expressed as follows:

$$\sigma_E = \frac{F}{A_0},\qquad \varepsilon_E = \frac{\Delta L}{L_0}, \tag{1}$$

where A_0 is the cross section area of the specimens. F and ΔL are the tensile load and the elongation of the specimens, respectively. Furthermore, the true stress σ_T and true strain ε_T are obtained by the following equations [25]:

$$\sigma_T = \sigma_E(1+\varepsilon_E),\qquad \varepsilon_T = \ln(1+\varepsilon_E). \tag{2}$$

Note that (2) is inapplicable after the onset of necking. Figure 1(b) shows the specimen fixed in the MTS. Three repeated tests are conducted to ensure the repeatability.

2.1.2. SHTB Tests. SHTB test system is used to study the dynamic behavior of 6005 Al alloy. The striker of SHTB is 20 mm in diameter and 500 mm in length. The incident bar is 20 mm in diameter and 3300 mm in length, while the transmitted bar is 20 mm in diameter and 1500 mm in length. To obtain a good signal of the strain and signal-to-interference ratio, both bars are made of LY12 Al alloy with ultrahigh hardness. The bullet, promoted by the launching system with high-pressure gas, impacts the flange of the incident bar. Then, a tensile loading wave generated through the incident bar. A strain gauge, affixed to the surface of the incident bar, records the incident signal and the reflected signal, while another one affixed to the surface of the transmitted bar records the transmission signal. All of signals simplified through a high dynamic strain meter are stored and recorded by an oscilloscope. Finally, a computer deals with the strain signals. Figure 2 shows the experimental device of the SHTB system.

Based on the hypothesis of the one-dimensional stress wave and strain and stress axial uniformity, the expression of

TABLE 1: The chemical composition of 6005 Al alloy.

Chemical element	Si	Fe	Cu	Mn	Mg	Cr	Zn	Ti	Al
Wt.%	0.5~0.9	0.35	0.3	0.5	0.4~0.7	0.3	0.2	0.1	Balanced

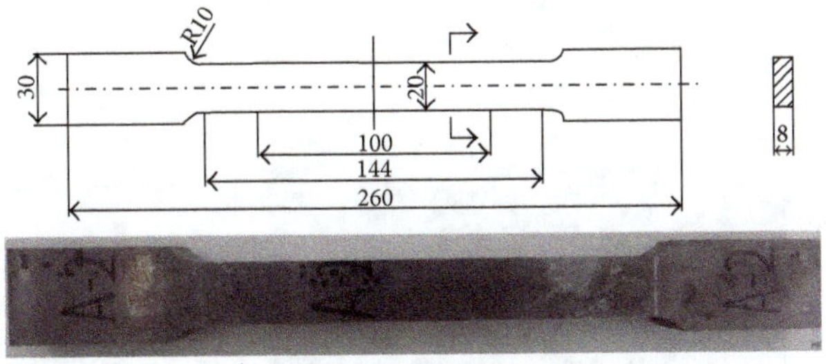

(a)

(b)

FIGURE 1: (a) Specimen for the conventional tension test; (b) MTS 647 hydraulic wedge grip used for the conventional tension test.

(a)

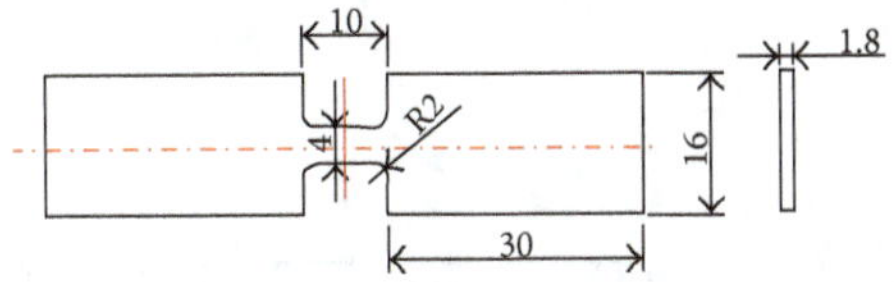

(b)

(c)

FIGURE 2: (a) Split-Hopkins tensile bar (SHTB) testing system; (b) specimen for the tension test under high strain rate; (c) specimen fixed in the SHTB system.

engineering stress, engineering strain, and strain rate is calculated by the following equations:

$$\sigma_e(t) = \frac{EA}{A_S}\varepsilon_t(t),$$
$$\varepsilon_e(t) = -\frac{2C_0}{L}\int_0^t \varepsilon_r(t)dt, \quad (3)$$
$$\dot{\varepsilon}(t) = -\frac{2C_0}{L}\varepsilon_r(t),$$

where E is the elastic modulus of the incident bar. A and A_S are the cross section areas of the incident bar and specimens, respectively. C_0 is the stress wave speed and L is the gauge length of the specimens. $\varepsilon_t(t)$ and $\varepsilon_r(t)$ are the signals of the transmission wave and reflection wave. $\dot{\varepsilon}(t)$ is the strain rate. The true stress and true strain of the dynamic tensile experiment can also be obtained by (2).

In this study, three different strain rates, that is, 700 s^{-1}, 1100 s^{-1}, and 3300 s^{-1}, are conducted to research the dynamic behavior at high strain rates. Three repeat measurements are performed in each test in order to improve the accuracy of results. The specimen is fixed in the SHTB as shown in Figure 2(c).

2.1.3. Tensile Shearing Tests. Tensile shearing tests have been carried out to investigate the relationship between stress triaxiality, double-notch specimens with different stress triaxialities are machined. The specimens are similarly used with the dimensions of 60 mm in overall length, 5 mm in gauge length, 3 mm in notch length, 20 mm in width, and 7 mm in thickness. The angles between loading direction and the bearing cross section between two notches, denoted as α, are designed as 0°, 30°, 45°, 60°, and 90°, respectively. Figure 3 shows the specimens of shearing tests. The tensile shearing tests of 0° specimens can be regarded as pure shear tests, while other tests are under the combined action of tension and shearing. For these tests, the normal stress and shearing stress can be obtained through the following equations:

$$\sigma = \frac{F \sin\alpha}{A_{\min}},$$
$$\tau = \frac{F \cos\alpha}{A_{\min}}, \quad (4)$$

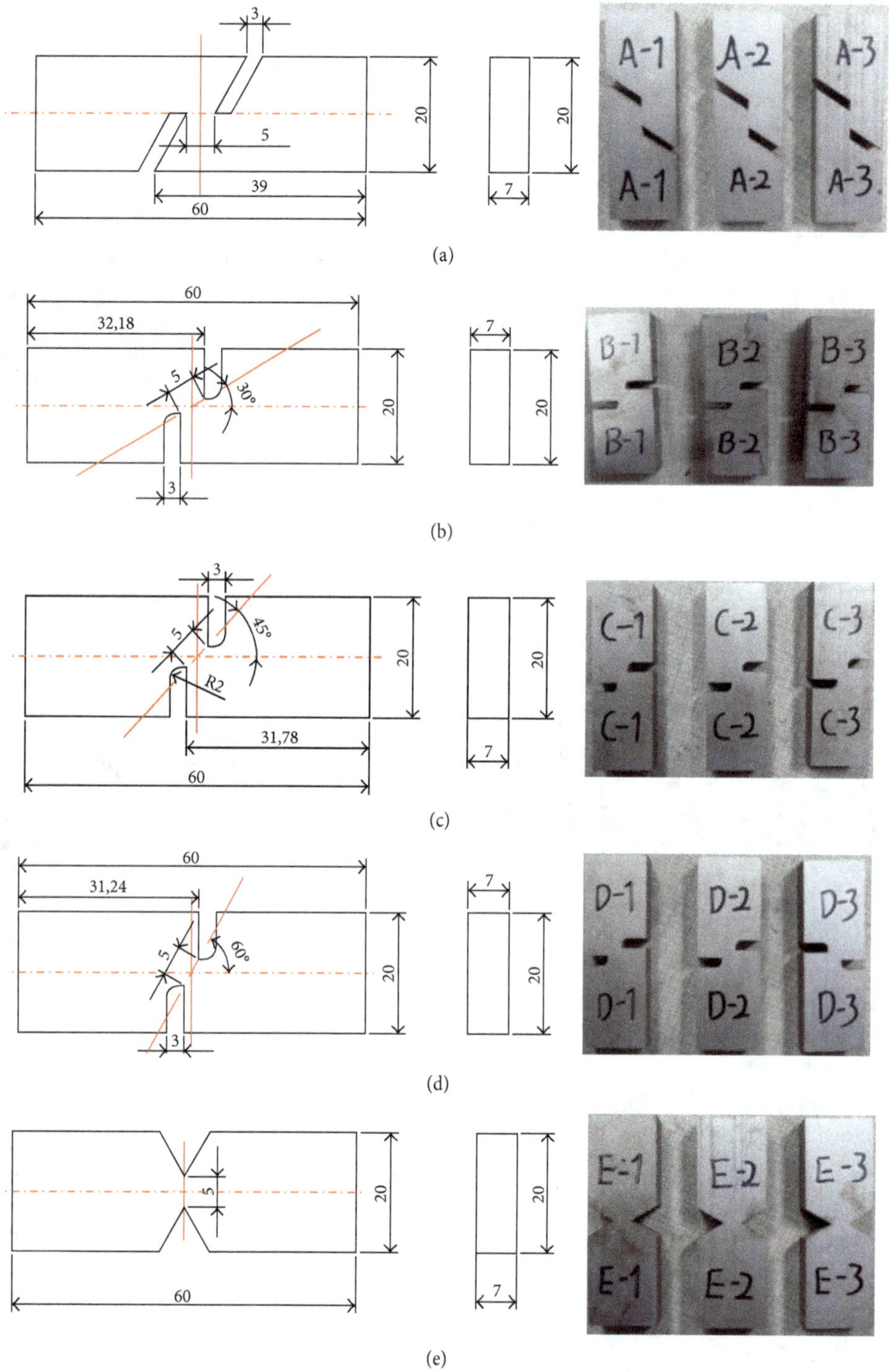

FIGURE 3: Specimens with different notch degrees: (a) 0°, (b) 30°, (c) 45°, (d) 60°, and (e) 90° for tensile shearing tests.

where A_{min} is the bearing area of the notched specimens given as follows:

$$A_{\text{min}} = tl, \tag{5}$$

where t and l are the thickness and length of the bearing cross section between two notches. For the plane stress state, principal stresses can be calculated with the following equation:

$$\left.\begin{matrix}\sigma_{\text{max}} \\ \sigma_{\text{min}}\end{matrix}\right\} = \frac{\sigma}{2} \pm \sqrt{\frac{\sigma^2}{4} + \tau^2}. \tag{6}$$

The stress triaxiality is defined as

$$\sigma^* = \frac{\sigma_{\text{h}}}{\sigma_{\text{m}}} = \frac{(\sigma_1 + \sigma_2 + \sigma_3)/3}{\sqrt{\left((\sigma_1 - \sigma_2)^2 + (\sigma_2 - \sigma_3)^2 + (\sigma_3 - \sigma_1)^2\right)/2}}, \tag{7}$$

Table 2: The theoretical initial stress triaxialities of different specimens.

Angle α	0°	30°	45°	60°	90°
σ^*	0	0.105	0.167	0.236	0.33

Table 3: Values of material parameters for 6005 Al alloy.

Parameter	ρ (g/cm^3)	C_p (J·(K·g)$^{-1}$)	χ	T_m (K)
Value	2.71	0.96	0.9	933

where σ_h is the hydrostatic stress, σ_m is the von Mises equivalent stress, and σ_1, σ_2, and σ_3 represent three principal stresses. According to (4)–(7), the stress triaxiality can be rewritten as

$$\sigma^* = \frac{\sin\alpha}{3\sqrt{\sin^2\alpha + 3\cos^2\alpha}}. \tag{8}$$

The initial stress triaxialities of different specimens are listed in Table 2. It can be found that the stress triaxiality increases from 0 to 0.33 with the increase of α. These tests are conducted by MTS, and the loading speed is 1 mm/min.

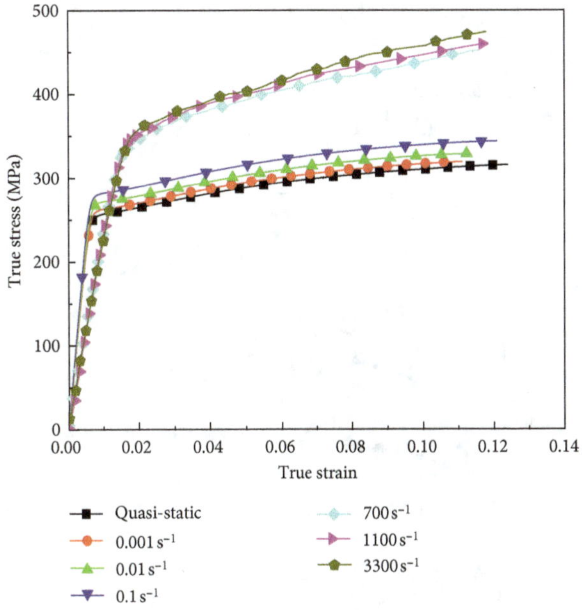

Figure 4: True strain-stress curves of 6005 Al alloy at different strain rates.

2.2. Constitutive Model. For tensile loading, to reflect the relationship between strain, stress, strain-rate hardening, and ambient temperature, the Johnson-Cook model [22], one of the widely used constitutive models, is commonly applied to describe the flow behavior, which is shown as follows:

$$\sigma = \left(A + B\varepsilon^n\right)\left(1 + C\ln\dot{\varepsilon}^*\right)\left(1 - \left(T^*\right)^m\right), \tag{9a}$$

where σ and ε are the stress and equivalent plastic strain, respectively. A is the yield stress of the material. B and n are the parameters of strain hardening. C is the strain-rate hardening parameter, determined by the material. $\dot{\varepsilon}^* = \dot{\varepsilon}/\dot{\varepsilon}_0$ is the dimensionless plastic strain rate, where $\dot{\varepsilon}$ is the strain rate and $\dot{\varepsilon}_0$ is the reference strain rate. $T^* = (T - T_r)/(T_m - T_r)$ is the dimensionless temperature indicating the thermal softening effect, where T_r is the room temperature and T_m is the melting point of the material. During high strain-rate loading, specimens are reasonable to be assumed in adiabatic conditions. In this case, the rate of temperature change is calculated through

$$\dot{T} = \chi\frac{\sigma\dot{\varepsilon}}{\rho C_p}, \tag{9b}$$

where ρ and C_p and are the density and heat capacity of the material. χ is the conversion factor of work into heat. The temperature rise can be estimated using the following equation:

$$\int_{T_0}^{T} \rho C_p\, dT = \chi\int_0^{\varepsilon_e} \sigma\, d\varepsilon, \tag{9c}$$

where ε_e is the strain corresponding to the maximum stress. ρ and C_p are assumed as constants, while χ is taken as 0.9 for metals [31]. The related parameters are given in Table 3. The equation can be rewritten as

$$\Delta T = \frac{0.9}{\rho C_p}\int_0^{\varepsilon_e} \sigma\, d\varepsilon. \tag{9d}$$

All tensile tests in this study are conducted at room temperature. Furthermore, strain hardening and strain-rate hardening are only considered without the effect of temperature for low strain-rate tests. Therefore, the JC model can be simplified as the following equation:

$$\sigma = \left(A + B\varepsilon^n\right)\left(1 + C\ln\dot{\varepsilon}^*\right). \tag{9e}$$

In this study, $10^{-4}\,\mathrm{s}^{-1}$ is specified as the reference strain rate, that is, $\dot{\varepsilon}_0 = 10^{-4}\mathrm{s}^{-1}$. The constitutive behavior model can be rewritten as

$$\sigma = A + B\varepsilon^n. \tag{9f}$$

Then, take logarithm on both sides of the equation, and the equation can be rewritten as

$$\ln\left(\sigma - A\right) = \ln B + n\ln\varepsilon. \tag{9g}$$

Unknown parameters A, B, and n can be determined with the least squares method. In B and n represent the intercept of the straight line and the slope of the straight line, respectively.

Next, to obtain the value of C, the equation can be transformed as follows:

$$\frac{\sigma_2(\dot{\varepsilon})}{\sigma_1} - 1 = C\ln\dot{\varepsilon}^*, \tag{9h}$$

where $\sigma_1 = A + B\varepsilon^n$, σ_1 represents the stress of the material at $10^{-4}\,\mathrm{s}^{-1}$ strain rate, and $\sigma_2(\dot{\varepsilon})$ represents the yield stress at a strain rate of $\dot{\varepsilon}$.

Table 4: The yield and ultimate stresses under different strain rates for 6005 Al alloy.

Strain rate $\dot{\varepsilon}$ (s^{-1})	Quasi-static	0.001	0.01	0.1	700	1100	3300
Yield stress σ_y (MPa)	255.27	260.42	269.53	279.56	325.03	338.08	346.38
Ultimate stress σ_u (MPa)	315.82	319.70	329.03	344.21	452.29	455.34	474.58

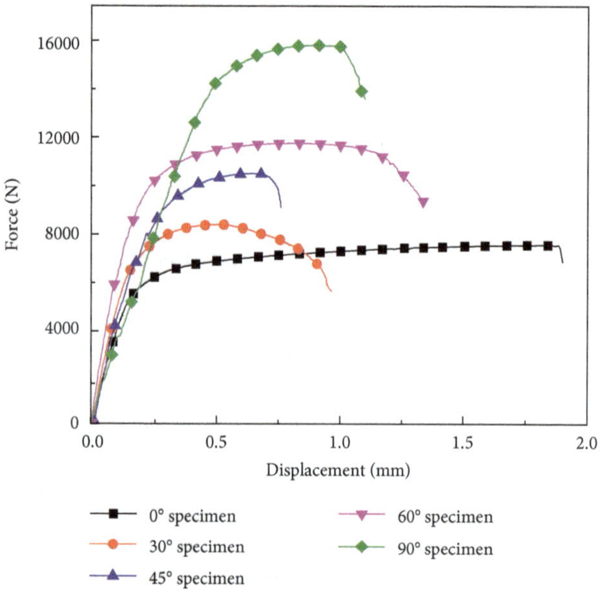

Figure 5: Experimental results for different specimens.

Table 5: Parameters of the Johnson-Cook model of 6005 Al alloy.

Parameter	A (MPa)	B (MPa)	n	C
Value	255.28	411.69	0.82	0.0186

3. Results

3.1. Flow Stress at Different Strain Rates. The true stress-strain curves under different tensile strain rates before the onset of necking are plotted in Figure 4. The tension process can be generally divided into two stages: elastic stage and plastic stage. One thing to note about the difference of Young's modulus between low and high strain rates at the elastic stage is the fact that, during Hopkinson tensile tests, the stress wave in the specimen needs two or three reflections to achieve uniformity, resulting in the inhomogeneity of stress in the specimen. Therefore, the stress-strain relation at high strain rates is inaccurate. For the plastic stage, the strain hardening effect can be seen under each condition. A stronger strain hardening effect is observed at higher strain rates ($700\,s^{-1}$–$3300\,s^{-1}$) in the plastic stage. In addition, a clear strain-rate hardening effect is observed when comparing curves of different strain rates. The yield strength and ultimate strength for 6005 Al alloy under different strain rates are all listed in Table 3.

From Table 4, it can be seen that the yield stress and ultimate stress gradually increase when the strain rate increases. The yield stress experiences an increase of 35.7% from quasi-static to $3300\,s^{-1}$, while the ultimate stress increases by 50.3%. Evidently, 6005 Al alloy is a material with great dependence on the strain rate.

3.2. Results of Tensile Shearing Tests. For this group of tests, the central cross section of specimens is under tension-shear combined loading, which is a complex stress state. The true stress and true strain cannot be obtained directly due to the different deformation mode for each test. The experimental results are thus exhibited with the curves of force and displacement, as shown in Figure 5. As the comparison shows, the maximum tensile force for different specimens differs a lot. As the notch angle increases from 0° to 90°, the maximum tensile force increases monotonously from 7718 N to 16301 N experiencing an increase of 111.21%. The fracture displacements scatter from 0.76 mm to 1.9 mm, and no clear correlation to notch angle is observed.

3.3. Material Parameter Identification. The experimental data of the quasi-static tensile are used to determine the three material constants A, B, and n, and the yield stress data at the strain rate range from quasi-static to $0.1\,s^{-1}$ to determine C of the simplified Johnson-Cook model. The temperature soft parameter m can be obtained by comparing the results from experiments and the simplified JC model. The four parameters for 6005 Al alloy at different strain rates can be obtained with the method mentioned in Section 2.2 as shown in Table 5. The relationship between ln $(\sigma - A)$ and ln ε is shown in Figure 6.

To compare the predicted results from the JC constitutive model and experimented results, both of the two data groups are plotted in Figure 7. However, the values of yield stress at high strain rates are underestimated. Considering the temperature rise caused by the adiabatic process at high strain rates is limited (30 K calculated by (9d)), the temperature effect is neglected. It could be observed that the prediction agrees well with the experiments at the quasi-static strain rate, and with the increase of strain rate, the prediction deviates from experimented results increasingly. This deviation may be caused by neglecting the variations of the strain-rate coefficients. That is, the parameter C in the JC model is supposed to be a variable instead of a constant. Apparently, the original JC constitutive model cannot depict the flow stress behavior at dynamic loading adequately. Therefore, it is necessary to modify the original JC constitutive model.

3.4. The Modification of the JC Constitutive Model. For the simplified JC constitutive model neglecting the temperature effect, flow stress is a function of strain and strain rate. The relationship between flow stress and ln $\dot{\varepsilon}^*$ is considered as linear in the original JC model, which is an ideal case. Actually, the strain-rate coefficient may also correlate with strain. In other words, for a material under a certain strain-rate loading, the strain-rate coefficient may vary with the

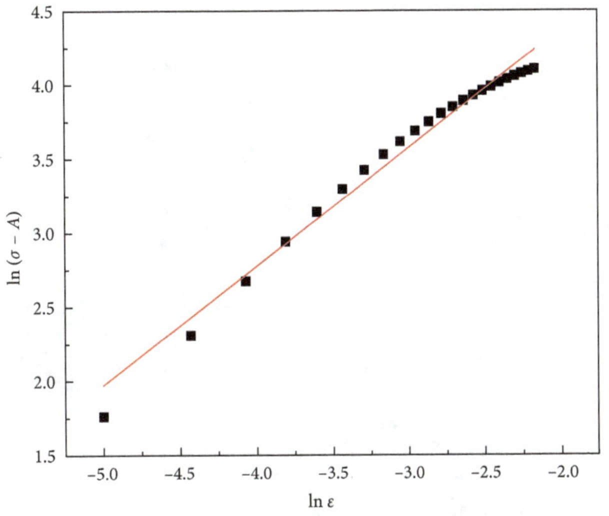

FIGURE 6: Relationship between $\ln(\sigma - A)$ and $\ln \varepsilon$.

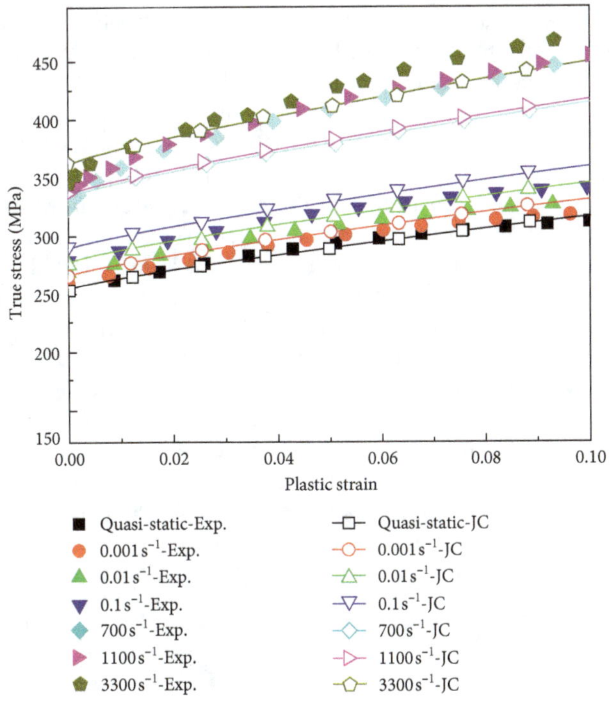

FIGURE 7: Comparison between predicted results of the JC model and experimental results at different strain rates.

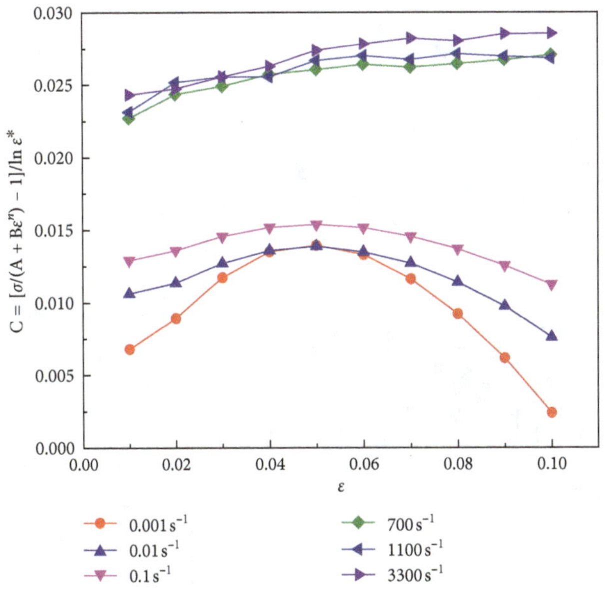

FIGURE 8: Relationship between C and ε.

development of plastic deformation. Figure 8 exhibits the relationship between the strain-rate coefficient and strain. On one hand, a nonlinear relationship can be found between C and ε. On the other hand, for different strain rates, the relationship is different. A nonlinear relationship can be observed as well when it comes to C and $\ln \dot{\varepsilon}^*$, as shown in Figure 9. Therefore, it can be assumed that C is a function of strains and strain rates. The relationship is given as follows:

$$C = f(\varepsilon, \ln \dot{\varepsilon}^*). \quad (10)$$

From Figures 8 and 9, considering the nonlinear relationship between C and the two variables, C can be described as a binary quadratic polynomial of ε and $\ln \dot{\varepsilon}^*$ independently. In addition, the interaction effect between the two variables should be taken into consideration. A quadratic polynomial regression model can thus be used to describe the relationship between C and ε and between C and $\ln \dot{\varepsilon}^*$ shown as follows:

$$C = C_0 + C_1\varepsilon + C_2\varepsilon^2 + C_3 \ln \dot{\varepsilon}^* + C_4(\ln \dot{\varepsilon}^*)^2 + C_5\varepsilon \ln \dot{\varepsilon}^*, \quad (11)$$

where C_0–C_5 are coefficients needed to be solved. The relationship between C and the two independent variables can be described with a surface in the three-dimensional space coordinate system. The six coefficients can be obtained directly by nonlinear surface fitting (shown in Figure 10). The blue surface represents experimental data, and the red one is the fitting surface.

The parameters of the modified JC constitutive model of 6005 Al alloy are listed in Table 6. Comparison between the predicted and experimented results is shown in Figure 11. Obviously, the modified JC model predicts the flow stress more accurately compared with the original JC model. It is indicated that the modification using the C function is effective to depict the flow behavior of the investigated material.

3.5. FEA Modeling of Tensile Shearing Tests. In this paper, numerical simulations are used to obtain the actual initial stress triaxiality and equivalent plastic strain to fracture. ABAQUS/Standard, as a common FEA code, is employed to simulate tensile shearing tests. Five tensile shearing tests of different angle specimens are simulated. The material property is defined as elastic-plastic and isotropic hardening. The stress-strain data of the plastic stage from the quasi-static tensile test are input. The element type is the three-dimensional hexahedral element with reduced integration. The 3D finite element model meshes of specimens are all shown in Figure 12. The minimum size of elements is 0.2. The nodes on one end of the 3D finite element model are fixed, while displacement

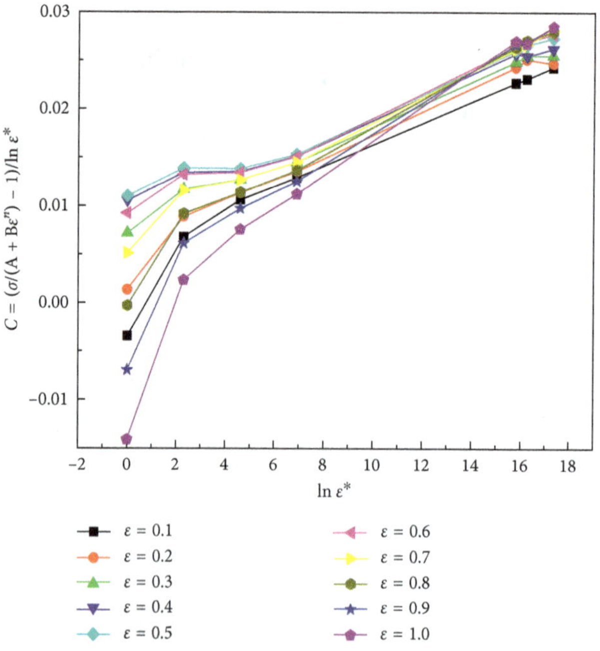

FIGURE 9: Relationship between C and In $\dot{\varepsilon}^*$.

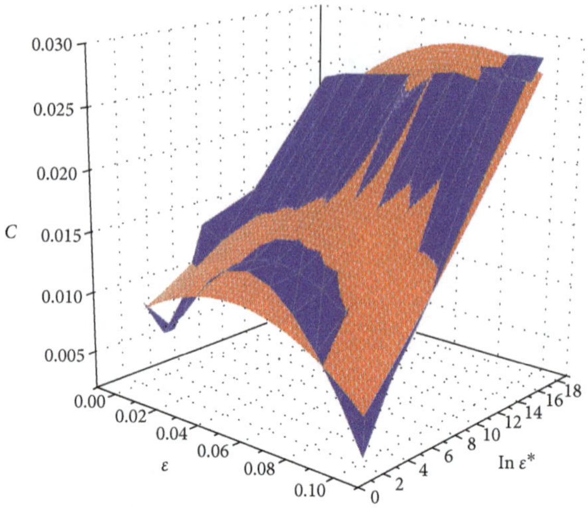

FIGURE 10: Nonlinear surface fitting of the quadratic polynomial regression model.

boundary conditions are imposed on the ones on the other end. It is worth stressing that fracture criterion is not applied in modeling for the reason that the numerical simulations are performed to validate the numerical analysis results.

Results obtained from the numerical simulations and experiments are compared in Figure 13. The two groups of force-displacement curves are in good agreement before the onset of fracture, as can be seen in Figure 13. Because no failure criterion is employed in the simulations, the simulations do not reflect the fracture behavior. However, due to the good agreement before the onset of fracture, it is still reliable to obtain the equivalent plastic strain to fracture and stress triaxiality from numerical simulations. The FEA model can thus be used to formulate the fracture strain.

3.6. Fracture Criterion for 6005 Al Alloy. In a certain triaxiality region, the coupling of the triaxiality and fracture strain constitutes a fracture criterion. The model built in Section 3.5 is used to obtain the equivalent plastic strain to fracture. By comparing the force-displacement curves of simulations and experiments, the fracture time step can be determined. The equivalent plastic strain at this time step is the fracture strain. It should be noted that the object to study is the critical elements where the fracture begins. The point is to determine the location where the fracture begins. It is observed from the finite element models that the maximum equivalent plastic strain at the fracture time step locates in the root of notches, which means that the fracture begins there. Figure 14 shows the distribution of equivalent plastic strain at the fracture time step in a typical numerical simulation.

In Section 2.1.3, a theoretical calculation method based on the assumption of the plane stress state for the initial stress triaxiality is proposed. However, specimens under tensile loading are actually in the three-dimensional stress state. Therefore, it is necessary to acquire the actual initial stress triaxiality from the numerical simulations. The two sets of initial stress triaxialities and corresponding fracture strains are listed in Table 7.

According to Table 7, the actual initial stress triaxialities, within the range from 0.04 to 0.741, are obviously greater than the theoretical values. It may be caused due to the three-dimensional stress instead of the plane stress state in which the specimen is. The actual initial stress triaxialities coupled with equivalent plastic strains to fracture are plotted in Figure 15. Results indicate that the equivalent plastic strain to fracture presents nonmonotonic dependence on the initial stress triaxiality. In the stress triaxiality range from 0.04 to 0.369, which is a low triaxiality level, when σ^* increased, the fracture strain increases as well from 0.767 to 1.423. A drop shows up as the stress triaxiality exceeds 0.369, a relatively high triaxiality level. According to the study by Bao and Wierzbicki [23], for low triaxialities (0.04–0.369), a parabolic fit is found to be satisfactory. The fitting result is shown as (12). A plot of the equation is shown in Figure 16.

$$\varepsilon_p^f = 0.76 * \sigma^{*2} - 1.12 * \sigma^* + 1.85. \quad (12)$$

4. Discussion

4.1. Evaluation of the Modified JC Model. In Section 3.2, a modified JC constitutive model has been proposed in which C is considered to be relative to strain and strain rate. Previous studies [32, 33] have modified JC constitutive model for high-strength alloys by changing the form of the strain hardening factor and for AA6005-T6 by multiplying the strain-rate factor with a true stress. The method used in the present study has been successfully employed to modify the JC model for 7050-T745 aluminum alloy in [34].

To accurately evaluate the forecasting error of two constitutive models of 6005 aluminum alloy, three evaluating indices are investigated. They are the correlation coefficient, normalized mean bias value, and average absolute relative error denoted as R, NMBE, and AARE, respectively, which are expressed as follows:

TABLE 6: The material constants of the modified JC constitutive model of 6005 Al alloy.

Parameter	A (MPa)	B (MPa)	n	C_0	C_1	C_2	C_3	C_4	C_5
Value	255.3	411.7	0.8230	0.002238	0.1890	0.001622	0.007192	−2.480	-3.836×10^{-5}

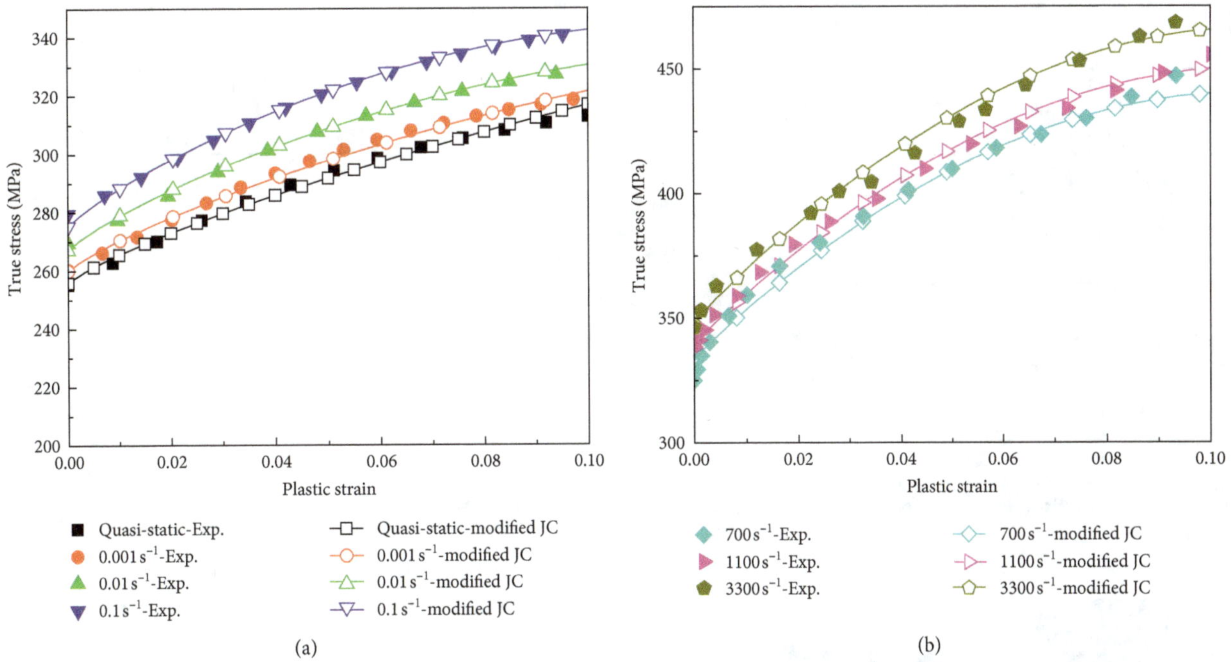

FIGURE 11: Comparison between the predicted results and experimented results: (a) low strain rates and (b) high strain rates.

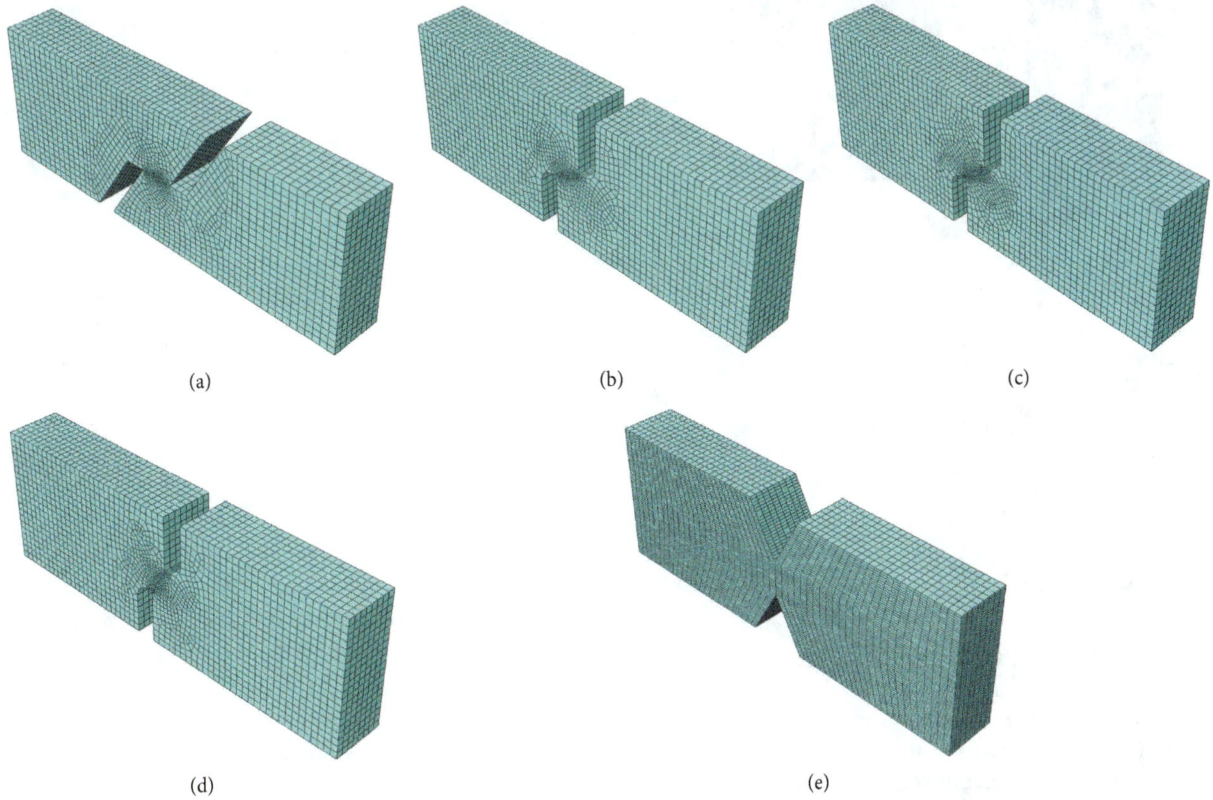

FIGURE 12: The 3D finite element model meshes of (a) 0°, (b) 30°, (c) 45°, (d) 60°, and (e) 90° specimens.

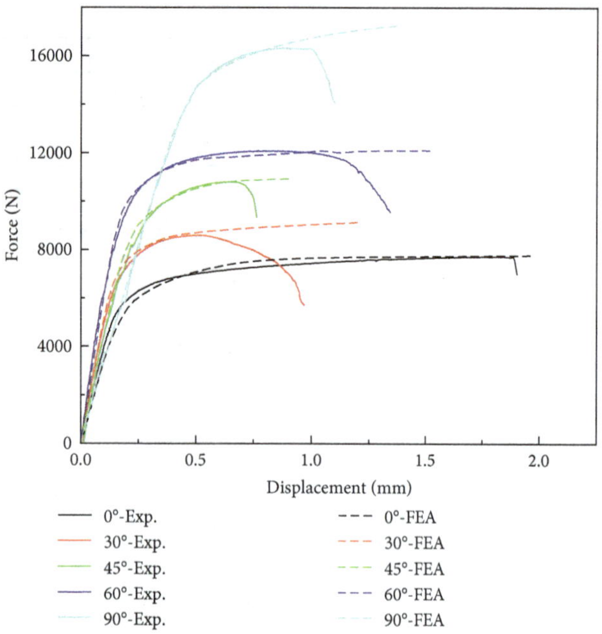

FIGURE 13: Comparison between numerical simulations and experiments.

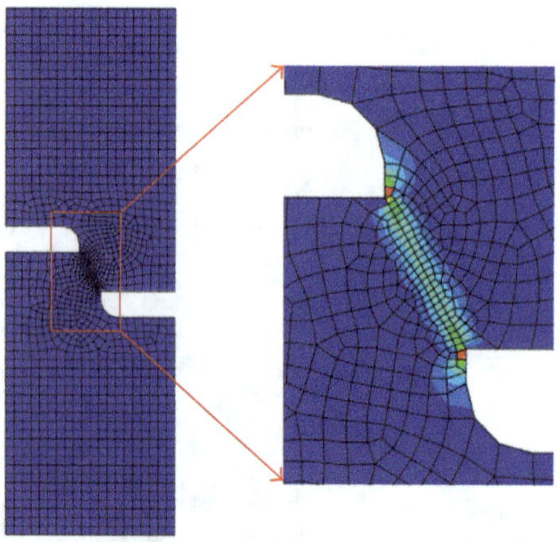

FIGURE 14: The distribution of equivalent plastic strain at the fracture time step in a typical numerical simulation.

$$R = \frac{\sum_{i=1}^{N}(E_i - \overline{E})(P_i - \overline{P})}{\sqrt{\sum_{i=1}^{N}(E_i - \overline{E})^2 (P_i - \overline{P})}}, \quad (13)$$

$$\text{NMBE} = \frac{\sum_{i=1}^{N}(E_i - P_i)}{\sum_{i=1}^{N} E_i}, \quad (14)$$

$$\text{AARE} = \frac{1}{N}\sum_{i=1}^{N}\left|\frac{E_i - P_i}{E_i}\right| \times 100\%, \quad (15)$$

where E_i and P_i are the experimental and predicted values, while $\overline{E}$ and $\overline{P}$ are the mean values of them, respectively. N is the total number of data used in this study. The correlation coefficient is used to characterize the degree of a linear correlation relationship between two variables. The NMBE represents the mean bias value of predicted values, and a positive NMBE value means overprediction, while negative one means underprediction from a model. AARE is an unbiased statistic that estimates the accuracy of predictions by comparing the relative error term by term.

TABLE 7: The calculated and actual initial stress triaxialities and corresponding fracture strains.

α	0°	30°	45°	60°	90°
σ^*_{cal}	0	0.105	0.167	0.236	0.33
σ^*_{sim}	0.04	0.159	0.221	0.369	0.741
ε^f_p	0.767	0.948	1.140	1.423	0.732

σ^*_{cal}: the initial stress triaxiality calculated with (8); σ^*_{sim}: the initial stress triaxiality obtained from the numerical simulations.

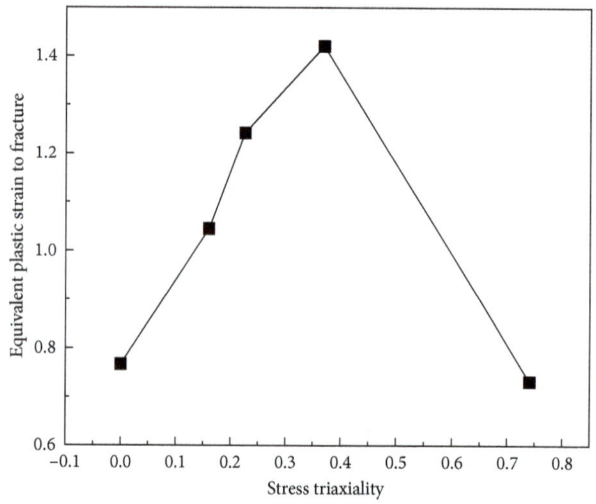

FIGURE 15: Actual initial stress triaxialities coupled with equivalent plastic strains to fracture.

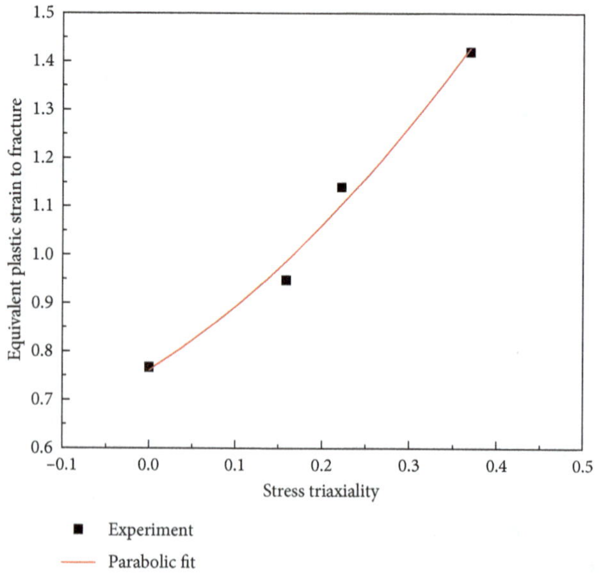

FIGURE 16: Parabolic fitting for low triaxialities.

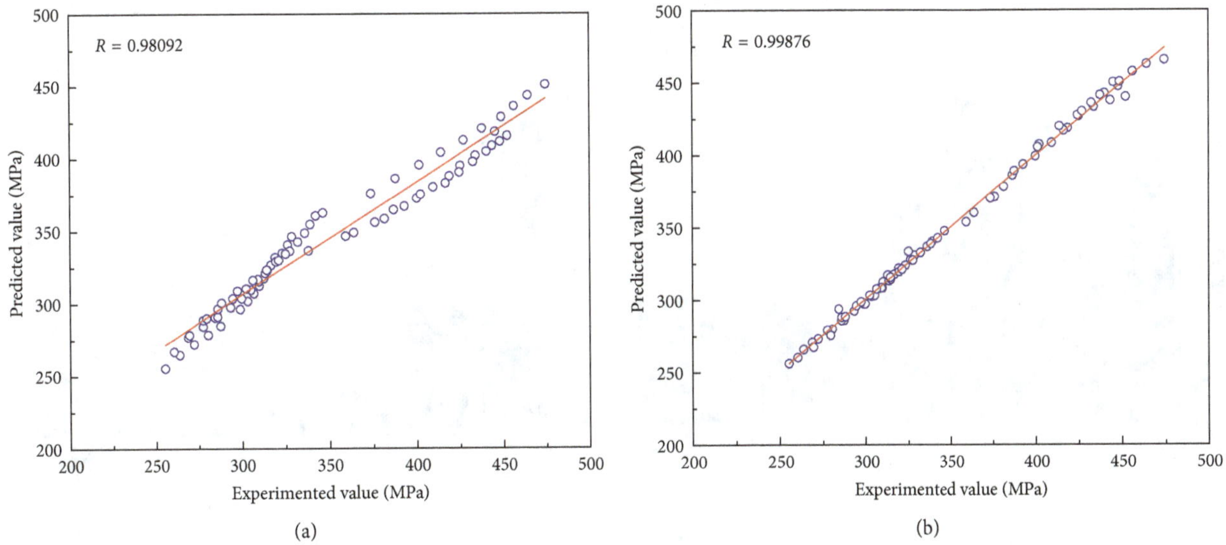

FIGURE 17: Correlation analysis between experimental values and predicted values: (a) the original JC model and (b) the modified JC model.

Experimental values of flow stress and corresponding predicted values obtained from the original JC model as well as modified JC model are plotted in the coordinate system as shown in Figure 17. It is can be seen that the data points for the modified JC model are more close to the best line compared with the original one. The comparison indicates that the predicted values of the modified JC model correlate with experimental values better than the original JC model.

The performance of the original Johnson-Cook model and modified Johnson-Cook model is shown in Table 8. The average absolute relative errors of the original JC model and modified JC model are 3.71% and 1.34%, while the normalized mean bias errors are 6.8% and −0.049%, respectively. These results show that the modified JC model features higher prediction accuracy than the original JC model.

In addition, the absolute relative error is employed to analyze the performance of the two models. The absolute relative error can be calculated with the following equation:

$$\text{Absolute relative error} = \left|\frac{E_i - P_i}{E_i}\right| \times 100\%. \tag{16}$$

It can be obtained that the absolute relative error given by the original JC model varies from 0.0031% to 8.14%, while the value is in the range of 0.05% to 3.26% for the modified JC model. The distribution of the absolute relative errors is plotted in Figure 18. For almost 95% data sets, the relative error of the modified JC model is concentrated in the range of 0 to 2%, while the relative error scatters from 0 to 9% erratically for the original JC model. In conclusion, the modified Johnson-Cook model can describe the flow behavior of 6005 aluminum alloy more accurately than the original Johnson-Cook model.

TABLE 8: The performance of the two models.

Indicator	R	AARE (%)	NMBE (%)
Original JC	0.9809	3.71	6.80
Modified JC	0.9987	1.34	-0.049

4.2. The Fracture Behavior under Low Stress Triaxiality. In this paper, a parabolic equation is established to determine the fracture strain, in which stress triaxiality is the only independent variable. Several studies [10, 23, 24] have investigated the fracture behavior of aluminum alloy, while most of them focus on the high stress triaxiality. The Johnson-Cook fracture model is the most commonly used failure criterion. Bao and Wierzbicki [23] have studied the fracture ductility of AA2024-T351 in the entire range of stress triaxialities (−1/3 to 0.95). Three expressions were given for negative (−1/3 to 0), low (0–0.4), and high (0.4–0.95) triaxialities, respectively. A similar tendency of fracture strain is observed for AA6005 in the range from 0.04 to 0.741.

With an elongation beyond 10%, 6005 Al alloy can be regarded as a ductile material. To further investigate the fracture mode of 6005 Al alloy during tensile, pure shear, and tensile-shear combined tests, some magnified fracture surfaces are collected by a scanning electron microscope (SEM). Figure 19 shows the typical fracture surfaces of tensile shearing specimens with the notch degree from 0° to 90°. Figure 19(a) shows the pure shearing fracture surface, where there are typical "ripple waves" and the direction of the shear fracture surface is parallel to that of the maximum shear stress. It indicates that the fracture mechanism is shear fracture. During the pure shear test, the stress triaxiality is about zero. The shear band appears in the center of the specimen. Cracks are produced within the shear band and then coalesced. When the equivalent plastic strain reaches the critical value, the fracture takes place. Figure 19(e) shows the fracture surface of 90° specimen, where there exist a number of dimples with different size. During the tensile test on 90° specimen, there is a high stress triaxiality in the root of the notch. The specimen fractures after experiencing the process of nucleation, growth, and coalescence of voids. For tensile shearing tests, dimple and shear combination of different degrees can be seen in Figures 19(b)–(19d). In the

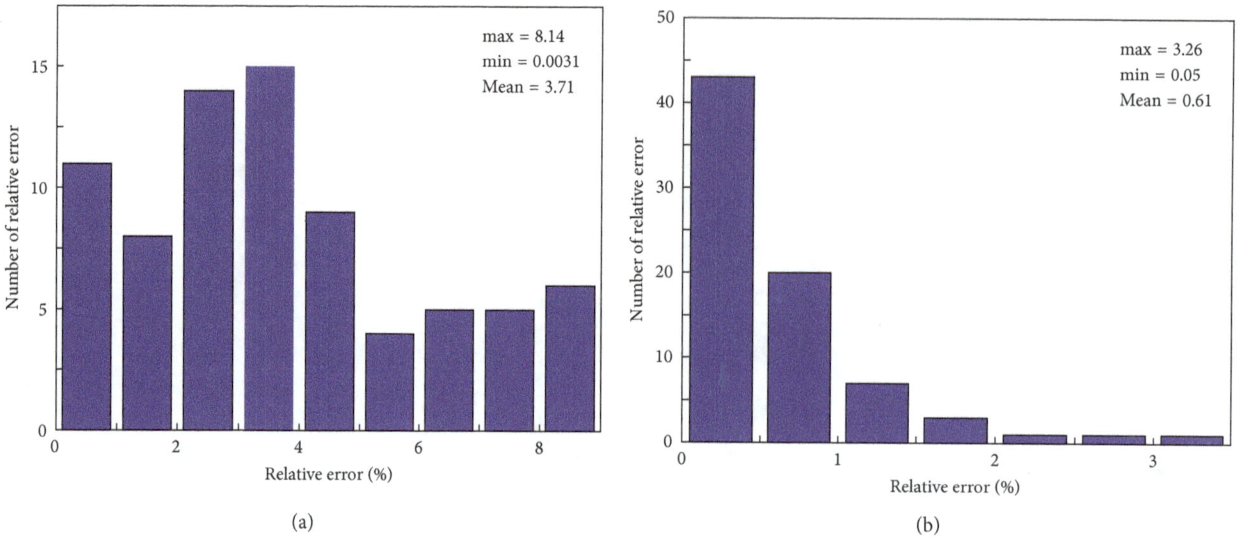

FIGURE 18: The distribution of the absolute relative errors of (a) the original JC model and (b) modified JC model.

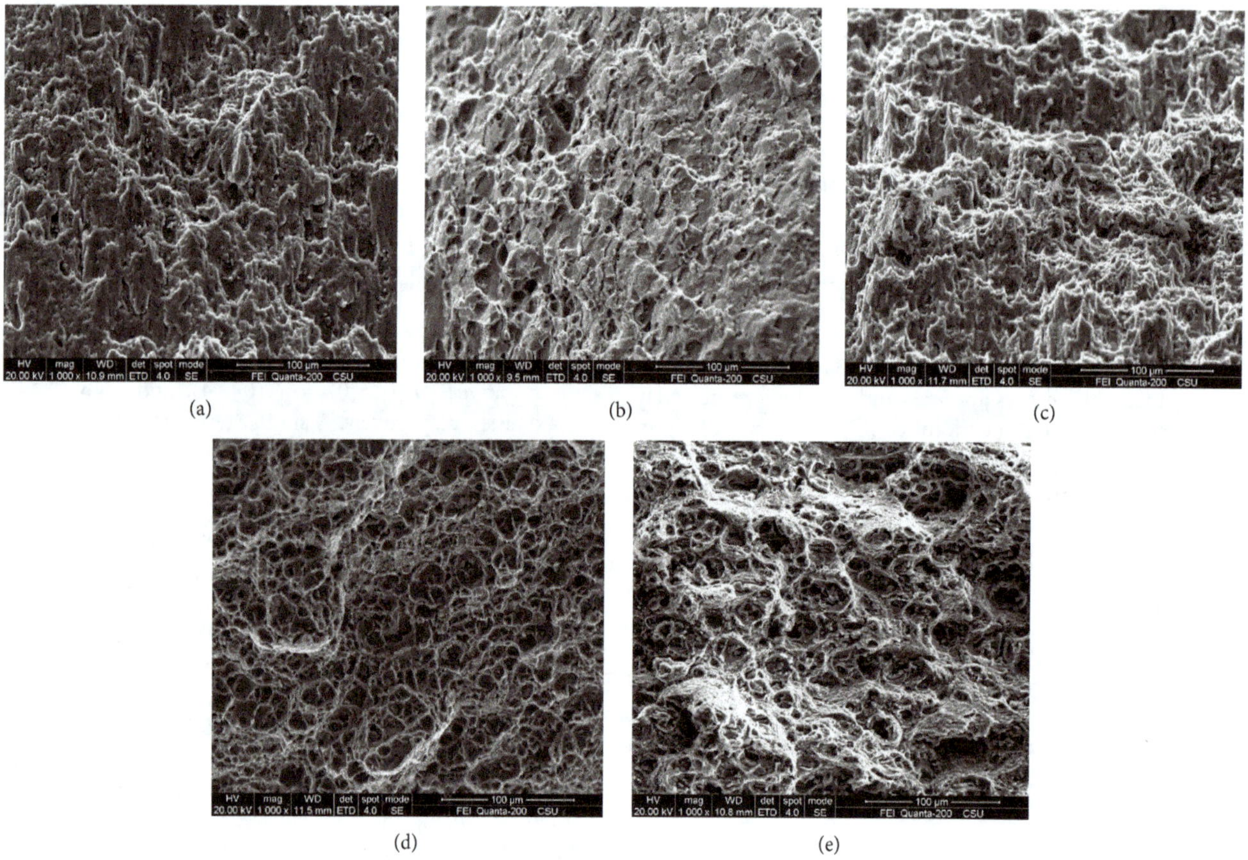

FIGURE 19: Fracture surfaces of (a) 0°, (b) 30°, (c) 45°, (d) 60°, and (e) 90° notch specimens.

range of low stress triaxialities, fracture is caused by a combined function of shear and void growth.

5. Conclusion

In this study, the dynamic constitutive behavior of 6005 Al alloy is investigated by means of experiments and numerical simulations. Tensile tests under different strain rates indicate that the flow stress of the investigated material is rather sensitive to strain rate. What is more, a clear effect of strain hardening is observed. Based on the tensile test data, a simplified Johnson-Cook model was established to describe the flow stress of 6005 Al alloy at various strain rates. Furthermore, the original JC model was modified using the

C function. The modified JC model is proved to be effective in predicting the flow stress of 6005 Al alloy.

A set of tensile shearing tests were carried out to study the fracture behavior under low stress triaxialities. Numerical simulations based on FEA were used to obtain the actual initial stress triaxialities and equivalent plastic strain to fracture. The tendency of the fracture strain is non-monotonic in the triaxiality range of 0.04 to 0.74. A parabolic fracture criterion was thus proposed in the triaxiality range of 0.04 to 0.369. Finally, the fractography analysis for 6005 Al alloy indicates that the material has a typical ductile fracture mechanism including the shear fracture under pure shear and the dimple fracture under uniaxial tensile.

Conflicts of Interest

The authors declare that there are no conflicts of interest regarding the publication of this paper.

Acknowledgments

The authors thank Modern Analysis and Testing Center of Central South University for experimental assistance in performing the mechanical tests and electron microscope scan experiment. This work was supported by the National Natural Science Foundation of China (nos. 51405517 and U1334208) and the Natural Science Foundation of Hunan (no. 2015JJ3155).

References

[1] I. N. Fridlyander, V. G. Sister, O. E. Grushko, V. V. Berstenev, L. M. Sheveleva, and L. A. Ivanova, "Aluminum alloys: promising materials in the automotive industry," *Metal Science and Heat Treatment*, vol. 44, no. 10, pp. 365–370, 2002.

[2] W. Brocker and R. D. Rosenberger, "Light weight potentials in coach body structures of high-speed trains," in *Proceedings of WCRR*, vol. 97, pp. 713–721, Florence, Italy, November 1997.

[3] Z. Lu, B. Li, C. Yang et al., "Numerical and experimental study on the design strategy of a new collapse zone structure for railway vehicles," *International Journal of Crashworthiness*, vol. 22, no. 5, pp. 1–15, 2017.

[4] J. Yan, S. Yao, P. Xu, Y. Peng, H. Shao, and S. Zhao, "Theoretical prediction and numerical studies of expanding circular tubes as energy absorbers," *International Journal of Mechanical Sciences*, vol. 105, pp. 206–214, 2015.

[5] P. Xu, C. Yang, Y. Peng, S. Yao, D. Zhang, and B. Li, "Crash performance and multi-objective optimization of a gradual energy-absorbing structure for subway vehicles," *International Journal of Mechanical Sciences*, vol. 107, pp. 1–12, 2016.

[6] Y. Peng, X. Wang, X. Xiong, and P. Xu, "Crashing analysis and multi-objective optimisation of duplex energy-absorbing structure for subway vehicle," *International Journal of Crashworthiness*, vol. 21, no. 4, pp. 338–352, 2016.

[7] Y. Peng, S. Wang, S. Yao, and P. Xu, "Crashworthiness analysis and optimization of a cutting-style energy absorbing structure for subway vehicles," *Thin-Walled Structures*, vol. 120, pp. 225–235, 2017.

[8] G. Liu, J. Xie, and S. Xie, "Experimental and numerical investigations of a new U-shaped thin plate energy absorber subjected to bending and friction," *Thin-Walled Structures*, vol. 115, pp. 215–224, 2017.

[9] S. M. Mirfalah-nasiri, A. Basti, and R. Hashemi, "Forming limit curves analysis of aluminum alloy considering the through-thickness normal stress, anisotropic yield functions and strain rate," *International Journal of Mechanical Sciences*, vol. 117, pp. 93–101, 2016.

[10] G. Haugou, E. Markiewicz, J. Fabis, and G. Gary, "Contribution to the definition of a partial overlapping plastic strain rates domain for moderate loadings–application to tensile testing on metallic materials," *International Journal of Crashworthiness*, vol. 9, no. 2, pp. 187–194, 2004.

[11] X. Chen, Y. Peng, S. Peng, S. Yao, C. Chen, and P. Xu, "Flow and fracture behavior of aluminum alloy 6082-T6 at different tensile strain rates and triaxialities," *PLoS One*, vol. 12, no. 7, article e0181983, 2017.

[12] F. Delannay, T. Pardoen, D. Lassance, and D. Fabre, "Micromechanics of room and high temperature fracture in 6xxx Al alloys," *Progress in Materials Science*, vol. 52, no. 1, pp. 62–129, 2007.

[13] B. Milkereit, N. Wanderka, C. Schick, and O. Kessler, "Continuous cooling precipitation diagrams of Al–Mg–Si alloys," *Materials Science and Engineering: A*, vol. 550, pp. 87–96, 2012.

[14] W. Yang, L. Huang, R. Zhang, M. Wang, Z. Li, and Y. Jia, "Electron microscopy studies of the age-hardening behaviors in 6005A alloy and microstructural characterizations of precipitates," *Journal of Alloys and Compounds*, vol. 514, pp. 220–233, 2012.

[15] O. S. Lee, H. Choi, and H. Kim, "High-temperature dynamic deformation of aluminum alloys using SHPB," *Journal of Mechanical Science and Technology*, vol. 25, no. 1, pp. 143–148, 2011.

[16] H. Ying, X. Feng, J. Ying, and J. Jiao, "A modified Johnson Cook model for elevated temperature flow behavior of T24 steel," *Materials Science and Engineering: A*, vol. 577, pp. 138–146, 2013.

[17] L. Berkovic, A. Chabotier, F. Coghe, and L. Rabet, "Measuring and modeling of low temperature Hopkinson tests," *Procedia Engineering*, vol. 10, pp. 1645–1650, 2011.

[18] G. T. Gray, S. R. Chen, and K. S. Vecchio, "Influence of grain size on the constitutive response and substructure evolution of MONEL 400," *Metallurgical and Materials Transactions A*, vol. 30, no. 5, pp. 1235–1247, 1999.

[19] K. Y. Huh and A. D. Gosman, "A phenomenological model of diesel spray atomization," in *Proceedings of the International Conference on Multiphase Flows*, Tsukuba, Japan, September 1991.

[20] M. A. Mccarthy, J. R. Xiao, and C. T. Mccarthy, "Modelling of bird strike on an aircraft wing leading edge made from fibre metal laminates–part 2 ," *Applied Composite Materials*, vol. 11, no. 5, pp. 317–340, 2004.

[21] J. Qin, R. Chen, X. Wen, Y. Lin, M. Liang, and F. Lu, "Mechanical behaviour of dual-phase high-strength steel under high strain rate tensile loading," *Materials Science and Engineering: A*, vol. 586, pp. 62–70, 2013.

[22] G. R. Johnson and W. H. Cook, "A constitutive model and data for metals subjected to large strains, high strain rates and high temperatures," in *7th International Symposium on Ballistic*, pp. 541–547, Hague, Netherlands, April 1983.

[23] Y. Bao and T. Wierzbicki, "On fracture locus in the equivalent strain and stress triaxiality space," *International Journal of Mechanical Sciences*, vol. 46, no. 1, pp. 81–98, 2004.

[24] B. Liu, R. Villavicencio, and C. Guedes Soares, "Experimental and numerical plastic response and failure of pre-notched transversely impacted beams," *International Journal of Mechanical Sciences*, vol. 77, pp. 314–332, 2013.

[25] G. R. Johnson and W. H. Cook, "Fracture characteristic of three metals subjected to various strains, strain rates,

temperatures and pressures," *Engineering Fracture Mechanics*, vol. 21, pp. 31–48, 1985.

[26] A. L. Gurson, "Continuum theory of ductile rupture by void nucleation and growth: part I—yield criteria and flow rules for porous ductile media," *Journal of Engineering Materials and Technology*, vol. 99, no. 1, pp. 297–300, 1977.

[27] Y. Bao and T. A. Wierzbicki, "Comparative study on various ductile crack formation criteria," *Journal of Engineering Materials and Technology*, vol. 126, no. 3, pp. 314–324, 2004.

[28] T. Wierzbicki, Y. Bao, Y. W. Lee, and Y. Bai, "Calibration and evaluation of seven fracture models," *International Journal of Mechanical Sciences*, vol. 47, no. 4-5, pp. 719–743, 2005.

[29] Y. Bao and T. A. Wierzbicki, "Application of extended Mohr-Coulomb criterion to ductile fracture," *International Journal of Fracture*, vol. 161, no. 1, pp. 1–20, 2010.

[30] A. Gilioli, A. Manes, M. Giglio, and T. Wierzbicki, "Predicting ballistic impact failure of aluminium 6061-T6 with the rate-independent Bao-Wierzbicki fracture model," *International Journal of Impact Engineering*, vol. 76, pp. 207–220, 2015.

[31] F. Feng, S. Huang, Z. Meng et al., "Experimental study on tensile property of AZ31B magnesium alloy at different high strain rates and temperatures," *Materials & Design*, vol. 57, pp. 10–20, 2014.

[32] Y. C. Lin, X. M. Chen, and G. Liu, "A modified Johnson-Cook model for tensile behaviors of typical high-strength alloy steel," *Materials Science and Engineering: A*, vol. 527, no. 26, pp. 6980–6986, 2010.

[33] T. Børvik, A. H. Clausen, M. Eriksson, T. Berstad, O. Sture Hopperstad, and M. Langseth, "Experimental and numerical study on the perforation of AA6005-T6 panels," *International Journal of Impact Engineering*, vol. 32, no. 1–4, pp. 35–64, 2006.

[34] J. Q. Tan, M. Zhan, S. Liu, T. Huang, J. Guo, and H. Yang, "A modified Johnson-Cook model for tensile flow behaviors of 7050-T7451 aluminum alloy at high strain rates," *Materials Science and Engineering: A*, vol. 631, pp. 214–219, 2015.

7

Influence of Grain Boundary on the Fatigue Crack Growth of 7050-T7451 Aluminum Alloy based on Small Time Scale Method

Weihan Wang, Weifang Zhang, Hongxun Wang, Xiaoliang Fang, and Xiaobei Liang

Science & Technology on Reliability & Environmental Engineering Laboratory, Beihang University, Beijing 100191, China

Correspondence should be addressed to Weifang Zhang; 08590@buaa.edu.cn

Academic Editor: Liviu Marsavina

Based on the small time scale method, the influence of grain boundary on the fatigue crack growth of 7050-T7451 has been investigated. The interaction between fatigue crack and grain boundary was investigated by in situ SEM testing. Results showed that the fatigue crack growth will be retarded by grain boundary when the angle between fatigue crack and grain boundary is greater than 90 degrees. Mechanism analysis showed that the fatigue crack tip would not be able to open until the loading reached the 55% of maximum load, and the fatigue crack had been closed completely before the loading was not reduced to the minimum value, which led to the crack growth retardation. When the 7050-T7451 aluminum alloy suffered from fatigue loading with constant amplitude, a behavior of unstable fatigue crack growth could be observed often, and results indicated that the bridge linked mechanism led to the behavior. The grain boundary was prone to fracture during fatigue loading, and it became the best path for the fatigue crack growth. The fatigue crack tip would be connected with fractured grain boundary eventually, which led to the fast crack growth in different loading stage.

1. Introduction

7050-T7451 is a kind of high strength aluminum alloy which has been widely used in aerospace fields for its excellent properties, such as high specific strength, good fracture toughness, and stress corrosion resistance [1, 2]. Due to the fact that the aircraft bears complicated loadings in service, including maneuver loading, gust loading, and ground loading, the accumulation of fatigue damage will appear in the structure of 7050-T7451 aluminum alloy, which may lead to the fatigue fracture and cause disastrous accidents.

In essence, the fatigue fracture of 7050-T7451 aluminium alloy is the process of microdefect's initiation and growth. Microstructure such as grain boundary and second phase play an important role in the fatigue crack growth. Researchers found that the plastic deformation of metallic materials was seriously influenced by the grain boundary. Kamp et al. [3] observed that the transgranular propagation of persistent slip bands was impeded by the grain boundary and the fatigue crack tended to grow along grain boundaries or subgrain boundaries. Scharnweber et al. [4] analyzed the effect of grain boundary on small crack growth in stainless steel, and results showed that the hindering effect of grain boundary on fatigue crack growth is mainly affected by the plastic area of crack tip. Holzapfel et al. [5] found that the subsurface orientation of grain boundaries is an important factor for interpreting crack growth rate at the surface. Terentyev and Gao [6] demonstrated that both low and high angle grain boundaries exhibit resistance to brittle crack propagation. Wang and Dong [7] indicated that the precipitation of carbide at grain boundary strengthened the grain boundary and decreased the crack growth rate.

At present, the researches for the effect of grain boundary on the fatigue crack growth are mainly focused on the fracture analysis and in situ testing [8–10]. The fracture analysis could be used for analyzing the damage mechanism; however, the deformation process during the loading period is hardly reflected and it limits the recognition of damage mechanism. The in situ testing is able to observe the whole process of fatigue crack growth in real time; however, existing methods mainly focused on the research of damage mechanism between hundreds or thousands of loading cycles, and seldom studies focused on the behavior of fatigue crack growth within one loading cycle. Recently, a method of small time

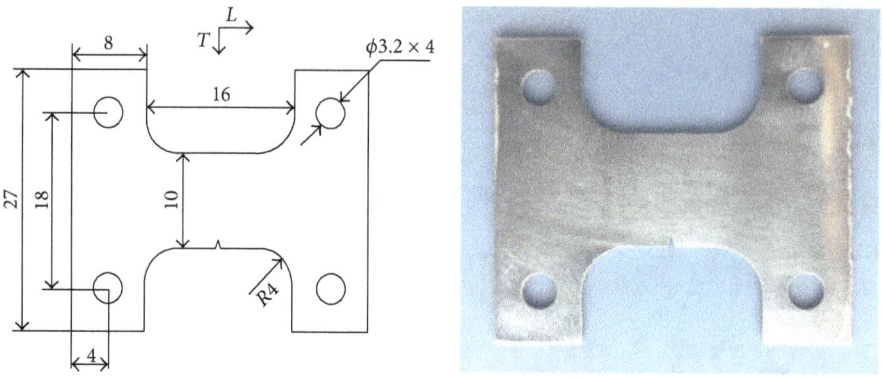

(a) (b)

Figure 1: Geometry of the sample: (a) the sample design. (b) Manufactured sample (unit: mm).

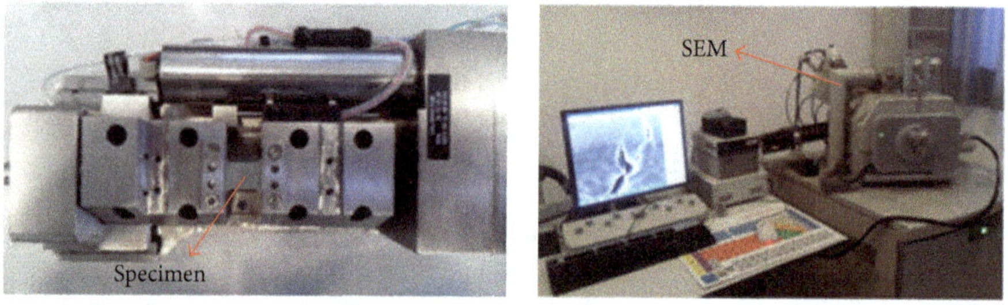

(a) (b)

Figure 2: Experimental platform: (a) in situ testing stage. (b) Quanta 650 SEM.

scale used for the analysis of fatigue crack growth [11] has been developed, and it is different with existing cycle-based method for improving the accuracy. The advantage of this method [12] is that it could be used for analyzing the crack growth behavior at any time instant during one loading cycle.

With regard to the high strength aluminium alloy, the fatigue fracture is the result of damage accumulation formulated by different loading cycles [13–15], and the behavior of fatigue crack growth within one loading cycle determines the service life of 7050-T7451 aluminum alloy. Therefore, researching the damage deformation in smaller time scale is important for mastering the fracture mechanism of 7050-T7451 aluminum alloy, and it is beneficial for predicting the residual life of the material.

In this paper, the influence of grain boundary on fatigue crack growth of 7050-T7451 aluminum alloy has been investigated based on the small time scale method. The interaction between grain boundary and fatigue crack growth was researched in the experiment, such as retardation effect on the fatigue crack growth, as well as the behavior of unstable fatigue crack growth within one loading cycle; the result could provide the basis for the optimization and life prediction of 7050-T7451 aluminum alloy.

2. Experiment

As shown in Figure 1, the in situ sample with width $W = 27$ mm, length $L = 32$ mm, and thickness $T = 0.8$ mm was prepared, and L-T is the sampling direction.

Table 1: Mechanical properties of 7050-T7451 aluminum alloy.

Tensile strength (MPa)	Yield strength (MPa)	Young's modulus (GPa)
541	476	72

The material used in the experiment is 7050-T7451 aluminum alloy, and the material mechanical properties are given in Table 1.

Before the in situ testing, the fatigue precracking was accomplished in the hydraulic testing machine, INSTRON 8801, and the maximum length of precrack is 1 mm. After the precracking testing, the metallographic treatment was applied on the specimen, and the etching agent is Graff Sargent solution (1 mL HF, 16 mL HNO_3, 3 g CrO_3, and 83 mL H_2O).

3. Results and Discussion

As shown in Figure 2(a), the sample was clamped in the in situ testing stage (Deben 2000), and the maximum loading of stage is 2 KN. The testing stage was installed in Quanta 650 scanning electron microscope (SEM) as shown in Figure 2(b), and the chamber pressure is less than 3×10^{-2} Pa.

Fatigue loading with constant amplitude was applied on the specimen. The stress ratio (R) is 0.1 and the maximum loading is 1750 N. From the SEM, it was observed that the fatigue crack began to grow stably when the loading cycles

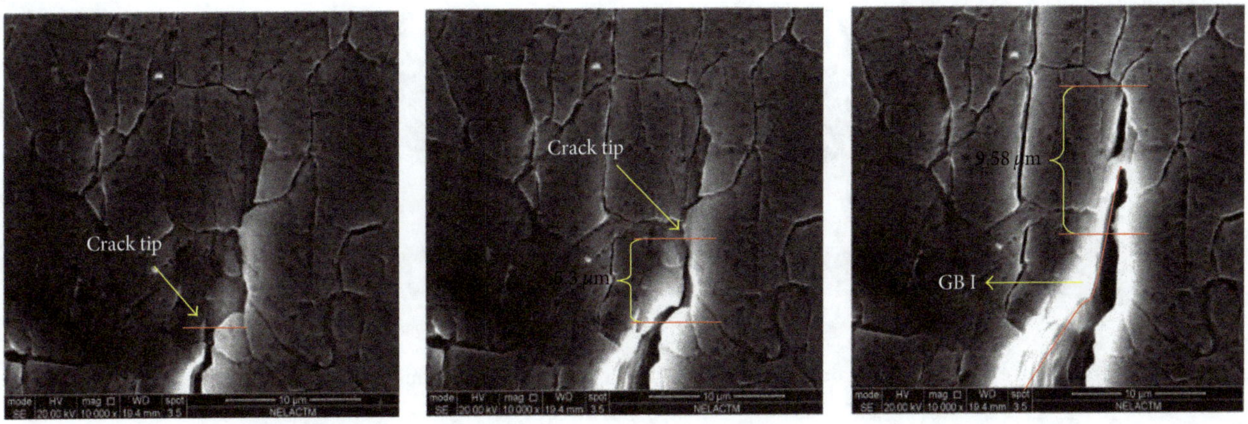

(a) (b) (c)

FIGURE 3: Micromorphology of fatigue crack: (a) at 40 cycles, (b) at 50 cycles, and (c) at 65 cycles.

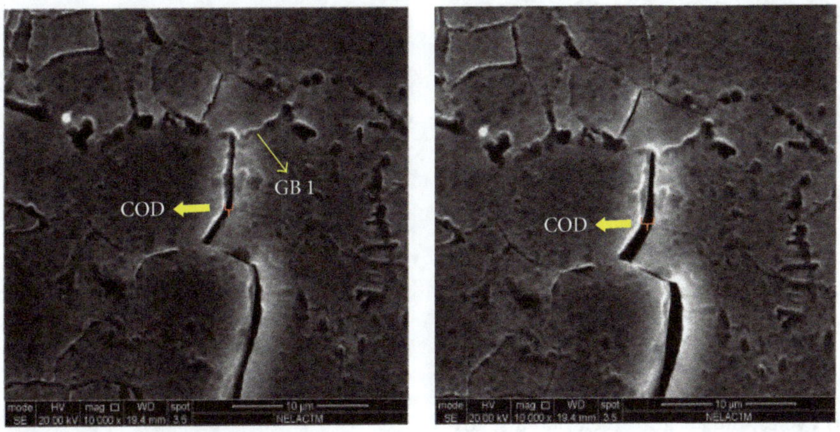

(a) (b)

FIGURE 4: Micromorphology of fatigue crack: (a) at 77 cycles and (b) at 99 cycles.

arrived at 40, and the micromorphology of fatigue crack from 40 cycles to 65 cycles is shown in Figure 3.

It can be seen that the fatigue crack kept stable growth from 40 cycles to 65 cycles from Figure 3. When the fatigue crack propagates along grain boundary I (GB I), the mean fatigue crack growth rate is 5.85×10^{-4} mm/N. As shown in Figure 4(a), the fatigue crack tip approached grain boundary 1 (GB 1) at 77 cycles.

From Figure 4(a), it can be seen that the angle between fatigue crack and GB 1 is more than 90 degrees at 77 cycles. When the loading cycle arrives at 99 cycles, the fatigue crack tip still stayed under GB 1 as shown in Figure 4(b). From Figure 4(b), it was observed that the crack opening displacement (COD) further increased at 99 cycles, and the retardation effect of GB 1 on the fatigue crack growth is more obvious.

In the loading stage of 99 cycles, micromorphology of fatigue crack at different loading level is shown in Figure 5.

As shown in Figure 5(a), the fatigue crack tip maintained closure when the loading increased to 540 N. The fatigue crack tip started to open when the loading reached 960 N as shown in Figure 5(b). From Figures 5(c) and 5(d), it can be seen that the COD increased with the increasing of loading. As shown in Figure 5(d), when the loading reached 1750 N, the COD also achieved the maximum.

From Figure 5, it can be seen that some microcracks appear in front of the main crack, and it mainly initiated in the grain boundary. Under the tensile loading, the front of main crack will become a high stress concentration area. Due to the uncoordinated deformation existing between grain and grain boundary, some grain boundary with weak performance will initiate microcrack in front of the main crack under tensile loading.

In the unloading stage of 99 cycles, micromorphology of fatigue crack at different loading level is shown in Figure 6.

As shown in Figure 6, when the loading decreased from 1380 N to 540 N, the COD also decreased, and the fatigue crack tip recovered from the opening status to the closure status. From Figure 6(d), it can be seen that the fatigue crack tip had been already closed until the loading was decreased to the minimum. It was obtained that the loading value of crack closure is 540 N which is lower than that of crack opening.

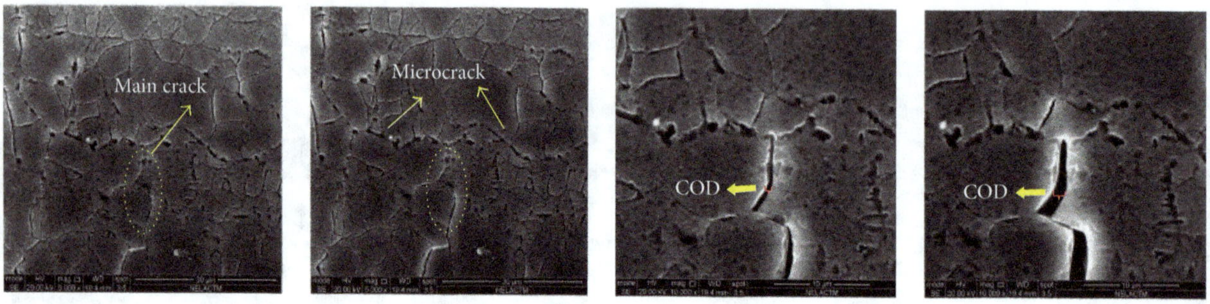

(a) (b) (c) (d)

FIGURE 5: In the loading stage of 99 cycles, micromorphology of fatigue crack at different loading level: (a) at 540 N, (b) at 960 N, (c) at 1380 N, and (d) at 1750 N.

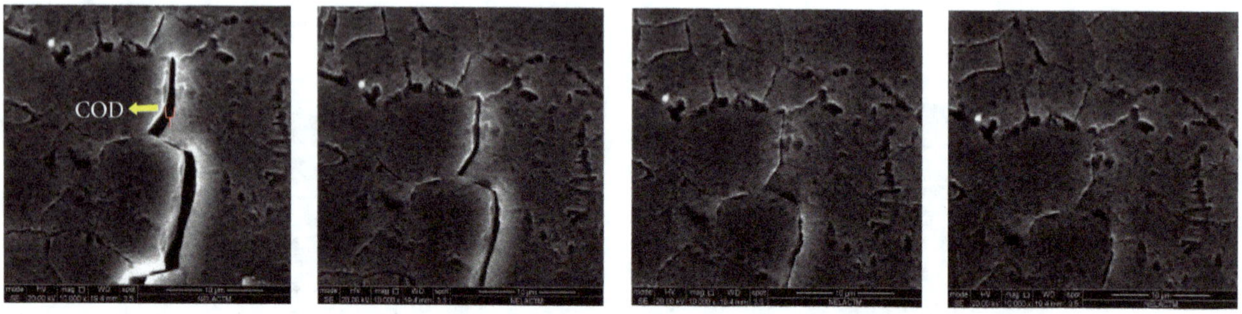

(a) (b) (c) (d)

FIGURE 6: In the unloading stage of 99 cycles, micromorphology of fatigue crack at different loading level: (a) at 1380 N, (b) at 960 N, (c) at 750 N, and (d) at 540 N.

When it reached 111 cycles, the fatigue crack had already been grown rapidly along GB 1 as shown in Figure 7. It can be seen that the fatigue crack growth was retarded by GB 1 for 33 cycles since the loading cycles of 77.

As shown in Figure 7, the fatigue crack had been grown along GB 1 and the crack tip approached Grain 1 at 111 cycles. Meanwhile, two new cracking paths had been developed: the first path is the crack grew along GB 2, and the second path is the crack that passed through GB 3 and grew into Grain 1.

Because the fatigue crack had been retarded by GB 1 from 77 cycles for several loading cycles, large amounts of deformation energy were concentrated in front of the crack tip. When the energy reached the maximum, the fatigue crack tip will be suddenly opened and the fatigue crack grew along GB 1 promptly. The fatigue crack was able to pass through GB 3 and grow into Grain 1 by the driving of released energy; meanwhile, the remaining energy drove the fatigue crack to grow along GB 2.

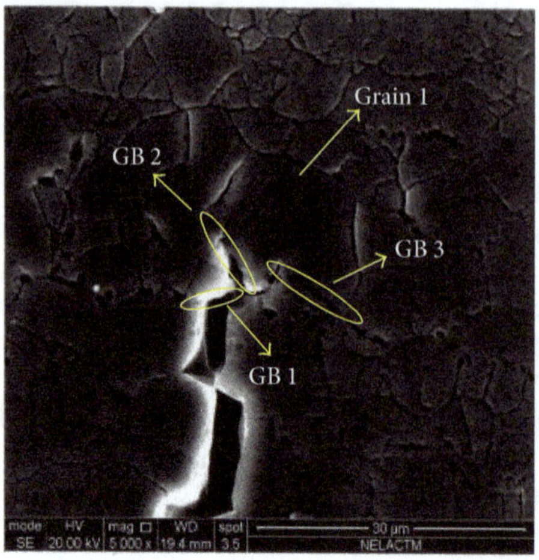

FIGURE 7: Micromorphology of fatigue crack at 111 cycles.

In the loading stage of 112 cycles, micromorphology of fatigue crack at different loading level is shown in Figure 8.

From Figures 8(a) and 8(b), it can be seen that the fatigue crack tip remained closed when the loading value is less than 700 N. As shown in Figure 8(c), when the loading reached 700 N, the fatigue crack tip started to open. From Figures 8(d) and 8(e), it can be seen that the COD increased with the increment of the loading, and the fracture of GB 4 could be observed in front of the crack tip.

As shown in Figures 8(f) and 8(g), when the loading reached 1225 N, the tendency that the crack would grow along GB 2 and connect with GB 4 is more obvious. From Figure 8(h), it can be seen that the COD had been achieved to the maximum when the loading reached 1750 N, and it presented a tendency that the crack tip would be soon connected with GB 4.

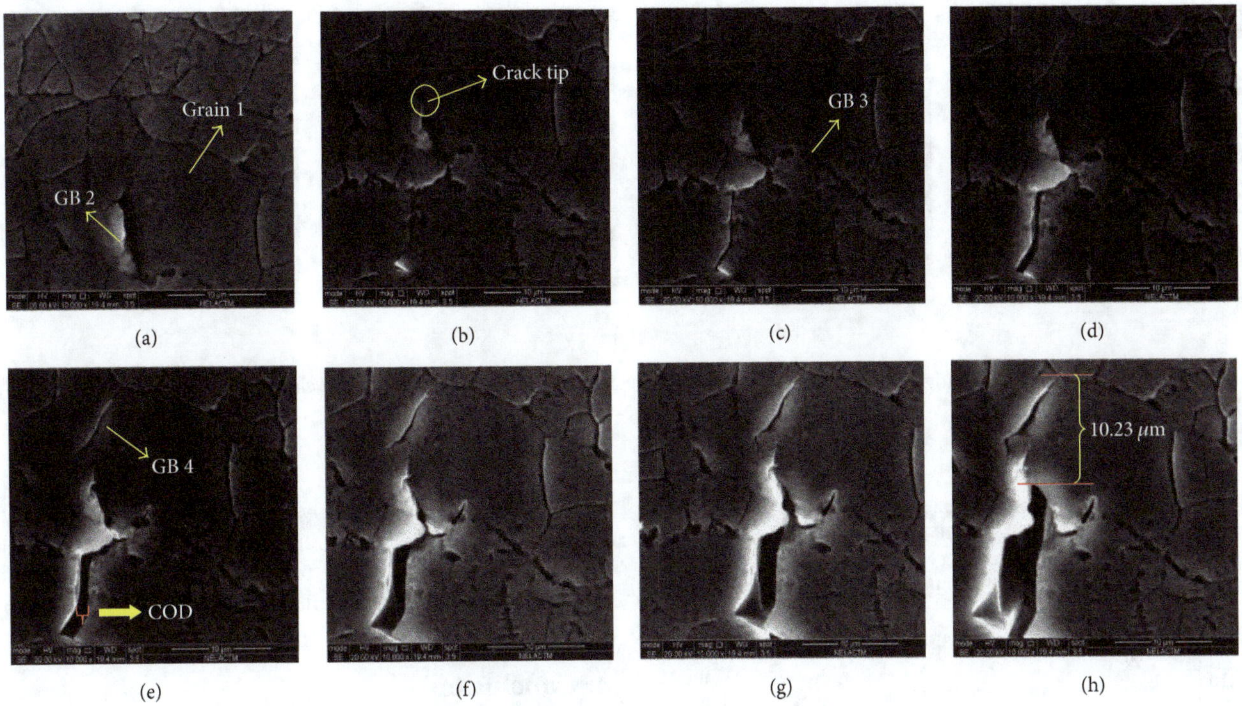

FIGURE 8: In the loading stage of 112 cycles, micromorphology of fatigue crack at different loading level: (a) at 175 N, (b) at 525 N, (c) at 700 N, (d) at 875 N, (e) at 1050 N, (f) at 1225 N, (g) at 1400 N, and (h) at 1750 N.

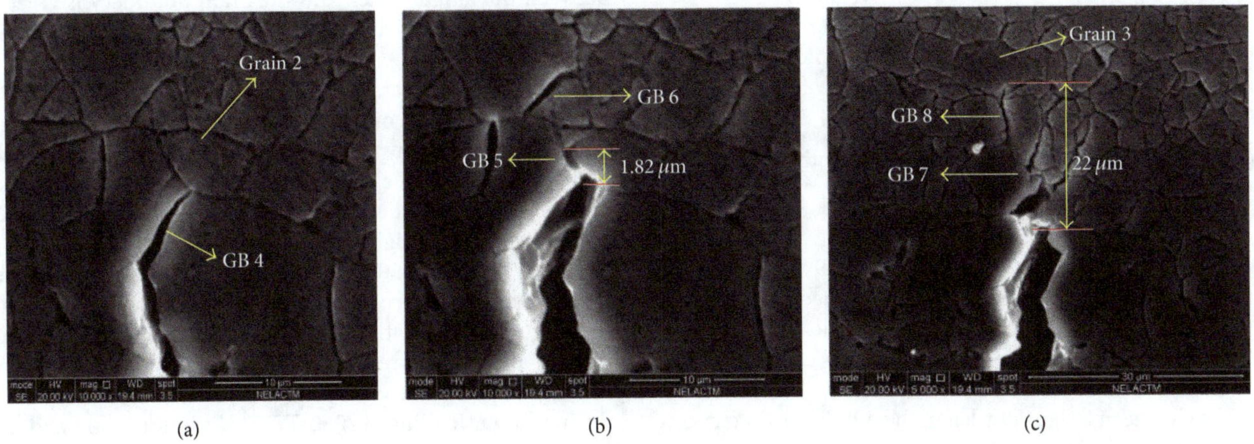

FIGURE 9: Micromorphology of fatigue crack at different loading cycles: (a) at 114 cycles, (b) at 116 cycles, and (c) at 119 cycles.

After two loading cycles, the micromorphology of fatigue crack at 114 cycles is shown in Figure 9(a).

As shown in Figure 9(a), the fatigue crack tip had been connected with GB 4 at 114 cycles, and the crack tip approached Grain 2 at present. It was obtained that the fatigue crack growth rate is 3.41×10^{-3} mm/N when the fatigue crack propagates along GB 4. When it reached 116 cycles, it can be seen that the fatigue crack grew along GB 5 as shown in Figure 9(b). The increment of the length of the fatigue crack is 1.82 um between 114 and 116 cycles, and the fracture of GB 6 could be observed. It was obtained that the fatigue crack growth rate is 9.1×10^{-4} mm/N when the fatigue crack propagates along GB 5. From Figure 9(c), it can be seen that the fatigue crack had been connected with GB 6 when the loading cycle reached 119, and the fatigue crack tip entered into Grain 3 at present. It is obvious that from 116 to 119 cycles, the increment of the length of the fatigue crack is 22 um. It was obtained that the fatigue crack growth rate is 7.3×10^{-3} mm/N when the fatigue crack propagates along GB 6 and GB 8.

Micromorphology of fatigue crack from 126 cycles to 131 cycles is shown in Figure 10.

As shown in Figure 10(a), the increment of fatigue crack is 5 um in Grain 3 from 119 cycles to 126 cycles. Meanwhile, the fracture of GB 9 could be observed in front of Grain 3. As shown in Figure 10(b), the fatigue crack further grew at 129 cycles, and the fatigue crack tip opening increased.

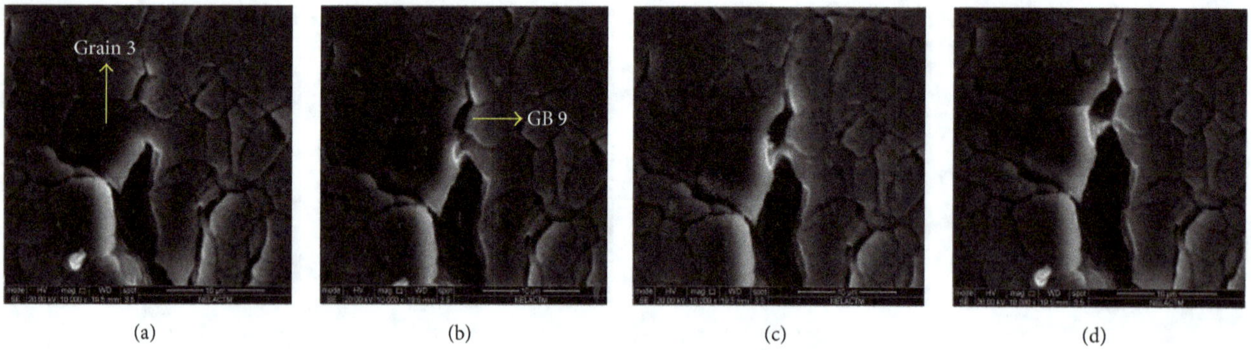

FIGURE 10: Micromorphology of fatigue crack at different loading cycles: (a) at 126 cycles, (b) at 129 cycles, (c) at 130 cycles, and (d) at 131 cycles.

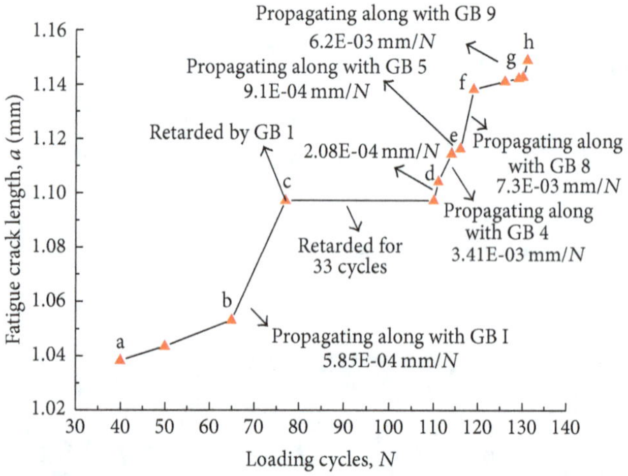

FIGURE 11: Increment of fatigue crack length with the change of loading cycles.

When the loading cycle reached 130, GB 9 was fractured further and the fatigue crack tip approached GB 9. From Figure 10(d), it can be seen that the fatigue crack tip had completely passed through Grain 3 and it was connected with GB 9.

The change of fatigue crack length with the loading cycles from 40 cycles to 131 cycles is shown in Figure 11.

It can be seen that the fatigue crack growth rate is different between different loading cycles. The increment of fatigue crack is the minimum from 77 cycles to 111 cycles in the experiment and the fatigue crack growth rate is 2.08×10^{-4} mm/N. The retardation of fatigue crack growth since 77 cycles is mainly caused by the impediment of grain boundary (GB I). The fatigue crack growth will be retarded by the grain boundary when the angle between fatigue crack and grain boundary is greater than 90 degrees in 7050-T7451 aluminum alloy.

The fatigue crack growth is the fastest from 116 cycles to 119 cycles, and the fatigue crack growth rate is 7.3×10^{-3} mm/N. The following is 6.2×10^{-3} mm/N from 130 cycles to 131 cycles. The rapid growth of fatigue crack is mainly caused by the bridge linked mechanisms between grain boundary and fatigue crack. During the fatigue loading of 7050-T7451 aluminum alloy, the grain boundary is prone to fracture. When the fatigue crack tip connected with the grain boundary, the fatigue crack presented fast unstable growth, which led to the life reduction of 7050-T7451 aluminum alloy.

4. Conclusions

(1) When 7050-T7451 aluminum alloy is under the fatigue loading with constant amplitude, the fatigue crack growth will be retarded by the grain boundary. When the angle between the fatigue crack and the grain boundary is greater than 90 degrees, the retardation phenomenon is more obvious.

(2) The fatigue crack tip is not able to open until the loading reached 55% of the maximum value when the fatigue crack was retarded by the grain boundary. The fatigue crack had been completely closed when the loading is not reduced to the maximum value, and it is observed that the crack opening loading is higher than the crack closure loading.

(3) When the 7050-T7451 aluminum alloy suffered from fatigue loading with constant amplitude, a behavior of unstable fatigue crack growth could be observed in 7050-T7451 aluminum alloy, which related to the bridge linked mechanism of fatigue crack. The grain boundary was prone to fracture during fatigue loading of 7050-T7451 aluminum alloy, and the fractured GB became the best path for the fatigue growth. During the fatigue loading stage, the fatigue crack tends to grow in the direction of cracking grain boundary, and it will be connected with fractured GB eventually, which led to the fast growth of fatigue crack.

Competing Interests

The authors declare that they have no competing interests.

Acknowledgments

This research is supported by two National technical infrastructure projects, the first Project no. is JSZL 2014601B001 and the second Project no. is JSZL 2014601B004. Special thanks go to Professor Dai for his assistance in the revision of paper.

References

[1] A. W. Mello, A. Nicolas, R. A. Lebensohn, and M. D. Sangid, "Effect of microstructure on strain localization in a 7050 aluminum alloy: comparison of experiments and modeling for various textures," *Materials Science and Engineering: A*, vol. 661, pp. 187–197, 2016.

[2] D. C. C. Magalhães, M. F. Hupalo, and O. M. Cintho, "Natural aging behavior of AA7050 Al alloy after cryogenic rolling," *Materials Science and Engineering A*, vol. 593, pp. 1–7, 2014.

[3] N. Kamp, N. Gao, M. J. Starink, and I. Sinclair, "Influence of grain structure and slip planarity on fatigue crack growth in low alloying artificially aged 2xxx aluminium alloys," *International Journal of Fatigue*, vol. 29, no. 5, pp. 869–878, 2007.

[4] M. Scharnweber, W. Tirschler, V. Mikulich, S. Jacob, C.-G. Oertel, and W. Skrotzki, "Influence of crack length and grain boundaries on the propagation rate of short cracks in austenitic stainless steel," *Scripta Materialia*, vol. 67, no. 7-8, pp. 677–680, 2012.

[5] C. Holzapfel, W. Schäf, M. Marx, H. Vehoff, and F. Mücklich, "Interaction of cracks with precipitates and grain boundaries: understanding crack growth mechanisms through focused ion beam tomography," *Scripta Materialia*, vol. 56, no. 8, pp. 697–700, 2007.

[6] D. Terentyev and F. Gao, "Blunting of a brittle crack at grain boundaries: an atomistic study in BCC Iron," *Materials Science and Engineering A*, vol. 576, pp. 231–238, 2013.

[7] P. Wang and J. X. Dong, "Effect of the grain boundary carbide on the crack propagation rates of GH864 alloy," *Rare Metal Materials and Engineering*, vol. 43, no. 11, pp. 2723–2727, 2014.

[8] W. Y. Zhang, "Study on the low cycle fatigue fracture in TA5 titanium alloy plates," *Rare Metal Materials and Engineering*, vol. 27, no. 3, pp. 156–160, 1998.

[9] K. Dám, P. Lejček, and A. Michalcová, "In situ TEM investigation of microstructural behavior of superplastic Al-Mg-Sc alloy," *Materials Characterization*, vol. 76, pp. 69–75, 2013.

[10] C. Meng, D. Zhang, L. Zhuang, and J. Zhang, "Correlations between stress corrosion cracking, grain boundary precipitates and Zn content of Al-Mg-Zn alloys," *Journal of Alloys and Compounds*, vol. 655, Article ID 35434, pp. 178–187, 2016.

[11] Z. Lu and Y. Liu, "Small time scale fatigue crack growth analysis," *International Journal of Fatigue*, vol. 32, no. 8, pp. 1306–1321, 2010.

[12] W. Zhang and Y. Liu, "Investigation of incremental fatigue crack growth mechanisms using in situ SEM testing," *International Journal of Fatigue*, vol. 42, pp. 14–23, 2012.

[13] A. Pineau, D. L. McDowell, E. P. Busso, and S. D. Antolovich, "Failure of metals II: fatigue," *Acta Materialia*, vol. 107, pp. 484–507, 2015.

[14] B. Sun, Y.-L. Xu, and Z. Li, "Multi-scale model for linking collective behavior of short and long cracks to continuous average fatigue damage," *Engineering Fracture Mechanics*, vol. 157, pp. 141–153, 2015.

[15] J. Schijve, "The application of small overloads for fractography of small fatigue cracks initiated under constant-amplitude loading," *International Journal of Fatigue*, vol. 70, pp. 63–72, 2015.

8

Effect of Equal Channel Angular Pressing on the Surface Roughness of Solid State Recycled Aluminum Alloy 6061 Chips

Adel Taha Abbas,[1] Mohamed Adel Taha,[2] Adham Ezzat Ragab,[3] Ehab Adel El-Danaf,[1] and Mohamed Ibrahim Abd El Aal[2]

[1] *Department of Mechanical Engineering, College of Engineering, King Saud University, Riyadh, Saudi Arabia*
[2] *Department of Mechanical Design and Production, College of Engineering, Zagazig University, Sharkia, Egypt*
[3] *Department of Industrial Engineering, College of Engineering, King Saud University, Riyadh, Saudi Arabia*

Correspondence should be addressed to Adel Taha Abbas; atabbas1954@yahoo.com

Academic Editor: Pavel Lejcek

Solid state recycling through hot extrusion is a promising technique to recycle machining chips without remelting. Furthermore, equal channel angular pressing (ECAP) technique coupled with the extruded recycled billet is introduced to enhance the mechanical properties of recycled samples. In this paper, the surface roughness of solid state recycled aluminum alloy 6061 turning chips was investigated. Aluminum chips were cold compacted and hot extruded under an extrusion ratio (ER) of 5.2 at an extrusion temperature (ET) of 425°C. In order to improve the properties of the extruded samples, they were subjected to ECAP up to three passes at room temperature using an ECAP die with a channel die angle (Φ) of 90°. Surface roughness (R_a and R_z) of the processed recycled billets machined by turning was investigated. Box-Behnken experimental design was used to investigate the effect of three machining parameters (cutting speed, feed rate, and depth of cut) on the surface roughness of the machined specimens for four materials conditions, namely, extruded billet and postextrusion ECAP processed billets to one, two, and three passes. Quadratic models were developed to relate the machining parameters to surface roughness, and a multiobjective optimization scheme was conducted to maximize material removal rate while maintaining the roughness below a preset practical value.

1. Introduction

In order to reduce energy consumption and CO_2 emissions in manufacturing processes, recycling of wastes became very important. The main methods used for scrap recycling are conventional and solid state recycling [1]. The conventional recycling, which involves remelting of scrap, has many disadvantages such as high energy consumption, high production and labor cost, and large number of operations. Furthermore, in case of aluminum chips recycling, the losses are about 50% of the recycled scrap due to high surface area to volume ratio of the chips and contamination with machining oils [2, 3]. The need for a recycling method in which the scrap can be recycled without remelting to avoid the disadvantages associated with the conventional method was the motivation of many recent investigations [1, 4–7] of solid state recycling techniques. It was found that recycling of aluminum alloys chips by solid state method through hot extrusion can produce approximately fully dense products [4]. Moreover, about 95% of the primary material can be used by avoiding metal loss during the remelting process [4]. Furthermore, solid state recycling process can be applied to produce finished or semifinished products directly [5–7]. In order to improve the mechanical properties of the extruded recycled billets, it can be subjected to severe plastic deformation to refine and homogenize the grain structure. Recently, several studies on solid state recycling of machining chips through severe plastic deformation (SPD) have been conducted [8–10]. SPD techniques induce remarkably high strains that are capable of refining bulk materials with coarse grain structures.

Equal channel angular pressing (ECAP) technique being the most prominent of SPD techniques has been widely used for producing ultra-fine-grained (UFG) metals and alloys. ECAP involves pressing the sample in a die having intersecting equal channels. The deformation is mainly carried out by simple shear while the cross-section is maintained. This means that the samples can be repeatedly pressed to accumulate large amount of plastic strain that is capable of breaking down the coarse grains into submicron grains. During deformation the grain develop subgrains with low angle boundaries along with high dislocation density. Further deformation causes the subgrains to develop into grains with high angle boundaries. The associated enhancement in mechanical properties has been demonstrated in many previous works [11–15].

Abbas et al. [16] investigated the surface roughness for five starting material conditions (annealed and ECAP processed to 1, 2, 3, and 4 passes) for 1050 aluminum alloy. Quadratic models that related the four surface roughness (R_a, R_q, R_t and R_z) values to the three cutting parameters (cutting speed, feed rate, and depth of cut) were developed, while the fourth parameter (number of ECAP passes) was considered as categorical factor. The results showed that optimum (minimum) surface roughness was achieved at minimum depth of cut (0.15 mm) and minimum feed rate (0.0 5 mm/rev) with slightly different cutting speeds that ranged between 193 m/min for R_z and 207.5 m/min for R_a. The surface quality of the ECAP processed material was found to be much superior compared to its annealed counterpart, for the same cutting conditions.

Khamis et al. [17] investigated the mechanical properties of recycled AA6061 chip that was produced from high speed milling. ANOVA analysis showed that holding compaction time was the most influential parameter whereas chip size had the least effect. The desirability function concluded that longer holding time and more precompaction with large chip size will give the higher tensile strength (UTS) value.

Chan and Lajis [18] investigated the cold compaction of aluminum 6061 chips to conclude that the optimum pressure is 650 MPa. The associated density was reported to be about 98.5% of the reference sample without going into any sintering process. They concluded that pressureless sintering breaks the interlocking which increases the gaps that increase the overall volume which causes the density of the compressed specimen to be reduced. They concluded that the mechanical properties of the compacted samples are still lower than their cast counterpart.

Duflou et al. [19] investigated the solid state recycling techniques of aluminum alloys directly from production scrap. They used three different techniques, namely, hot extrusion, screw extrusion, and spark plasma sintering (SPS), and compared them with the conventional remelting route for reference. It was concluded that aluminum recycling reduced the environmental impact with preferential material and energy savings. An overall impact reduction with a factor of 2 for the SPS route and 3-4 for the extrusion routes was reported.

Chiba and Yoshimura [7] investigated the solid state recycling of Al-Si alloy chips produced by milling and turning. Cold compaction into columnar billets that were subsequently profile-extruded into equilateral c-channels at 600 K, under extrusion ratios of 10 and 18, was performed. Uniaxial tensile tests showed that the dense recycled material had a higher ductility than the original ingot, with a reduction of around 30% in the ultimate tensile strength. The material recycled from the turning chips exhibited insufficient bonding, as compared to that recycled from the milling chips under the same conditions, thereby resulting in inferior mechanical properties.

The objective of the current work is to investigate the surface roughness, as an indicator of product surface quality, of solid state recycled 6061 aluminum chips produced by dry turning. Solid state recycling involved three steps comprising cold compaction followed by hot extrusion and finally for improving mechanical properties the extruded billet was further deformed in an equal channel die up to one, two, and three passes. Cutting parameters, namely, depth of cut, feed rate, and cutting speed, will be the process variables considered in a quadratic model that relates surface roughness to experimental machining conditions.

Table 1: Chemical composition of Aluminum Alloy AA6061.

Si	Mg	Fe	Cu	Mn	Others	Al
0.8	0.85	0.48	0.26	0.051	0.329	Balance

2. Materials and Methods

The material used in this paper is aluminum alloy AA6061 (as-received extruded rod). Table 1 presents the chemical composition of this alloy. The machined chips were prepared by dry turning of the as-received rod under the following cutting conditions: cutting speed of 88 m/min (V), feed of 1 mm/rev (f_r), and depth of cut of 1 mm (d). The chips were placed in a die with a diameter of 25 mm, cold compacted at room temperature, and then hot extruded at a temperature of 425°C with an extrusion ratio of 5.2. In order to improve the properties of the extruded samples, the samples were subjected to ECAP process up to three passes at room temperature using an ECAP die with a channel die angle (Φ) equal 90°.

The microstructure of as-received and extruded specimens was examined using optical microscopy on Zeiss Axio Imager. On the other hand, after ECAP, the microstructure was examined using FEI INSPECT S50 scanning electron microscope (SEM), because of the expected fine microstructure. Metallographic specimens were ground, polished, and etched using kellers reagent. Vickers microhardness was measured under an applied loading of 100 gf and a dwell time of 15 sec at room temperature using Buehler Micromet 5100 device. The tensile test was conducted at room temperature using LR300 K (300 kN) universal testing machine. Tensile tests were performed at an initial strain rate of $8.33 \times 10^{-4}\ s^{-1}$ at room temperature. Tensile test specimens were fabricated parallel to extrusion and ECAP direction in accordance with ASTM-B557-06.

The final processed billets were machined using Emco Concept Turn 45 CNC lathe. This machine is equipped with Sinumeric 840-D. All test runs were done using CNC part

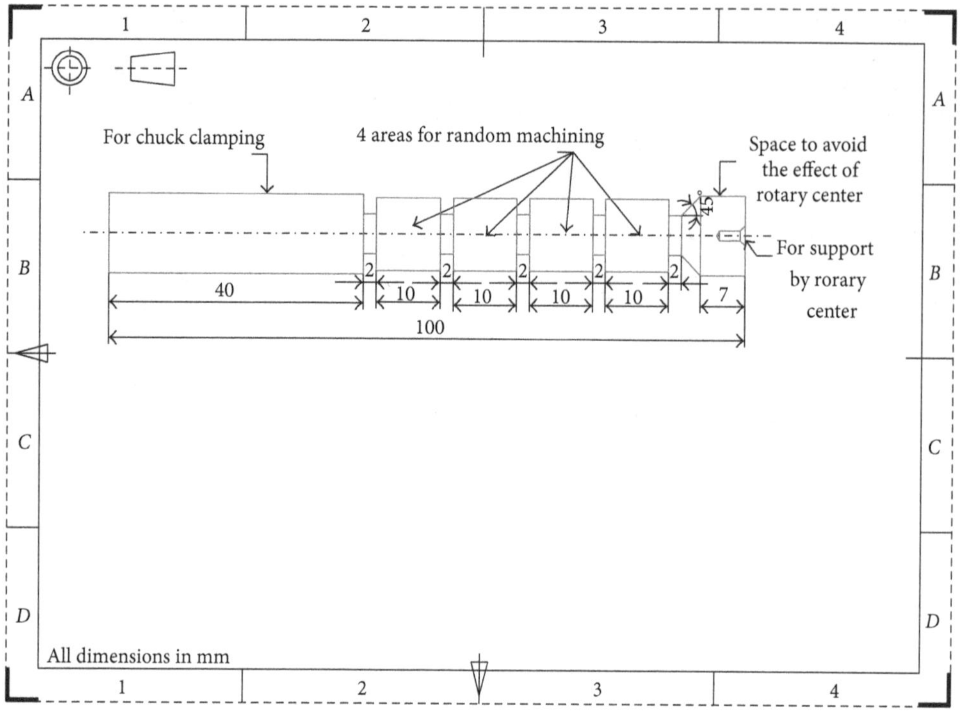

FIGURE 1: Test specimen drawing.

TABLE 2: The cutting parameters and their levels used in the Box-Behnken design.

Designation	Process parameter	Level 1	Level 2	Level 3	Level 4
A	V	100	150	200	—
B	d	0.10	0.25	0.40	—
C	f_r	0.012	0.081	0.15	—
D	Specimen condition	No ECAP (1)	One pass ECAP (2)	Two passes ECAP (3)	Three passes ECAP (4)

program. The specifications for tool holder and insert are SVJCL2020K16 and VCGT160404 FN-ALU, respectively. The clearance angle, cutting edge angle, and nose radius are maintained at 5°, 35° and 0.4 mm, respectively. All cutting parameters were controlled via CNC part program and all the experiments were conducted in wet conditions. The surface roughness was evaluated using surface roughness tester TESA. The test specimens had an initial diameter of 12 mm and a length of 100 mm. The test specimens were divided into six areas. The first area with a length of 40 mm was used for chuck clamping, and the other four areas with 10 mm for each and separated by a groove of 2 mm were used for random machining with different cutting parameters and the last area was used for centering. The test specimen drawing is shown in Figure 1. The test specimens had four different starting conditions: hot extruded at 425°C and ECAP up to three passes. The cutting parameters and values of surface roughness (R_a, and R_z) and the calculated material removal rate were recorded in the Appendix. Where R_a is the arithmetic average deviation of the assessed profile (μm) and R_z is the average maximum height of the profile (μm).

3. Design of Experiments

Box-Behnken design is one of response surface methodology (RSM) designs that is used extensively in studying and optimizing surface roughness of different machining processes through several publications [20–23]. RSM is a collection of statistical and mathematical techniques used to optimize a considered output variable, or response (represented in the model by the dependent variable) that is influenced by a set of controllable input variables, or factors (represented in the model by the independent variables) [24].

The Box-Behnken design was first introduced in 1960 by Box and Behnken [24]. In this research a rotatable Box-Behnken design consisting of 60 experiments was built including four factors, namely, cutting speed, depth of cut, feed rate, and specimen condition. Table 2 summarizes the studied factors and their levels. The specimen condition was treated as a categorical variable, while the other factors were treated as continuous variables. The Appendix illustrates the full list of experiments with measured response values (R_a and R_z) in micrometers.

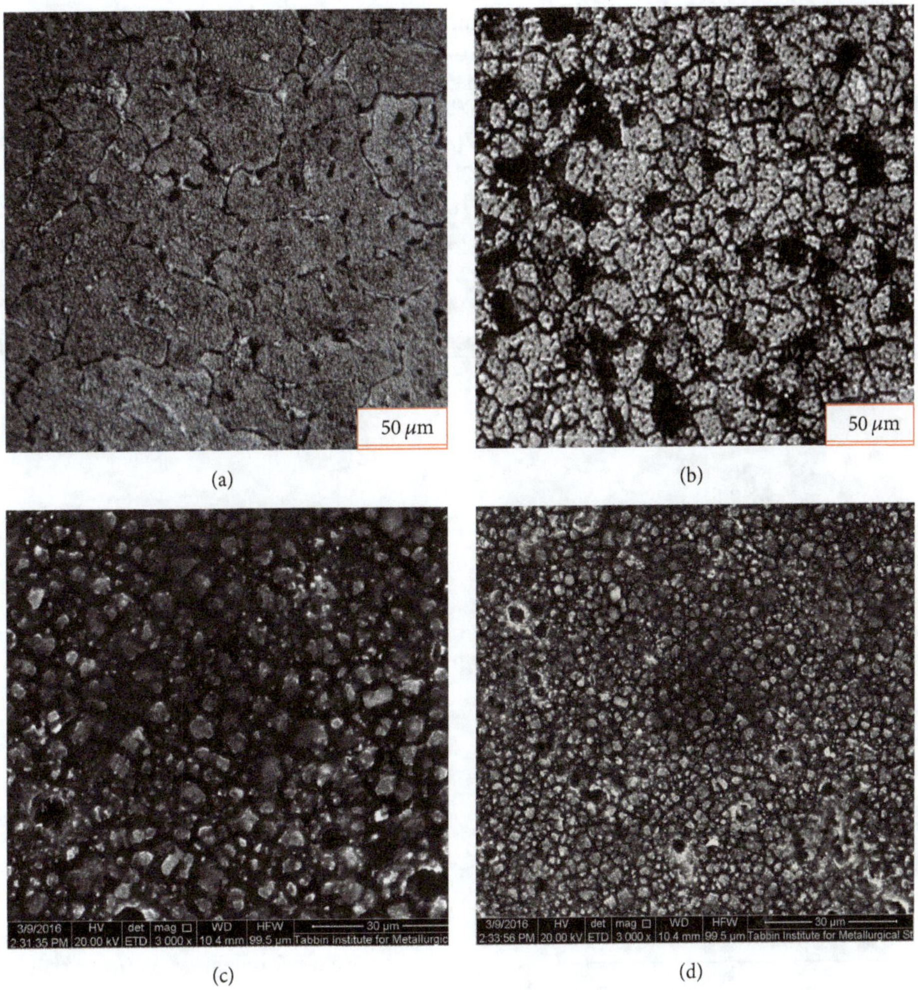

FIGURE 2: Microstructure of (a) as-received, (b) ET = 425°C, (c) ECAP-1, and (d) ECAP-3.

4. Results and Discussions

The microstructures of the as-received sample are shown in Figure 2(a). It consists of coarse-grained structure with grain size of about 48 μm. On the other hand, the microstructure of the hot extruded specimen consists of fine grains of about 13.5 μm, as shown in Figure 2(b). The finer grained microstructure is caused by severe plastic strain imposed during the recycling processes and the positive effect of chip boundaries which can act as barriers between grains to prevent grain growth [25–28]. The decrease in grain size was congruent with previous work done by Haase et al. [25] where the grain size of the as-received AA-6060 alloy was 385 μm, while it was 31.8 μm for the hot extruded chips at 450°C. Also, Selmy et al. [26, 27] recycled AA-6061 chips through hot extrusion; the grain size was 48 μm for as-received samples while it was 15.8 and 10.9 μm for the recycled samples produced by hot extrusion at 500 and 350°C, respectively. Moreover in previous work by Hu et al. [28], the Mg chips were classified into three different groups based on its size. The Mg chips were cold compacted and followed by hot extrusion at 400°C. The grain size of the as-received sample was 200 μm, while it was 22, 25, and 11 μm for the three different groups of Mg chips recycled by hot extrusion. SEM micrographs of ECAPed samples after 1 and 3 passes are shown in Figures 2(c) and 2(d), respectively. It is obvious that additional grain refinement is exhibited with increasing number of ECAP passes. The grain size for the ECAPed samples processed to 1 and 3 passes is 5 and 3.25 μm, respectively.

TABLE 3: Effect of ECAP process on the Solid Sate Recycled Aluminum Alloy 6061 chips after Extrusion temperature 425°C.

Description	ET = 425°C	ECAP-1	ECAP-2	ECAP-3
Relative density %	96.4	99	99.8	99.8
Microhardness, Vickers	43	76.25	80.75	82.75
Ultimate T. Strength, MPa	114.7	198	230.8	306.4
Elongation to failure, mm	9.85	10.3	11.2	12.3
Grain size, μm	13.5	5.22	4	3.25

Table 3 presents the relative density, microhardness, and ultimate tensile strength; elongation to failure and grain size of the solid state recycled aluminum alloy 6061 chips after extrusion and after the extruded billet were processed by

TABLE 4: Quadratic models for R_a, and R_z.

Response	D	Equation
R_a	1	$1.334 - 0.00971A - 2.101B - 9.972C + 0.000036A^2 + 4.468B^2 + 120.55C^2 + 4.56B*C$
	2	$1.191 - 0.00971A - 2.101B - 9.972C + 0.000036A^2 + 4.468B^2 + 120.55C^2 + 4.56B*C$
	3	$1.153 - 0.00971A - 2.101B - 9.972C + 0.000036A^2 + 4.468B^2 + 120.55C^2 + 4.56B*C$
	4	$1.130 - 0.00971A - 2.101B - 9.972C + 0.000036A^2 + 4.468B^2 + 120.55C^2 + 4.56B*C$
R_z	1	$7.04 - 0.01399A - 27.97B - 51.62C + 36.90B^2 + 496.6C^2 + 0.0771A*B + 0.1211A*C$
	2	$5.77 - 0.01399A - 27.97B - 51.62C + 36.90B^2 + 496.6C^2 + 0.0771A*B + 0.1211A*C$
	3	$5.57 - 0.01399A - 27.97B - 51.62C + 36.90B^2 + 496.6C^2 + 0.0771A*B + 0.1211A*C$
	4	$5.45 - 0.01399A - 27.97B - 51.62C + 36.90B^2 + 496.6C^2 + 0.0771A*B + 0.1211A*C$

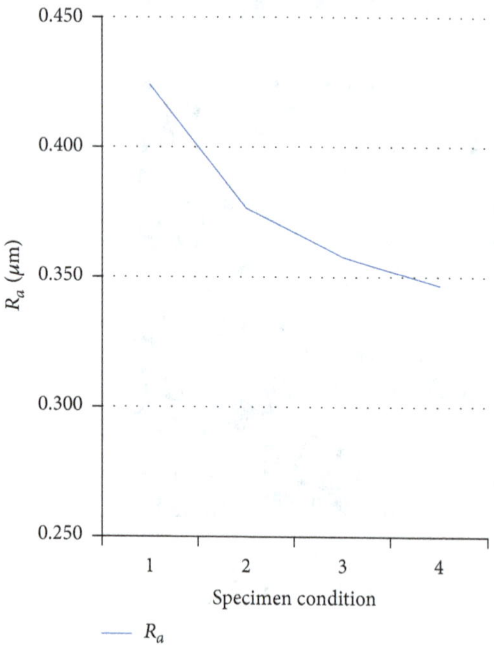

FIGURE 3: Surface roughness, R_a (μm) versus specimen condition.

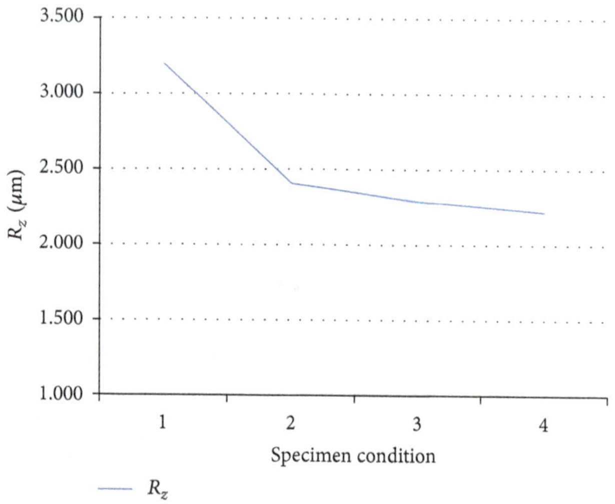

FIGURE 4: Surface roughness, R_z (μm) versus specimen condition.

ECAP to one, two, and three passes. The values are presented in Table 3 for the properties of the solid state recycled billets point towards the successfulness of recycling the chips into a consolidated bulk materials via cold compaction, followed by hot extrusion and finally ECAP. Table 3 shows the influence of number of ECAP passes on the microhardness of recycled samples. The microhardness has increased from 43 HV after extrusion to about 83 HV when extrusion was coupled with ECAP to 3 passes, which means that the hardness is almost doubled. Also, the ultimate tensile strength has increased from 115 to 307 MPa, which means that the tensile strength is almost tripled. This remarkable increase in mechanical strength is mainly due to the high dislocation density and the grain refinement associated with severe deformation accompanied with ECAP. Room temperature ECAP deformation has also resulted in improving the relative density. The increase in mechanical strength is expected to have significant effect on surface roughness.

Figures 3 and 4 show the variation of both R_a and R_z versus the specimen condition, respectively. The roughness was measured at the same machining conditions (cutting speed of 150 m/min, depth of cut of 0.25 mm, and feed rate of 0.081 mm/rev.) which embrace the center point of the Box-Behnken design. The figures illustrate that the roughness values of studied specimens decrease with the number of ECAP passes. The change between extruded specimens (no ECAP) and one pass ECAP specimens is significantly high. Extra passes reduce the value of surface roughness but with less slope.

A quadratic model of the form given in (1) was fitted to the data using Minitab 17 software depending on the Box-Behnken design. Stepwise technique was used to reduce the model into significant terms only. Stepwise is an automated technique that adds terms to the model one by one. The added term is kept in the model if it has a significant effect (p value < 0.05); otherwise, it is removed. The process is repeated until the model includes only all of the significant term.

$$y = \beta_o + \sum_{i=1}^{k} \beta_i X_i + \sum_{i=1}^{k} \beta_{ii} X_i^2 + \sum_{i<j} \sum \beta_{ij} X_i X_j, \qquad (1)$$

where β_o is the constant term, β_i represents the linear effects, β_{ii} represents the pure quadratic effects, and β_{ij} represents the second level interaction effects. Table 4 illustrates the

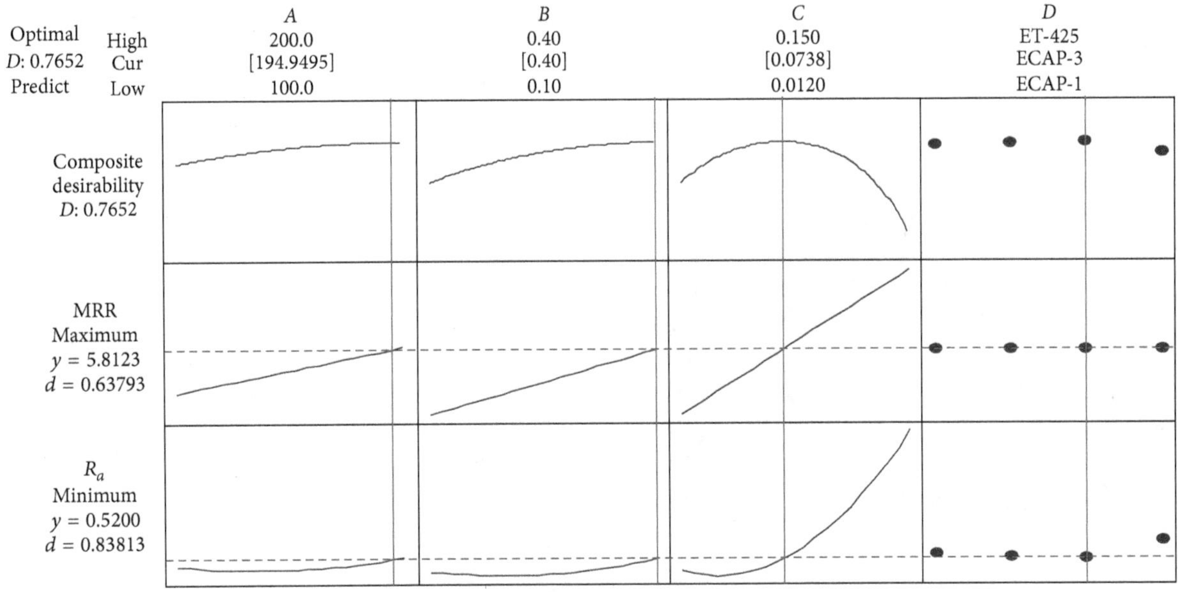

FIGURE 5: Optimization plot for R_a.

quadratic models developed for R_a and R_z. The specimen condition (factor D) and its significant interactions are not included in the equation since it is treated as a categorical factor as mentioned before. The adequacy of the quadratic models was tested through calculating the coefficient of determination. Adjusted R-squared and predicted R-squared were calculated as 99% and 98% for R_a, respectively, and 95% and 93% for R_z, respectively. Anderson-Darling test was conducted to check the normality of residuals with a result of p value = 0.695 for R_a and 0.169 for R_z assuming normally distributed residuals. The null hypothesis of such a test is that the data are normal and a p value < 0.05 proves nonnormality.

Material removal rate (MRR) was calculated for each run using (2), where V is surface cutting speed (m/min.), f_r is feed rate (mm/rev.), and d is depth of cut (mm). Optimization process was run using the desirability function approach to maximize MRR while maintaining R_a below 0.8 μm, which is practically adopted in industry.

$$\text{MRR} = V * d * f_r \quad (2)$$

Desirability function approach was used to estimate the values of studied process parameters that maximize the MRR keeping the R_a at levels not exceeding a practical value of 0.8 μm. The optimization plot, illustrated in Figure 5, shows that an optimum MRR of 5.812 cm^3/min with R_a of 0.520 μm is obtained at cutting speed of 195 m/min., depth of cut of 0.4 mm, and feed rate of 0.073 mm/rev, for three-pass ECAP alloy (specimen condition 4).

ANOVA was used to examine the effect of studied factors, in both linear and quadratic forms, and their interactions for their significance in the model. The results for R_a, illustrated in Table 5, show that all factors has significant effect on R_a with their both linear and quadratic terms. The feed rate, by far, had the greatest influence on the surface roughness in both its linear and quadratic terms.

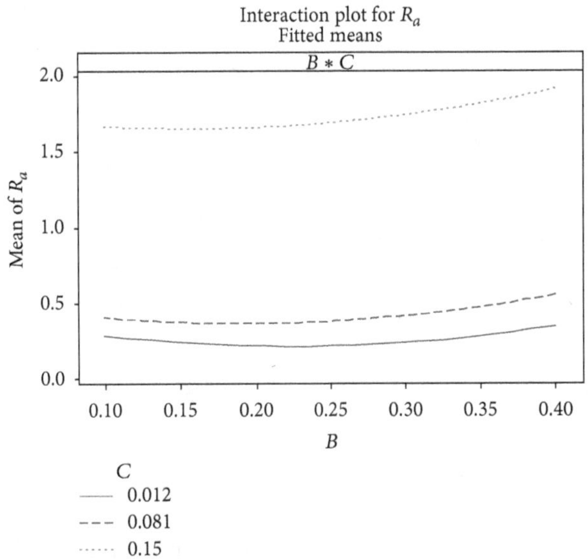

FIGURE 6: Interaction plot of depth of cut and feed rate for R_a.

Figure 6 shows the interaction plot for depth of cut and feed rate which is the only significant interaction affecting R_a. Although the interaction is significant, it has little effect on the variation of R_a which is illustrated by almost parallel curves in the plot. The same pattern is illustrated in the response surface given in Figure 7. The ANOVA results for R_z showed the same patterns except that the significant interactions are cutting speed/depth of cut and cutting speed/feed rate.

To validate the model, extra 12 experimental runs were conducted. Three cutting conditions were selected for the four starting specimen conditions (ET-425°C, ECAP-1, ECAP-2, and ECAP-3). Two among these three conditions

Table 5: ANOVA results for R_a.

Source	DF	Adj SS	Adj MS	F-value	p value
Model	10	23.0431	2.3043	475.18	0
Linear	6	18.0698	3.0116	621.05	0
A	1	0.0761	0.0761	15.7	0
B	1	0.1812	0.1812	37.37	0
C	1	17.436	17.436	3595.57	0
D	3	0.3765	0.1255	25.88	0
Square	3	4.9376	1.6459	339.4	0
$A * A$	1	0.1172	0.1172	24.17	0
$B * B$	1	0.1492	0.1492	30.77	0
$C * C$	1	4.8654	4.8654	1003.32	0
2-way interaction	1	0.0356	0.0356	7.35	0.009
$B * C$	1	0.0356	0.0356	7.35	0.009
Error	49	0.2376	0.0048		
Lack-of-fit	41	0.2373	0.0058	169.84	0
Pure error	8	0.0003	0		
Total	59	23.2807			

Table 6: Extra runs for valediction and verification of the predicted equations in Table 4.

Run Order	Condition	Surface Speed (m/min)	Depth of cut (mm)	Feed (mm/rev)	R_a, (μm)	
					Experimental	Predicted
1	ET-425	125	0.325	0.125	1.378	1.294
2	ET-425	195	0.400	0.073	0.758	0.732
3	ET-425	250	0.500	0.20	4.88	4.507
4	ECAP-1	125	0.325	0.125	1.267	1.151
5	ECAP-1	195	0.400	0.073	0.654	0.589
6	ECAP-1	250	0.500	0.20	4.589	4.364
7	ECAP-2	125	0.325	0.125	1.190	1.113
8	ECAP-2	195	0.400	0.073	0.610	0.551
9	ECAP-2	250	0.500	0.20	4.303	4.326
10	ECAP-3	125	0.325	0.125	1.082	1.090
11	ECAP-3	195	0.400	0.073	0.502	0.520
12	ECAP-3	250	0.500	0.20	4.223	4.303

were taken within the data range, one of which was the predicted optimum condition. The third condition was taken out of the data range. For all these 12 conditions, the measured surface roughness values were compared to the predicted ones as shown in Table 6. Figure 8 shows a scatter plot for both measured and predicted values. It is clear from the figure that the relation between them is linear with calculated R^2 of 0.99. It is worth mentioning that the machining parameters that corresponded to the optimum condition of R_a and MMR (cutting speed of 195 m/min., depth of cut of 0.40 mm, and feed rate of 0.073 mm/rev) yielded an experimental value of R_a of 0.502 compared to the predicted value of 0.520 which exhibits less than 5% error.

5. Conclusion

Solid state recycling of 6061 aluminum chips was conducted. The final bulk material was produced via cold compaction followed by hot extrusion and finally equal channel angular pressing for one, two, and three passes. Mechanical properties revealed the effectiveness of the compaction route in producing a densified product. Surface roughness was measured for four materials conditions comprising as extruded sample as well as samples extruded and ECAP processed to one, two, and three passes. The sample ECAP processed to three passes exhibited the highest hardness and consequently gave the lowest surface roughness. Box-Behnken experimental design was used to investigate the effect of three machining parameters (cutting speed, feed rate, and depth of cut) on the surface roughness of the machined specimens. A quadratic model was developed that related R_a and R_z to the machining parameters, with the material condition being a categorical variable. Feed rate was concluded to be the most influential parameter affecting surface roughness. Multiobjective optimization results showed that the minimum value of MRR was calculated as 5.812 cm^3/min for surface roughness

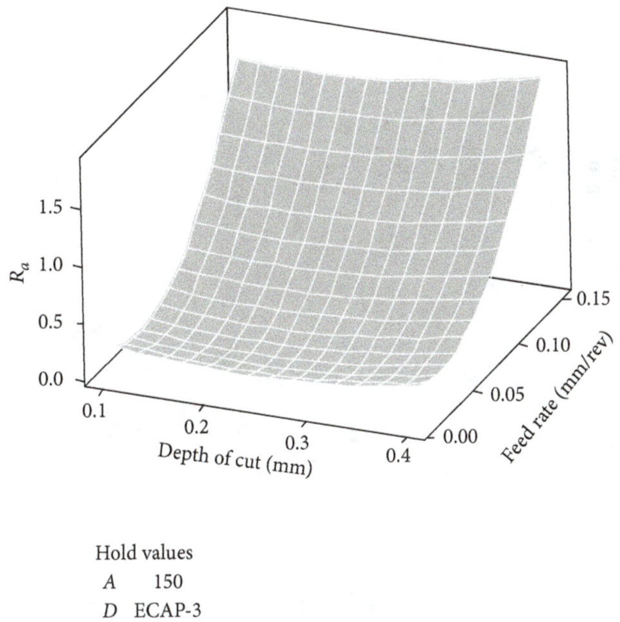

FIGURE 7: Response surface plot for R_a versus depth of cut and feed rate at a mean cutting speed of 150 m/min.

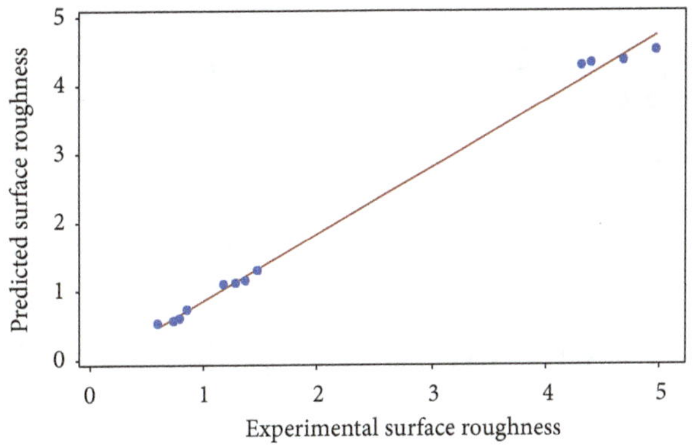

FIGURE 8: Scatter plot for the predicted versus experimental R_a values for the (12) validation runs.

of 0.520 μm at cutting speed of 195 m/min., depth of cut of 0.40 mm, and feed rate of 0.073 mm/rev. (12) validation experimental tests were conducted with three new cutting conditions. The difference between the experimental and predicted values of R_a for the new 12 experiments was in the range of 3–10%, which proves the applicability of the developed models.

Nomenclature

d: Depth of cut, mm
ECAP: Equal channel angular pressing
ER: Extrusion ratio
ET: Extrusion temperature, °C
Φ: Channel die angle
f_r: Feed rate, mm/rev
MRR: Material removal rate, cm^3/min
N: Spindle speed, rpm
R_a: The arithmetic average deviation of the assessed profile, μm
R_z: The average maximum height of the profile, μm
Status: (1) for extrude temperature 425°C, (2) one pass ECAP, (3) two passes ECAP, and (4) Three passes ECAP
V: Surface speed, meter/min.

Competing Interests

The authors declare that they have no competing interests.

Table 7

Run order	Condition	Surface speed (M/min)	Depth of cut (mm)	Feed (mm/rev)	R_a (Mm)	R_z (Mm)	Material removal rate (Cm^3/min)
1	ET-425	100	0.4	0.081	0.862	3.237	3.240
2	ET-425	200	0.4	0.081	0.907	6.696	6.480
3	ET-425	150	0.4	0.012	0.396	2.853	0.720
4	ET-425	150	0.4	0.15	2.034	11.016	9.000
5	ET-425	100	0.25	0.012	0.338	2.952	0.300
6	ET-425	200	0.25	0.012	0.595	4.109	0.600
7	ET-425	100	0.25	0.15	1.959	9.350	3.750
8	ET-425	200	0.25	0.15	2.058	11.010	7.500
9	ET-425	150	0.25	0.081	0.419	3.161	3.038
10	ET-425	150	0.25	0.081	0.425	3.217	3.038
11	ET-425	150	0.25	0.081	0.427	3.212	3.038
12	ET-425	100	0.1	0.081	0.509	3.695	0.810
13	ET-425	200	0.1	0.081	0.577	3.381	1.620
14	ET-425	150	0.1	0.012	0.353	2.699	0.180
15	ET-425	150	0.1	0.15	1.870	10.418	2.250
16	ECAP-1	100	0.4	0.081	0.574	2.837	3.240
17	ECAP-1	200	0.4	0.081	0.645	5.493	6.480
18	ECAP-1	150	0.4	0.012	0.256	2.053	0.720
19	ECAP-1	150	0.4	0.15	1.985	8.808	9.000
20	ECAP-1	100	0.25	0.012	0.296	1.725	0.300
21	ECAP-1	200	0.25	0.012	0.329	2.113	0.600
22	ECAP-1	100	0.25	0.15	1.617	6.083	3.750
23	ECAP-1	200	0.25	0.15	1.824	8.624	7.500
24	ECAP-1	150	0.25	0.081	0.382	2.444	3.038
25	ECAP-1	150	0.25	0.081	0.378	2.398	3.038
26	ECAP-1	150	0.25	0.081	0.369	2.378	3.038
27	ECAP-1	100	0.1	0.081	0.435	2.272	0.810
28	ECAP-1	200	0.1	0.081	0.504	3.021	1.620
29	ECAP-1	150	0.1	0.012	0.242	1.694	0.180
30	ECAP-1	150	0.1	0.15	1.750	10.046	2.250
31	ECAP-2	100	0.4	0.081	0.545	2.695	3.240
32	ECAP-2	200	0.4	0.081	0.613	5.218	6.480
33	ECAP-2	150	0.4	0.012	0.243	1.950	0.720
34	ECAP-2	150	0.4	0.15	1.886	8.368	9.000
35	ECAP-2	100	0.25	0.012	0.281	1.639	0.300
36	ECAP-2	200	0.25	0.012	0.313	2.007	0.600
37	ECAP-2	100	0.25	0.15	1.536	5.779	3.750
38	ECAP-2	200	0.25	0.15	1.733	8.193	7.500
39	ECAP-2	150	0.25	0.081	0.363	2.322	3.038
40	ECAP-2	150	0.25	0.081	0.359	2.278	3.038
41	ECAP-2	150	0.25	0.081	0.351	2.259	3.038
42	ECAP-2	100	0.1	0.081	0.413	2.158	0.810
43	ECAP-2	200	0.1	0.081	0.479	2.870	1.620
44	ECAP-2	150	0.1	0.012	0.230	1.609	0.180
45	ECAP-2	150	0.1	0.15	1.663	9.544	2.250
46	ECAP-3	100	0.4	0.081	0.529	2.614	3.240

TABLE 7: Continued.

Run order	Condition	Surface speed (M/min)	Depth of cut (mm)	Feed (mm/rev)	R_a (Mm)	R_z (Mm)	Material removal rate (Cm^3/min)
47	ECAP-3	200	0.4	0.081	0.594	5.062	6.480
48	ECAP-3	150	0.4	0.012	0.236	1.892	0.720
49	ECAP-3	150	0.4	0.15	1.829	8.117	9.000
50	ECAP-3	100	0.25	0.012	0.273	1.590	0.300
51	ECAP-3	200	0.25	0.012	0.303	1.947	0.600
52	ECAP-3	100	0.25	0.15	1.490	5.605	3.750
53	ECAP-3	200	0.25	0.15	1.681	7.947	7.500
54	ECAP-3	150	0.25	0.081	0.352	2.252	3.038
55	ECAP-3	150	0.25	0.081	0.348	2.210	3.038
56	ECAP-3	150	0.25	0.081	0.340	2.191	3.038
57	ECAP-3	100	0.1	0.081	0.401	2.094	0.810
58	ECAP-3	200	0.1	0.081	0.464	2.784	1.620
59	ECAP-3	150	0.1	0.012	0.223	1.561	0.180
60	ECAP-3	150	0.1	0.15	1.613	9.257	2.250

Acknowledgments

This Project was supported by King Saud University, Deanship of Scientific Research, College of Engineering Research Center. Authors would like to thank Twasol Research Excellence Program (TRE Program), King Saud University, Riyadh, Saudi Arabia, for support.

References

[1] A. E. Tekkaya, M. Schikorra, D. Becker, D. Biermann, N. Hammer, and K. Pantke, "Hot profile extrusion of AA-6060 aluminum chips," *Journal of Materials Processing Technology*, vol. 209, no. 7, pp. 3343–3350, 2009.

[2] J. Z. Gronostajski, J. W. Kaczmar, H. Marciniak, and A. Matuszak, "Production of composites from Al and AlMg2 alloy chips," *Journal of Materials Processing Technology*, vol. 77, no. 1–3, pp. 37–41, 1998.

[3] W. H. Macintosh, "Induction furnaces for melting secondary aluminium," *Conservation and Recycling*, vol. 6, no. 1-2, pp. 41–48, 1983.

[4] J. Z. Gronostajski, H. Marciniak, and A. Matuszak, "Production of composites on the base of AlCu4 alloy chips," *Journal of Materials Processing Technology*, vol. 60, no. 1-4, pp. 719–722, 1996.

[5] M. Haase and A. E. Tekkaya, "Cold extrusion of hot extruded aluminum chips," *Journal of Materials Processing Technology*, vol. 217, pp. 356–367, 2015.

[6] W. Tang and A. P. Reynolds, "Production of wire via friction extrusion of aluminum alloy machining chips," *Journal of Materials Processing Technology*, vol. 210, no. 15, pp. 2231–2237, 2010.

[7] R. Chiba and M. Yoshimura, "Solid-state recycling of aluminium alloy swarf into c-channel by hot extrusion," *Journal of Manufacturing Processes*, vol. 17, pp. 1–8, 2015.

[8] T. Aida, N. Takatsuji, K. Matsuki, S. Kamado, and Y. Kojima, "Homogeneous consolidation process by ECAP for AZ31 cutting chips," *Journal of Japan Institute of Light Metals*, vol. 54, no. 11, pp. 532–537, 2004.

[9] T. Takahashj, Y. Kume, M. Kobashi, and N. Kanetake, "Solid state recycling of aluminum machined chip wastes by compressive torsion processing," *Keikinzoku/Journal of Japan Institute of Light Metals*, vol. 59, no. 7, pp. 354–358, 2009.

[10] T. Peng, Q. D. Wang, and J. B. Lin, "Microstructure and mechanical properties of Mg-10Gd-2Y-0.5Zr alloy recycled by cyclic extrusion compression," *Materials Science and Engineering A*, vol. 516, no. 1-2, pp. 23–30, 2009.

[11] E. A. El-Danaf, "Mechanical properties, microstructure and texture of single pass equal channel angular pressed 1050, 5083, 6082 and 7010 aluminum alloys with different dies," *Materials and Design*, vol. 32, no. 7, pp. 3838–3853, 2011.

[12] M. S. Soliman, E. A. El-Danaf, and A. A. Almajid, "Effect of equal-channel angular pressing process on properties of 1050 Al alloy," *Materials and Manufacturing Processes*, vol. 27, no. 7, pp. 746–750, 2012.

[13] M. Furukawa, Z. Horita, M. Nemoto, and T. G. Langdon, "The use of severe plastic deformation for microstructural control," *Materials Science and Engineering A*, vol. 324, no. 1-2, pp. 82–89, 2002.

[14] K. Nakashima, Z. Horita, M. Nemoto, and T. G. Langdon, "Influence of channel angle on the development of ultrafine grains in equal-channel angular pressing," *Acta Materialia*, vol. 46, no. 5, pp. 1589–1599, 1998.

[15] R. Z. Valiev and T. G. Langdon, "Principles of equal-channel angular pressing as a processing tool for grain refinement," *Progress in Materials Science*, vol. 51, no. 7, pp. 881–981, 2006.

[16] A. T. Abbas, A. E. Ragab, E. A. El-Danaf, and E. A. Al Bahkali, "Effect of equal-channel angular pressing on the surface roughness of commercial purity aluminum during turning operation," *Proceedings of the Institution of Mechanical Engineers, Part B: Journal of Engineering Manufacture*, 2016.

[17] S. S. Khamis, M. A. Lajis, and R. A. O. Albert, "A sustainable direct recycling of aluminum chip (AA6061) in hot press forging employing response surface methodology," *Procedia CIRP*, vol. 26, pp. 477–481, 2015.

[18] B. L. Chan and M. A. Lajis, "Recycling of aluminium 6061 ChipThrough cold compression," *International Journal of Engineering & Technology*, vol. 15, no. 4, 2015.

[19] J. R. Duflou, A. E. Tekkaya, M. Haase et al., "Environmental assessment of solid state recycling routes for aluminium alloys: can solid state processes significantly reduce the environmental impact of aluminium recycling?" *CIRP Annals—Manufacturing Technology*, vol. 64, no. 1, pp. 37–40, 2015.

[20] H. Tebassi, M. A. Yallese, R. Khettabi, S. Belhadi, I. Meddour, and F. Girardin, "Multi-objective optimization of surface roughness, cutting forces, productivity and Power consumption when turning of Inconel 718," *International Journal of Industrial Engineering Computations*, vol. 7, no. 1, pp. 111–134, 2016.

[21] K. Kadirgama, M. M. Noor, N. M. Zuki et al., "Optimization of surface roughness in end milling on mould aluminium alloys (AA6061-T6) using response surface method and radian basis function network," *Jourdan Journal of Mechanical and Industrial Engineering*, vol. 2, no. 4, pp. 209–213, 2008.

[22] K. Kadirgama, M. M. Noor, M. M. Rahman, M. R. M. Rejab, C. H. C. Haron, and K. A. Abou-El-Hossein, "Surface roughness prediction model of 6061-T6 aluminium alloy machining using statistical method," *European Journal of Scientific Research*, vol. 25, no. 2, pp. 250–256, 2009.

[23] D. C. Montgomery, *Design and Analysis of Experiments*, John Wiley & Sons, New York, NY, USA, 2013.

[24] G. E. P. Box and D. W. Behnken, "Some new three level designs for the study of quantitative variables," *Technometrics*, vol. 2, no. 4, pp. 455–475, 1960.

[25] M. Haase, N. Ben Khalifa, A. E. Tekkaya, and W. Z. Misiolek, "Improving mechanical properties of chip-based aluminum extrudates by integrated extrusion and equal channel angular pressing (iECAP)," *Materials Science and Engineering A*, vol. 539, pp. 194–204, 2012.

[26] A. I. Selmy, A. M. El-Gohry, M. I. Abd El Aal, and M. A. Taha, "Characteristics of solid state recycling of aluminum alloy (AA6061) chips by hot extrusion," in *Proceedings of the International Conference of Engineering Sciences and Applications*, vol. 1, pp. 316–323, Aswan, Egypt, 2016.

[27] A. I. Selmy, M. I. Abd El Aal, A. M. El-Gohry, and M. A. Taha, "Solid-state recycling of aluminum alloy (AA-6061) chips via hot extrusion followed by equal channel angular pressing (ECAP)," *The Egyptian International Journal of Engineering Sciences and Technology*, vol. 21, pp. 33–42, 2016.

[28] M. Hu, Z. Ji, X. Chen, and Z. Zhang, "Effect of chip size on mechanical property and microstructure of AZ91D magnesium alloy prepared by solid state recycling," *Materials Characterization*, vol. 59, no. 4, pp. 385–389, 2008.

9

Repairing 2024 Aluminum Alloy via Electrospark Deposition Process

G. Renna,[1] P. Leo,[1] G. Casalino,[2] and E. Cerri[3]

[1]*Innovation Engineering Department, University of Salento, Via per Arnesano s,n, 73100 Lecce, Italy*
[2]*DMMM, Politecnico di Bari, Viale Japigia 182, 70126 Bari, Italy*
[3]*Department of Engineering and Architecture, University of Parma, v.le G. Usberti 181/A, 43124 Parma, Italy*

Correspondence should be addressed to P. Leo; paola.leo@unisalento.it

Academic Editor: Antonio Riveiro

The electrospark deposition (ESD) technique has been studied as a potential method to repair locally damaged 2024 rolled sheets supplied in natural-aged (T4) and artificial-aged (T6) conditions. The 2024-T4 and 2024-T6 tensile samples were first notched, and then the notches were filled (repaired) by ESD with the same aluminum alloy. The effect of process parameters on the microstructure of the filling material and the substrate properties was studied by optical and scanning electron microscopy. Tensile and hardness tests were performed. The tensile test showed that T4 and T6 as-repaired specimens had low tensile properties, which was due to defectiveness and residual stress caused by high cooling rate during reparation. However, the as-repaired specimens were heat-treated at either 135°C or 190°C to improve the mechanical properties. A better yield strength was observed for the T4 heat-treated alloy. The ductility and ultimate tensile strength did not change, being mainly affected by voids and microcracks.

1. Introduction

The need to reduce operating, maintenance, and waste-related costs of an aircraft has prompted aeronautical industries to develop new repair procedures for aeronautical components. Besides, because of the extreme cost of materials and labor for fabricating high-value component, it is also necessary to repair all those components which are affected milling ball indentation and cutter pull out, which occur during the manufacturing process [1, 2]. New repairing technologies should overcome the limitations of the actual ones and be able to operate in the maintenance facility, which would avoid transporting it in a repairing workshop or sending the worn-out part to the original manufacturer [3].

Aluminum alloy 2024 was a forerunner of a variety of 2xxx series alloy that is widely applied in the aerospace industry for its superior properties such as high strength-to-weight ratio, very good resistance to fatigue crack growth, and good fracture toughness [4, 5]. It is commonly supplied in the T4 or T6 temper since such treatments give good mechanical properties. Generally, the 2024 parts used in aviation for damage tolerance applications, such as the lower wing skins and fuselage structure of commercial aircraft, usually easily undergo localized corrosion (i.e., pitting and exfoliation) especially in chloride-containing environment [6, 7].

Unfortunately, the conventional fusion welding methods (i.e., tungsten inert gas welding (TIG), plasma transferred arc and electron beam welding, and laser cladding) are not suitable for repairing this alloy [1, 5]. It has been proved that several problems like solidification cracking, grain boundary melting in the base metal (liquation cracking), and changes in metallurgical structure would arise mainly due to the high thermal process input [8–14].

Electrospark deposition (ESD) technology is an effective method for the deposition of anticorrosion and wear-resistance coatings. It is a low heat input pulsed micro-bonding process that uses high-energy density and short duration of electrical pulses, typically ranging from a few

Table 1: Chemical composition (wt. %) for the alloy AA2024.

Composition	Ti	Zi	Fe	Cr	Si	Cu	Mn	Mg	Al
AA2024	<0.15	<0.25	<0.50	<0.5	<0.5	3.8–4.9	0.3–0.9	1.2–1.8	Bal.

microseconds to milliseconds, to deposit the electrode material onto the component's surface. It appears to be a very interesting and economic solution for the restoration and refurbishment of worn or damaged high-valued parts, especially those materials ordinarily considered poor-weldable by conventional repairing processes [2, 15–21].

In this technique, an electrical arc is pulsed between a rotating consumable electrode (anode) and the workpiece/substrate (cathode) to produce the deposit by detaching small droplets of material from the electrode and ejecting towards substrate. During the process, the electrical arc is pulsed with a frequency varying between 0.1 and 4 KHz using a high-frequency power supply [22, 23].

The short pulse duration and high-pulse frequency induce both very fast cooling rate and a rapid solidification process in combination with a small amount of material transferred during each pulse ESD technology. This rapid solidification rate, believed to approach 10^5-$10^{6\circ}$C/s, results in the refinement of the microstructure to either the nanostructured or amorphous level [23]. The solidification mode of the deposited alloy, as well as the resulting grain morphology, is controlled mainly by the thermal condition that exists at the beginning of the solidification. Otherwise, the fine scale microstructure is controlled primarily by the postsolidification cooling rate [17].

This innovative technology has gained much interest in the repairing field since it can offer a variety of advantages including low pollution, technical feasibility for metal deposition in atmospheric pressure, and welding equipment handling and transportability. Moreover, the reduction of residual stress and distortion compared to the other fusion processes renders this technique highly applicable to those materials possessing a high tendency to crack during welding, such as superalloys, MCrAlY, and thermally sensitive materials (2000 series aluminum alloy) [22–24].

Many works have studied the mass transfer that occurs during the ESD process [25–27], which is paramount to produce a high wear-resistant coating [28–32]. The relationship between process parameters and the deposit structure has been also investigated [33–36]. Only few studies are related to the utilisation of ESD as a repair technology. Farhat [24] determined that the repair of a locally damaged NiCoCrAlY or CoNiCrAlY coating can be done by ESD to further extend the lifetime of coatings. While Johnson showed the possibility of repairing a single crystal turbine blade with low net heat input free from distortion [37].

Besides, to date investigations focused on assessment of ESD as repair technology, not concerning thermally sensitive materials such as the 2024 alloy.

In this paper, a comparative study between deposits prepared by ESD on AA2024 standard dog-bone substrates both in natural-aged (T4) and artificial-aged (T6) states has been presented. Detailed analyses of the microstructure modification of the filling alloy and the effect of process parameters on the substrate properties and microstructure were performed.

Furthermore, the as-repaired specimens were heat treated at 135°C and 190°C to reduce the detrimental effect of residual stresses on the mechanical properties. Hence, also the effect of postrepair heat treatment (PRHT) on the microstructure and mechanical properties of the repaired specimens was investigated.

2. Experimental Methods

AA2024 aluminum alloy was used for the electrode and the substrate materials. ESD substrate was supplied as T4 and T6 rolled plates. Chemical components (wt. %) of the AA2024 alloy are listed in Table 1.

Dog-bone specimens were realized from 4 mm thick plates (Figure 1(a)). A notch was machined in the midsize of the dog-bone specimens to be later filled up with the electrode material, as shown in Figure 1(b). Deposition was carried out using a hand-held gun at room temperature with argon (Ar) protection, which avoids contamination of the deposit zone by interstitial elements such as oxygen or nitrogen. During deposition, the electrode rotated at 1200 rpm speed. The electrospark deposition process parameters are shown in Table 2, where E is the value of the electric arc pulse energy in Joule. These parameters are based on a previous study of Leo et al. [36].

The substrate was kept in a fixed position, while the hand-held gun was moving according to three orthogonal XYZ axes. At the beginning, the semicircular shape of the notch was coated by adjacent depositions. The depositions had circular shape to fit the semicircular shape of the notch. The close semicircular depositions were realized to cover the entire surface of the notch along the y-axis. The filling was obtained by superimposing several depositions along the Z direction (Figure 1(b)). Then, the samples were postrepair heat treated (PRHT) for 2 h at 135°C and for 6 h at 190°C. All samples were air-cooled. Holding times and temperatures are displayed in Table 3. Samples were sectioned across the deposit center, embedded in epoxy resin and then polished to a mirror finish. Analyses were performed on the XZ section in the as-polished condition to assess the metallurgical health within the deposit and in the bonding zone. The microstructure was revealed by etching with Keller's reagent. Three complementary methods were used to study the microstructure: stereo microscopy (SM), optical microscopy (OM), and scanning electron microscopy (SEM) with EDS. The grain structure of substrate was revealed by electropolishing (20% perchloric acid and 80% ethanol at 0°C; electropolishing parameter: 15 V and 60 s), anodic

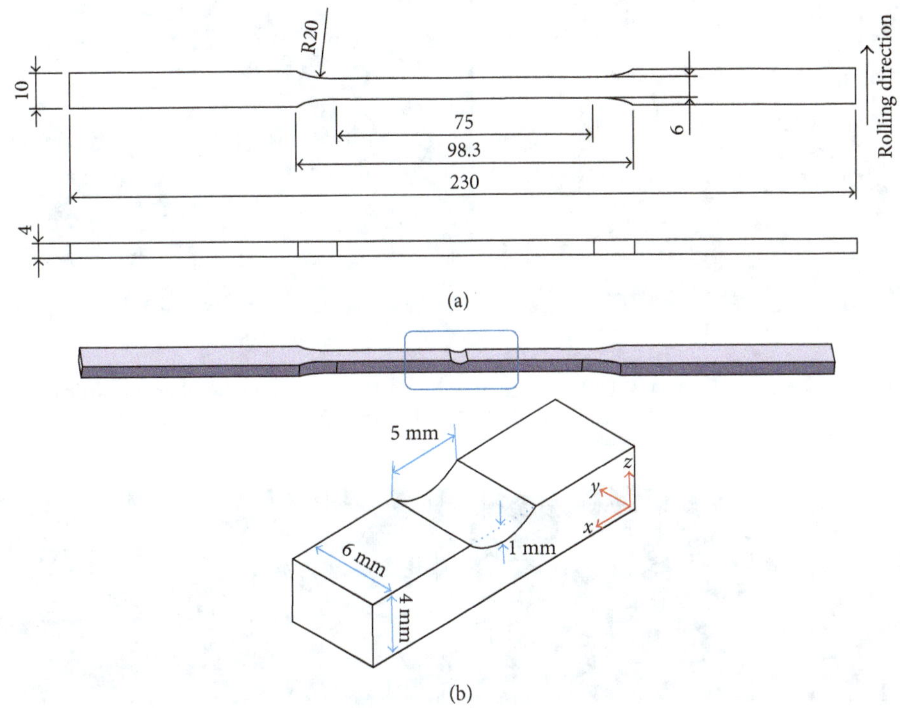

FIGURE 1: (a) Sizes of the dog-bone specimen for tensile tests and (b) draft of defect dimensions machined into dog-bone specimens. These defects will be machined into one-half of the specimens and subsequently filled with ESD.

TABLE 2: Process parameters used in the experiments.

E (J)	Ar (l/min)	Electrode rotation speed (rpm)
0.9	20	1200

TABLE 3: PRHT on dog-bone specimens of AA2024 in both T4 and T6 conditions.

	Temperature (°C)	Holding time (h)	Cooling
T_{135_2}	135	2	Air cooling
T_{190_6}	190	6	Air cooling

oxidation (Barker etching; anodizing parameters: 20 V and 80 s), and subsequent investigation under polarized light in OM.

The defect areas inside the deposit and the length of cracks at the substrate/deposit interface on the XZ plane (Figure 1(b)) were obtained using the NIS software for imaging analysis. NIS-Elements is a NIKON software supplied with Epiphot 200 OM. Defects area percentage was measured as the sum of the defect areas divided by the area of the repaired zone. Moreover, using ImageJ software, both Feret's diameter and the shape factor (f_{circle}) of the filling defects have been evaluated. The shape factor comes from the following equation:

$$f_{\text{circle}} = \frac{4\pi A}{P^2}, \tag{1}$$

where P is the perimeter and A is the area of defect. For a perfect spherical defect, the shape factor is one, while for a strongly elongated defect, it is very close to zero. Particularly, two main morphological classes of defects were considered in this study:

(i) Defects with $0 < f_{\text{circle}} \leq 0.5$

(ii) Defects with $0.5 < f_{\text{circle}} \leq 1$

Microhardness tests were performed with the load of 100 g for 15 s. Microhardness measurements were carried out on the XZ plane following a grid arrangement parallel to the top surface of the deposit in accordance with ISO 6507-1: 2005. The distance between the indentations both in the horizontal and vertical direction was 100 μm. Then, the average values of microhardness were calculated considering all the indentation made in the deposit zone and in the BM, respectively. These measures were carried out on both as-repaired and PRHT samples. Tensile tests were carried out, at room temperature, in a universal testing machine in accordance with ISO 6892-1:2009 perpendicularly to the rolling direction (RD). The displacement control, at an initial strain rate of 0.25 mm/min, was used. Strains were measured with an axial extensometer with a gauge length of 12.5 mm.

3. Results and Discussion

3.1. Base Material Characterization. In AA2024-T4 alloy, the hardening mechanism is due to the Guinier–Preston zone (G–P zone) in the grains that hinder dislocation movement. While the AA2024-T6 reaches the peak hardness due to a finer transitional precipitates distribution that remains coherent with the solid-solution matrix, it contributes to precipitation strengthening. Both G–P zone (~10 nm thick

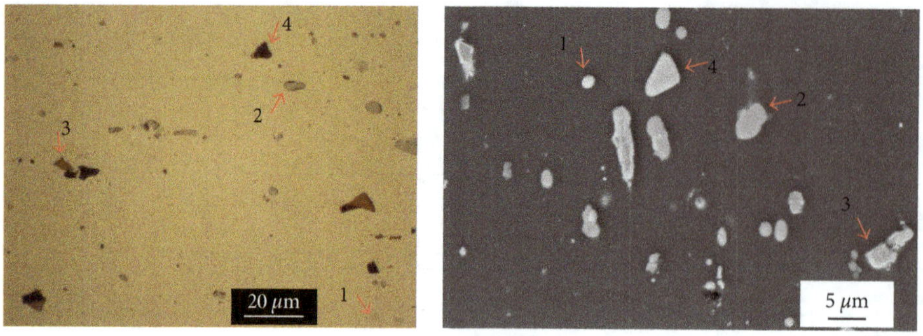

(a) (b)

FIGURE 2: (a) OM and (b) SEM image of AA2024 BM after chemical etchant: (1) mainly roughly spherical particles Al_2CuMg, (2) light grey particles $CuAl_2$, (3) light brown particles (Mn, Fe)$_3SiAl_{12}$, and (4) dark brown phase Al(Cu, Fe, Mn).

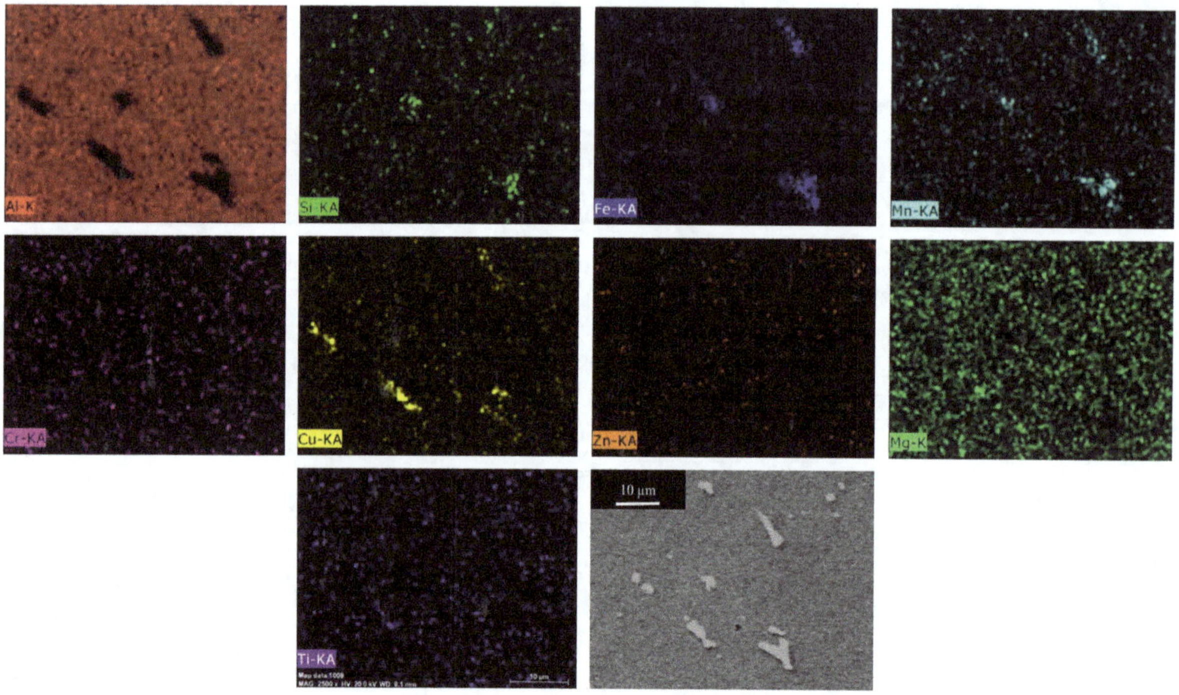

FIGURE 3: Elemental map of 2024 BM. The analysis confirms the chemical composition of typical intermetallic particles.

and 100 nm in diameter) and the precipitate particles (larger than G–P zone) are too small to be clearly resolved by OM or SEM microscopy [4, 5, 38]. Figure 2(a) presents the microstructure of the AA2024 alloy base material (BM) in the perpendicular direction to rolling manufacturing. OM and SEM micrographs of 2024, in T4 (Figure 2(a)) and T6 (Figure 2(b)) states, respectively, show the aluminum matrix together with a series of intermetallic particles of different morphologies and sizes. According to the literature and the SEM/EDS analysis, the main spherical intermetallic particles are Al_2CuMg (S-phase) (1, Figures 2(a) and 2(b)), while the light grey particles are Al_2Cu (θ phase) (2, in Figures 2(a) and 2(b)). Moreover, (Mn, Fe)$_3SiAl_{12}$ (light brown, 3, Figures 2(a) and 2(b)) and Al(Cu, Fe, Mn) phases (dark brown, 4, Figures 2(a) and 2(b)) were detected. They were as large as 10 μm and had an irregular shape. Furthermore, according to the literature, the fine dispersoids in the Al matrix can be either $Cu_2Mn_3Al_{20}$ or Al_2Cu [5, 39]. The element mapping (Figure 3) of the BM confirmed the presence of those phases.

The mechanical properties of the base material both in the T4 and T6 conditions, at room temperature, are presented in Table 4 [5].

3.2. Defectiveness and Microstructure within the Fillings. Figures 4(a) and 4(b) show the SM pictures of the notch after filling with the ESD process. Figure 4(c) is an OM micrograph of a *XZ* plane of the electrospark filling. It can be clearly seen that the top surface was continuous and without visible cracks or other defects (Figures 4(a) and 4(b)). However, some defects are shown in Figure 4(c). The main defects detected in the analyzed fillings were voids with different morphologies and microcracks that extended both perpendicularly and parallel to the substrate/filling interface. In particular, the typical voids identified within the fillings had

TABLE 4: Mechanical properties of base material (BM) in T4 and T6 conditions.

Material	Yield strength (MPa)	Ultimate tensile strength (MPa)	Elongation (%)	Vickers hardness ($HV_{0.1/15}$)
AA2024-T4	339	485	29	141 ± 1.1
AA2024-T6	482	516	16,4	158 ± 2.6

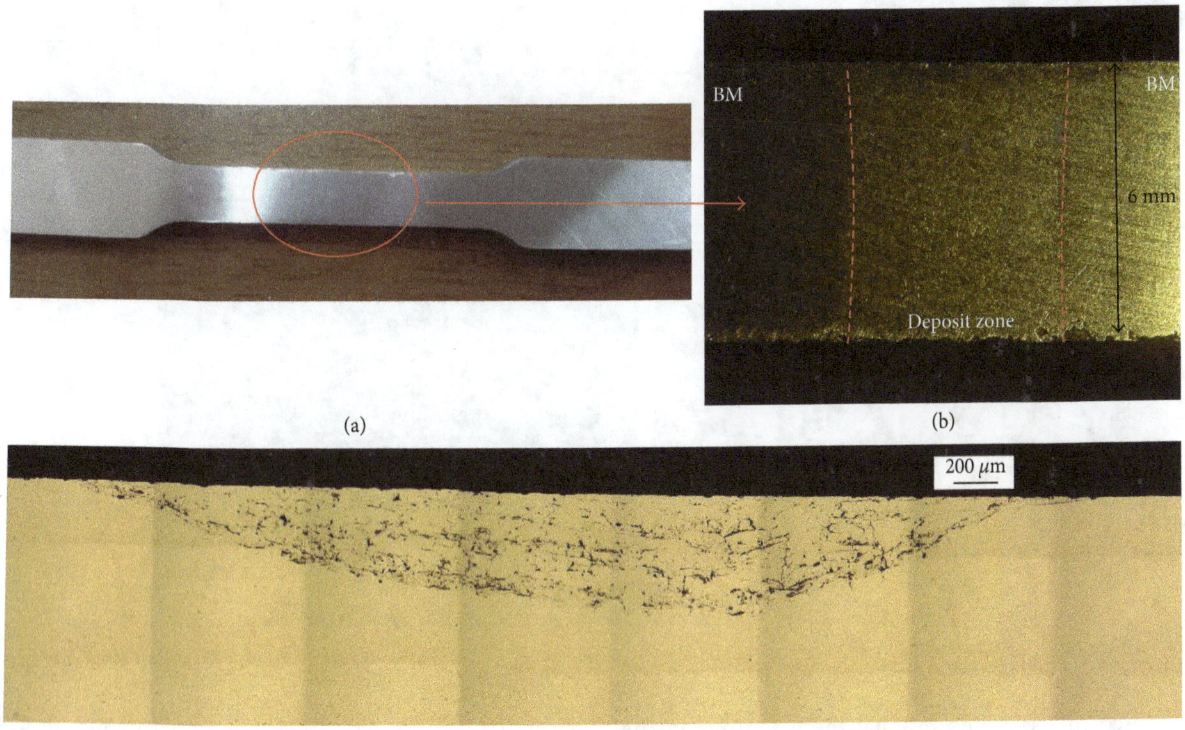

FIGURE 4: (a, b) The repair from top view and (c) typical cross section micrograph of electrospark deposit.

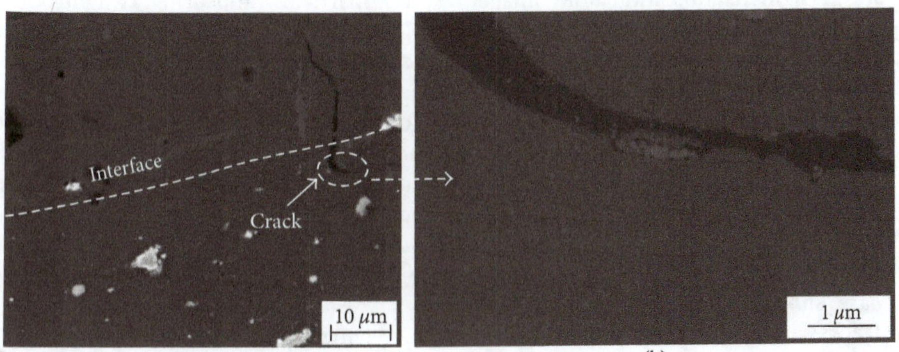

...s showing cracking behavior in the deposit/base metal interface for the sample repaired and (b) SEM image ...ack paths by resolidified products, typical of liquation cracking.

... (gas porosity), ... and thin ... these

electrode did not remain constant during the deposition process [19, 36]. All the previous aspects changed the quantity of the metal that melted and/or evaporated. Therefore, peaks and valley formed on the surface, which appeared corrugated and showed voids [19, 30, 31]. Cracking ...in the filling is believed to be the result of the thermal ...hich builds up during cooling and solidification of

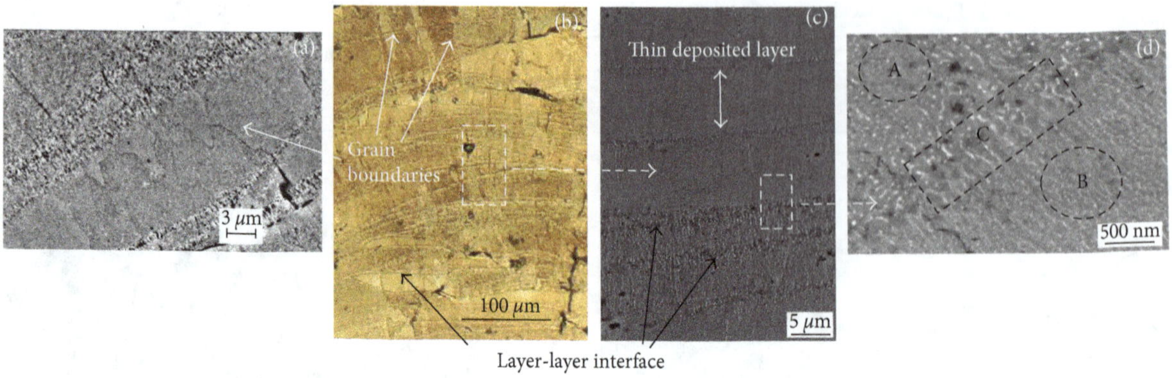

FIGURE 6: OM and SEM images of microstructure parallel to build direction: (b, c) layered deposit (morphology cross section), (a) fine microstructure to layer-layer interface and within of the layers, and (d) long, slender columnar-like grains.

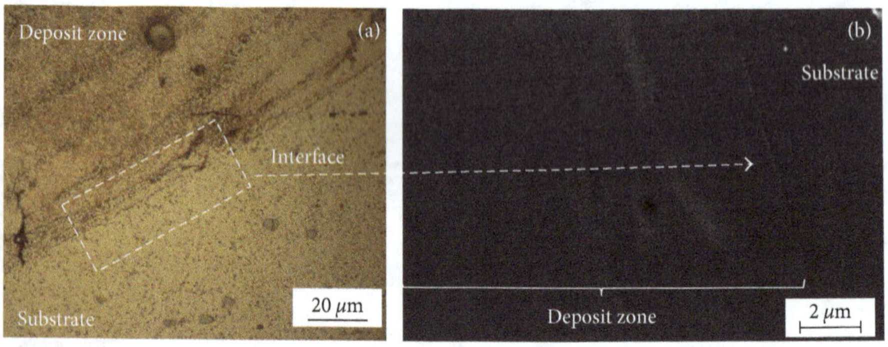

FIGURE 7: (a) OM and (b) SEM micrographs showing the deposit/base metal interface for the sample repaired.

the deposited material. The residual thermal stress in the filling was always tensile [40, 41]. Moreover, from the analysis of the substrate/filling interface, it was observed also that both the substrates in T4 and T6 states were characterized by cracks that preferentially extended along the grain boundaries (Figure 5(a)). High magnification SEM image outlined the crack path by the resolidified products, which are typical of the liquation cracking phenomenon (Figure 5(b)). Simultaneously, the rapid cooling from the ESD process induced a great thermal strain that opened the liquated zones. In fact, 2024 aluminum alloy is susceptible to solidification and liquation cracking having large range of solidification, high thermal expansion, large shrinkage, and many alloying elements [10, 41].

Optical and scanning electron micrographs on the *XZ* plane of the filling are shown in Figure 6. Numerous superimposed linear tracks can be observed clearly (Figures 6(b) and 6(c)) due to the remelting which was caused by the superimposed layer (layer-layer interface). Moreover, these layers differed in size and overlapped one over the other. The random size layers just revealed the instability in the manual deposition process. The average thickness of the several deposited layers was calculated through the application of intercept method. It was then determined to be $\sim 12.4 \pm 6.1\,\mu$m. In addition, columnar grain structure grew perpendicularly to the substrate (Figures 6(a) and 6(b)). These columnar grains grew epitaxially along the crystallographic direction of previously solidified grain, through multiple layers, producing microstructure normal to the deposition layers [10, 36, 38, 42]. The coating microstructure was fine due to the high cooling rate [15, 16]. Moreover, some layers were characterized by a cellular dendritic microstructure (highlighted by the circle B in Figure 6(d)), and some other layers showed a uniform distribution of fine small second phase particles (appearing as light point in the circle A of Figure 6(d)). Moreover, the interlayer thin zone exhibited a coarse microstructure, generally equiaxed (highlighted by the box C in Figure 6(d)). These different microstructure morphologies in thin deposit layers can be attributed to the variable solidification conditions in the volume of the molten splat (liquid phase) [16, 43]. The interface between the filling and substrate is shown in Figure 7. The filling formed a metallurgical bound with the substrate (Figure 7).

3.3. Analysis of Defects. The defects in the filling material and at the filling/substrate interface were examined in detail. In the filling, for all the samples obtained, some randomly distributed defects with spherical and laminar shapes were observed in the deposit volume. The average area fraction of these defects was 2.5%, which can be considered acceptable. Figures 8(a) and 8(b) display the plots of the Feret diameter as a function of the shape factor (f_{circle}) and the
distribution of the defects having shape f
tween 0 and 0.5 and 0.5 and 1
Figure 8(a), the defect
ones and th

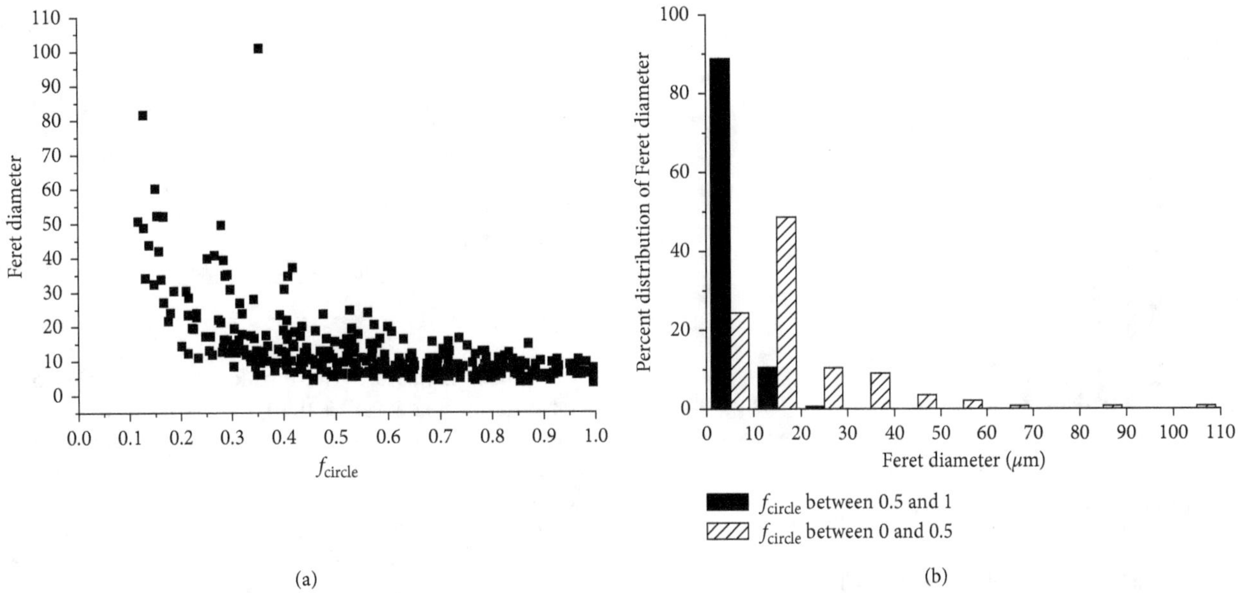

Figure 8: Analysis of defects within the volume of the deposit obtained with the electrical parameters is reported in Table 2. (a) Feret diameter versus f_{circle} of defects and (b) Feret diameter distribution of the defects having shape factor (f_{circle}) between 0 and 0.5 and 0.5 and 1.

Feret diameter distribution of the defects shown in Figure 8(b) illustrated how the almost totality of defects with f_{circle} between 0 and 0.5 had length shorter than 20 μm. In addition, 50% of them had diameters between 10 μm and 20 μm, but with defects of considerable extension. The tensile thermal stress increased during the growth of the filling, causing large cracks in the filling and delamination at the weak interface bond. The small size voids were created by the formation of asperities or trapped gas. Figure 8(b) shows that spherical defects (f_{circle} 0.5–1) were mostly distributed (90%) in the lower class of the Feret diameters (up to 10 μm) while the remaining had Feret diameters between 10 and 30 μm. Regarding the defects at the substrate/filling interface, for the two different treatment conditions of the BM (T4 and T6 states), the values of the average crack length and the ratio between the crack number (N°_{tot}) and interface length (mm) examined are obtained and reported in Table 5. In addition, Figure 9 shows the distribution of the length of cracks that propagated in the substrate from the substrate/filling interface in the BM state T4 and T6. It is noteworthy that the ESD process generated the same N°_{tot}/mm value regardless of the BM state (T4 or T6) and a considerable rise of the average crack length in the T6 BM. The increase in the average crack length observed in 2024-T6 sample was due to the high hardness of the substrate. Once triggered, the cracks propagated more easily thanks to the brittleness of the region. In fact, observing the crack length distribution plots, the deposition on the 2024-T6 sample produced lower crack percentage for length fields below 15 μm and a rise in the percentage of longer cracks with consequent rise in the crack length field variation. Therefore, the depositions made on the 2024-T6 induced greater defects at the interface than the depositions made on the 2024-T4.

Table 5: Average crack length and ratio between the cracks number and interface length examined both in the case in which the filling was performed on the base material in T4 and T6 states.

	Average crack length (μm)	N°_{tot} (mm)
2024-T4	11.3 ± 7.2	10.7
2024-T6	26.5 ± 16	11.14

3.4. Mechanical Characterization

3.4.1. Microhardness. Figure 10 shows the microhardness profiles of the fillings and AA2024 substrate both in T4 and T6 conditions. The graph shows the average microhardness values detected at increasing distances from the substrate/deposit interface. Table 6 shows the average microhardness. In addition, the average microhardness values of the substrate measured at 100 μm from the coating interface are indicated. The hardness of the filling was generally lower than that of the substrate in all the samples examined (Figure 10 and Table 6). The decrease of the hardness was due to the aging in the base material and to some defects, which formed inside the coating. Moreover (Figure 10), in the repaired zone, the average microhardness values remained almost constant, which confirm that there were no significant microstructural changes along the thickness of the repaired zone. The results in Table 6 suggest that the microhardness of the substrate close to the interface did not vary compared to that of the BM zone. Therefore, the mechanical properties of the substrate were not altered. Furthermore, the results showed that the postrepair heat treatment (PRHT) performed on samples had no effect on the hardness of the filling. In fact, it remained almost constant regardless of the heat treatment temperature employed.

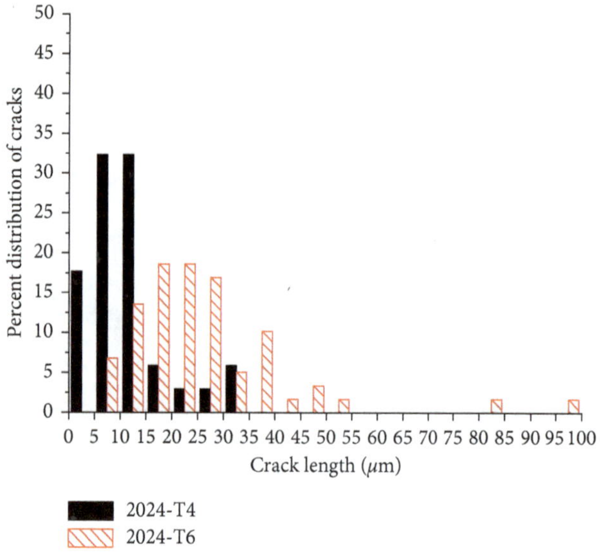

Figure 9: Analysis of defects to substrate/filling interface: distribution of the crack length that propagates in the substrate from the interface both in the case in which the filling was performed on the base material in T4 and T6 states.

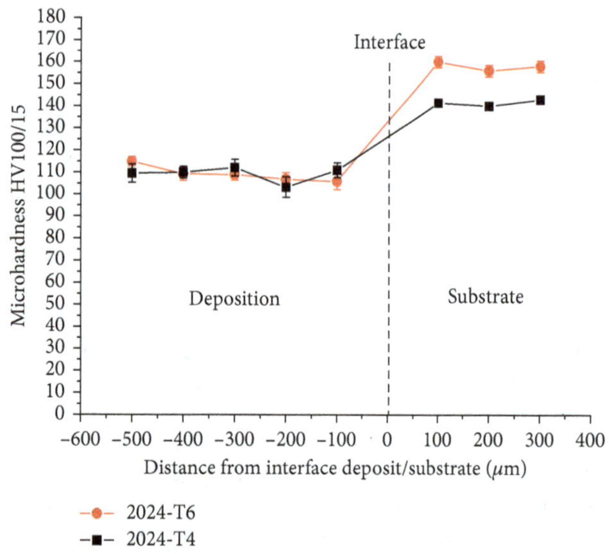

Figure 10: Microhardness measurement through the filling layer and AA2024 substrate both in T4 and T6 conditions.

The hardness of 2024-T4 substrate, that was underaged, increased after the T_{190_6} PRHT due to the precipitation of more hardening particle [38]. Due to overaging, the hardness value of the 2024-T6 substrate, that was peak aged, decreased with the increasing time and temperature of the heat treatment. In fact, it is known that if precipitate particles are large and widely spaced, they can be readily bypassed by moving dislocations which bow out between them and rejoin by a mechanism first proposed by Orowan [4].

Table 6: Average hardness value ($HV_{100/15}$) with the locations each sample.

Sample	Filling	Substrate	Interface
2024-T4 repaired	109.5 ± 5	141 ± 1.1	140 ± 1.1
2024-T4 repaired + T_{135_2}	107 ± 5	141 ± 2.8	140 ± 2.8
2024-T4 repaired + T_{190_6}	104 ± 4	148 ± 1.8	149 ± 1.8
2024-T6 repaired	109.5 ± 5	158 ± 2.6	157 ± 2.6
2024-T6 repaired + T_{135_2}	107.7 ± 5	156 ± 3.4	156 ± 3.4
2024-T6 repaired + T_{190_6}	107.3 ± 6.9	151 ± 1.2	150 ± 1.2

3.4.2. Tensile Behavior. The tensile test was carried out to determine the behavior of the repaired specimens. Table 7 contains the values of the yield stress (σ_y), ultimate tensile strength (UTS), the load at fracture, strain at fracture (elongation e_f), and cross-sectional area (S_0) for as-received BM in T4 and T6 states, as-received BM notched, and repaired specimens both in untreated and treated T_{135_2} and T_{190_6} states. The stress-strain curves obtained are shown in Figure 11. From the results of the tensile tests, both as-repaired and PRHT samples displayed a decay of the mechanical properties compared BM (both notched and unnotched 2024 samples). This was probably caused by the presence of widespread defects inside the deposit and at the substrate/deposit interface as well as by different microstructural characteristics of the deposit and the BM. The yield stress (σ_y) and the ultimate tensile strength (UTS) of the as-repaired samples and PRHT samples were lower with respect to the notched samples. At the same time, a reduction in ductility owing to the defects present within the deposit and at the interface was observed. Porosity in the deposit led to stress concentration, so the crack easily started and propagated throughout the sample thickness. Moreover, the cracks at the interface induced the low tensile strength rupture. An increase of the mechanical properties (Table 7) was obtained by the high temperature and long duration PRHT treatment. The yield strength and UTS of the PRHT samples were greater with respect to the as-repaired samples. The T_{135_2} heat treatment increased the tensile strength of the 2024-T4 samples of 17 MPa and the yield strength of 34 MPa, while the T_{135_2} heat treatment increased the tensile strength of the 2024-T6 samples of 16 MPa and the yield strength of 39 MPa. Likewise, the T_{190_6} heat treatment increased the tensile strength of the 2024-T4 samples of 16 MPa and the yield strength of 72 MPa, while the T_{190_6} heat treatment increased the tensile strength of the 2024-T6 samples of 18 MPa and the yield strength of 46 MPa. The improvements of σ_y and UTS were primarily due to the reduction of the residual stresses and/or to the increasing of hardness due to heat treatment. The role of the residual stress reduction was particularly effective in the case of the PRHT performed at 190°C on the 2024-T6 sample in which the rise in the yield strength was associated with the overaging of precipitated particles (Table 6), which leads to softening of microstructure. So, the PRHT at 190°C was more effective when carried out on the 2024-T4 sample since, in addition to a residual stress reduction, the heat treatment also led to an increase of the substrate hardness.

Table 7: Tensile test for the as-received BM in T4 and T6 states, as-received BM notched, the untreated repaired samples, and heat-treated T_{135_2} and T_{190_6} repaired samples.

Sample	σ_y (MPa)	F_y (N)	UTS (MPa)	Maximum load (N)	e_f (%)	S_0 (mm^2)
2024-T4	339	8315.7	485	11,893	29	24.53
2024-T4 notched	346	7480.52	503	10,877	13.3	21.62
2024-T4 notched T_{135_2}	213.3	4645.7	487.2	10610.9	12	21.78
2024-T4 repaired	257	6198.84	402	9696.24	9	24.12
2024-T4 repaired + T_{135_2}	291	7245.9	419	10,440	10.1	24.9
2024-T4 repaired + T_{190_6}	329.2	7966.64	442.5	10,202.2	5.51	24.2
2024-T6	482	11,852.4	516	12,691	16.4	24.59
2024-T6 notched	482	10,488.3	532	11,585	4.5	21.76
2024-T6 repaired	372	8965.2	439	10,579.9	4	24.1
2024-T6 repaired + T_{135_2}	411	9913.32	455	10,973	4.2	24.12
2024-T6 repaired + T_{190_6}	418.8	10,302.5	457.29	11,238.6	3.24	24.6

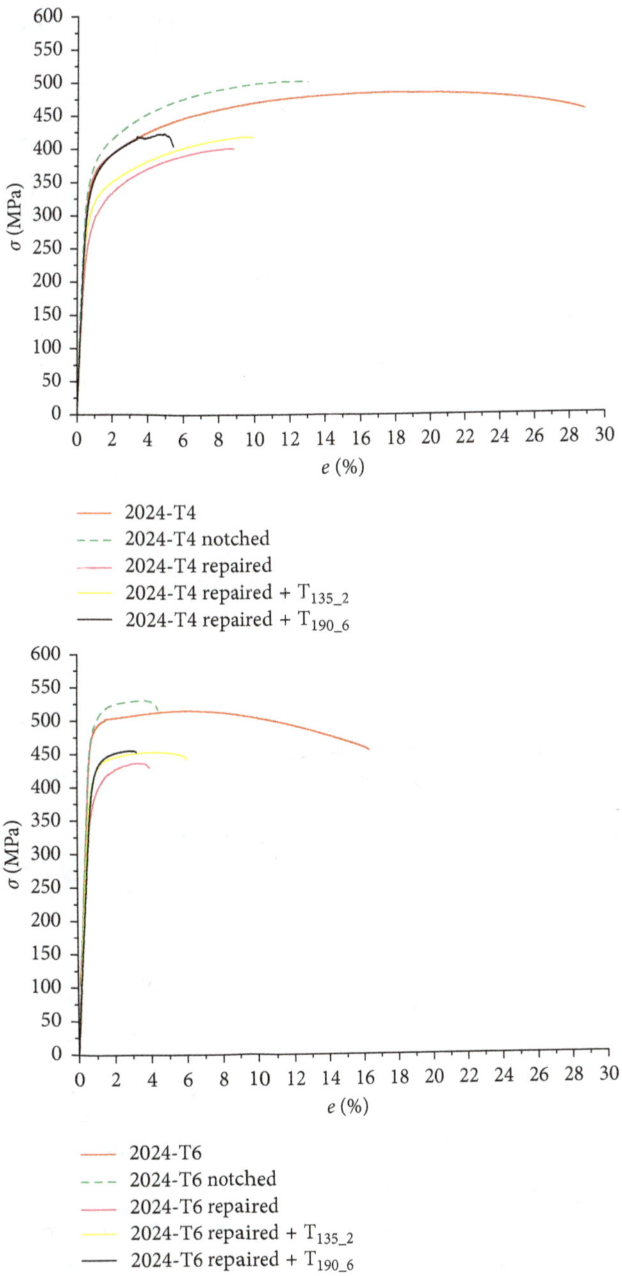

Figure 11: Tensile stress-strain curves.

4. Conclusion

The study of the 2024 aluminum alloy electrospark deposition process on a similar substrate brought the following evidences:

(1) The filling microstructure was layer by layer, very fine, and mixed with both cellular and equiaxed dendrites. In the filling layers, grains formed long columnar grains with a preferential growth along the build-up direction. The microstructure was strongly inhomogeneous due to the different cooling rates during the deposition process. The hardness of the deposited metal was lower than that of the substrate. The substrate did not present a heat affected zone.

(2) The reparation showed the presence of some defectiveness. Those defects, randomly distributed in the microstructure had spherical, casual, and laminar shapes. The average area fraction of these defects was 2.5% which is considered acceptable. It was estimated that the fillings had 63% of laminar and 37% spherical shape defects. The formation of defects with laminar morphology was mainly attributed to the tensile residual stresses. The length size of the cracks to interface was higher for the artificial aged deposition.

(3) The tensile properties of the repaired samples were lower compared to the notched ones (for both T6 and T4 states) due to the residual stress and defectiveness. The PRHT improved the yield strength of the repaired specimen by reducing residual stress. However, the ductility of the 2024 alloy was strongly reduced by the reparation process. PRHT at 190°C was most effective when carried out on 2024-T4 since, in addition to a residual stress reduction, the heat treatment also led to a rise in the substrate hardness.

Conflicts of Interest

The authors declare that they have no conflicts of interest.

References

[1] M. Brandt, *Laser Additive Manufacturing: Materials, Design, Technologies, and Applications*, Woodhead Publishing, Cambridge, UK, 2016.

[2] H. Paydas, A. Mertens, R. Carrus, J. Lecomte-Beckers, and J. Tchoufang Tchuindjang, "Laser cladding as repair technology for Ti–6Al–4V alloy: influence of building strategy on microstructure and hardness," *Materials and Design*, vol. 85, pp. 497–510, 2015.

[3] N. Price, *AESF/EPA Conference for Environmental and Process Excellence*, Environmental Protection Agency, Lake Buena Vista, FL, USA, 2004.

[4] I. J. Polmear, *Light Alloys: Metallurgy of the Light Metals*, Butterworth Heinemann, Oxford, UK, 1995.

[5] J. R. Davis, *ASM Specialty Handbook: Aluminum and Aluminum Alloy*, ASM International, Russell Township, OH, USA, 1993.

[6] F. C. Campbell, *Manufacturing Technology for Aerospace Structural Materials*, Elsevier, Amsterdam, Netherlands, 2006.

[7] C. M. Abreu, M. J. Cristòbal, P. Freitas et al., "Microstructure of the passive layer formed on AA2024-T3 aluminum alloy surface implanted with nitrogen," *Surface and Interface Analysis*, vol. 40, no. 3-4, pp. 290–293, 2008.

[8] R. Onjukka, "Welding vs riveting–which has the fatigue life for airplanes," *Welding International*, vol. 75, pp. 29–33, 1996.

[9] S. Kou, "Welding Metallurgy and Weldability of High Strength Aluminum Alloys," *Welding Research Council Bulletin*, vol. 320, pp. 1–20, 1986.

[10] M. F. Ghaini, M. Sheikhi, M. J. Torkamany, and J. Sabbaghzadeh, "The relation between liquation and solidification cracks in pulsed laser welding of 2024 aluminium alloy," *Materials Science and Engineering: A*, vol. 519, pp. 167–171, 2009.

[11] A. F. Norman, V. Drazhner, and P. B. Prangnell, "Effect of welding parameters on the solidification microstructure of autogenous TIG welds in an Al–Cu–Mg–Mn alloy," *Materials Science and Engineering: A*, vol. 259, pp. 53–64, 1999.

[12] W. R. Oates, *Materials and Applications*, vol. 3, AWS, Miami, FL, USA, 8th edition, 1972.

[13] H. Zhao, D. R. White, and T. DebRoy, "Current issues and problems in laser welding of automotive aluminium alloys," *International Materials Reviews*, vol. 44, pp. 238–266, 1999.

[14] X. Cao, W. Wallace, J. P. Immarigeon, and C. Poon, "Research and progress in laser welding of wrought aluminum alloys. II. Metallurgical microstructures, defects, and mechanical properties," *Materials and Manufacturing Processes*, vol. 18, no. 1, pp. 23–49, 2003.

[15] Y. J. Xie, M. C. Wang, and D. W. Huang, "Comparative study of microstructural characteristics of electrospark and Nd: YAG laser epitaxially growing coatings," *Applied Surface Science*, vol. 253, no. 14, pp. 6149–6156, 2007.

[16] Y. J. Xie and M. C. Wang, "Epitaxial MCrAlY coating on a Ni-base superalloy produced by electrospark deposition," *Surface and Coatings Technology*, vol. 201, no. 6, pp. 3564–3570, 2006.

[17] Y. J. Xie and M. C. Wang, "Microstructural morphology of electrospark deposition layer of a high gamma prime superalloy," *Surface and Coatings Technology*, vol. 201, no. 3-4, pp. 691–698, 2006.

[18] C. J. Chen, M. C. Wang, D. S. Wang, and Y. M. Liu, "Study on corrosion characteristic of high-energy micro-arc alloying of Al-Y electrode on AZ31 magnesium alloy," *Transactions of Materials and Heat Treatment*, vol. 28, pp. 106–110, 2007.

[19] S. Frangini and A. Masci, "A study on the effect of a dynamic contact force control for improving electrospark coating properties," *Surface and Coatings Technology*, vol. 204, no. 16-17, pp. 2613–2623, 2010.

[20] S. Frangini, A. Masci, and A. Di Bartolomeo, "Cr7C3-based cermet coating deposited on stainless steel by electrospark process: structural characteristics and corrosion behavior," *Surface and Coatings Technology*, vol. 149, no. 2-3, pp. 279–286, 2002.

[21] S. K. Tang, T. C. Nguyen, and Y. Zhou, "Materials transfer in electro-spark deposition of TiCp/Ni metal-matrix composite coating on Cu substrate," *Welding Research*, vol. 89, 2010.

[22] R. N. Johnson, "Robust coatings for corrosion and wear: the electrospark deposition process," in *Proceedings of the Tri-service conference on corrosion*, Battelle, Columbus, OH, USA, pp. 341–410, 1999.

[23] M. Brochu, D. W. Heard, J. Milligan, and S. Cadney, "Bulk nanostructure and amorphous metallic components using the electrospark welding process," *Assembly Automation*, vol. 30, no. 3, pp. 248–256, 2010.

[24] R. Farhat, *Repair of Damaged MCrAlY Coatings Targeting Petroleum Industry Applications*, McGill University, Montreal, QC, Canada, 2012.

[25] I. V. Galinov and R. B. Luban, "Mass transfer trends during electrospark alloying," *Surface and Coatings Technology*, vol. 79, no. 1–3, pp. 9–18, 1996.

[26] A. Lesnjajk and J. Tusek, "Processes and properties of deposits in electrospark deposition," *Science and Technology of Welding and Joining*, vol. 7, no. 6, pp. 391–396, 2002.

[27] C. Changjun, W. Maocai, L. Yiming, W. Dongsheng, and J. Ren, "Mass transfer trends and the formation of a single deposition spot during high-energy micro-arc alloying of AZ31Mg alloy," *Journal of Materials Processing Technology*, vol. 198, no. 1–3, pp. 275–280, 2008.

[28] W. F. Wang, M. C. Wang, F. J. Sun, Y. G. Zheng, and J.-M. Jiao, "Microstructure and cavitation erosion characteristics of Al-Si alloy coating prepared by electrospark deposition," *Surface & Coatings Technology*, vol. 202, no. 21, pp. 5116–5121, 2008.

[29] D. Liu, W. Gao, Z. Li, H. Zhang, and Z. Hu, "Electro-spark deposition of Fe-based amorphous alloy coatings," *Materials Letters*, vol. 61, no. 1, pp. 165–167, 2007.

[30] S. Cadney and M. Brochu, "Formation of amorphous Zr41.2Ti13.8Ni10Cu12.5Be22.5 coatings via the electrospark deposition process," *Intermetallics*, vol. 16, no. 4, pp. 518–523, 2008.

[31] S. Cadney, G. Goodall, G. Kim, A. Moran, and M. Brochu, "The transformation of an Al-based crystalline electrode material to an amorphous deposit via the electrospark welding process," *Journal of Alloys and Compounds*, vol. 476, no. 1-2, pp. 147–151, 2009.

[32] Y. J. Xie and M. C. Wang, "Isothermal oxidation behavior of electrospark deposited MCrAlX-type coatings on a Ni-based superalloy," *Journal of Alloys and Compounds*, vol. 480, no. 2, pp. 454–461, 2009.

[33] L. E. Brown, S. N. Hammond, and M. C. Nordin, "Method and apparatus for electrospark alloying," US Patent 6417477, 2002.

[34] A. Lešnjak and J. Tušek, "Processes and properties of deposits in electrospark deposition," *Science and Technology of Welding and Joining*, vol. 7, no. 6, p. 391, 2002.

[35] J. Liang, W. Gao, Z. Li, and Y. He, "Hot corrosion resistance of electrospark-deposited Al and Ni Cr coatings containing dispersed Y_2O_3 particles," *Materials Letters*, vol. 58, no. 26, p. 3280, 2004.

[36] P. Leo, G. Renna, and G. Casalino, "Study of the direct metal deposition of AA2024 by electrospark for coating and reparation scopes," *Applied Sciences*, vol. 7, no. 12, p. 945, 2017.
[37] R. N. Johnson, "Alternative coatings for wear and corrosion: the electrospark deposition process," in *Proceedings of the AESF Conference for Environmental Excellence*, Lake Buena Vista, FL, USA, January 2002.
[38] D. A. Porter and K. E. Easterling, *Phase Transformations in Metals and Alloys*, Chapman & Hall, London, UK, 1992.
[39] L. F. Mondolfo, *Aluminium Alloys: Structure and Properties*, Butterworth, Oxford, UK, 1976.
[40] Z. Chen and Y. Zhou, "Surface modification of resistance welding electrode by electro-spark deposited composite coatings: part I. Coating characterization," *Surface & Coatings Technology*, vol. 201, no. 3-4, pp. 1503–1510, 2006.
[41] W. Jian-sheng, M. Hui-min, Y. Hong-ying, F. Zi-shuan, and S. Dong-bai, "Characterization and wear behavior of WC-0.8Co coating on cast steel rolls by electro-spark deposition," *International Journal of Minerals, Metallurgy and Materials*, vol. 16, p. 707, 2009.
[42] R. W. Messler, *Principles of Welding*, John Wiley & Sons Inc., New York, NY, USA, 1999.
[43] W. Kurz and D. J. Fisher, *Fundamentals of Solidification*, Trans Tech Publications, Switzerland, 4th edition, 1998.

10

Effect of Modified Pin Profile and Process Parameters on the Friction Stir Welding of Aluminum Alloy 6061-T6

J. C. Verduzco Juárez,[1] G. M. Dominguez Almaraz,[1] R. García Hernández,[2] and J. J. Villalón López[1]

[1] *Facultad de Ingeniería Mecánica, Universidad Michoacana de San Nicolás de Hidalgo (UMSNH), Santiago Tapia No. 403, Col. Centro, 58000 Morelia, MICH, Mexico*

[2] *Instituto de Investigaciones Metalúrgicas, UMSNH, Edificio U, Ciudad Universitaria, Gral. Francisco J. Mugica S/N, Felicitas del Rıo, 58030 Morelia, MICH, Mexico*

Correspondence should be addressed to G. M. Dominguez Almaraz; dalmaraz@umich.mx

Academic Editor: Hongchao Kou

This work deals with the effect of a new "bolt-head" pin profile on the friction stir welding performance of the aluminum alloy 6061-T6, compared to traditional pin profiles. Friction stir welding parameters such as the tool rotation speed and the welding speed were investigated together with the different pin profiles; the results show that the new "bolt-head" pin profile leads to better mechanical properties of welded specimens. The pin profiles used in this work were the straight square (SS), straight hexagon (SH), taper cylindrical (TC), and the straight hexagon "bolt-head" (SHBH). It was found that the last pin profile improves the material flow behavior and the uniform distribution of plastic deformation and reduces the formation of macroscopic defects on the welded zone. Mechanical tensile tests on welded specimens were performed to determine the tensile strength: the specimens welded with the SHBH pin profile have shown the highest mechanical properties. An approach is presented for material flow on this aluminum alloy using the SHBH pin profile, which is related to the improvement on the resulting mechanical properties.

1. Introduction

The alloy 6061-T6 is the most popular and used of all aluminum alloys; it presents an adequate balance of mechanical properties, cost, and weight. This alloy is used for aerospace and automotive industries under joining and welding requirements [1–4]. The traditional fusion welding methods, such as MIG and TIG, are characterized by high heat dissipation and the transformations of microstructural phases, leading to a number of defects (microporosities, less fine equiaxed recrystallized grain) and thermic instabilities (hardening precipitates), on the joining elements [5, 6]. For aeronautical application of the 6061-T6 alloy, the laser welding leads to high mechanical properties (reduction of FZ and HAZ, even if porosity may be promoted) and good finishing [7, 8]; however, the cost is a principal constraint for this welding method.

Friction stir welding is a solid state welding technique which has been developed at the end of 1991 by "The Welding Institute" in the United Kingdom [9]. A nonconsumable rotating tool communicates energy to the two facing surfaces, which may be of the same metallic alloy or different metallic alloys. The rotating movement and the related friction generate heat dissipation between the two facing surfaces, leading to softening of this region and facilitating the high shear plastic deformation and mixing of the material [10–12].

It has been recognized that the principal parameters affecting the bonding strength of FSW and the resulting microstructure are the rotational speed of tool, the welding speed, the axial tool force, and the angle of tool in regard to the welding surface (tilt angle) [13, 14].

The application of this joining technique has been widely used for the aluminum alloy 6061-T6 [15–19]; nevertheless, in this work is presented the effect of a new profile of the pin tool to improve the mechanical properties on this material, after the friction stir welding.

The pin profile is a parameter which plays important role during the FSW process, affecting the welding temperature, plastic deformation, the material mixing, the grain size, the welding defects, and so forth [20–22]. Some different pin profiles have been used to investigate their effects on the mechanical properties of welded materials using FSW [23–25]. Furthermore, recent investigations are oriented to the determination of the physical and mechanical effects of pin profiles: a model has been proposed for a taped cylindrical pin profile [22], which proposes equations for the heat generation: (a) from the shoulder surface, (b) from the taper probe surface, and (c) from the pin tip surface. The heat generation integral expression for the taper probe surface is

$$Q_2 = \int_0^{2\pi} \int_0^l \omega \cdot r^2 \cdot \tau_{\text{contact}} \cdot dz \cdot d\theta \tag{1}$$

with the solution

$$Q_2 = \frac{\pi \cdot \omega \cdot \tau_{\text{contact}}}{2} \frac{H_{\text{probe}}}{\cos\alpha} \left(R_{\text{PS}} + R_{\text{PT}}\right)^2, \tag{2}$$

where ω is the angular velocity, τ_{contact} is the contact stress, H_{probe} is the vertical length of taper cylindrical pin, α is the taper angle, R_{PS} is the bottom radius of the pin, R_{PT} is the top radius of the pin, and l is the inclined length of the pin: $l = H_{\text{probe}}/\cos\alpha$.

Equation (1) is obtained considering r independent of z (along the vertical axis of the pin), which is not the case. Here, we propose r as a function of z: $r = R_{\text{PT}} - z\tan\alpha$; the integral equation (1) becomes

$$Q_2 = \int_0^{2\pi} \int_0^{H_{\text{probe}}} \omega \cdot \tau_{\text{contact}} \cdot \left(R_{\text{PT}}^2 - 2R_{\text{PT}} \cdot z\tan\alpha + z^2 \cdot \tan^2\alpha\right) dz \cdot d\theta. \tag{3}$$

With $H_{\text{probe}}\tan\alpha = R_{\text{PT}} - R_{\text{PS}}$, the solution is

$$Q_2 = 2\pi \cdot \omega \cdot \tau_{\text{contact}} \frac{H_{\text{probe}}}{3} \left(R_{\text{PT}}^2 + R_{\text{PT}} \cdot R_{\text{PS}} + R_{\text{PS}}^2\right). \tag{4}$$

Equation (4) gives values slightly lower compared to (2) for the taper probe surface. The principal conclusion of [22] is that temperature registered during FSW process decreases with the taper angle.

Analytical models for heat generation were obtained for the pins: triangular, square, pentagonal, and hexagonal profiles [26], with the following principal conclusions: (a) heat generation increases from the triangular to square profile; then, it decreases to hexagonal pin profile; (b) increase on the tool rotation speed under constant weld speed induces heat generation increase; (c) the increase on the weld speed, under constant tool rotation speed, leads to decrease on heat generation. These conclusions are not in the same direction in regard to the results obtained for straight and taper cylindrical pin profiles [27], in which heat generation increases with angular speed ω (the tool rotation speed) and with the tangential speed V (welding speed). Recently, small tool pin profiles have been investigated using FSW on the 6061 aluminum alloy [28], with better results on mechanical properties for the triangular pin profile compared to the threaded tapered cylindrical and square pin profiles.

Table 1: Chemical composition in weight of aluminum alloy AISI 6061-T6.

Al	95.8 to 98.6
Cr	0.04 to 0.35
Cu	0.15 to 0.4
Fe	Max. 0.7
Mg	0.8 to 1.2
Mn	Max. 0.15
Si	0.4 to 0.8
Ti	Max. 0.15
Zn	Max. 0.25
Other	Each max. 0.05 Total max. 0.15

Table 2: Principal mechanical properties of aluminum alloy AISI 6061-T6.

Density (kg/m3)	2700
Hardness, Brinell	95
σ_y (MPa)	270
σ_u (MPa)	310
E (GPa)	68.9
Poisson ratio	0.33
Elongation at break (%)	17

2. Experimental Set-Up and Procedure

The chemical composition in weight and the principal mechanical properties of aluminum alloy 6061-T6 are shown in Tables 1 and 2, respectively.

Plates of dimensions 200 mm × 150 mm × 9.5 mm were used for the FSW purpose. The classification of the FSW tests according to the welding parameters for this work is shown in Table 3, whereas in Table 4 are listed the pins profiles and their dimensions.

Heat generation is increased with the shoulder diameter because the area of friction is increased; the effect of this parameter was not studied in this work. Figure 1 presents the four pin profiles used in this investigation; the PSHBH nomination represents the "bolt-head" pin, previous to manufacture of the bolt-head at the end of the pin. The dimensions of the "bolt-head" pin are shown in Figure 2, whereas in Figure 3 are depicted the welding plates rigidly clamped to the machine bed without any obstruction for the rotation and translation motion of the pin. The alignment of welding line and pin center was maintained close to zero deviation and all tests were carried out with 0 degrees for the tool tilt angle. The FSW process was implemented in a vertical machining center: Dyna Myte/DM-4800 equipped

TABLE 3: Classification of FSW tests, according to the welding parameters.

Nomination test	Tool rotational speed (rpm)	Welding speed (mm/min)	Penetration speed (mm/min)	Axial force (kN)	Used pin profile
C1	800	40	3	8	SS, SHBH
C2	1000	90	9	8	TC, SHBH
C3	1200	90	9	8	SH, SHBH

TABLE 4: Models for the pins according to the tool profile, the material, and dimensions.

Models	Pin profile	Shoulder surface	Material	Pin length (mm)	Shoulder diameter (mm)
M1	SS	Flat	Steel 4140	9	19
M2	TC	Flat	Steel 4140	9	19
M3	SH	Flat	Steel 9840	9	25.4
M4	SHBH	Flat	Steel H-13	9	25.4

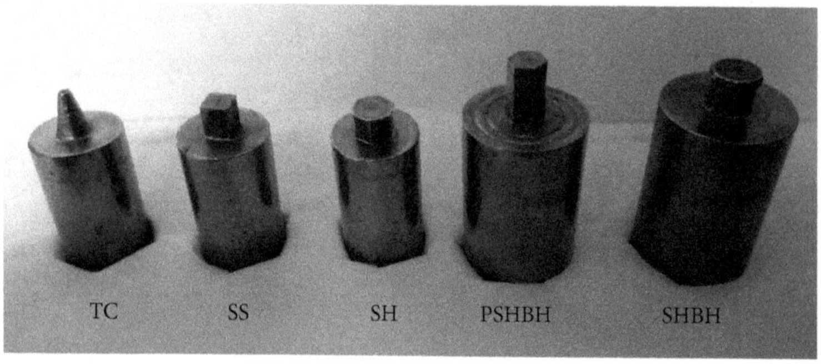

FIGURE 1: Manufactured pin profiles for the FSW process.

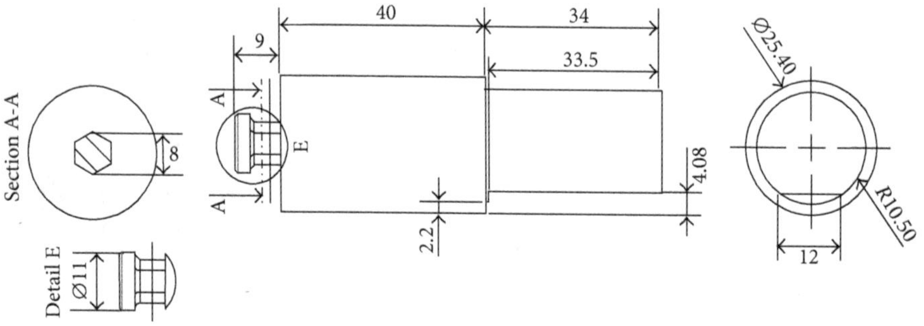

FIGURE 2: Dimensions (mm) for the "bolt-head" pin profile.

with Mitsubishi controls and a maximum tool rotation of 6000 rpm.

The testing specimens were obtained from the welded plates in perpendicular direction in regard to the FSW line for the tests: tensile, hardness, and the microstructural observation by optical microscope. Tensile specimens were manufactured according to the norm ASTM-E8M, with a gauge length of 64 mm, a gauge width of 6 mm, and a thickness of 9.5 mm, as shown in Figure 4.

All tensile tests were carried out on a tension machine model Zwick/Roell Z100 with 100 KN of maximum test load and at constant test displacement of 6 mm/min. The microhardness Mitutoyo model HM 200 was used to obtain the hardness profiles along the welding zone with an applied load of 0.1 Kg and 10 sec of time application. In addition, the FSW welding process was obtained under the following condition: the distance between the tip of pin and the bottom of the welding workpiece was 0.3 mm.

3. Results and Discussions

Plates of thickness 9.5 mm of aluminum alloy 6061-T6 were joined by FSW technique using classical pin profiles (SS, SH, and TC) and the new pin profile "bolt-head" to compare the results on welding and mechanical properties [29]. In Figure 5 are illustrated the different zones, from 1 to 8, along the welding line from which the tensile specimens have been machined. This figure shows the welding direction as well as

Table 5: Tensile tests for specimens welded by the SHBH pin profile and base material (BM) specimens.

Nomination test-tool	Specimen ID-section (Figure 5)	Stress at 0.2% plastic strain (MPa)	Ultimate tensile stress (MPa)	dL at break (%)	Fracture localization
C1-M1	1-3	113	146	4.7	HAZ
C1-M1	1-4	118	150	6.2	HAZ
C2-M2	2-3	115	128	3.8	TMAZ
C3-M3	3-2	110	125	3.9	TMAZ
C3-M3	3-3	106	108	4.0	TMAZ
C3-M3	3-4	108	129	6.6	TMAZ
C1-M4	4-2	114	161	10.2	NHAZ
C1-M4	4-4	115	163	10.8	NHAZ
C3-M4	5-2	114	176	10.5	NHAZ
C3-M4	5-4	124	177	9.5	NHAZ
C1-M4	6-2	111	162	11.1	NHAZ
C1-M4	6-4	111	162	11.3	NHAZ
C2-M4	7-2	119	161	8.7	NHAZ
C2-M4	7-3	116	172	5.8	NHAZ
C2-M4	7-4	115	176	9.0	NHAZ
C2-M4	8-2	116	169	4.3	NHAZ
C2-M4	8-3	115	171	9.6	NHAZ
C2-M4	8-4	116	176	10.0	NHAZ
—	BM-1	289	309	14.2	—
—	BM-2	296	316	11.5	—
—	BM-3	286	308	14.2	—

Figure 3: The two welding plates clamped to machine bed and the welding tool at the end of the FSW process.

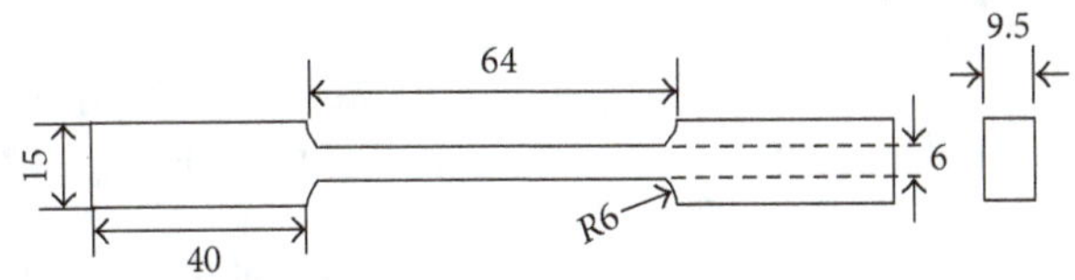

Figure 4: Tensile specimen dimensions (mm), following the norm ASTM-E8M.

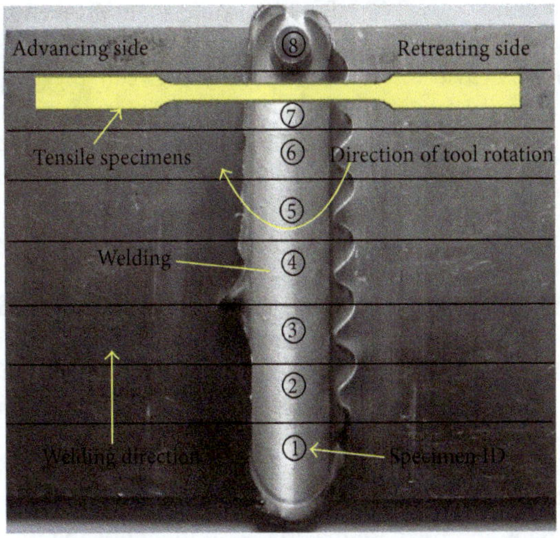

Figure 5: Zones for the tensile specimen along the welding line. Direction of welding, advancing and retreating sides, and direction of tool rotation.

the advancing and retreating sides and the direction of tool rotation.

The tensile specimens were obtained from the zones 2 to 4; the zones 1 and 8 were eliminated for this purpose. Figure 6 presents the macrostructure of the transverse welding zone for the different used pin profiles in function of the rotational speed of tool and the welding speed. The tunnel defect is observed for the pin profiles, SS, TC, and SH, for the three rotating speeds of tool (800, 1000, and 1200 rpm) and for both welding speeds (40 and 90 mm/min). The use of low rotational speed and low welding speed with the pin profile SS induces middle ultimate tensile stress on the tensile specimens (146–150 MPa, Table 5); whereas in increasing the

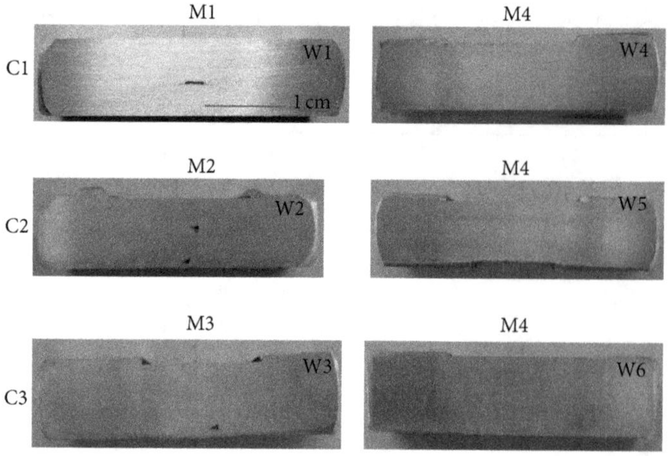

FIGURE 6: Tunnel defect for welding specimens using different pin profiles, for three tool rotating speeds and two welding speeds.

last two velocities and using the pin profiles TC and SH, the ultimate tensile stress decreases as shown on the same table. The maximum ultimate tensile stress is obtained with the SHBH pin profile (176-177 MPa), using high values for the two weld speeds.

Concerning the tunnel defect in the welded specimens, it is present for the three pin profiles, SS, TC, and SH, as shown in Figure 6; nevertheless, this defect is eliminated using the SHBH pin profile for all rotating and welding used speeds, as depicted in the same figure. An attempt to explain the causes that induce the elimination of tunnel defect with the use of SHBH pin profile is undertaken in the last section. These results seem to agree with the physical conditions close to the contact zone between the pin and the welding material: the heat generation increases when the tool rotation speed increases; then, the flow of material across the welding thickness increases too. For tapered pin profiles particularly, the contact area of pin decreases with the depth of the welding thickness and the flow of material increases across the thickness, contributing to tunnel defect.

In Table 5 are listed the tensile tests for the welded specimens using the four pin profiles, as well as the tensile tests on the base material (last three specimens in this table). The ultimate tensile stress for the welded specimens ranges from 108 to 177 MPa, whereas for the base material these values range from 308 to 316 MPa [30, 31].

Figure 7 presents the tensile tests graphs for the welded specimens and the base material with the following identification in regard to Table 5: C1-M1 = W1-3 (146 MPa UTS); C2-M2 = W2-3 (128 MPa UTS); C3-M3 = W3-3 (108 MPa UTS); C1-M4 = W4-3 (163 MPa UTS); C2-M4 = W5-4 (176 MPa UTS); C3-M4 = W6-4 (177 MPa UTS). In all cases for tensile tests on welded specimens, the higher values for UTS were obtained with the SHBH pin profile. Furthermore, the fracture sites under tensile tests of welded SHBH specimens were located outside the welding zone, as shown in Figure 8.

Figure 9 presents the Vickers microhardness measured along the weld cross-section comprising the zones: the nugget zone (NZ), the thermomechanical heat zone (TMHZ), the heat affected zone (HAZ), and the base metal (BM), for the welded SHBH specimens. The higher values of this mechanical property have been obtained for low rotation speed during the FSW process and intermediate values for welding speed; this tendency has been mentioned by some authors [32, 33], for aluminum alloys.

Fracture surfaces for the tensile specimens are shown in Figure 10: the corresponding fracture surface for specimens containing the tunnel defect is depicted in Figure 10(a), whereas the fracture surface for specimen without the tunnel defect is presented in Figure 10(b). Figure 10(c) shows the fracture surface for the base material. Low plastic deformation characterizes the separation of fracture surfaces for specimens with tunnel defect, revealing that low resistance is opposed to tensile loading: this behavior is represented by the tensile tests W2-3 and W3-3 in Figure 7.

The increase of plastic strain is related to the ductile fracture and the microvoids coalescence, which is observed on this aluminum alloy for tensile specimens presenting higher resistance, as illustrated on Figures 10(b) and 10(c). In this case, the peripheral plastic deformation zone increases with the UTS of tensile test on this material, as it is observed in the last two figures. The central zone of fracture surfaces for specimens without tunnel defect is typified by low or null plastic deformation and a granular nondeformed microstructure [34].

The material flow during the FSW process is a principal factor influencing the presence of welding defects, such as cavities, root flaws, and kissing bond [35]. Generally speaking, the material flow in FSW is determined by three different motions: (a) the material flow beneath and deeper of the tool shoulder, from and to the root of the probe, (b) the flow of material along the pin surface, downward in the vertical direction, and (c) the material flow from the bottom of pin towards the shoulder, away from the pin surface; this flow interacts with the two precedents flows [36, 37].

A swirl is present in the advancing side which increases with the welding speed and induces the radial motion illustrated, whereas in the retreating side the radial motion is considerably reduced and may change direction according to the welding parameters and the testing material [38, 39].

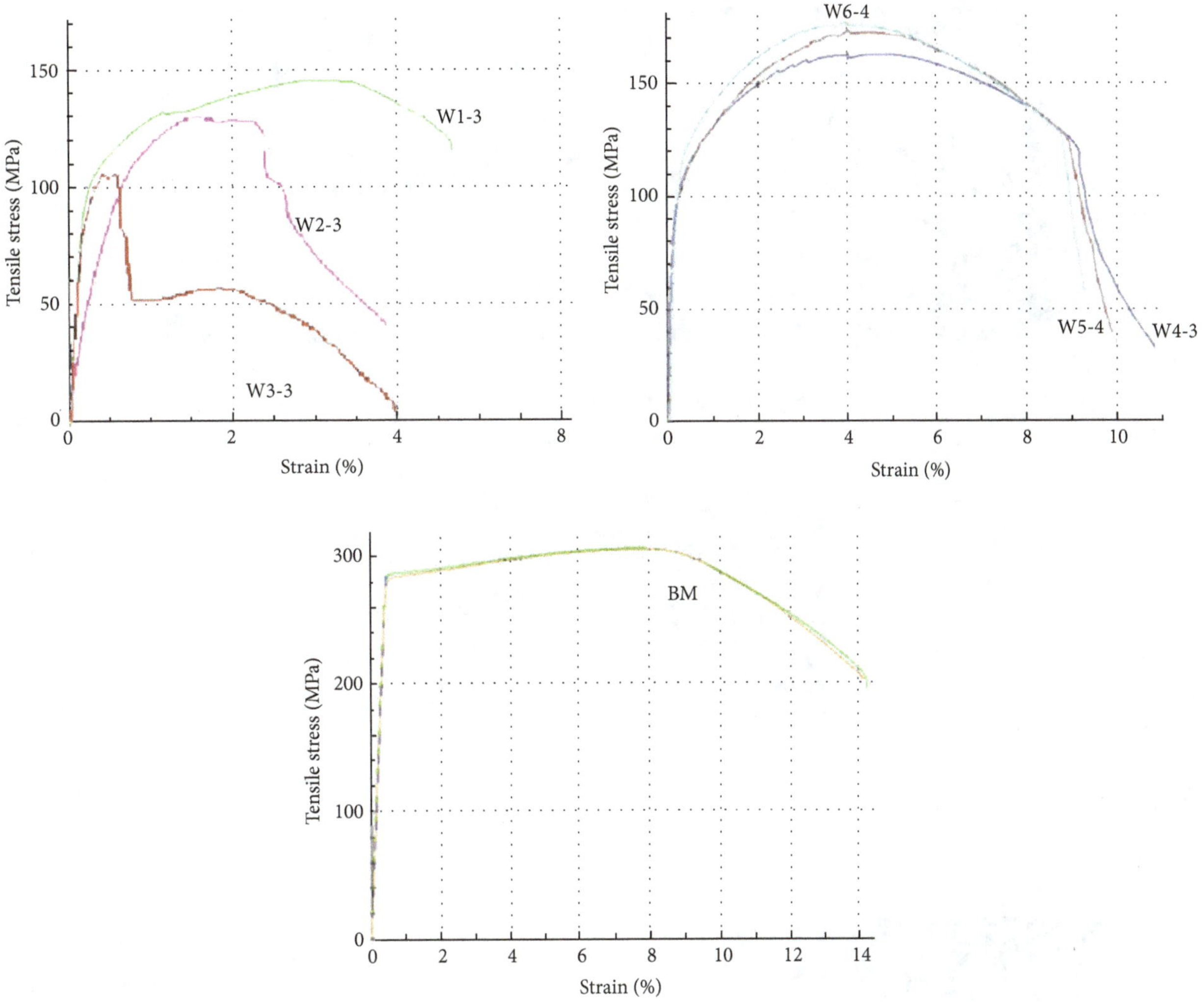

FIGURE 7: Tensile tests for welded and base material specimens.

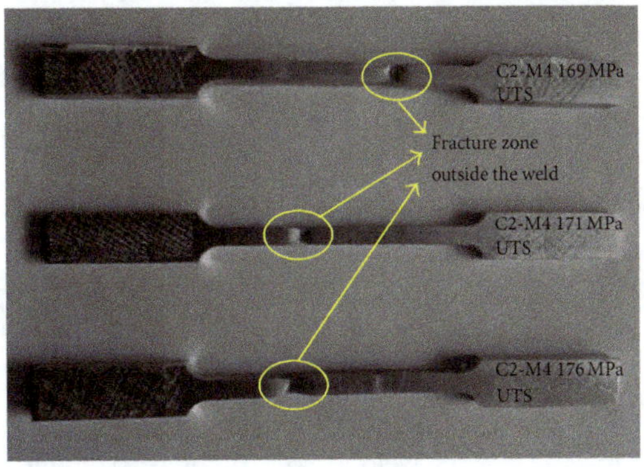

FIGURE 8: Localization of fracture under tensile tests for the SHBH welded specimens.

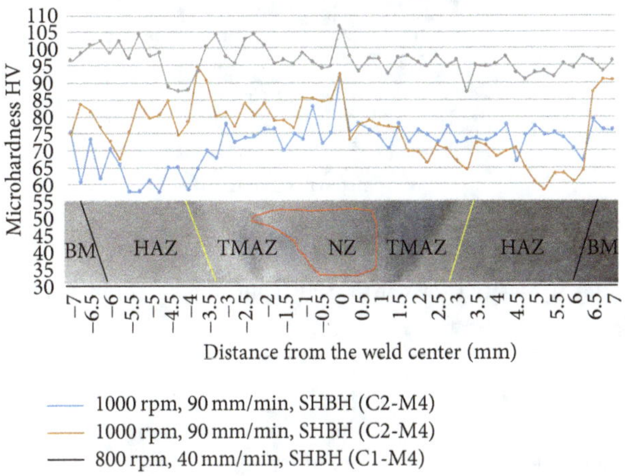

FIGURE 9: Vickers microhardness along the welding zone for the SHBH welded specimens.

In Figure 11 are represented the three mentioned flows, contributing to the flow of material during the welding process by FSW.

The flow of material along the pin surface and in the downwards vertical direction is reduced using the SHBH pin profile, whereas the flow from the bottom of pin towards the shoulder is moved more away from the pin surface. The

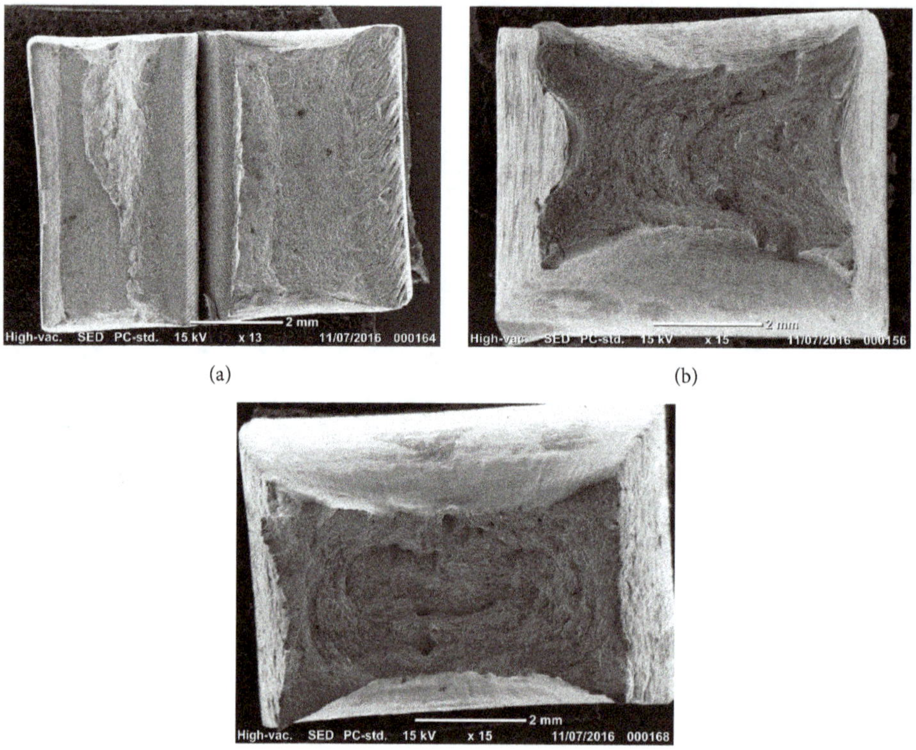

FIGURE 10: Fracture surfaces for tensile specimens: (a) with tunnel defect, (b) without tunnel defect, and (c) for the base material.

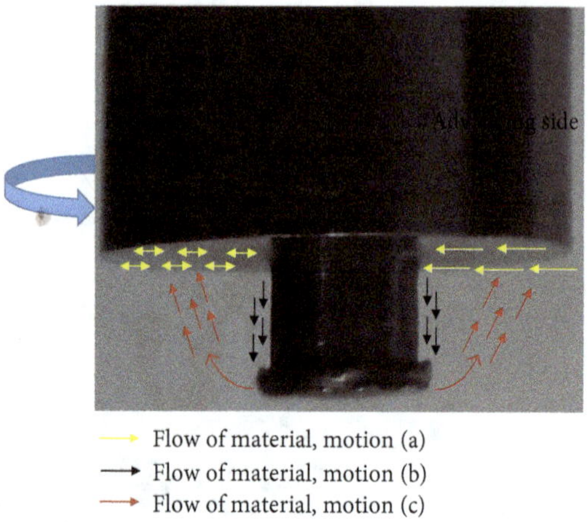

FIGURE 11: Material flows on the SHBH pin profile according to three principal motions.

bolt-head at the end of the pin reduces considerably the material flow in vertical direction: the mechanical effect is the reduction of discontinuities on the material with the rotating and the advancing motion, which leads to reduction of tunnel and other defects in this aluminum alloy. The vertical flow is associated with the continuity of material along the vertical direction during local FSW stirring; the reduction of this flow by the pin bolt-head induces the reduction of material discontinuity and hence welding defects.

4. Conclusion

In this work is presented a new pin profile denominated bolt-head, for the friction stir welding of the aluminum alloy 6061-T6. The tensile and hardness tests show that this pin profile improves such mechanical properties of the welded alloy. Additional conclusions related to this work are as follows:

(a) No tunnel defect was observed on the welded specimens using the SHBH pin profile, for the three rotating speeds and two welding speeds of experimental testing.

(b) A modified expression is proposed to calculate heat generation from the taper probe surface.

(c) The fracture for tensile specimens, using the SHBH pin profile, took place systematically outside the welding zone.

(d) The Vickers microhardness along the welding zone for the SHBH welded specimens was higher when using low rotational speeds of tool, for the testing welding speeds [40].

(e) Low plastic deformation is observed for the fracture surfaces of specimens with tunnel defect, indicating that low resistance is opposed during the tensile tests. The fracture surface of tensile specimens without tunnel defect shows appreciable plastic deformation around the peripheric rectangular area, indicating resistance of material to tensile loading.

(f) The bolt-head at the end of the pin reduces the material flow downward, contributing to reduction or elimination of the tunnel and other defects.

(g) Further investigations will be undertaken in the future to assess the effect of the bolt-head profile on the mechanical properties of joining materials by the FSW method, particularly, the heat generation, the material flow, modification of the bolt-head profile, the welding of other materials than aluminum alloys, and so forth.

Competing Interests

The authors declare that they have no competing interests.

Acknowledgments

The authors express their gratitude to the University of Michoacan in Mexico for the facilities received in the development of this work. A special mention of gratitude was made to CONACYT (The National Council for Science and Technology, Mexico) for the financial support destined to this study by the Program Grant CB-241117-2014.

References

[1] F. Fadaeifard, F. Gharavi, K. A. Matori, A. R. Daud, M. K. Ariffin, and M. Awang, "Investigation of microstructure and mechanical properties of friction stir lap welded AA6061-T6 in various welding speeds," *Journal of Applied Sciences*, vol. 14, no. 3, pp. 221–228, 2014.

[2] L. Liu, H. Nakayama, S. Fukumoto, A. Yamamoto, and H. Tsubakino, "Microscopic observations of friction stir welded 6061 aluminum alloy," *Materials Transactions*, vol. 45, no. 2, pp. 288–291, 2004.

[3] W. S. Miller, L. Zhuang, J. Bottema et al., "Recent development in aluminium alloys for the automotive industry," *Materials Science and Engineering: A*, vol. 280, no. 1, pp. 37–49, 2000.

[4] A. Heinz, A. Haszler, C. Keidel, S. Moldenhauer, R. Benedictus, and W. S. Miller, "Recent development in aluminium alloys for aerospace applications," *Materials Science and Engineering A*, vol. 280, no. 1, pp. 102–107, 2000.

[5] S. Jannet, K. Mathews, and R. Raja, "Comparative investigation of friction stir welding and fusion welding of 6061 T6-5083 O aluminum alloy based on mechanical properties and microstructure," *Bulletin of the Polish Academy of Sciences: Technical Sciences*, vol. 62, no. 4, pp. 791–795, 2014.

[6] R. Adalarasan and R. M. Santhanakumar, "Parameter design in fusion welding of AA 6061 aluminium alloy using desirability grey relational analysis (DGRA) method," *Journal of The Institution of Engineers (India): Series C*, vol. 96, no. 1, pp. 57–63, 2015.

[7] A. El-Batahgy and M. Kutsuna, "Laser beam welding of AA5052, AA5083, and AA6061 aluminum alloys," *Advances in Materials Science and Engineering*, vol. 2009, Article ID 974182, 9 pages, 2009.

[8] D. Narsimhachary, R. N. Bathe, G. Padmanabham, and A. Basu, "Influence of temperature profile during laser welding of aluminum alloy 6061 T6 on microstructure and mechanical properties," *Materials and Manufacturing Processes*, vol. 29, no. 8, pp. 948–953, 2014.

[9] W. M. Thomas, E. D. Nicholas, J. C. Needham, M. G. Murch, P. Temple-Smith P, and C. J. Dawes, "Friction-stir butt welding, GB Patent No. 9125978.8," International patent application No. PCT/GB92/02203, 1991.

[10] W. Woo and H. Choo, "Softening behaviour of friction stir welded Al 6061-T6 and Mg AZ31B alloys," *Science and Technology of Welding and Joining*, vol. 16, no. 3, pp. 267–272, 2011.

[11] P. Zhang, N. Guo, G. Chen et al., "Plastic deformation behavior of the friction stir welded AA2024 aluminum alloy," *The International Journal of Advanced Manufacturing Technology*, vol. 74, no. 5–8, pp. 673–679, 2014.

[12] T. Debroy, A. De, H. K. D. H. Bhadeshia, V. D. Manvatkar, and A. Arora, "Tool durability maps for friction stir welding of an aluminium alloy," *Proceedings of the Royal Society A: Mathematical, Physical and Engineering Sciences*, vol. 468, no. 2147, pp. 3552–3570, 2012.

[13] M. Indira Rani, R. N. Marpu, and A. C. S. Kumar, "A study of process parameters of friction stir welded AA 6061 aluminum alloy in O and T6 conditions," *ARPN Journal of Engineering and Applied Sciences*, vol. 6, no. 2, pp. 61–66, 2011.

[14] R. Nandan, T. DebRoy, and H. K. D. H. Bhadeshia, "Recent advances in friction-stir welding—process, weldment structure and properties," *Progress in Materials Science*, vol. 53, no. 6, pp. 980–1023, 2008.

[15] S. Rajakumar, C. Muralidharan, and V. Balasubramanian, "Establishing empirical relationships to predict grain size and tensile strength of friction stir welded AA 6061-T6 aluminium alloy joints," *Transactions of Nonferrous Metals Society of China (English Edition)*, vol. 20, no. 10, pp. 1863–1872, 2010.

[16] W. Woo, H. Choo, D. W. Brown, Z. Feng, and P. K. Liaw, "Angular distortion and through-thickness residual stress distribution in the friction-stir processed 6061-T6 aluminum alloy," *Materials Science and Engineering A*, vol. 437, no. 1, pp. 64–69, 2006.

[17] W. H. Jiang and R. Kovacevic, "Feasibility study of friction stir welding of 6061-T6 aluminium alloy with AISI 1018 steel," *Proceedings of the Institution of Mechanical Engineers, Part B: Journal of Engineering Manufacture*, vol. 218, no. 10, pp. 1323–1331, 2004.

[18] L. E. Murr, G. Liu, and J. C. McClure, "A TEM study of precipitation and related microstructures in friction-stir-welded 6061 aluminium," *Journal of Materials Science*, vol. 33, no. 5, pp. 1243–1251, 1998.

[19] S. Rajakumar, C. Muralidharan, and V. Balasubramanian, "Predicting tensile strength, hardness and corrosion rate of friction stir welded AA6061-T6 aluminium alloy joints," *Materials and Design*, vol. 32, no. 5, pp. 2878–2890, 2011.

[20] R. Palanivel, P. Koshy Mathews, N. Murugan, and I. Dinaharan, "Effect of tool rotational speed and pin profile on microstructure and tensile strength of dissimilar friction stir welded AA5083-H111 and AA6351-T6 aluminum alloys," *Materials and Design*, vol. 40, pp. 7–16, 2012.

[21] K. Elangovan, V. Balasubramanian, and M. Valliappan, "Influences of tool pin profile and axial force on the formation of friction stir processing zone in AA6061 aluminium alloy," *International Journal of Advanced Manufacturing Technology*, vol. 38, no. 3-4, pp. 285–295, 2008.

[22] V. S. Gadakh and K. Adepu, "Heat generation model for taper cylindrical pin profile in FSW," *Journal of Materials Research and Technology*, vol. 2, no. 4, pp. 370–375, 2013.

[23] M. K. Sued, D. Pons, J. Lavroff, and E. H. Wong, "Design features for bobbin friction stir welding tools: development

of a conceptual model linking the underlying physics to the production process," *Materials and Design*, vol. 54, pp. 632–643, 2014.

[24] M. Ilangovan, S. Rajendra Boopathy, and V. Balasubramanian, "Effect of tool pin profile on microstructure and tensile properties of friction stir welded dissimilar AA 6061-AA 5086 aluminium alloy joints," *Defence Technology*, vol. 11, no. 2, pp. 174–184, 2015.

[25] K. Ramanjaneyulu, G. Madhusudhan Reddy, A. Venugopal Rao, and R. Markandeya, "Structure-property correlation of AA2014 friction stir welds: role of tool pin profile," *Journal of Materials Engineering and Performance*, vol. 22, no. 8, pp. 2224–2240, 2013.

[26] V. S. Gadakh, A. Kumar, and G. J. Vikhe Patil, "Analytical modeling of the friction stir welding process using different pin profiles," *Welding Journal*, vol. 94, no. 4, pp. 115–124, 2015.

[27] S. Tikader, P. Biswas, and A. B. Puri, "A study on tooling and its effect on heat generation and mechanical properties of welded joints in friction stir welding," *Journal of The Institution of Engineers (India): Series C*, vol. 2016, 12 pages, 2016.

[28] H. I. Dawood, K. S. Mohammed, A. Rahmat, and M. B. Uday, "Effect of small tool pin profiles on microstructures and mechanical properties of 6061 aluminum alloy by friction stir welding," *Transactions of Nonferrous Metals Society of China*, vol. 25, no. 9, pp. 2856–2865, 2015.

[29] G. M. Dominguez Almaraz, J. C. Verduzco Juarez, R. García Hernández, and J. J. Villalón López, "Friction stir welding on aeronautical aluminum alloy 6061-T6," in *Proceedings of the 25th International Materials Research Congress (IMRC '16)*, Cancún, Mexico, August 2016.

[30] D. Ortiz, M. Abdelshehid, R. Dalton et al., "Effect of cold work on the tensile properties of 6061, 2024, and 7075 Al alloys," *Journal of Materials Engineering and Performance*, vol. 16, no. 5, pp. 515–520, 2007.

[31] T. Yokoyama, K. Nakai, and K. Katoh, "Tensile properties of AA6061-T6 friction stir welds and constitutive modeling in transverse and longitudinal orientations," *Journal of Light Metal Welding and Construction*, vol. 53, no. 8, pp. 19–28, 2015.

[32] N. Rajamanickam and V. Balusamy, "Effects of process parameters on mechanical properties of friction stir welds using design of experiments," *Indian Journal of Engineering and Materials Sciences*, vol. 15, no. 4, pp. 293–299, 2008.

[33] A. Karam, T. S. Mahmoud, H. M. Zakaria, and T. A. Khalifa, "Friction stir welding of dissimilar A319 and A413 cast aluminum alloys," *Arabian Journal for Science and Engineering*, vol. 39, no. 8, pp. 6363–6373, 2014.

[34] A. Dorbane, G. Ayoub, B. Mansoor, R. Hamade, G. Kridli, and A. Imad, "Observations of the mechanical response and evolution of damage of AA 6061-T6 under different strain rates and temperatures," *Materials Science and Engineering: A*, vol. 624, pp. 239–249, 2015.

[35] H.-B. Chen, K. Yan, T. Lin, S.-B. Chen, C.-Y. Jiang, and Y. Zhao, "The investigation of typical welding defects for 5456 aluminum alloy friction stir welds," *Materials Science and Engineering A*, vol. 433, no. 1-2, pp. 64–69, 2006.

[36] C. Hamilton, S. Dymek, and M. Blicharski, "A model of material flow during friction stir welding," *Materials Characterization*, vol. 59, no. 9, pp. 1206–1214, 2008.

[37] Q. Yang, S. Mironov, Y. S. Sato, and K. Okamoto, "Material flow during friction stir spot welding," *Materials Science and Engineering A*, vol. 527, no. 16-17, pp. 4389–4398, 2010.

[38] X. He, F. Gu, and A. Ball, "A review of numerical analysis of friction stir welding," *Progress in Materials Science*, vol. 65, pp. 1–66, 2014.

[39] R. S. Mishra and Z. Y. Ma, "Friction stir welding and processing," *Materials Science and Engineering R: Reports*, vol. 50, no. 1-2, pp. 1–78, 2005.

[40] J. S. Leon and V. Jayakumar, "Investigation of mechanical properties of aluminium 6061 alloy friction stir welding," *American Journal of Mechanical Engineering and Automation*, vol. 1, no. 1, pp. 6–9, 2014.

11

Effect of a Deep Cryogenic Treatment on Wear and Microstructure of a 6101 Aluminum Alloy

Volker Franco Steier,[1] Edgar Sobral Ashiuchi,[2] Lutz Reißig,[3] and José Alexander Araújo[2]

[1]*Department of Automotive Engineering, University of Brasilia, 72444-240 Brasilia, DF, Brazil*
[2]*Department of Mechanical Engineering, University of Brasilia, 70910-900 Brasilia, DF, Brazil*
[3]*Fraunhofer Institute for Mechanics of Materials IWM, Woehlerstrasse 11, 79108 Freiburg, Germany*

Correspondence should be addressed to Volker Franco Steier; vfs@unb.br

Academic Editor: Francesco Delogu

The aim of this work is to evaluate the effect of a deep cryogenic treatment (DCT) on the wear behavior and on the microstructure of an aluminum alloy. In order to compare the level of improvement on the wear resistance provided by the DCT with a more traditional technique, a test matrix which included DCT, CrN coated specimens, and combinations of both modification methods was conducted. The wear behavior was investigated using microabrasive wear tests. The cryogenic treated specimens proved to have similar low wear rates as the specimens coated with CrN. The most distinct improvement was reached with a combination of both techniques. In the case of the DCT, the performed microstructural analysis identified the generation of additional GP-zones as the reason for the improved wear resistance.

1. Introduction

Especially in the automotive sector, the use of aluminum is steadily increasing in order to reduce the weight of cars, which improves the efficiency and reduces the emission of CO_2. The drawbacks of aluminum are its low yield strength and the low wear resistance (wear resistance is 1/wear rate). One possibility to enhance the mechanical properties of aluminum is to add alloying elements like silicon, manganese, or magnesium [1–3]. However, the wear resistance of aluminum alloys is still insufficient for many applications.

One possibility to improve the wear resistance of aluminum alloys is to apply deep cryogenic treatments (DCTs). One of the advantages of DCTs is that the microstructural changes occur in the entire bulk material and not just on the surface (as it is the case for coatings) [4]. DCTs have been widely used in many industries such as aerospace and automotive, electronic, and mechanical engineering to improve the wear resistance and the dimensional stability of steel components [5]. However, no industrial application of DCTs for aluminum components has so far been established and there are just few published works on the effect of DCTs on aluminum and aluminum alloys.

A first investigation regarding the effect of DCTs on residual stresses of an aerospace aluminum alloy was carried out in 2000 [6]. In the following years, several researchers applied DCTs on different aluminum alloys [7–10]. In most cases, it was reported an improvement of the mechanical properties after the application of DCTs. However, these studies did not try to explain the possible causes associated with such behavior. More recent work has suggested that the formation of hard second phases (like Mg_2Si or $Al_xFe_xSi_x$) [11–13], grain size effects [8], and the generation of dislocations [14] could be related to the reported improvements, but more firm conclusions were not provided. In this setting, the aim of this work is to comprehensively investigate the effect of DCTs on the wear resistance and on the microstructure of an aluminum alloy.

2. Methods

To compare the impact of DCTs on the wear behavior, five different groups of specimens were tested. First, a group of unmodified specimens, which act as a reference material, was tested. The second group of specimens was modified with a DCT. The third batch consisted of specimens that were coated with a chromium nitride (CrN) layer. This batch was added to the test matrix to compare the effect of the DCT with a more traditional technique known to improve the wear resistance [15–17]. Zhou et al. showed that CrN coatings exhibit an excellent sliding wear resistance, when deposited on aluminum alloys [18, 19]. For this reason, CrN was chosen as surface coating for this work. The fourth and fifth batches were modified with a combination of DCT and a CrN coating. The difference between both batches was the order of application of the modification methods.

All five batches were tested in comparative microabrasive wear tests, and, to identify the effect of DCTs on the microstructure of the aluminum alloy, a comprehensive series of analysis was carried out. It included classical metallographic methods supported by light microscopy, Scanning Electron Microscopy (SEM), Transmission Electron Microscopy (TEM), hardness tests, energy-dispersive spectroscopy (EDS), X-Ray Diffraction (XRD) analysis, and differential scanning calorimetry (DSC). Details on the test procedures, on the application of these analyses techniques and on the material/specimens considered, are provided next.

2.1. Material. The raw material used for the tests was an aluminum alloy 6101-T4. To operate as reference material, the alloy was tested as received (further named as AR). More precisely, the state of the alloy should be defined as naturally aged, since the aluminum specimens were not tested just after production. The nominal chemical composition of this material is given in Table 1.

2.2. Modification Methods. The DCT was applied by the company Kyros Tecnologia. The process starts at room temperature and goes down to −190°C within a period of 12 h. The temperature of −190°C is kept constant for further 12 h. After this dwell time, the temperature is increased to room temperature within 12 h. Before the liquid nitrogen is injected in the chamber, it was evaporated to enable a controlled cooling and to prevent a severe gradient of temperature between the surface and the core of the specimens. The deep cryogenically treated specimens are henceforth named CTed.

Physical vapor deposition was used to apply the CrN coating by an external contract. As target, a 99.9% pure chromium was used. The deposition was carried out in a N_2 atmosphere. Besides a thoroughly cleaning of the specimens with alcohol in an ultrasonic bath, no further treatment was applied before the coating process. Due to the microstructural changes, which can be generated in aluminum alloys, a low temperature during the coating process is important. CrN was selected for the tests as it could be deposited at comparable low temperatures. These specimens are henceforth named CrNC.

To investigate if a combination of both modification methods could lead to a superposition of effects, two additional test series were carried out. One was a series where the CTed specimens were subsequently coated (combination of DCT and coating-1, Comb-1) and the other consisted of coated specimens that were later submitted to DCT (combination of coating and DCT-2, Comb-2). To improve the readability of the paper, a list of abbreviations for the specimens in reported in Table 2.

Table 1: Nominal chemical composition of the aluminum alloy 6101-T4.

Elements	Al	Si	Mg	Fe	Cu	Mn	Cr	Zn
Percentage	98.8	0.3–0.7	0.3–0.8	0.5	0.1	0.03	0.03	0.01

Table 2: Specimens abbreviations.

Treatment applied to the specimens	Abbreviation
As received	AR
Deep cryogenic treatment	CTed
Coated with a CrN layer	CrNC
Cryogenically treated and afterwards coated with CrN	Comb-1
First coated with CrN and afterwards cryogenically treated	Comb-2

To guarantee equal surface conditions, all specimens were polished under equal conditions (before the coating process and the DCT). The polishing was carried out with abrasive paper (grit 1200), until the surface was optically equal.

2.3. X-Ray Diffraction. X-Ray Diffraction (XRD) spectroscopy was used to investigate the effect of the DCT on the aluminum alloy. For the measurements, a high resolution Rigaku Ultima IV, with a Cu Kα source and a wavelength of 1.54 A was used. The scan was performed with a step size of 0.5°/min from an angel of 0° to 120°. The voltage and current settings were 35 kV and 15 mA, respectively.

2.4. Hardness Measurements. To examine the effect of the cryogenic treatment on the hardness, Vickers indentations were carried out (Zwick ZHU 250). Ten indentations were performed with HV1 (10 N). The force was applied for 10 s and measures were taken with an optical microscope.

For coated specimens, the coating was removed to conduct hardness tests. This was done by polishing the specimens with an abrasive aluminum suspension (mean particle size 0.5 μm). The polishing process was executed by hand with very low forces to avoid the creation of residual stresses. The process was frequently interrupted and visually controlled to remove only the surface coating and maintain the near surface substrate material.

To avoid that microstructural differences between the specimens affect the hardness analysis of the CTed and the AR specimens, the hardness measurements were carried out on the same sample. This means that the hardness was first examined in the AR state. The same sample was then treated cryogenically. Subsequently, the hardness was measured again close to the previously performed indentations.

2.5. Microstructure Analysis. To investigate the microstructure of the specimens before and after the DCT, polished and etched surfaces were photographed. The polishing process was executed using different diamond suspensions down to a grain size of 1 μm. For the final polish, an oxide polishing suspension was used (Struers). The etching was carried out using a Barker etchant. The microscopic images were taken with an optical microscope using polarized light.

The TEM was performed in the material engineering department in the Pontifical Catholic University of Rio de Janeiro (JEOL JEM 2100F). The acceleration voltage used was 200 kV. The preparation of the specimens was carried out in two steps. First, the specimens size was reduced to a thickness of 100 μm. Subsequently, a common electropolishing process was applied to reduce the specimens height to the needed thickness. The electrolyte consisted of nitric acid and methanol in a mass fraction of 30% and 75%, respectively. The used voltage was 20 V at a temperature of 20°C.

2.6. Differential Scanning Calorimetry (DSC). Due to the difficulties to observe ultrafine precipitates like Guinier-Preston-zones (GP-zones) by using SEM or TEM, DSC was used to analyze the microstructural differences generated by the DCT (Perkin Elmer, DSC 8500). The measurement was executed with a constant Argon purge, to avoid oxidation reactions. After an initial dwell time (2 min, 20°C) to homogenize the samples temperature, a heating rate of 20 K/min was considered. A further 2 min dwell time was applied at the final temperature of 330°C. As usual for DSC measurements, an empty specimen container, made of aluminum, was used as a reference. The test was carried out with two samples. Both samples (AR and CTed) had the same mass of 28.5 mg. To better visualize the differences between the CTed and the AR aluminum alloy, the DSC signals were subtracted one from the other.

2.7. Abrasive Wear Tests. The abrasive wear tests were performed with a Calowear testing machine (CSM Instruments SA) (Figure 1). The samples were mounted in an angle of 80° versus the horizontal. As rotating sphere, a steel ball (Inox 420) with a diameter of 20 mm was used. The ball was preroughened by preliminary wear tests to enable a good entrainment of the slurry in the contact zone [20]. The rotation direction of the steel ball was adjusted so that the abrasive slurry was directly fed in the abrasive contact. The tests were executed with a ball speed of 0.1 m/s. The contact force between the contact parts depends on the weight, the distance, and the angle between the ball and the specimens. Those parameters were kept constant for all tests. In order to archive three-body wear, the contact force was kept at a comparable low level of 0.05 N for all tests [20, 21]. As abrasive medium, a mixture of deionized water and Al_2O_3 was used (weight% 31 : 69, resp., $Al_2O_3 : H_2O$). The specification of the ceramic powder given by the manufacturer is D20-0.35 μm, D50-0.6 μm, and D90-1.4 μm (CR6, Baikowski).

To investigate the volume loss as a function of the sliding distance, six craters were generated on each sample. After a predefined sliding distance was reached, the tests were interrupted. The defined sliding distances were 22, 32, 49, 65, 81, and 105 m. To avoid interaction between the single spheres, a distance of at least 3 mm was maintained.

The diameter of the crater was measured with an optical microscope using a circular caliber whose size was adapted until the distinct edges of the crater were congruent (Figure 2, red line).

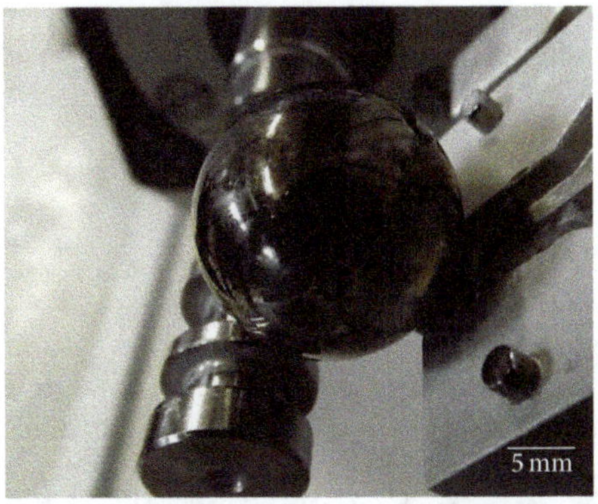

FIGURE 1: Microabrasive wear test.

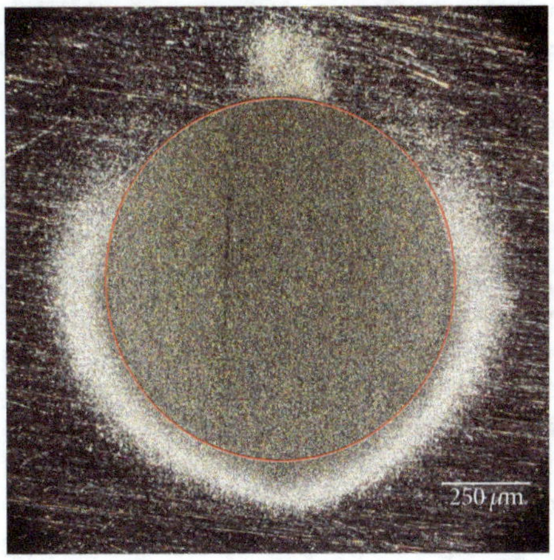

FIGURE 2: Crater on an AR sample with scuffed area around the crater.

Deviating from the common measurement method where two 90° tilled measurements are used, this method avoids the common problem that one of the distance measurements needs to be done in the scuffed edges, where a clear definition of the edge is impossible.

In this research, no separate analysis of the wear coefficient of the substrate (κ_s) and the wear coefficient of the coating (κ_c) was carried out, as only the collective wear coefficient was used to evaluate the effect on the wear resistance. This means that the results of the coated specimens always show a mean wear rate consisting of κ_s and κ_c (Figures 5, 6, and 7).

3. Calculation

The modified Archard equation was used [22] to calculate the wear rate of all uncoated specimens. According to this equation, the wear rate κ can be calculated by dividing the removed volume V by the sliding distance S and the applied normal load N:

$$\kappa = \frac{V}{S \cdot N}. \quad (1)$$

The total wear volume, V, can be calculated using the diameter of the generated crater b and the radius R of the steel ball [22]:

$$V \approx \frac{\pi \cdot b^4}{64 \cdot R} \quad \text{for } b \ll R. \tag{2}$$

4. Results

4.1. Wear Tests. As described in the methodology, the thickness of the coating was determined using a SEM. One representative image of a coated sample is shown in Figure 3. The coated surface can be clearly identified by the bright layer.

A mean coating thickness of 660 nm was measured.

To identify the wear mechanisms generated by the microabrasive wear test, a surface investigation of the generated craters was carried out. In Figure 4, two images of the crater surface are shown to illustrate the arising wear mechanism. This wear mechanism was observed on the surfaces of all tested specimens.

Figure 4(a) shows one crater generated with many continuous vertically oriented grooves. The grooves were generated by the microabrasive wear tests due to microcutting [23]. Their regular and continuous appearance indicates a two-body wear behavior. In this case, the abrasive wear particles are attached to the rotating ball and therefore generate straight grooves. The measured width of the grooves was between 0.6 μm and 1.5 μm. This fits well to the particle size of the abrasive slurry (see Section 2.7).

The SEM image in Figure 4(b) shows two types of grooves. First, the continuous groves which are related to particles attached to the rotating ball. Second, grooves show the repetitive occurrence of indentations. An explanation for the appearance of such indentations can be the combination of two- and three-body wear mechanisms. The cyclic indentations in the grooves can be generated by big, rotating particles. The distance between the indentations is approximately 1.9 μm, which is in accordance with the specification of the abrasive powder. It is therefore assumed that the cyclic indentations are generated by particles which have been pushed into the substrate. After the particles penetrated in the substrate, the ball rotation provoked anew displacement of the particle. The cyclic indentations indicate that such particles are not moved out of the contact but just pushed only one revolution further, before a new penetration into the substrate occurs. Since the particles are not permanently attached to the rotating ball, the wear mechanism for the scratches with cyclic indentations can be categorized as a sort of three-body wear.

Both sorts of scratches were found in all substrate materials, for which reason a mixed wear regime can be assumed in all tested specimens.

The results of the wear tests are shown in Figures 5–7. Each point shown in Figures 5 and 6 represents the average of three measurements carried out on three different equally treated specimens after a prescribed sliding distance. The error bars illustrate the deviation of the single measurement points. Figure 5 depicts the wear volume versus the sliding distance, S, multiplied by the contact force, N. Since N was kept constant, the diagram can also be understood as a plot of the removed volume against the sliding distance. From this figure, one can notice that the specimens tested under AR conditions show the highest amount of removed volume, while the Comb-2 specimens show the lowest one. The curves for the CrNC, the Comb-1, and the CTed specimens are similar.

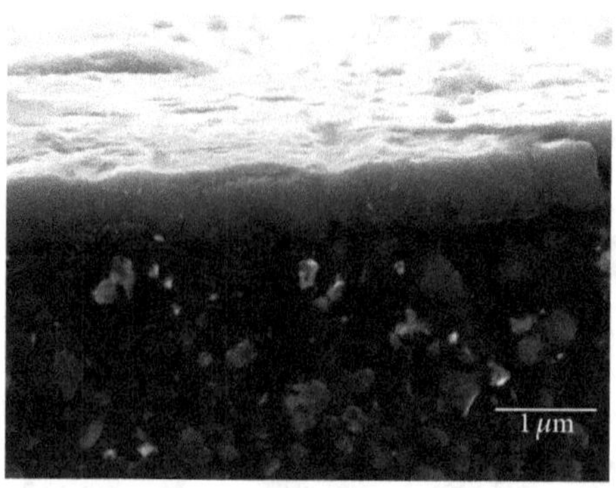

FIGURE 3: CrN surface coating.

Figure 6 shows the calculated wear rate of the same tests. One can notice that the AR specimens presented a relatively high wear rate while all modified specimens showed clearly lower wear rates. Focusing on the curve progression of the AR specimens, the wear rates decrease drastically in the beginning of the tests. A similar but less pronounced behavior was also found for the CTed specimens at low sliding distances. At high sliding distances, no further decrease occurred. In contrast, an increase in wear rate was found for all types of coated specimens. These comparable low wear rates were found at the first- or/and the second-measurement points. After the initial rise of the wear rate, which ends at S·N value of around 2.5 (third checkpoint), the curve remained constant. The mean wear rate for all tests conducted and the results for the substrate hardness measurements are shown in Figure 7. The error bars illustrate the standard deviation of each test series. The black circles over the columns represent the substrate hardness of each batch. Since the dispersion of the hardness measurements was below 2% for all test series, the error bars do not appear. To compare the results more precisely, the wear and hardness values of all batches are listed in Table 3.

The values in Table 3 were organized according to the wear rate of each batch to make the analysis of the results easier. The highest wear resistance was found for the coated specimens. Hence, the first raw (AR specimens) showed the highest and the last raw (Comb-2 specimens) shows the lowest wear rate.

Table 3 further displays that the substrate hardness is not following the pattern of the wear rate. The highest wear rate was found for the hardest material. Furthermore, it was found that the DCT did not lead to a hardness increase. It rather resulted in a slight reduction of the hardness. A significant hardness reduction varying from 30% up to 36% was found for all coated specimens. As described in the methodology, these values represent the hardness of the substrate measured after the removal of the surface coating.

TABLE 3: Wear and hardness results.

Test series	Mean wear rate (10^{-13} m^2/N)	Percentage of wear rate reduction	Hardness (HV1)	Percentage of hardness reduction
AR	6.91	—	69.0	—
CTed	5.06	27	66.6	3
Comb-1	4.79	31	44.9	35
CrNC	4.74	31	44.5	36
Comb-2	3.71	46	48.1	30

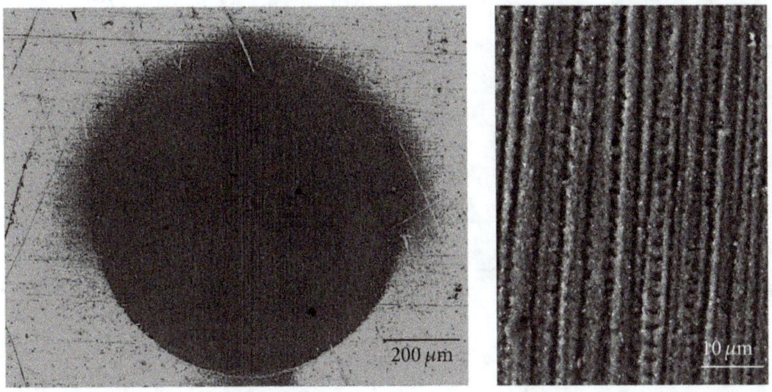

(a) (b)

FIGURE 4: (a) Crater in an AR specimen with continuous grooves. (b) SEM image of the grooves in the crater (AR specimen). (b) is a zoom in the crater shown in (a).

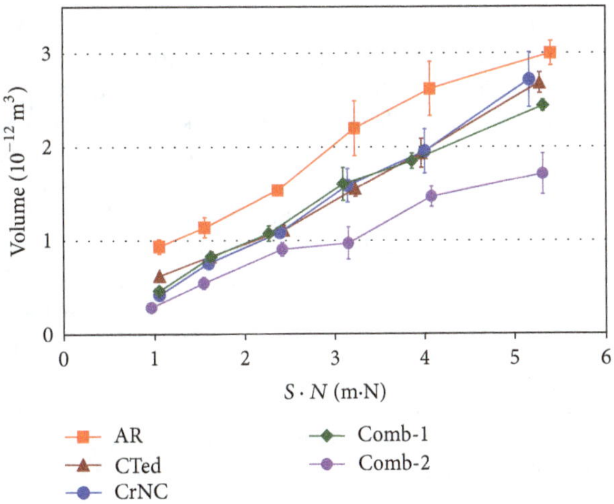

FIGURE 5: Wear volume versus sliding distance multiplied by the contact force.

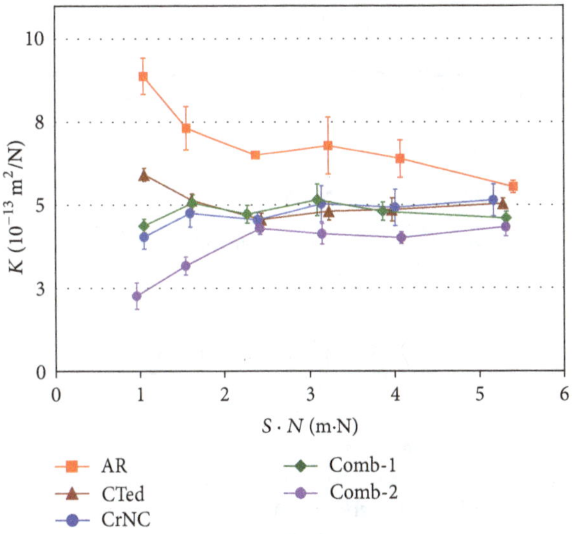

FIGURE 6: Wear rate versus sliding distance multiplied by the contact force.

4.2. X-Ray Diffraction. An X-Ray Diffraction analysis was carried out to investigate whether secondary phases or precipitates of the alloying elements could have been formed by the DCT. The occurrence of such phenomenon was mentioned by others [11–13] as the cause for an increase in wear resistance. Therefore, a comparative analysis between AR and CTed specimens was executed. The results of this analysis are shown in Figure 8.

In order to visualize both spectra in one diagram, the signal of the CTed specimens was numerically shifted (by the addition of a constant number to all values). While all distinct peaks occur in both spectra, the corresponding peak intensities differ significantly. The peak intensity can be correlated to the amount of each material or, more specifically, to the amount of the crystalline orientation of the material. The numbers over the peaks in the diagram represent the orientation of the aluminum crystals. Therefore, it needs to be mentioned that all identified peaks can be attributed to

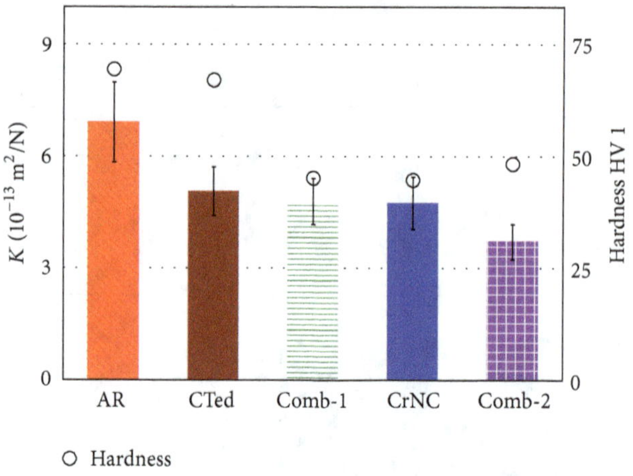

FIGURE 7: Wear rate and substrate hardness of all specimens types.

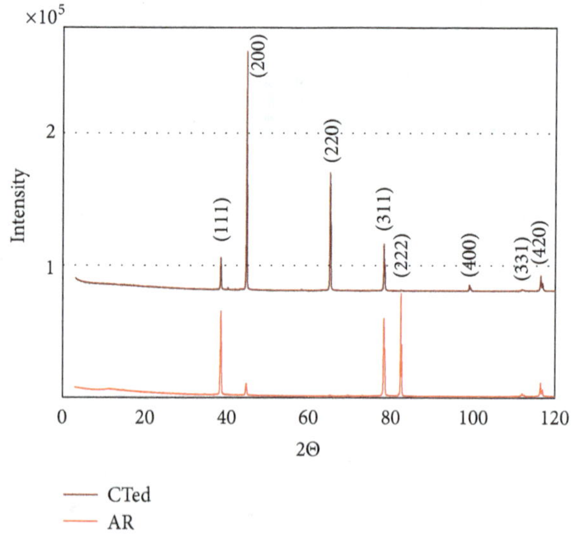

FIGURE 8: XRD spectrum of an AR and a CTed specimen.

aluminum. No secondary phases could be detected either in the CTed or in the AR specimens.

4.3. Microstructure. The microstructure investigation was carried out in order to try to find possible differences between the CTed and the AR specimens. The initial study included two basic metallographic methods: first, an analysis using optical microscopy of the polished surface (Figure 9) and second an analysis of the polished and etched surface (Figure 10). As the microstructure of the aluminum is influenced by the conformation process, it was ensured that the extracted samples had the same orientation. As a reference, the rolling direction of the primary aluminum block was taken. All shown microstructure images were taken from surface whose axial orientation can be described as longitudinal and lateral to the rolling direction.

In Figures 9(a) and 9(b), the polished surfaces of an AR and a CTed samples are shown. The bright surface indicates the aluminum structure. The dark marks are precipitates.

TABLE 4: Precipitations of the aluminum alloy.

Elements	Al (%)	Si (%)	Mg (%)	Fe (%)	O (%)
$Al_xFe_xSi_x$	92.4	3.2	0.5	3.9	—
Mg_2Si	90.8	4.6	4.3	0.3	—
SiO_2	73.3	7.9	0.1	—	18.7

Two types of precipitates were found with this metallographic analysis. The gray rod-shaped marks are characteristic for Mg_2Si and $Al_xFe_xSi_x$ precipitates. The smaller round shaped black precipitates are SiO_2. Energy disperse analysis (EDS) executed with a SEM confirmed this analysis. The percentage of concentrations found for all elements are listed in Table 4. Similar particles were also found in other 6XXX aluminum alloys [24]. As the precipitations are very small, the EDX beam always incorporated the substrate around the precipitates. Therefore, aluminum is always present in all the measurements. The presence of Mg and of Fe in the $Al_xFe_xSi_x$ and the Mg_2Si precipitates, respectively, is related to this same measurement procedure.

In Figure 10, the etched samples are shown. Similar to the unetched specimens, no distinctly higher or lower number of precipitates were found. The etched specimens further allow an evaluation of the grain size. Neither the orientation of the grains nor the grain size itself showed remarkable differences. Therefore, it can be mentioned that the DCT did not have any visually detectable effect with optical investigation methods.

4.4. Transmission Electron Microscope. To complete the microstructure study, a TEM analysis was carried out. Besides the big precipitates, detectable with the optical microscope, which have been identified via SEM/EDS, smaller precipitates with a size of approximately 50 nm and 4 nm were found using TEM (Figure 11(a)). The EDS/TEM analysis showed a concentration of copper in the bigger particles (Cu_xAl_x 50 nm, Figures 11(b) and 11(c)). Copper precipitates have also been found by others [25, 26] in a 6XXX aluminum alloy.

Furthermore, a large number of GP-zones (4 nm) were found arbitrarily distributed in the treated and the untreated specimens (Figures 12(a) and 12(b)).

Due to the very low size of the GP-zones, their chemical elements could not be distinctively identified. The EDX analysis revealed a random distribution of all alloying elements. The GP-zones were found in the AR and CTed specimens. A statement regarding the quantitative appearance of GP-zones could not be done once the distribution and the size of such zones were very disperse.

4.5. Differential Scanning Calorimetry (DSC). The graph in Figure 13 represents the difference of the DSC measurements between the CTed and the AR specimens. To obtain such a curve, two separate measurements were carried out. In both of them, an empty specimen container was taken as reference. Afterwards, the two curves were subtracted from each other to emphasize the differences between the modified and unmodified sample.

In order to clarify the nomenclature, the precipitation sequence for this type of aluminum alloy is presented. It is

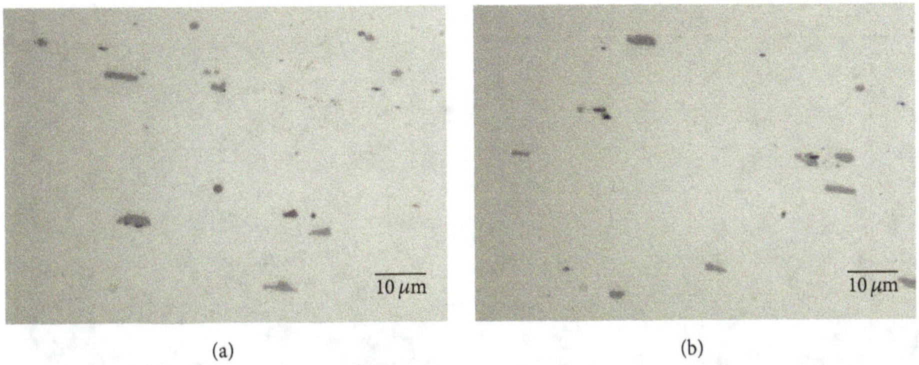

FIGURE 9: (a) Polished AR sample. (b) Polished CTed sample.

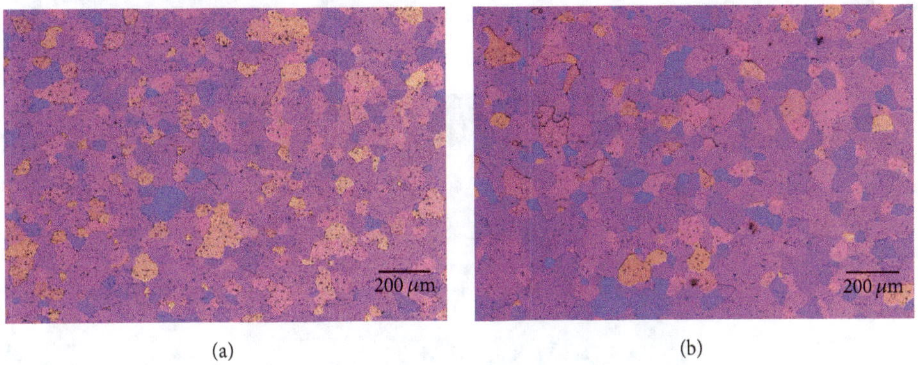

FIGURE 10: (a) Etched, AR sample. (b) Etched, CTed sample.

generally agreed that the precipitation process in Al-Mg-Si alloys can be divided into the following steps [27]:

$$\alpha\,(\text{sss}) \longrightarrow \text{GP-zones} \longrightarrow \beta'' \longrightarrow \beta' \longrightarrow \beta, \quad (3)$$

where α(sss) $\rightarrow$ is the supersaturated solid solution. GP-zones occur in form of irregular structures (Figure 12). β'' precipitates are usually needle-shaped and occur along $\langle 100 \rangle_{Al}$ plane, with a monoclinic structure. β' precipitates are rod-shaped and occur in the same plane. β phase (Mg_2Si) is usually formed as platelets along {100} plane of aluminum.

A wide endothermic peak (80°C–160°C) with the center at 115°C can be observed in Figure 13. This peak can be related to dissolution of GP-zones [28]. At higher temperatures, two exothermic peaks were revealed at approximately 235°C and 310°C. The peak centered at 235°C indicates the formation of β'' and the one at 310°C is a sign for β' precipitations [28, 29]. The endothermic ramp above 325°C is attributed to the start of a dissolution process for just formed precipitates.

5. Discussion

5.1. X-Ray Diffraction. The first method used to determine the formation of precipitates was XRD. The spectra depicted that the crystal orientations of the scanned specimens were different. It is unclear if the different orientations were caused by the cryogenic treatment or if they are related to the fact that two different specimens were examined. In this case, the difference in crystal orientation is related to inhomogeneity of the material.

However, no other elements nor second phases were found in the CTed or in the AR specimens. Although this result may seem surprising, as the substrate contains alloying elements, the reason for that is associated with the detection limit of the equipment (between 0.5% and 3%). Small particles may hence be hindering their detection using XRD. As shown in Table 1, the maximum alloy element concentration is specified with 0.5%.

5.2. Microstructure. No significant differences between the CTed and the AR specimens were found. Difficulties associated with this microstructural analysis method are the small sizes of the sections investigated. As the distribution of precipitates is not homogeneous, differences can be easily found (or not) depending on the section that was photographed. Therefore, often-found assumption that Mg_2Si, $Al_xFe_xSi_x$, or similar precipitates are generated by DCTs could not be confirmed. It is further unlikely that such big precipitates are directly generated by the DCT, as precipitates are usually formed by diffusion. The low temperatures during the DCT do not favor their formation.

The TEM/EDS analysis showed an appearance of smaller precipitates (size of 50 nm) with a high concentration of cupper. This led to the assumption that such precipitates consisted of metastable Cu_2Al [29]. Furthermore, small GP-zones were found with the TEM.

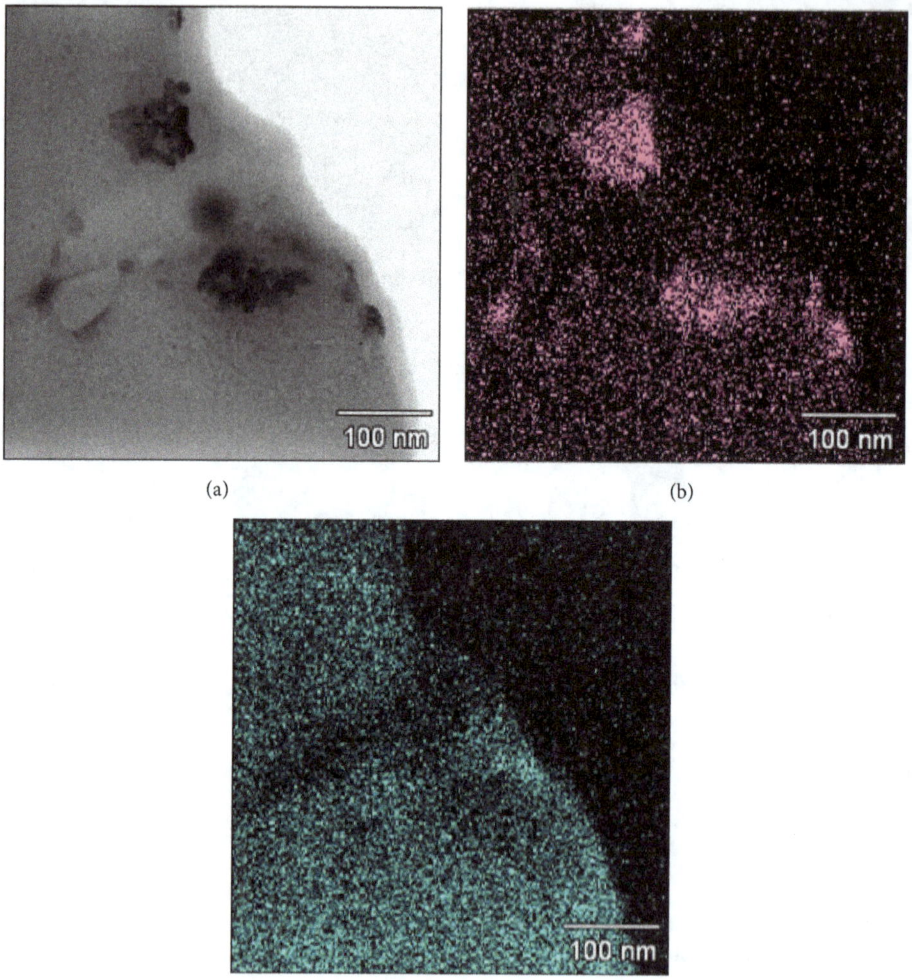

(a) (b) (c)

FIGURE 11: (a) Bright field TEM image of an AR sample (the small dark dots represent GP-zones). (b) Element analysis of the GP-zones (pink dots indicate the appearance of Cu). (c) Element analysis of the GP-zones (green dots indicate the appearance of Al).

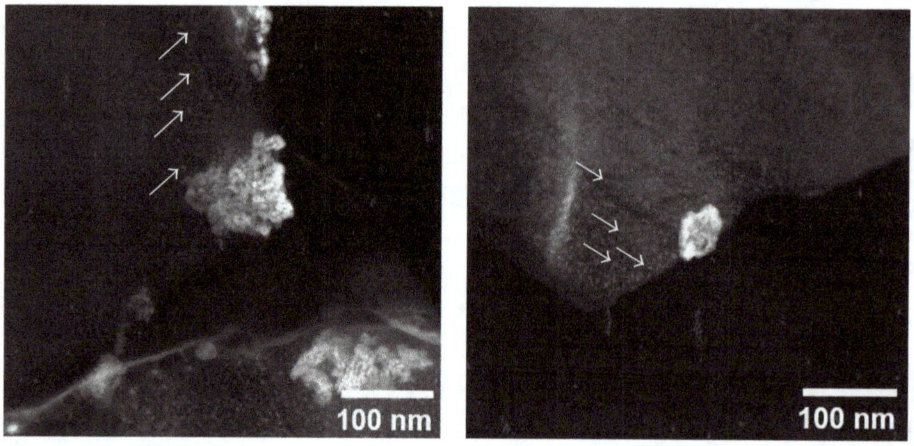

(a) (b)

FIGURE 12: (a) Dark field TEM image of an AR sample (small, widely distributed white dots in the size of approximately 4 nm are GP-zones). (b) Dark field TEM image of a DCT sample (small, widely distributed white dots in the size of approximately 4 nm are GP-zones).

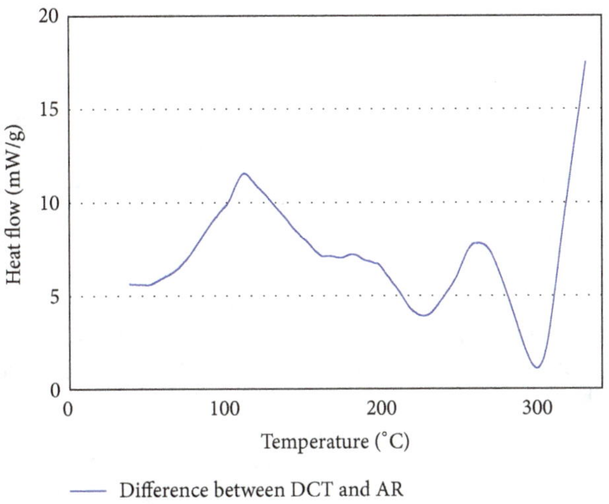

FIGURE 13: Differential scanning calorimetry (signal of the CTed sample with the AR material as reference).

The decreasing solubility of the alloying elements at lower temperatures is likely to be responsible for the generation of the GP-zones. At low temperatures, the movement of the atoms, as well as the distance between them, is decreasing. Therefore, it is energetically more favorable to emit chemical impurities from the solid aluminum solution, because the difference in the atomic radius disturbs the crystalline structure. This can lead to the emission of alloying elements, which accumulate in nanosized GP-zones. Whether the contraction of the aluminum structure is strong enough to emit alloying elements depends on the discrepancy between the atomic radii of the aluminum atoms and the alloying elements.

As shown with the comparative images in Figure 12, the small sections scanned with the TEM do not allow a serious proof that the DCT increases the number of GP-zones. Anyhow, GP-zones as well as the small precipitates (size 50 nm) were found in the AR and as well in the CTed specimens. Their occurrence in both types of specimens is not surprising, as the natural aging process also generates GP-zones [29]. A robust and conclusive screening of the microstructure with the TEM would be too cost- and time-consuming. For this reason, another analysis method, which includes a larger volume fraction, was used to confirm the generation of GP-zones by DCTs.

5.3. Differential Scanning Calorimetry. As explained above, the curve presented in Figure 13 shows the difference between the AR and the CTed specimens. The found peaks reveal a clear difference generated by the DCT. The curve depicted a wide endothermic peak between 80°C and 160°C, which can be associated with the dissolution of GP-zones. The applied temperature regime of the DSC analysis dissolved GP-zones that have been generated by the DCT. The volume under the peak indicates that the number of dissolved GP-zones is extensive. Therefore, an improved wear resistance due to the generation of GP-zones seems reasonable. The peak-width indicates that the size and presumably the components of the GP-zones are different, which confirms the assumption that the GP-zones cannot be related to only one type of alloying element.

The results illustrated in Figure 13 also showed two exothermic peaks (235°C and 310°C), attributed to the generation of β'' and β' precipitates. These peaks indicate an increased amount of precipitate generation in the CTed specimens. As explained before, it is unlikely that DCTs are generating big precipitates. The diffusion-controlled accumulation of alloying elements is not likely to appear at such low temperatures. It is more probable that the just ended dissolution process of the GP-zones (in between 80°C and 160°C) supported the generation of bigger precipitates. Other authors mentioned the possibility that larger GP-zones might not be dissolved completely due to fast heating rate of 20 K/min. Considering this, not completely dissolved GP-zones can act as nuclei for the precipitates and accelerate their formation [28, 29].

5.4. Hardness. To improve the understanding of the hardness results, it is again mentioned that the *hardness of the substrates* was measured and compared. The surface coating was smoothly removed (by polishing) before the hardness tests were carried out. This procedure was carried out to measure the hardness of the substrate and thereby uncover changes in the microstructure of the substrate.

As shown in Figure 7, the hardness measurements revealed a significant effect of the coating process on the hardness of the substrate. All coated specimens showed a significant decrease in the hardness. Therefore, the discussion of the coated and uncoated specimens is done separately.

5.4.1. Uncoated Specimens. The hardness of the CTed specimens showed a hardness reduction of 3% compared to the AR specimens. This result was not expected, since the CTed specimens contain a significantly higher number of GP-zones. The results of other researchers might explain these conflicting results. It was found that DCTs reduced residual stresses in aluminum alloys [6, 30]. An effect of residual stresses on the hardness of materials is widely known and was already proven for similar aluminum alloys [31, 32]. Taking into account that the reduction of residual stresses decreases the hardness and that the generation of GP-zones increases the hardness, the sum of both effects can be either an increase or a decrease in the hardness of the aluminum alloy.

5.4.2. Coated Specimens. The batches CrNC, Comb-1, and Comb-2 were coated with CrN. One reason for the selection of the CrN coating was its comparably low deposition temperature. As evidenced by these results, the temperature of the deposition process, even though lower than for other coatings, affected the microstructure of the alloy. The exact temperature during the coating process was not provided by the coating company. The surface temperature was estimated to be between 100°C and 150°C.

As explained in Section 5.3, a wide endothermic peak, which indicated that the GP-zones were dissolved, appeared between 80°C and 160°C during the DSC measurements. Assuming the deposition time and the temperature during the coating process, a dissolution of GP-zones and small precipitates in the near surface area is highly possible. Such behavior was found by other researchers [29], who applied a preaging process (at 100°C), that also led to dissolved GP-zones and a reduced hardness. The fact that the measured hardness is only slightly higher than the hardness of pure aluminum supports this assumption [33].

Considering this in the actual discussion, all coated batches were unintentionally heat-treated during the coating process, for which reason their hardness is comparably low. The hardness of the batches Comb-1 and CrNC is nearly equal. The hardness of the Comb-2 batch increased 6%, compared with its coated partners. This increase can be contributed to the increased number of GP-Zones, generated by the subsequently executed DCT.

5.5. Wear Results. As shown in Figure 5, the AR specimens had the highest wear rate for all sliding distances. Especially at a low sliding distances (the first-two checkpoints), the wear rates are markedly high. The same behavior was found for the CTed specimens. Some percentage of this initial high wear rate can be associated with an overestimation of the removed volume at low sliding distances. It occurs due to the difficulty in measuring the diameter of the crater [22]. Therefore, a slight overestimation can be considered for the first points. Notice that all uncoated specimens showed a pronounced scuffed area around the crater (Figure 2). This turned the measurement of its diameter and consequently the determination of the volume loss and of the wear rate into a less precise task. Other authors found similar behavior especially for uncoated soft materials [20]. In the case of a coated surface, the scuffed area around the crater was much smaller and this type of measurement inaccuracy could be neglected.

In contrast, a low wear rate was found for the first-measurement points of all coated specimens. It is easy to perceive that this effect is generated by the wear resistant coating. Subsequent tests carried out to detect the sliding distance necessary to perforate the surface coating indicated a breakthrough after a sliding distance of 0.3 m. Therefore, the first-measurement point represents a mixture of the surface coating and the substrate material and does not represent the wear rate of the pure coating. As the coating is only present on the surface, the biggest wear resistance can hence be expected for low sliding distances. After the perforation of the coating, the volume ratio between coating and substrate is continuously decreasing. Hence, the wear resistance slightly decreased with an increase in sliding distance. This effect can be seen in Figure 6. In general, the wear rate reduction is relatively high considering that the height of the deposited coating was only 660 nm (Figure 3).

The CTed specimens led to a similar reduction in wear rate as the one observed for the CrNC specimens. As shown by the DSC tests, additional GP-zones were generated due to the DCT. Such zones are responsible for the improved wear resistance. As widely known, GP-zones and precipitates exhibit a higher wear resistance than the pure aluminum. As reported in Table 1, several alloying elements are present in the aluminum alloy, for which reason it is expected that the GP-zones consisted of diverse alloying elements.

In order to generate an aluminum alloy with superior wear resistance, one batch of specimens was first modified cryogenically and afterwards coated with CrN (batch Comb-1). Since the DCT changes the microstructure in the entire material and the CrN mainly acts in near surface regions, a superposition of both effects was expected for the specimens treated with both modification methods. However, such superposition did not occur for the Comb-1 batch. As explained before, the coating process had a side effect on the substrate material, which dissolved GP-zones and small precipitates. It therefore also affected the wear resistance. For this reason, the results of the Comb-1 specimens were similar to the CrNC specimens.

To confirm that a superposition of both effects is possible, an additional test series was carried out. The Comb-2 batch was first coated with CrN and afterwards DCT. During the coating process, the GP-zones were dissolved in the same way. However, the subsequent DCT generated new GP-zones and therefore reversed the effects of the coating process. Using this modification sequence, the combination of an increased number of GP-zones with a surface coating generated a significant improvement (of 48%) in wear resistance (Figure 7).

6. Conclusion

The results of the wear tests showed that DCT led to an improved wear resistance of the aluminum alloy. CrN surface coating showed a likewise improvement of the wear behavior. If both modification methods are executed in the correct order (first CrN coating and second DCT), a superposition of such effects can be reached. A reversed order does not lead to similar effects, since the temperature rise, which occurs during the coating process, eliminates the microstructural changes and dissolves the GP-zones.

The microstructural analyses revealed that DCT did not generate large precipitates or other significant microstructural changes like in the grain size. Such effects are diffusion controlled and therefore require higher temperatures. However, a distinct generation of small and widely dispersed GP-zones was detected with DSC. The results showed a clearly pronounced dissolution of GP-zones for the cryogenically treated specimens. The authors claim that the formation of GP-zones generated during the DCT was the main reason for the improved wear resistance.

Competing Interests

The authors declare that there are no competing interests regarding the publication of this paper.

Acknowledgments

The authors would like to acknowledge the Brazilian program Science Without Borders from CNPq for the financial support of the project (401671/2013-5 and 314034/2013-7). They also would like to thank Kyros Tecnologia Ltda. for carrying out deep cryogenic treatments and the Department of Materials Engineering in the Pontifícia Universidade

Católica do Rio de Janeiro (PUC-Rio) for conducting TEM measurements.

References

[1] R. J. Arsenault, "The strengthening of aluminum alloy 6061 by fiber and platelet silicon carbide," *Materials Science and Engineering*, vol. 64, no. 2, pp. 171–181, 1984.

[2] S. Seifeddine, S. Johansson, and I. L. Svensson, "The influence of cooling rate and manganese content on the β-Al5FeSi phase formation and mechanical properties of Al-Si-based alloys," *Materials Science and Engineering A*, vol. 490, no. 1-2, pp. 385–390, 2008.

[3] H. Torabian, J. P. Pathak, and S. N. Tiwari, "Wear characteristics of Al-Si alloys," *Wear*, vol. 172, no. 1, pp. 49–58, 1994.

[4] B. Podgornik, F. Majdic, V. Leskovsek, and J. Vizintin, "Improving tribological properties of tool steels through combination of deep-cryogenic treatment and plasma nitriding," *Wear*, vol. 288, pp. 88–93, 2012.

[5] D. Mohan Lal, S. Renganarayanan, and A. Kalanidhi, "Cryogenic treatment to augment wear resistance of tool and die steels," *Cryogenics*, vol. 41, no. 3, pp. 149–155, 2001.

[6] P. Chen, L. Malone, R. Bond, and P. Torres, "Effects of cryogenic treatment on the residual stress and mechanical properties of an aerospace aluminum alloy (NASA)," 2000.

[7] K. E. Lulay, K. Khan, and D. Chaaya, "The effect of cryogenic treatments on 7075 aluminum alloy," *Journal of Materials Engineering and Performance*, vol. 11, no. 5, pp. 479–480, 2002.

[8] X.-Q. Jiang, N. Li, H. He, X.-J. Zhang, C.-C. Li, and H. Yang, "Effect of cryogenic treatment on mechanical properties and microstructures of 3102 Al-alloy," *Materials Science Forum*, vol. 546–549, no. 2, pp. 845–848, 2007.

[9] W. D. Zhang, J. Y. Yang, J. Z. Dang, Y. Liu, and H. Xu, "Effects of cryogenic treatment on mechanical properties and corrosion resistance of LC4 aluminum alloy," *Advanced Materials Research*, vol. 627, pp. 694–697, 2013.

[10] J. Wang, X. Xue, Z. Yang, H. Zhang, and Y. Zhou, "Effect of cryogenic treatments on mechanical properties of 2A11 aluminum alloy," *Advanced Materials Research*, vol. 146-147, pp. 1646–1650, 2011.

[11] S. Zhirada, *Effect of cryogenic treatment on the mechanical properties of steel and aluminum alloys [M.S. of Applied Science]*, Mechanical and Industrial Engineering, Library and Archives Canada, Concordia University, Montreal, Canada, 2005.

[12] G. Li, H. Wang, Y. Cai, and X. Yuan, "Behavior of $\beta(Mg_{17}Al_{12})$ phases in mg-al alloy subject to cycling cryogenic treatment," *Key Engineering Materials*, vol. 575-576, pp. 390–393, 2014.

[13] S. Karabay, M. Yilmaz, and M. Zeren, "Investigation of extrusion ratio effect on mechanical behaviour of extruded alloy AA-6101 from the billets homogenised-rapid quenched and as-cast conditions," *Journal of Materials Processing Technology*, vol. 160, no. 2, pp. 138–147, 2005.

[14] K. Gu, J. Wang, and Y. Zhou, "Effect of cryogenic treatment on wear resistance of Ti-6Al-4V alloy for biomedical applications," *Journal of the Mechanical Behavior of Biomedical Materials*, vol. 30, pp. 131–139, 2014.

[15] G. Y. Liang, T. T. Wong, J. M. K. MacAlpine, and J. Y. Su, "A study of wear resistance of plasma-sprayed and laser-remelted coatings on aluminium alloy," *Surface and Coatings Technology*, vol. 127, no. 2-3, pp. 233–238, 2000.

[16] I. P. Mertsalo, V. T. Yavors'kyi, M. D. Klapkiv, and R. S. Mardarevych, "Wear resistance of anodic-spark coatings on aluminum alloys," *Materials Science*, vol. 39, no. 1, pp. 136–139, 2003.

[17] S. Guanhong, H. Xiaodong, J. Jiuxing, and S. Yue, "Parametric study of Al and Al_2O_3 ceramic coatings deposited by air plasma spray onto polymer substrate," *Applied Surface Science*, vol. 257, no. 17, pp. 7864–7870, 2011.

[18] F. Zhou, C.-M. Suh, S.-S. Kim, and R.-I. Murakami, "Tribological behavior of CrN coating on aluminum alloys deposited by arc ion plating," *Journal of Materials Research*, vol. 17, no. 12, pp. 3133–3138, 2002.

[19] F. Zhou, C.-M. Suh, S.-S. Kim, and R.-I. Murakami, "Sliding-wear behavior of TIN- and CrN-coated 2024 aluminum alloy against an Al_2O_3 ball," *Tribology Letters*, vol. 13, no. 3, pp. 173–178, 2002.

[20] R. I. Trezona and I. M. Hutchings, "Three-body abrasive wear testing of soft materials," *Wear*, vol. 233-235, pp. 209–221, 1999.

[21] K. Adachi and I. M. Hutchings, "Wear-mode mapping for th micro-scale abrasion test," *Wear*, vol. 255, no. 1–6, pp. 23–29, 2003.

[22] M. G. Gee, A. J. Gant, I. M. Hutchings et al., "Results from an interlaboratory exercise to validate the micro-scale abrasion test," *Wear*, vol. 259, no. 1–6, pp. 27–35, 2005.

[23] K.-H. Z. Gahr, "Wear by hard particles," *Tribology International*, vol. 31, no. 10, pp. 587–596, 1998.

[24] W. Q. Song, P. Krauklis, A. P. Mouritz, and S. Bandyopadhyay, "The effect of thermal ageing on the abrasive wear behaviour of age-hardening 2014 Al/SiC and 6061 Al/SiC composites," *Wear*, vol. 185, no. 1-2, pp. 125–130, 1995.

[25] J. Ouyang, E. Yarrapareddy, and R. Kovacevic, "Microstructural evolution in the friction stir welded 6061 aluminum alloy (T6-temper condition) to copper," *Journal of Materials Processing Technology*, vol. 172, no. 1, pp. 110–122, 2006.

[26] H. Kaçar, E. Atik, and C. Meriç, "The effect of precipitation-hardening conditions on wear behaviours at 2024 aluminium wrought alloy," *Journal of Materials Processing Technology*, vol. 142, no. 3, pp. 762–766, 2003.

[27] G. B. Burger, A. K. Gupta, P. W. Jeffrey, and D. J. Lloyd, "Microstructural control of aluminum sheet used in automotive applications," *Materials Characterization*, vol. 35, no. 1, pp. 23–39, 1995.

[28] W. F. Miao and D. E. Laughlin, "Differential scanning calorimetry study of aluminum alloy 6111 with different pre-aging treatments," *Journal of Materials Science Letters*, vol. 19, no. 3, pp. 201–203, 2000.

[29] W. F. Miao and D. E. Laughlin, "Effects of Cu content and preaging on precipitation characteristics in aluminum alloy 6022," *Metallurgical and Materials Transactions A: Physical Metallurgy and Materials Science*, vol. 31, no. 2, pp. 361–371, 2000.

[30] Q.-C. Wang, L.-T. Wang, and W. Peng, "Thermal stress relief in 7050 aluminum forgings by uphill quenching," *Materials Science Forum*, vol. 490-491, pp. 97–101, 2005.

[31] G. P. Dolan and J. S. Robinson, "Residual stress reduction in 7175-T73, 6061-T6 and 2017A-T4 aluminium alloys using

quench factor analysis," *Journal of Materials Processing Technology*, vol. 153-154, no. 1–3, pp. 346–351, 2004.

[32] W. R. LaFontaine, C. A. Paszkiet, M. A. Korhonen, and C.-Y. Li, "Residual stress measurements of thin aluminum metallizations by continuous indentation and x-ray stress measurement techniques," *Journal of Materials Research*, vol. 6, no. 10, pp. 2084–2090, 1991.

[33] Y. Ito and Z. Horita, "Microstructural evolution in pure aluminum processed by high-pressure torsion," *Materials Science and Engineering A*, vol. 503, no. 1-2, pp. 32–36, 2009.

12

Simulation of Solidified Microstructure and Experimental Comparative Study of Twin-Roll Casting Aluminum Alloys

Teng Ma,[1,2] Junting Zhang,[3] Xiaochao Cui,[1,3] and Xiaosi Sun[1]

[1]*School of Materials Science and Engineering, Taiyuan University of Science and Technology, Taiyuan 030024, China*
[2]*Huake Institute, Taiyuan University of Science and Technology, Taiyuan 030024, China*
[3]*School of Applied Science, Taiyuan University of Science and Technology, Taiyuan 030024, China*

Correspondence should be addressed to Xiaochao Cui; cuixiaochao@sohu.com

Academic Editor: Paolo Ferro

A coupled macro-micro mathematical model of twin-roll casting was established in the study. The continuous solidification process of Al-10Mg aluminum alloys in front of the nip point was numerically simulated by the cellular automaton method and the solidification microstructure, dendritic grain radius, and secondary dendrite arm spacing were obtained through simulation to predict the mechanical property of wedge strips. In order to verify the reliability of the simulation results, the metallographic examination and tensile tests were performed with the as-cast specimens. The results showed that the gain size, the distribution characteristics of various grain regions, dendrite arm spacing, and the yield strengths obtained from simulation were consistent with experimental results.

1. Introduction

Twin-roll casting is characterized by the short process, low energy consumption, and low manufacturing cost and widely used to produce aluminum alloy strips with good mechanical properties [1–4]. The solidification structure prepared by continuous casting affects alloy properties as well as the quality and mechanical properties of alloy strips [5–7]. Therefore, it is necessary to develop new rolling technologies and deeply explore the relationship between the casting process, solidification structure, and metal performance.

The solidification process of strips can be simulated with the computer simulation technology to study nucleation and grain growth and analyze solidification structure and geometric features. The formation process of the solidification structure can be displayed directly with the simulation technology. Several mathematical models were developed for the twin-roll casting process. Some micro mathematical models were established to simulate the evolution process of the solidification microstructure in a micro region by Liu et al. [8], S. D. Chen and J. C. Chen [9], and Huang et al. [10]. However, the complete solidification process of the strip or the prediction of mechanical properties of the as-cast alloy strip was not reported. The casting-state microstructure of aluminum alloy or other alloys had been explored in the twin-roll casting experiments [11–13]. However, the solidification evolution was seldom explored.

According to cellular automata theory, the mathematical model of the solidification structure of twin-roll continuous casting was established in this study. With Al-10Mg aluminum alloys as the study target, the continuous solidification process in front of the nip point was numerically simulated in Procast. The dendritic grain radius and the secondary dendrite arm spacing were obtained to predict the mechanical properties of wedge strips. Finally, the mathematic model was verified through metallographic examination and tensile tests.

2. Mathematic Models

2.1. Macroscopic Model. Macro temperature field is the basis for simulating microstructures. According to the heat transfer process between the metal liquid, casting roller, and the environment, Xiong et al. proposed the mathematical model based on unsteady heat conduction [14]:

$$\rho c_p \frac{\partial T}{\partial t} = \frac{\partial}{\partial x}\left(\lambda \frac{\partial T}{\partial x}\right) + \frac{\partial}{\partial y}\left(\lambda \frac{\partial T}{\partial y}\right) + \frac{\partial}{\partial z}\left(\lambda \frac{\partial T}{\partial z}\right) - \rho L \frac{\partial f_s}{\partial t}, \quad (1)$$

where ρ is alloy density; c_p is a specific heat at constant pressure; λ is thermal conductivity; L is latent heat; and f_s is solid fraction.

2.2. Microscopic Model

2.2.1. Nucleation Model. During the twin-roll continuous casting, the undercooling of the liquid phase is generally far less than the undercooling required for uniform nucleation and liquid-phase surface oxidation, broken dendrites, and heterogeneous nucleation exist in the liquid metal. Therefore, the heterogeneous nucleation model is adopted to describe the solidification process of liquid metal. At a given undercooling ΔT, the continuous nucleation model based on Gaussian distribution and proposed by Gandin and Rappaz [15] is adopted:

$$\frac{dn}{d(\Delta T)} = \frac{n_{\max}}{\sqrt{2\pi}\Delta T_0} \exp\left[-\frac{1}{2}\left(\frac{\Delta T - \Delta T_N}{\Delta T_0}\right)^2\right], \quad (2)$$

where ΔT_0 is the standard deviation; $n_{\max}$ is the maximum density of nuclei; and ΔT_N is the mean nucleation undercooling.

2.2.2. Growth Model. Twin-roll continuous casting is a near-rapid solidification process. Therefore, the modified KGT model is adopted to describe the constrained dendrite tip growth. The dynamic effect in the dendrite growth is not significant [16], so compositional undercooling and curvature undercooling other than the kinetic degree of the undercooling of dendrite tip are considered in the model. Assuming that the equilibrium distribution coefficient of alloy and diffusion coefficients of solute remain unchanged, the modified KGT model is given as [8]

$$\Delta T = \Delta T_c + \Delta T_R,$$
$$\Delta T_c = mc\left[1 - \frac{1}{1 - Iv(\mathrm{pe})(1-k)}\right], \quad (3)$$
$$\Delta T_R = \frac{2\Gamma}{R},$$

where c is the solute concentration at the liquid interface; k is the solute partition coefficient; (pe) is the Ivantsov function; and Γ is the Gibbs-Thomson coefficient. According to the stability principle of the dendrite tip, the growth velocity of dendrite tip v and the growing radius of dendrite tip R can be calculated as follows:

$$R = 2\pi\sqrt{\frac{\Gamma}{mG_c\zeta_c - G}},$$
$$\mathrm{pe} = \frac{VR}{2D_L}, \quad (4)$$
$$v^2\left(\frac{\pi^2\Gamma}{\mathrm{pe}^2 D_L^2}\right) + v\left\{\frac{m_L C_0(1-K_0)}{D_L[1-(1-K_0)Iv(\mathrm{pe})]}\right\} + G = 0.$$

The secondary dendrite spacing is an important parameter to describe characteristics of solidification structure and can reflect the micro segregation and other casting defects. The secondary dendrite spacing, λ_2, is given as [17]

$$\lambda_2 = 5.5\left(Mt_f\right)^{1/3}, \quad (5)$$

where

$$t_f = \frac{\Delta T_0}{G_L R},$$
$$M = \frac{T_m \Gamma D_L \ln\left(C_f/C_0\right)}{m_L(k-1)\left(C_f/C_0\right)}, \quad (6)$$

where C_f is the final composition of liquid metal of dendrite root. The growth velocity of columnar crystals and dendrite tip as well as the dendrite spacing in the process of twin-roll continuous casting can be expressed by the modified KGT model.

3. Experiment and Model

Table 1: Chemical composition of Al-10Mg alloy.

Al	Mg	Si	Fe	Cu	Zn	Other components (Mn + Ni)
89.70	9.90	0.11	0.18	0.01	0.03	0.01

3.1. Experimental. The alloy material used in the study is Al-10Mg aluminum alloy (Table 1). Aluminum ingot and magnesium ingot were melted in an airtight furnace according to the proportion of Al-10Mg aluminum alloy and inert gases were injected into the furnace to protect the melting process. The casting roller size is Ø400 mm/Ø300 mm; the roll gap width is 2–4 mm; the gap between nozzle and roll is 0.10~0.15 mm. The chamber was preheated to 350~400°C and the temperature was maintained. Then the twin-roll caster was started and molten metal was poured into the chamber at the pouring temperature of 690°C. The casting speed of the twin-roll caster was 1 m/min.

When the aluminum alloy strip went through the rolling area, plastic deformation occurred and primary crystal shape was modified. In order to get initial solidification, the wedge strip was taken out after casting and processed to the length

Table 2: Thermal physical properties of Al-10Mg alloy.

Melting temperature (°C)	Specific heat ($J\cdot kgK^{-1}$)	Density of liquid ($kg\cdot m^{-3}$)	Thermal conductivity of liquid ($W\cdot mk^{-1}$)
607	1088.6	2.55	108.9

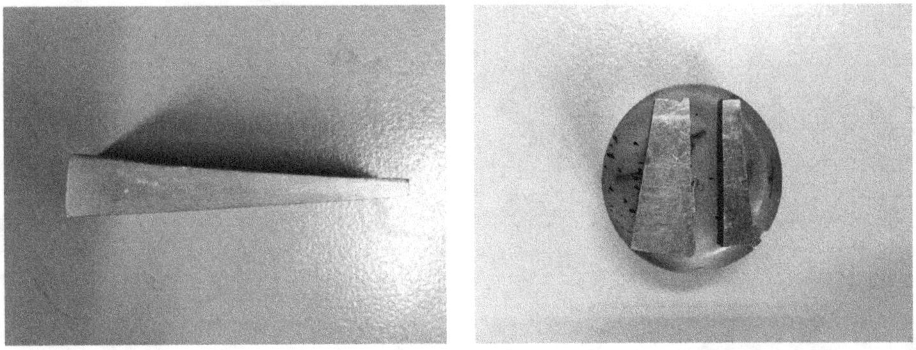

(a) (b)

Figure 1: Twin-roll cast Al-10Mg alloy wedge-shaped plate specimen: (a) wedge-shaped plate; (b) mounting specimen.

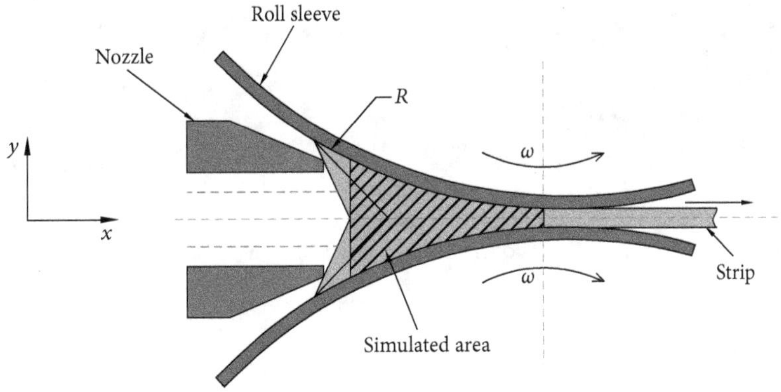

Figure 2: Diagram of the twin-roll casting process.

of 15 mm (Figure 1). After polishing, the strip was corroded in 50% HF. The rolled plane and the longitudinal section of the sample were observed by the metallographic microscope (LeiCADM2700M).

Three groups of standard samples (10 mm × 4 mm × 3 mm) were taken out along three directions: the horizontal direction, rolling direction, and the angle of 45°. The tensile tests were carried out with universal testing machine to obtain mechanical properties and stress-strain curve of Al-10Mg (strain rate $\varepsilon = 5 \times 10^{-4}\ s^{-1}$).

3.2. Geometrical Model. In the process of twin-roll continuous casting, the continuous nucleation of the molten metal occurred at the meniscus of the roller when the rollers were reversely rotated. The cast-rolled strip was thickened constantly in the growth of dendrite. When the strip was close to the nip point, the solidification process was completed and the initial structure was formed. Therefore, the continuous solidification process in front of the nip point is numerically simulated in the study. The calculation area is shown in Figure 2 and the thermal physical properties of Al-10Mg alloy are shown in Table 2 [18]. The microstructure was analyzed and the mechanical properties were predicted. The simulation parameters were consistent with the production parameters. The thickness of the strip is 3 mm. Figure 2 shows the schematic diagram of the twin-roll casting process. Under the conditions of continuous casting, the heat transfer in the roll barrel direction is far less than that in the other two directions. Therefore, the changes in the grain morphology mainly occurred in the longitudinal section and the rolled plane of the sample, but the grain morphology was the same in the strip width direction. Therefore, in order to simplify the model and reduce the calculation load, a certain section along the roll barrel direction was selected for analysis in the model under the assumption of adiabatic side seal. The strip width is 5 mm in the simplified model.

4. Results and Discussion

Figures 3 and 4, respectively, show the simulated and experimentally obtained microstructures of rolled plane and longitudinal section, which are 1 mm away from the roll in the continuous casting process. Figure 5 shows the schematic diagram of sample selection for microstructure

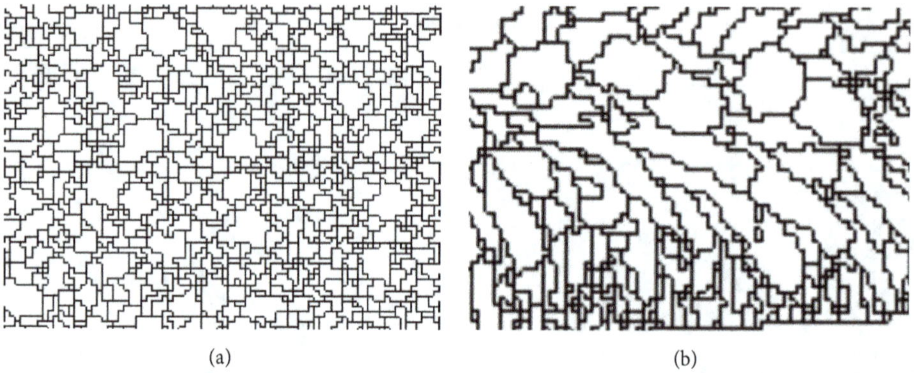

FIGURE 3: Solidification simulation of twin-roll cast Al-10Mg alloy wedge-shaped plate: (a) rolled plane; (b) longitudinal section.

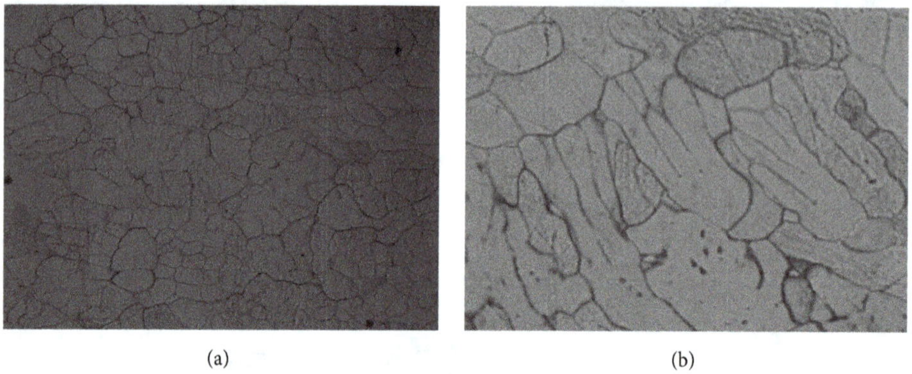

FIGURE 4: Solidification simulation of the cross section and lengthwise section of twin-roll cast Al-10Mg alloy wedge-shaped plate: (a) rolled plane; (b) longitudinal section.

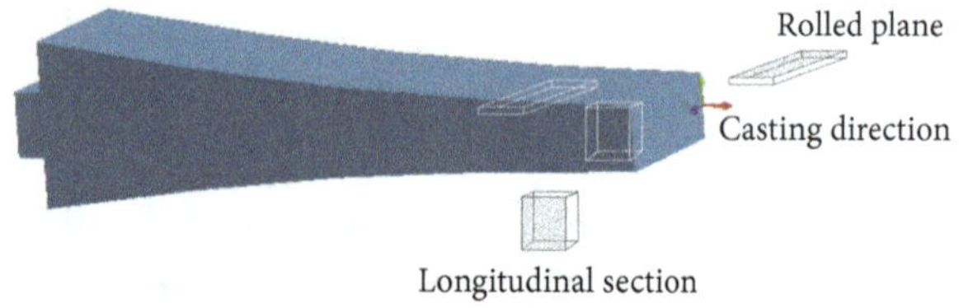

FIGURE 5: Schematic illustration of sample selection for microstructure and texture analysis.

and texture analysis. On the surface of rolled plane, the size of grains is uniform and equiaxed (Figure 3(a)). The grains near the casting zone are relatively larger (the left side of Figure 3(a)), whereas the grains near the rolling zone are relatively smaller (the right side of Figure 3(a)). In the simulation results, the grain size is $3.2 \times 10^{-8}\,m^2$–$4.9 \times 10^{-7}\,m^2$, whereas metallographic test results indicate that the grain size is $8.0 \times 10^{-8}\,m^2$–$8.1 \times 10^{-7}\,m^2$. The above results prove that the calculated results are well consistent with experimental results. The simulated grain size is a little smaller than the result by metallographic test because of the assumption made in the theoretical models. In addition, the solidification process is a complicated physical process and it is difficult to exactly determine the changing rule of thermal properties with temperature. In Figure 3(b), the solidification structure of longitudinal section from the roll surface into the strip center is, respectively, composed of the fine grain zone, columnar grain zone, and equiaxed crystal zone. The predicted results including the grain shape and grain size are consistent with the results of metallographic tests.

Figure 6(a) shows the fraction solid distribution of molten metal at the steady state. Figures 6(b) and 6(c), respectively, show primary dendritic radius distribution and secondary dendrite arm spacing distribution. The primary dendritic radius and the secondary dendrite arm spacing are important parameters for estimating the solidification structure and can reflect various roll casting defects. The solidification process of the solution reaches a stable state after the casting process becomes stable. The primary dendritic radius in the front of columnar grain growth zone reflects the growth condition of columnar grain growth. The smaller radius indicates the more obvious growth tendency. Figure 6(b) shows that the primary dendritic radius of strip surface is smaller than that of strip canter. As shown in Figure 6(c), the secondary dendrite arm spacing is gradually decreased from the outer zone to the inner zone along the thickness direction and is gradually increased from the outer zone to the inner zone along the rolling direction with the solidification. The largest secondary dendrite arm spacing occurs near the nip point and is about 19.55 μm. The measured result (23 μm) in Figure 4(a) is basically consistent

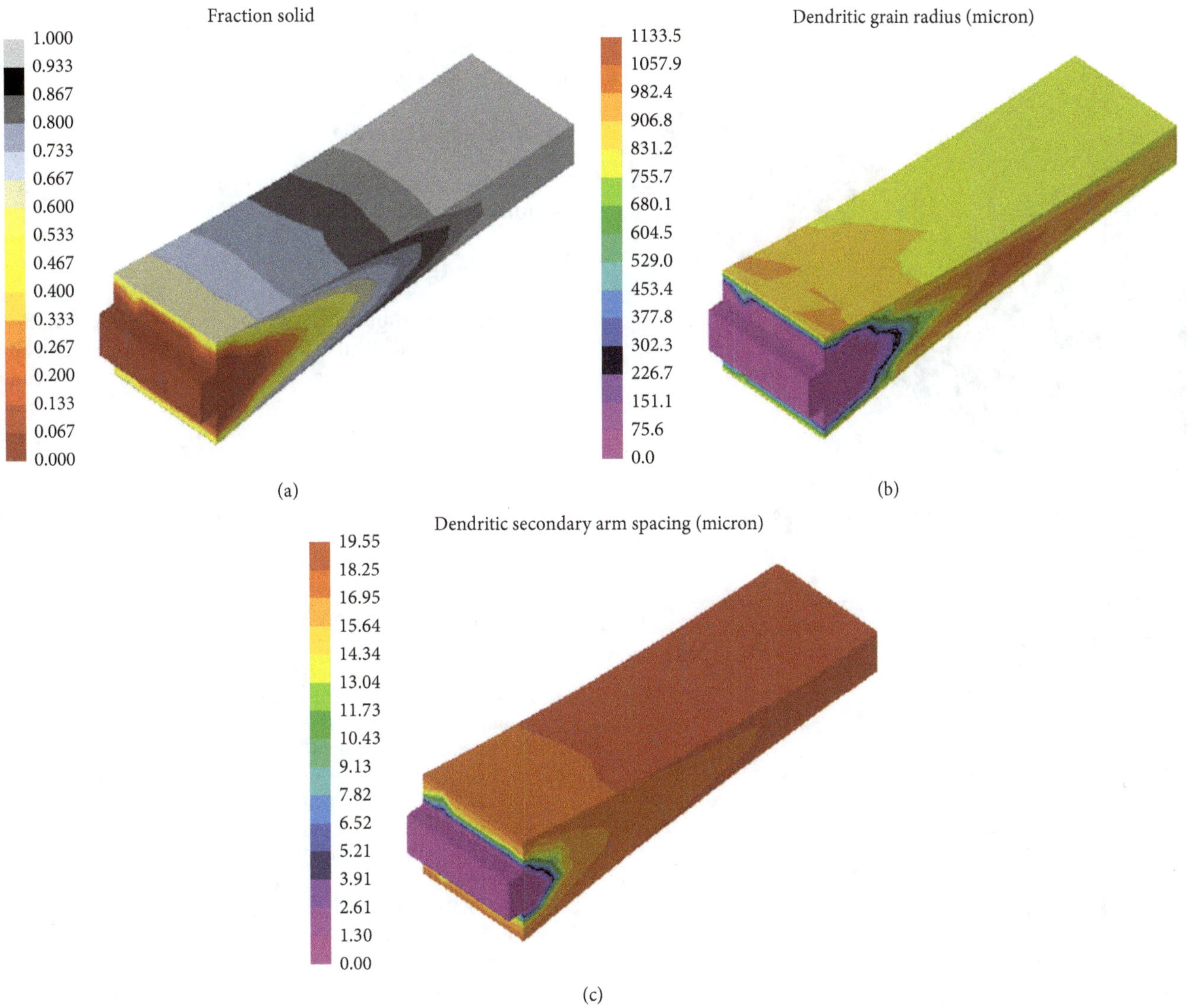

FIGURE 6: Solid fraction graph, dendritic grain radius graph, and secondary dendrite arm spacing graph of twin-roll cast Al-10Mg alloy wedge-shaped plate: (a) solid fraction graph; (b) dendritic grain radius graph; (c) secondary dendrite arm spacing graph.

with the simulation result, indicating that the model is applicable to predict the dendrite arm spacing.

Roll casting solidification is a sub-rapid solidification process. Compared with the traditional casting structures, the strip structure possesses the smaller grain size, lower segregation degree, and excellent mechanical properties. In this paper, the mechanical properties of the strip were predicted based on the solidification structure obtained by the mathematic model of the casting process. The predicted yield strength of twin-roll cast Al-10Mg strip was 158 Mpa, which was a little higher than the experimental result, 142 Mpa (Figure 7), due to the defects such as segregation, gas cavity, and the heterogeneity of grain size. The experimental results of uniaxial tensile mechanical properties of Al-10Mg aluminum alloy casting strip at room temperature are shown in Table 3. The corresponding engineering stress-strain curve is shown in Figure 8. According to the above results, the Al-10Mg casting strip shows the good mechanical properties, obvious yield phenomenon, and higher elongation.

TABLE 3: Tensile properties of twin-roll cast Al-10Mg plate.

Orientation	$\sigma_{0.2}$/MPa	σ_b/MPa	δ
0°	142.9	207.2	8.6
90°	121.1	183.3	7.4
45°	108.6	159.4	5.3

5. Conclusions

(1) A mathematical model of the solidification process on twin-roll continuous casting thin strip was developed based on the cellular automata (CA) theory. The continuous solidification process in front of the nip point was numerically simulated by Procast. The solidification microstructure, the

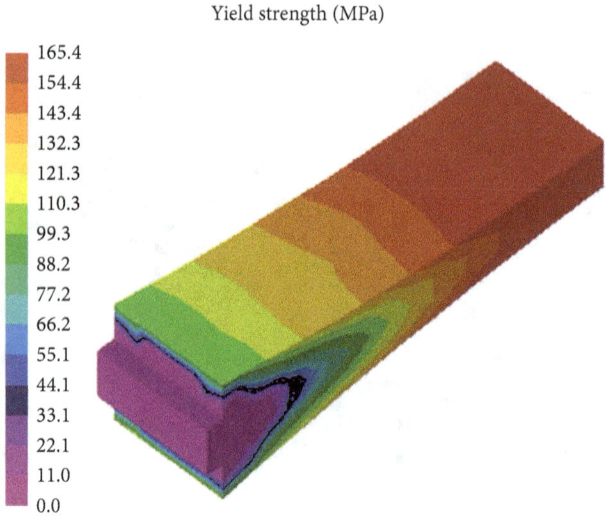

FIGURE 7: Simulation results of the yield strength of Al-10Mg alloy plate.

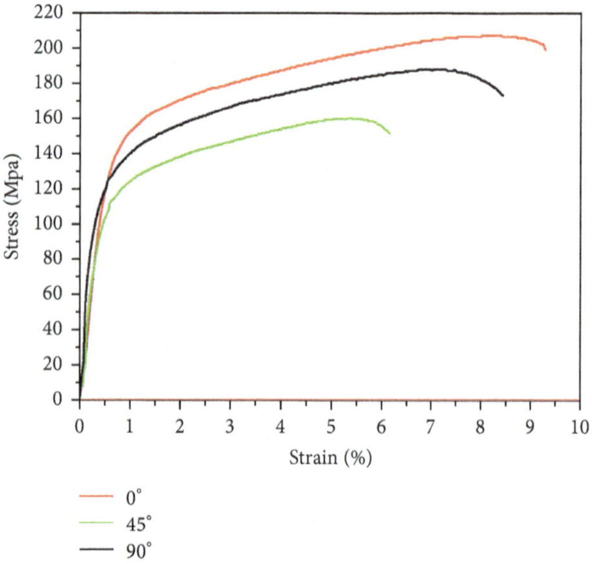

FIGURE 8: Engineering stress-strain curves of twin-roll cast Al-10Mg alloy plate at room temperature.

grains size, the dendritic grain radius, and the secondary dendrite arm spacing were obtained to predict the mechanical properties of the wedge strip.

(2) The simulation results were verified by metallographic examination and tensile tests. The simulation results of the microstructure on the wedge strip rolled plane and cross section, which was 1 mm away from the roll, were consistent with the experimental results. The simulation results of dendritic grain radius and the secondary dendrite arm spacing were, respectively, 1133 μm and 16 μm, which were consistent with the experimental results. The predicted yield strength of the aluminum alloy strip was 158 MPa, which was consistent with the experimental results (142.9 MPa).

(3) The solidification process of twin-roll cast alloy was described accurately with the established mathematical model. The model can be used to predict the solidification structure and mechanical properties of the strip under different process parameters and solidification parameters.

Conflicts of Interest

The authors declare that there are no conflicts of interest regarding the publication of this paper.

Acknowledgments

This work was supported by the National Natural Science Foundation of China (51574171 and 51641406) and the Natural Science Foundation of Shanxi Province of China (2015011002).

References

[1] P. D. Ding, B. Jing, C. M. Yang, and L. Fang, "Development and thought of thin-strip continuous casting," *The Chinese Journal of Nonferrous Metals*, vol. 14, no. S1, pp. s192–s196, 2004.

[2] H. S. Di, P. W. Bao, Y. C. Miao, G. D. Wang, and X. H. Liu, "Experimental study on the twin roll strip casting and analysis of processing stability," *Journal of Northeastern University (Natural Science)*, vol. 21, no. 3, pp. 274–277, 2000.

[3] Y. Birol, "Analysis of macro segregation in twin-roll cast aluminium strips via solidification curves," *Journal of Alloys and Compounds*, vol. 486, no. 1-2, pp. 168–172, 2009.

[4] C. Gras, M. Meredith, and J. D. Hunt, "Microstructure and texture evolution after twin roll casting and subsequent cold rolling of Al-Mg-Mn aluminium alloys," *Journal of Materials Processing Technology*, vol. 169, no. 2, pp. 156–163, 2005.

[5] H.-L. Mao, J.-H. Wang, G. Yang, Q.-N. Shi, J.-H. Yi, and L.-W. Chen, "Characterization of microstructure of 1235 aluminum alloy sheet produced by twin roll casting," *The Chinese Journal of Nonferrous Metals*, vol. 24, no. 4, pp. 863–869, 2014.

[6] M. C. Flemings and R. W. Cahn, "Organization and trends in materials science and engineering education in the US and Europe," *Acta Materialia*, vol. 48, no. 1, pp. 371–383, 2000.

[7] K. Shibuya, F. Kogiku, M. Yukumoto, S. Miyake, M. Ozawa, and T. Kan, "Development of a rapid solidification process with a double-roller method," *Materials Science and Engineering*, vol. 98, pp. 25–28, 1988.

[8] X.-B. Liu, Q.-Y. Xu, T. Jing, and B.-C. Liu, "Microstructure of aluminum twin-roll casting based on Cellular Automation," *Transactions of Nonferrous Metals Society of China (English Edition)*, vol. 18, no. 4, pp. 944–948, 2008.

[9] S. D. Chen and J. C. Chen, "Micro-model of simulation on twin-roll continuous casting thin strip solidification structure," *Rare Metal Materials and Engineering*, vol. 42, no. 1, pp. 14–18, 2013.

[10] F. Huang, H. S. Di, and G. S. Wang, "Numerical simulation on solidification microstructure evolution in twin-roll casting process," *Casting. Forging. Welding*, vol. 40, no. 5, pp. 57–60, 2011.

[11] J.-H. Cho, H.-W. Kim, C.-Y. Lim, and S.-B. Kang, "Microstructure and mechanical properties of Al-Si-Mg alloys fabricated by twin roll casting and subsequent symmetric and asymmetric rolling," *Metals and Materials International*, vol. 20, no. 4, pp. 647–652, 2014.

[12] H. Harada, S. Nishida, E. Masaki, and H. Watari, "Casting of high-aluminum-content Mg alloys strip by a horizontal twin-roll caster," *Metallurgical and Materials Transactions B: Process*

Metallurgy and Materials Processing Science, vol. 45, no. 2, pp. 427–437, 2014.

[13] S. J. Yao, X. K. Wang, J. Xu, N. M. Han, and S. K Guan, "Microstructure analysis of cast rolling 1235 aluminum alloy plates," *Light Alloy Fabrication Technology*, vol. 42, no. 4, pp. 30–34, 2014.

[14] S. M. Xiong, Q. Y. Xu, and J. W. Kang, *Modeling and Simulation Technology in Casting Process*, China Machine Press, Beijing, China, 2004.

[15] Ch.-A. Gandin and M. Rappaz, "A coupled finite element-cellular automaton model for the prediction of dendritic grain structures in solidification processes," *Acta Metallurgica et Materialia*, vol. 42, no. 7, pp. 2233–2246, 1994.

[16] M. B. Yang, F. S. Pan, X. D. Peng, and P. D. Ding, "Macroscopic and microscopic coupling mathematical model," *Journal of Chongqing University (Natural Science Edition)*, vol. 28, no. 5, pp. 39–44, 2005.

[17] B. Y. Sun, *Direct Strip Casting of Metals and Alloys Processing, Microstructure and Properties*, National Defense Industry Press, Beijing, China, 2014.

[18] B. Y. Hang, C. G. Li, L. K. Shi, G. Z. Qiu, and T. Y. Zuo, *Chinese Materials Dictionary Book 4*, China Material Engineering Canon, 2008.

13

Low-Temperature Brazing of Titanium using Al-Based Filler Alloys

A. Muhrat, H. Puga, and J. Barbosa

Centre for Micro-Electro-Mechanical Systems (CMEMS), The University of Minho, Campus of Azurém, 4800-058 Guimarães, Portugal

Correspondence should be addressed to A. Muhrat; eng.abdulsalam.edu@gmail.com

Academic Editor: Fantao Kong

Brazing of titanium using low melting temperature filler alloys is a preferred choice regarding cost and preserving its mechanical properties. However, brazing titanium at low temperature is still a challenge, especially regarding aluminum-based filler alloys. During the last years, several brazing methods and Al-based fillers were developed to meet industrial requirements; some of them might achieve some of those requirements. The use of ultrasound in brazing has gained increased attention recently, which helps to reduce the time and the necessity for a special brazing environment, subsequently, reducing cost and increasing applicability. Brazing titanium below the $\alpha \leftrightarrow \beta$ transformation temperature, using commercial and experimental Al-based fillers of different compositions, is presented in this review; including the procedures of traditional and ultrasound-assisted brazing methods. Correspondingly, the effects of brazing conditions and alloying elements on the mechanical properties and the intermetallic compounds formation are examined.

1. Introduction

Brazing of titanium at a temperature below $\alpha \leftrightarrow \beta$ transformation temperature has great importance in avoiding undesirable changes in its original microstructures [1, 2]. A low brazing temperature is particularly crucial for thin-wall brazing structures of aerospace components and heat exchangers [3]. The shorter brazing cycle/less brazing temperature helps to prevent the excess interaction between the joint components and growth of the intermetallic compounds at the joint interface [2]. Titanium has a strong chemical attraction of oxygen; therefore, even at room temperature, a passive amorphous oxide layer forms rapidly on its surface [4]. The oxidation of the surface makes titanium joining process challenging from the point of establishing the metal connection; furthermore, titanium should be protected from severe oxidation at temperatures exceeding 650°C [5]. In addition to susceptibility to grain coarsening and affecting strength/ductility of the parent components, high-temperature brazing requires special equipment, conditions, and longer processing time, leading to higher processing cost compared with low-temperature brazing.

Titanium joining has been widely studied [6]. Unlike brazing, welding is not used usually for titanium honeycomb sandwich structures [7–10] which require joining many points without distortion. Moreover, the welding process is not suitable for joining titanium aluminides [11]. A great number of brazing filler alloys were developed and studied for brazing titanium alloys; some of them were commercially used. In 1950s and 1960s, early work was focused upon Ag-, Au-, Cu-, and Pd-based filler alloys for ~(316–427°C) and Al-based filler alloys for ~204°C service temperatures. Ti- and Ti–Zr-based alloys were developed later, which showed better metallurgical compatibility with titanium [12] and produced higher joint strength when compared with the other systems of brazing fillers. Table 1 shows some examples of relatively high strength and temperature brazed joints using Ti- and Ti–Zr-based filler alloys. However, low-temperature brazing with more free conditions could find several applications: where the inert atmosphere cannot be applied, the parts to be joined cannot fit within limited and special environments such as the mass production by continuous brazing, the large brazing joints, or for in-place brazing. Furthermore, reducing the joining cost by lowering temperature is important for reducing the total fabrication cost and maintaining titanium share in the total structural weight of aerospace systems [13].

According to the main alloying elements, filler alloys for brazing titanium and its alloys can mainly be classified into five groups: titanium-based [1, 14, 16, 18–21], zirconium-based [16, 19, 22, 23], silver-based [24–27], aluminum-based [28–31], and nickel-based [19, 32] filler alloys. Attempts of brazing titanium using Al-based filler alloys and other filler systems started over than 60 years ago [33]. Many Al-based filler alloys have been developed since that time, and many attempts were made in pursuit of achieving suitable brazing conditions. Titanium was brazed to itself and to aluminum alloys by Al-based fillers using an oxyacetylene torch with flux [30, 33], in high-purity argon [10, 33], in vacuum [28–31, 34–36], and in air conditions: in fluxless process assisted by the ultrasonic vibration (USV) [37–39] or by additional pressure [40]. Investigating and improving the mechanical properties of the produced joint through a careful design of the brazing cycle and controlling the intermetallic compound formation/diffusion zone were among the primary explored objectives in brazing titanium and its alloys. Considerable efforts have been made to reduce brazing cycle time while still achieving a suitable strength by employing new techniques and developing new compositions of low-melting filler materials. Although significant work has been done so far, brazing of titanium by Al-based fillers still needs more development to increase its share in the industrial sector.

The developed microstructure and mechanical properties of brazing joints depend on the component compositions, temperature-time cycle, and the joint design. Choosing the right cycle parameters depends, in turn, on the rate and nature of the interaction at parent/filler interface. The short temperature-time brazing cycle protects the parent component from unwanted metallurgical changes, which may occur faster as the temperature increases; for instance, the aging effects which take place in 4 h at 537.8°C (1000°F) in Ti-6Al-4V would occur in ~10 min at 579.4°C (1075°F) [41]. However, brazing time could be increased according to the parent/filler materials' compositions to allow the interface element diffusion to take place for a certain limit which is necessary to achieve the optimal joint strength. The diffusion of some brazing processes, usually in high temperature, may proceed for a relatively long holding time until achieving the microstructure homogenization over all the joint, which is particularly known as diffusion brazing. Matsu et al., Ganjeh et al. and Chang et al. [16, 17, 42] investigated the diffusion brazing of several Ti- and Zr-based fillers, which led to specific microstructures and high shear strengths (Table 1). These processes involve isothermal solidification and homogenization of the joint [43].

In metal casting, the ultrasonic vibration (USV) provides multiple significant effects such as degassing [44–47], refining, and modifying the microstructures [48–51]. Using USV in brazing is an attractive assisting tool [37–39, 52–58]; however, studying the combined effects of USV with the technological parameters of brazing process and developing a joint interface more compatible with the USV have not yet received an adequate attention. The focus in this review will be on the titanium/aluminum interface which includes Al-based filler alloys for brazing titanium/titanium and titanium/aluminum parent (base) components, taking into consideration the technological parameters and procedures.

2. Aluminum-Based Filler Alloys

Various Al-based alloy systems have been tested for brazing titanium/titanium and titanium/aluminum parent components. The alloying additions to the filler materials serve several purposes, such as reducing melting temperature, improving wetting [28, 34], reducing intermetallic thickness [29], improving corrosion resistance [41], and improving static/dynamic strength at work temperature [29].

Lowering the brazing temperature and reducing the contact time of a molten filler alloy with a parent titanium alloy reduce the Ti erosion [26, 42] and increase the reliability of the thin-wall titanium joints [3]. Al-based filler alloys can be a suitable candidate for brazing titanium satisfying the following conditions: the melting temperature considerably below the $\alpha \leftrightarrow \beta$ transformation temperature of titanium alloys, a good metallurgical compatibility with titanium alloys, good wetting and spreading on titanium surface, good corrosion resistance [59], low density, and acceptable strength for a large joining surface [2]. Some commercial Al-based filler alloys of Al–Cu, Al–Mg, and Al–Si systems with addition of Cr and Ti were introduced [3, 19]. However, aluminum leads to fast formation of a brittle intermetallic layer at the joint interface with Ti and provides relatively low strength compared with other filler alloys; the activity of Al with Ti is considered the dominant challenge when designing Al-based fillers and optimizing the brazing process.

2.1. The Developing of Brazing Conditions/Methods in Brazing Ti/Ti and Ti/Al Using Aluminum-Based Fillers

2.1.1. Vacuum, Torch Brazing, and Other Methods. Lewis et al. [33] tested initially many of commercially available brazing fillers using careful sample preparation procedures. The obtained shear strength of the double-lap-brazed joints in a controlled-atmosphere furnace brazing was 73.77–115.69 MPa using 61S (6061) aluminum filler. Further torch-brazing tests were made by covering the specimen surfaces with a commercial titanium flux and using a slightly reducing flame. In general, the strength results were lower 44.82–88.94 MPa when compared with the controlled-atmosphere furnace brazing; the reason was explained by flux entrapment in the joint area or insufficient temperature control and oxidation protection in torch brazing. The use of torch brazing must always be accompanied by specific fluxes, choosing the suitable flux should consider the alloying elements in the molten filler and should be suitable for all joint elements. Fluxes for titanium are primarily mixtures of fluorides and chlorides of the alkali metals, sodium, potassium, and lithium [4].

Takemoto et al. [28] investigated brazing of pure Ti/Al3003 (Al–1Mn) joint in vacuum using three different Al-based filler alloys. Brazing time was short (3–5 min) and brazing temperature did not exceed 615°C. Table 2 shows the shear strength results.

TABLE 1: Examples of high-temperature brazing of titanium.

Shear strength (MPa)	Conditions	Filler	Parent metal	References
300	Vacuum + argon, 880°C, 60 min	37.5Ti – 37.5Zr – 15Cu – 10Ni	Commercially pure titanium (Grade 2)	[14]
261.7	Vacuum + argon, 870°C, 30 min	Ti-27Zr-14Cu-13Ni	Commercially pure titanium (Grade 2)	[15]
540	Argon, 950°C, 30 min	Ti – 26.8Zr – 13Ni – 13.9Cu	Ti – 6Al – 4V (Grade 5)	[16]
545	Vacuum, 1030°C, 3 min	Ti – 15Cu – 25Ni	Ti – 6Al – 4V (Grade 5)/Ti – 15V – 3Al – 3Cr – 3Sn	[17]

Pure Ti/Ti joints were investigated under the same level of vacuum using pure aluminum and Al–0.8%Si as filler materials at 680°C and 3 min holding time [28, 29]. The reported shear strength was ~92 MPa for both filler metals; anyway, increasing the holding time leaded to drastic decrement rate of the joint strength of the pure Al filler. It was found that the use of pure Al as a filler metal to braze Ti/Ti in 700°C leaded to self-separation of the joint after holding time of 30 min, and the thickness of the formed intermetallic compound layer ($TiAl_3$) was the greatest, which acts as an initiation site for cracks.

Sohn et al. [34] studied the diffusion between the Al–10Si–1Mg filler metal and the parent metals of CP-Ti/Al1050 by maintaining the temperature at 620°C and changing the holding time in 1×10^{-4} Torr vacuum as it is presented in Figure 1.

The best-achieved shear strength was 84 MPa for the brazing time of 25 min. For longer brazing time, bond strength decreased due to the formation of cavities in Al near Al/intermetallic interface. Shiue et al. [35] investigated a fast heating approach in joining titanium aluminide Ti50–Al50 by infrared brazing using pure aluminum and BAlSi–4 of the composition (Al– (11–13)Si–0.3Cu–0.1Mg–0.2Zn–0.15Mn–0.8Fe) as filler alloys. Brazing was done in the vacuum using a high-speed infrared source for heating of 900°C/min, and preceded by a preheating for 60 s/400°C. For the pure aluminum filler, a large formation of the brittle intermetallic phase $TiAl_3$ was found for all holding times, and the bonding strength was very weak. Figure 2 shows the place of the crack through the major phase of $TiAl_3$ in the joint.

While for BAlSi–4 filler alloy which contains Si, it was observed that the intermetallic Ti_2AlSi_3 was prominent at an early stage of infrared brazing. By increasing the time and temperature, the volume fraction of Ti_2AlSi_3 decreased dramatically and replaced by $Ti_5Al_{12}Si_3$, and then it disappeared. The highest strength 86.2 MPa was achieved for 900°C brazing temperature.

Khorunov et al. [31] investigated Al–Si and Al–Mg filler alloys in brazing OT4 (Ti–3Al–1.5Mn) titanium alloy in vacuum. The resulted strengths using Al–Mg filler alloys were relatively high (82–83 MPa), while the produced joint using Al–Si filler alloy was characterized by inferior quality with a continuous layer of silicide detected along the interface.

Some efforts have been made to simplify brazing process and carry out titanium brazing in normal ambient air conditions. These efforts were focused on developing low-

TABLE 2: Results of brazing Ti/Al3003 by Takemoto et al. [28].

Shear strength (MPa)	Brazing conditions	Fillers
38.25	615°C, 5 min, *vacuum*	Al–10Cu–8Sn
38.25	600°C, 5 min, *vacuum*	Al–20Ag–10Cu
69.88	600°C, 3 min, *vacuum*	Al–10Si–0.5Mg

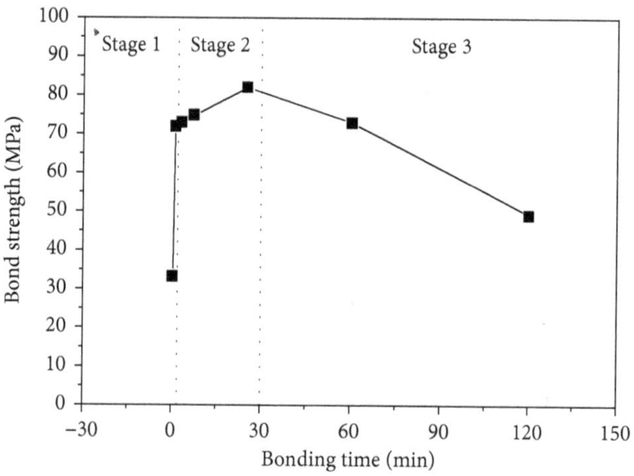

FIGURE 1: Shear strength variation of brazed joints CP-Ti/Al1050 at 620°C using Al–10Si–1Mg filler alloy. Reproduced from Sohn et al. [34].

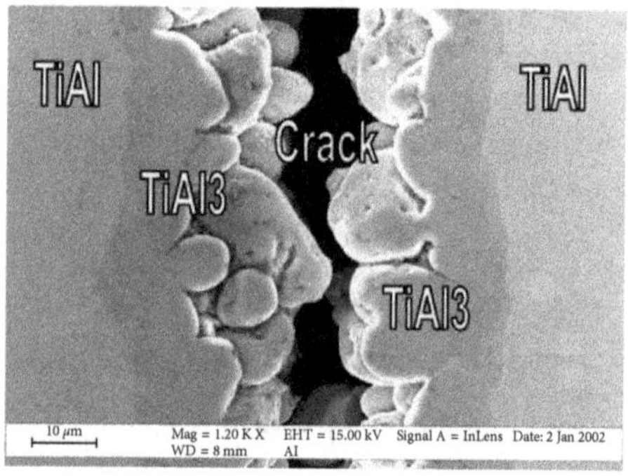

FIGURE 2: The crack location of the brazed joint Ti50Al50/Al/Ti50Al50 (900 C, 30s). Reproduced from Shiue et al. [35].

melting brazing fillers and finding a suitable technique to remove the oxide layer on the faying surfaces of the joint. In Germany, Bach et al. (cited in [2]) conducted an early attempt of Ti-6Al-4V brazing in the air by utilizing an alloy of

Table 3: Recent studies of ultrasound-assisted brazing of Ti/Ti and Ti/Al joints by Al-based fillers.

Reference	Materials	Heating method and best conditions (all in air)	Shear strength (MPa)
[37]	Parent metal: Ti Grade 5; filler metal: TiBraze Al-665(Al–2.5Mg–0.3Cr)/100 μm	Induction brazing, 660°C, (5 sec, 84 W) USV, no holding time	50–60
[39]	Parent metals: Ti Grade 5/Al1060; filler metal: Al–12Si	($A = 6\,\mu$m, 20 kHz) USV, 620°C, 4 sec USV + holding for 15 or 5 min + 4 or 0 sec USV	~68
[38]	Parent metal: Ti Grade 2; filler metal: TiBraze Al-665A (Al–2.5Mg–0.3Cr)/50 μm	Induction brazing, (25 kHz, 120 W) USV, 670°C, holding for 5 min at 560°C, then heating to 670°C and holding for 3 min, applying USV for 6 s	~64
[55]	Parent metals: Ti Grade 5/Al6061; filler metal: Al–9.4Si–10Zn–10Cu–2Ni	($A = 6\,\mu$m, 20 kHz) USV, 560°C, 4 sec USV + holding for 60 min + 4 sec USV	~43
[55]	Parent metals: Ti Grade 5/Al6061; filler metal: Al–15Sn–8.2Si–8.5Zn–8.5Cu–1.7Ni	($A = 6\,\mu$m, 20 kHz) USV, 560°C, 4 sec USV + holding for 15 min + 4 sec USV	~76

low-melting temperature, which was a thin Al foil coated with copper (33% Cu) for a brazing gap of 6–20 μm. The produced joint had a good quality; anyway, the shear strength did not reach more than 25 MPa, which serves in joining large areas only. This showed that the eutectic Al–Cu was active enough to react with titanium in the air. Karfoul and Muhrat [40] investigated Ti/Ti-brazed joint strength in relation to the filler metal compositions and brazing parameters in the air. New brazing filler alloys of Al–Cu–Ni–Ti group were studied to find the effects of the alloying elements, as well as investigating the possibility of brazing in the air. The highest average shear strength achieved was ~74–73 MPa using Al–8.54Cu–1.35Ni–0.08Ti and Al–6.44Cu–1.96Ni–0.92Ti filler alloys. Removing the oxide layers between the aluminum and titanium is the initiation key of the bonding process. Accordingly, adding pressure will lead, with the help of Ti surface roughness, to break the oxide layer of Al-based filler. Then, buckling proceeds through thickness cracking of the scale, and the stripped Al surface atoms will react with the Ti-oxide layer to reduce it to atomic Ti [40]. Using USV as an assistant tool in brazing, which leads to fast breaking of the oxide layer at the joint interface and makes the joining process possible in air conditions, is summarized in the following.

2.1.2. Ultrasound-Assisted Brazing of Ti by Al-Based Fillers. The sonochemical effects of the USV in liquids have been explored for over 80 years [60]. The interest in ultrasound-assisted brazing started more than 50 years ago. One of the early US patents was an ultrasound-assisted brazing unit using an ultrasonic tip surrounded by local heating flame [61].

Using USV for a short period results in very fast destroying of the oxide layer on the parent material and wetting it by the filler material [52], even it is effective in achieving a good wetting on ceramics by the filler materials in the air [53, 54, 62]. The acoustic cavitation generated in the liquid metal is considered as the most important physical effect of the ultrasonic oscillation propagation [63–66]. The formation of cavities starts with the half period of rarefaction at the weakest points of the melt [63]. Cavities continue to increase in size until they collapse by the compressing stresses during the next compression half period; the results of this collapse include high-intensity pressure pulses, temperature spikes, and high-velocity jets. The important physical and chemical processes due to the ultrasonic cavitation are melt degassing, filtration, wetting of the solid inclusions, and structure modifications. Moreover, the acoustic cavitation results in ultrasound energy adsorption and formation of acoustic streaming. The sound pressure must exceed a certain level to be able to initiate cavitation in the liquid metal, which is called cavitation threshold [63].

Employing USV in brazing results in short processing time, consequently, reducing the cost, and preserving the parent components from further metallurgical changes. The advantages include less specific atmosphere requirements due to the fast-breaking of the oxide layer [39, 52], enhanced spreading of the molten filler [55, 62], and homogenizing the microstructures of the joint [39]. Table 3 summarizes the technological parameters and the shear strength results of the ultrasound-assisted brazing of Ti/Ti and Ti/Al joints using Al-based fillers.

Chen et al. [52] explored the instantaneous breakage behavior of the oxide film of Ti–6Al–4V on its interface with the liquid Al–Si alloy and the chemical interactions during the process using 20 kHz frequency and 5.5–7 μm amplitude USV. The samples were heated to 620°C then the titanium sonotrode was quickly loaded on the titanium substrate with a constant pressure of 0.2 MPa. The applied time of USV was 2–16 s. Due to the USV effect, a large number of pits were formed on the titanium substrate as it is shown in Figure 3. The formation of the pit started by a small notch on the surface of the Ti oxide film due to "cavitation erosion" [65, 67] (N in Figure 4) which serves as a microchannel and allows direct interaction between the fresh Ti surface and molten Al alloy.

The suggested pit formation mechanism according to this study was firstly, the formation of a notch at the oxide layer of the titanium substrate and secondly, violent chemical dissolution of the solid phase Ti–6Al–4V into the liquid filler Al–Si, followed by further interfacial chemical reactions.

Chen et al. [39] conducted a further work in ultrasound-assisted brazing of Ti–6Al–4V/Al1060 in the air using Al–12Si filler. In this study, ultrasonic vibration was applied to Ti–6Al–4V plate for 4 s at 620°C followed by 0, 5,

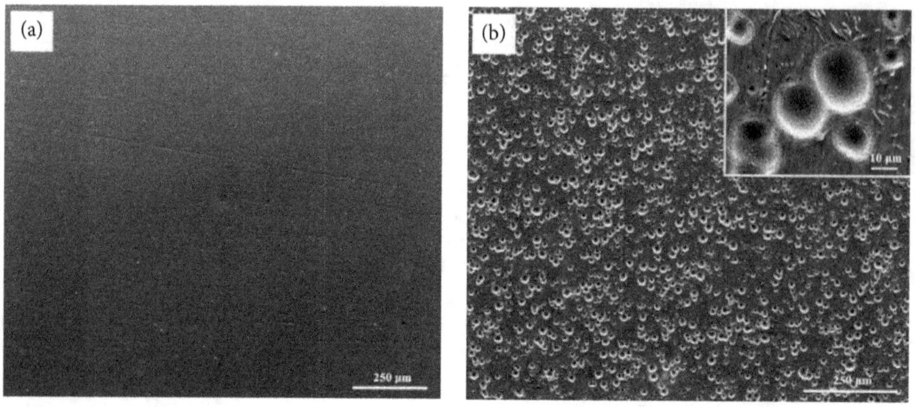

FIGURE 3: The morphology of Ti–6Al–4V substrate surface: (a) after interaction with the liquid Al–Si alloy for 10 min without ultrasonic treatment and (b) with ultrasonic treatment for 8 s. Reproduced from Chen et al. [52].

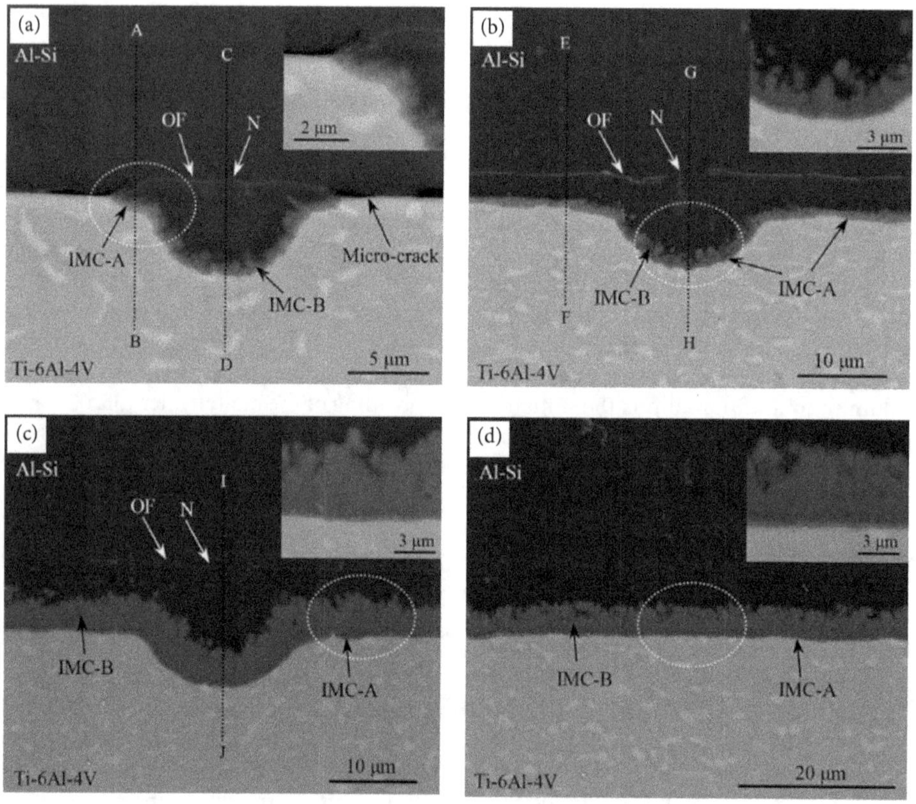

FIGURE 4: The microstructure (SEM) of Al–12Si/Ti–6Al–4V interface after: (a) 4 s; (b) 4 s + 5 min; (c) 4 s + 15 min; (d) 4 s + 15 min + 4 s. Reproduced from Chen et al. [39].

and 15 min holding time, and then by a second ultrasonic vibration period for 4 s. Applying the first ultrasonic vibration for 4 s caused formation of pits at the surface of the titanium alloy (Figure 4(a)). By applying a holding time of 5 or 15 min, the dissolving continued at Al1060 and Ti sides; consequently, the oxide film detached off the titanium substrate which is called "substrate oxide undermining" (Figures 4(b) and 4(c)). In addition to the pit formation, the cavitation and the stirring action of acoustic streaming caused the following effects during the ultrasound operation:

(i) Accelerating the dissolution process at Ti side once the pits had been formed on the Ti oxide film, as well as along Al1060 side.

(ii) During the second 4 s of USV, breaking the floating oxide film of the titanium alloy (Figure 4(d)). Simultaneously, breaking the Al dendrites and causing the Al phases to distribute uniformly in the joint.

Anyway, increasing holding time to 15 min or adding a second USV period did not improve the shear strength of the brazed joints.

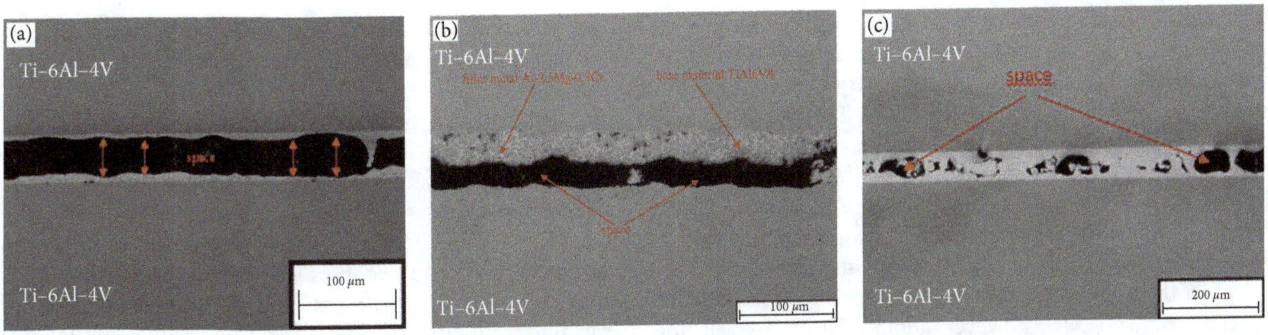

FIGURE 5: The formed voids and gaps in Ti–6Al–4V/Al–2.5Mg–0.3Cr/Ti–6Al–4V brazed joints by different ultrasound-assisted brazing parameters: (a) 100 μm, 670°C, 5 s, 60%, (b) 60 μm, 660°C, 5 s, 50%, and (c) 100 μm, 660°C, 5 s, 80%. Reproduced from Tillmann et al. [37].

Tillmann et al. [37] studied the ultrasound-assisted brazing of Ti–6Al–4V using Al–2.5Mg–0.3Cr filler and induction heating. The studied parameters were the layer thickness (60–200 μm), brazing temperature (660–680°C), and acoustic power (60–96 W), and then the sample of the best ultrasound-assisted brazing parameters was compared to a traditional vacuum brazing one. Increasing temperature over 660°C or reducing the thickness to less than 100 μm resulted in big voids in the brazed area. Increasing the acoustic power up to 70% (84 W) resulted in a considerable reduction of the oxide layer at the interface, and less intensities could not remove the oxide layer, while the highest one of 80% caused a significant cavitation leading to gap formation in the joint (Figure 5). The highest shear strength obtained using the sample of 100 μm, 660°C, 5 s, 70% did not reach more than 60 MPa and between 30 and 40 MPa for vacuum-brazed samples.

The lower strength values for the samples brazed in the vacuum were attributed to the low heating rate which resulted in an excessive formation of a low-ductility intermetallic at the interface.

Elrefaey et al. [38] conducted a preliminary investigation of the microstructure and the fracture behavior of ultrasound-assisted brazing of Grade 2 titanium using 50 μm TiBrazreAl-665A (Al–2.5Mg–0.3Cr) filler alloy. The results showed the formation of the thin interaction layer of Ti_3Al, and the joint failed mainly at Ti/Al interface due to the brittle nature of the intermetallic. Chen et al. [55] investigated the addition of Sn to Al–9.4Si–10Zn–10Cu–2Ni in ultrasound-assisted brazing of Ti/Al joint. In a similar way in other works [39, 62, 68], a simple overlap assembly was used with a preset joint clearance of 200 μm. The addition of Sn caused the widening of the melting range; before reaching the solidus temperature, the residual Sn-rich liquid filled the capillary gaps between the solidified $CuAl_2$ grains which significantly reduced the residual thermal stresses and enhanced the joint strength [55].

To guarantee the efficient transfer of USV to the joint, the fixing rig and the mechanism by which the USV is being transferred should be designed carefully. This includes the consideration of the acoustic properties of the fixing rig components, the applied force on the joining area, and the location of the sonotrode tip [69]. Elrefaey et al. [38] used, for instance, an apparatus where the horn was installed vertically, and the sample which was mounted into a steel holder was under 0.2 MPa pressure as a result of the horn weight (Figure 6). While, Tillmann et al. [37] used overlapped brazing joints with a torque of 8 Nm introduced by a screw connection.

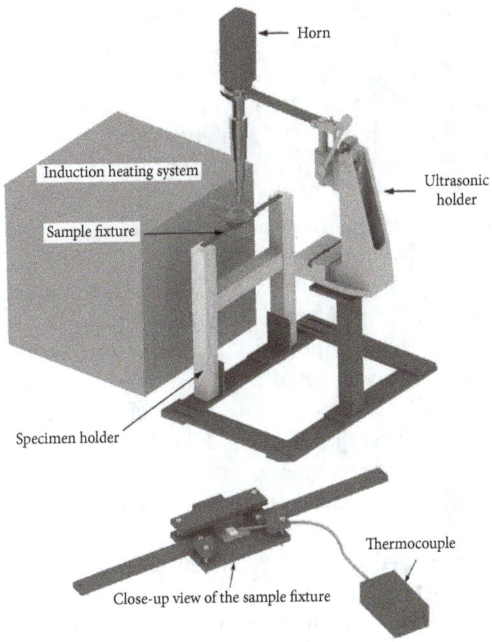

FIGURE 6: Schematic illustration of an ultrasonic-assisted brazing apparatus used by Elrefaey et al. Reproduced from Elrefaey et al. [38], with permission from American Welding Society (AWS).

Even though using USV is a valuable way to break up the oxide layer and improve the properties of the joint, the results of ultrasound-assisted brazing of Ti/Ti and Ti/Al confirm that the process needs more control and investigation on the brazing conditions/materials to achieve higher mechanical properties.

2.2. The Intermetallic Compound Formation at the Ti/Al Interface. Several intermetallic phases occur in the Ti–Al system such as α_2-Ti_3Al, γ-TiAl, $TiAl_3$, $TiAl_2$ and Ti_2Al_5 [70]. In fact, only Ti_3Al and TiAl are technically interesting [71].

Since almost all filler metals form intermetallic compounds with titanium [29], therefore, studying and controlling their formation/development at the interface largely determine the mechanical behavior of the brazed joints. The formation and growth of the intermetallic compound at the interface is related basically to the compositions of the filler materials, the parent material, and the temperature-time parameters. Brazing titanium with pure aluminum, for instance, mainly leads to the formation of $TiAl_3$ at the interface; adding Si (≥3%) to the aluminum filler alloy will lead to formation of $Ti_7Al_5Si_{12}$ and Ti_9Al_{23}; however, adding Cu or Ag will not change the probability of $TiAl_3$ formation [29]. The predicted phase by the effective heat of the formation rule does not necessarily form at the moving reaction interface if it is difficult to nucleate [72]. Many studies showed that $TiAl_3$ would form at the expense of the preexisting Ti_3Al or TiAl phases in the presence of excess Al. Sujata et al. [73] calculated the free energies of formation of different titanium aluminide compounds in the temperature range ~(0–1200°C), and they found that the free energy of $TiAl_3$ is less than that of Ti_3Al and TiAl but slightly higher than that of $TiAl_2$ and Ti_2Al_5. The preferred formation of $TiAl_3$ as the only product during solid Ti/liquid Al reaction was explained by two factors: the first one is the additional steps in $TiAl_2$ and Ti_2Al_5 formation which include TiAl as one of the starting phases and the second factor is the presence of the excess amount of liquid Al which minimizes the trend of TiAl or Ti-rich compound Ti_3Al formation. Wöhlert and Bormann [74] investigated the phase selection mechanism of intermetallic compounds in Ti/Al phase reaction. The predeposited layers of Ti_3Al and TiAl, between Al/Ti elemental layers after annealing for 50 and 20 h at 400°C, respectively, were disappeared and replaced by $TiAl_3$. Accordingly, the phase selection during the interface reaction was explained by the differences in growth velocities of the competing phases. Similar main phase selection reason was concluded by Michaelsen et al. [75] including the nucleation as a first step in $TiAl_3$ formation. Analogous results of $TiAl_3$ formation on a Ti_3Al-based alloy were reported by Zhang et al. [76] by using hot dip aluminizing (750°C-15 s) followed by interdiffusion treatment (600°C-30 h). It was found that the addition of Si, Mn, or Ag to the pure Al coating improved the thickness uniformity and reduced the formation of the transverse cracks. Further intermediate layers consisting of ($TiAl_3 + TiAl_2$) and/or TiAl were formed after high-temperture oxidation for all the coatings. Zhang et al. [77] reported in another work using similar aluminizing and interdiffusion procedures the growth of $TiAl_2$ and TiAl, as solid state reaction products at the interface of $TiAl_3$ and Ti-6Al-4V substrate after the subsequent oxidation at 700°C and 300h, whereas $TiAl_3 + TiAl_2$ and Ti_3Al layers were formed at 800°C and 300 h of oxidation in air. The mixed $TiAl_3$ + $TiAl_2$ layer resulted from O_2 reaction with $TiAl_3$. Xu et al. [78] applied vacuum hot pressing on Ti/Al laminates followed by annealing and found that the next reaction phase after $TiAl_3$ was TiAl at the Ti/Al interface after consuming Al layer and during a second-stage annealing treatment at higher temperatures. The diffusion at the Al/Ti interface can be accelerated with the amount of cold/hot rolling reduction [78, 79]. In brazing titanium alloy OT4 by Mg-containing aluminum alloys, the formation of $TiAl_2$ between $TiAl_3$ and titanium also could take place [31].

The diffusion of the chemical elements across the interface of the brazing joint is high at the beginning of the brazing process since the diffusion at the beginning is between liquid and solid phases. As soon as the formation of the intermetallic compounds takes place, the diffusion rate is reduced and becomes less and less by increasing the intermetallic layer thickness until the growth of the intermetallic compound is suppressed. Considering, in titanium-rich aluminide, titanium can diffuse faster than Al [80, 81]. The thickness of the intermetallic layer when using pure aluminum as a filler is the largest one, and any addition or impurity elements will suppress—but differently—the growth of the intermetallic layer [29].

Brazing titanium aluminide TiAl at 900°C for 30 s using pure Al filler produced very weak joint, and the primary phase in the joint was the brittle $TiAl_3$ phase [35]. On the other hand, for a different case with longer bonding cycle under pressure and without any additional filler, the "diffusion bonding" of two-phases gamma-based titanium aluminide Ti–45Al (in at.%) produced α_2-Ti_3Al phase at the bonding interface. It was concluded by Buque and Appel [82] that the unavoidable contamination of the diffusion couple with oxygen leads to the formation of a thin layer of α_2-Ti_3Al phase. Similar result was found by Herrmann and Appel [83] for various two-phase gamma-based titanium aluminides, even considerably higher vacuum was used.

2.3. *Main Aluminum-Based Filler Systems*

2.3.1. Pure Aluminum as a Filler Metal. Pure aluminum is highly active with titanium at brazing temperature causing high erosion level, consequently, unreliable brazed joints specially for the thin-wall titanium structures [2]. Takemoto et al. [28, 29] found that the highest thickness of the intermetallic compound $TiAl_3$ was caused by using pure aluminum as a filler. The small additions of alloying elements or impurities to the pure Al filler reduced the intermetallic layer thickness at the brazed interface. It was found that the addition of 0.8%Si greatly reduced the thickness and the growth rate of the intermetallic $TiAl_3$, and the produced Ti/Ti joints using Al-0.8Si filler maintained higher joint strengths after brazing for relatively higher temperatures and longer holding times comparing with pure Al filler [29]. In the same context, Shiue et al. [35] found that the bonding strength of TiAl/Al/TiAl joint was very weak to be evaluated due to the extensive presence of the stable $TiAl_3$ phase in the joint.

The minor modifications in the alloying elements affect the wetting and spreading behavior and should be designed according to the type of the Ti parent metal. For instance, comparing the fillers Al–4.5Mg–0.4Si–0.1Cr and Al–4Mg–0.5Si–0.4Mn [30], the former with more 0.5%Mg and 0.1Cr instead of 0.4Mn, has a lower contact angle and better spreading on Grade 5 titanium; by contrast, these

TABLE 4: Ti/Ti–Ti/Al-brazed joints using pure or low-alloyed Al fillers.

References	Materials				Conditions	Bonding		Shear strength (MPa)	Intermetallic compounds at Ti/filler interface
	Parent 1	Parent 2	Filler: Type	Filler: Thickness/gap width (μm)		Temperature (°C)	Time (min)		
[29]	Pure Ti	Pure Ti	Pure Al	100	Vacuum 2.7×10^{-3}	680	3	~92*	$TiAl_3$
						680	10	~42*	
			Al–0.8Si			680	3	~94*	$TiAl_3$
						680	10	~75*	
[31]	OT4 titanium	OT4 titanium	AD1 (Al–0.4Si–0.3Fe)	60	Vacuum 5×10^{-5} mmHg	685	3	76	
Cited in [2]	Ti Grade 5	Ti Grade 5	Al–0.3Si–0.8Fe–0.2Mg			>660		72	
[35]	Ti50–Al50	Ti50–Al50	Pure aluminum	75	Vacuum 8×10^{-5} mbar	900**	0.5	Very weak joint	$TiAl_3$

*The shear strength values calculated according to Figure 11 in [29]. **Infrared brazing 900°C/min, preheating in 400°C for 60 s.

modifications have a different effect in brazing Grade 2 titanium. Table 4 shows a summary of Ti/Ti–Ti/Al-brazed joints using pure or low-alloyed Al fillers.

TiBraze 665 which contains only 2.5% Mg [30]. Table 5 shows a summary of Ti/Ti–Ti/Al-brazed joints using Al–Mg-type filler alloys.

2.3.2. Al–Mg-Type Filler Alloys. Flom [84] reported the disappearing of magnesium from the interface layer between Al–Mg filler alloy and titanium parent alloy during vacuum brazing. Similarly, Khorunov et al. [31] found, after brazing using AMg6 alloy containing 6% of Mg, that only 1.5% of Mg remained in the joint and no Mg remained at the interface; the reduction of Mg was explained by the evaporation of Mg from the filler during the heating and melting in vacuum. The volatility of magnesium in elevated temperatures [85] and the reaction of the magnesium vapor with the oxide surface of aluminum make Mg an excellent wetting promoter [86]. Mg reacts with aluminum oxide forming a complex oxide $MgAl_2O_4$ [87] or reducing it to Al + MgO [34], which forms inclusions and induces bonding to occur; as a result, the content of Mg is reduced in the joint considerably [31]. The improved wetting of Al–Mg filler alloys was also confirmed by Khorunov et al. [31], where Al–Mg filler showed better wetting on titanium alloy OT4 for a temperature less than 700°C comparing with Al–Si. Raising brazing temperature more than 700°C resulted in reducing the contact angle for all the fillers containing Si or Mg. Mg content in the filler and the brazing conditions must be taken together into consideration when designing the brazing alloy, for example, Al–4Mg–0.5Si–0.4Mn alloy showed a good result of 117 MPa in torch brazing but produced only 61 MPa in vacuum brazing [30]. Another example, the vacuum brazing by Mg-containing filler alloys of 6% and 2.5% Mg showed almost identical strengths which were higher than the one using the Si-containing fillers [31]. Eckardt et al. [30] reported the formation of a variable and discontinuous intermetallic layer of $TiAl_3$ between the Al–Mg fillers and the titanium parent metal. In addition, it was reported that a thicker intermetallic layer was formed on Grade 5 titanium comparing with Grade 2 titanium. The high-Mg alloys did not show better fillet formation properties compared with the average-Mg braze alloy such as TiBraze 665.

2.3.3. Al–Si-Type Filler Alloys. Takemoto and Okamoto [29] reported that the small addition of 0.2–0.8%Si is highly effective in suppressing the intermetallic layer growth of $TiAl_3$. However, further additions ≥3% of Si increased the intermetallic thickness and promoted the formation of additional intermetallic compounds: Ti_9Al_{23} and $Ti_7Al_5Si_{12}$ at the interface, where $Ti_7Al_5Si_{12}$ was adjacent to the titanium parent side. Using Al–10Si–1Mg mainly produced $Ti_7Al_5Si_{12}$ and a small quantity of Ti_9Al_{23} intermetallic compounds. The best result of tensile test was for Al–10Si–1Mg comparing with the other two filler materials Al–10Cu–8Sn and Al–30Ag–10Cu. It was found that the fracture occurred within the filler metal when using Al–10Si–1Mg [29]. Under almost similar conditions, but for longer brazing time, Sohn et al. [34] investigated the diffusion brazing of Al/Ti joint by Al–10Si–1Mg. The bonding at the Ti/filler interface started as soon as the filler had been melted; two intermetallic compounds were observed, the continuous $Ti_7Al_5Si_{12}$ at the titanium side, which increased by the time due to the diffusion of Si into Ti, and the discontinuous one $Ti_5Al_{12}Si_3$ at the filler side and the fracture occurred in Al layer. In ultrasound brazing of titanium, two intermetallic compounds Ti_9Al_{23} and $Ti_7Al_5Si_{12}$ were formed at Ti-6Al-4V/Al-12Si filler interface [39]. The intermetallic IMC-A ($Ti_7Al_5Si_{12}$) formed at the pit edge during the first USV, while IMC-B (Ti_9Al_{23}) at the wall and the bottom of the pit. Applying holding time for 5 min caused the IMC-A ($Ti_7Al_5Si_{12}$) to form at the interface between IMC-B (Ti_9Al_{23}) and Ti–6Al–4V; by increasing the holding time to 15 min, the thickness of IMC-B (Ti_9Al_{23}) increased (Figure 4). Wells [41] reported that the Al–7.5%Si alloy cladded to a core of aluminum alloy 6951, which did not melt during brazing, produced stronger joint than 63Al–32Cu–5Ag–0.01Li for Grade 5 Ti. However, Khorunov et al. [31] found that Mg-containing filler alloys provided higher joint strength than the Si-containing filler alloy (5.3% Si), which was explained by the solidification of the silicide

Table 5: Ti/Ti–Ti/Al-brazed joints using Al–Mg-type filler alloys.

References	Materials		Filler		Conditions	Bonding		Shear strength (MPa)	Intermetallic compounds at Ti/filler interface
	Parent 1	Parent 2	Type	Thickness/gap width (μm)		Temperature (°C)	Time (min)		
[33]*	Ti	Ti	61S(Al (6061)) (Al–1Mg–0.6Si–0.25Cu–0.25Cr)	76	Argon (furnace)	~ 679–707**	5	73.8–115.7	
					Torch with flux			44.8–88.9	
[29]	Pure Ti	Pure Ti	Al–1Mg	100	Vacuum of 2.7×10^{-3} Pa	680	3		
[30]	Ti Grade 5, 2	Ti Grade 5, 2	Al–4.5Mg–0.4Si–0.1Cr	800	Vacuum/torch with flux	>638 liquidus temperature		Poor result	$TiAl_3$, Ti_9Al_{23}
			Al–4Mg–0.5Si–0.4Mn	500		>640 liquidus temperature		61 in vacuum and 117 by the torch	
[3]	Ti Grade 5, 2	Ti Grade 5, 2	TiBraze Al-665 (Al–2.5Mg–0.3Cr)	100	Vacuum/torch with flux	680		110.4 G5 97.6 G2	
			Al–4Mg–0.5Si–0.4Mn	320		650–700		98.8 G5 77.4 G2	
[31]	OT4 titanium	OT4 titanium	AMg6 (Al–6Mg–0.6Mn–0.4Si–0.4Fe–0.1Ti)	60	Vacuum 5×10^{-5} mmHg	685	3	82	$TiAl_2$, $TiAl_3$
			AMg6 (Al–6Mg–0.6Mn–0.4Si–0.4Fe–0.1Ti)	60		720		40	
			Al–2.5Mg–0.2Si–0.4Fe–0.2Cr	100		685		83	
[84]	Ti Grade 2/Grade 12	Ti Grade 2/Grade 12	TiBraze Al-645 (Al–5Mg–0.2Si–0.2Ti)	<25	EB + vacuum	>melting point	0.5–0.75		
[37]	Ti Grade 5	Ti Grade 5	TiBraze Al-665 (Al–2.5Mg–0.3Cr)	100	Ultrasound-assisted induction brazing in air	660	5 sec USV	50–60	
[38]	Ti Grade 2	Ti Grade 2	TiBraze Al-665A (Al–2.5Mg–0.3Cr)	50	Ultrasound-assisted induction brazing in air	670	5 min in 560°C + 3 min in 670°C + 6 s USV	~64	Ti_3Al

*A double-lap joint. **Considering 1205°F is the liquidus temperature [88].

Table 6: Ti/Ti–Ti/Al-brazed joints using Al–Si-type filler alloys.

References	Materials		Filler		Bonding			Shear strength (MPa)	Intermetallic compounds at Ti/filler interface
	Parent 1	Parent 2	Type	Thickness/gap width (μm)	Conditions	Temperature (°C)	Time (min)		
[34]*	CP-Ti	Al (1050)	Al–10Si–1Mg	100	Vacuum 10^{-4}Torr	620	25	84	$Ti_7Al_5Si_{12}$, $Ti_5Al_{12}Si_3$
[29]	Pure Ti	Pure Ti	Al–(0.1–10)Si	100	Vacuum 2.7×10^{-3}Pa	680	3	~94 for Al–0.8Si	$TiAl_3$**, Ti_9Al_{23}, $Ti_7Al_5Si_{12}$
[28]	Pure Ti	Al (3003)	Al–10Si–0.5Mg	100	Vacuum	600	3	~70	$Ti_7Al_5Si_{12}$, Ti_9Al_{23}***
[3]	Ti Grade 5, 2	Ti Grade 5, 2	AWS BAlSi-4 (Al–12Si–0.8Fe)	80	Vacuum/torch with flux	600–620		74.2 G5 69.6 G2	
[31]	OT4 titanium	OT4 titanium	Al–5.3Si–0.8Fe–0.3Cu–0.2Ti	100	Vacuum 5×10^{-5}mmHg	685	3	56	*Continuous silicide*
[89]	Ti Grade 5	Al (6060)	Al–8.4Si–20Cu–10Ge–0.1Re Al–8.4Si–20Cu–10Ge	200	Vacuum + argon	530 530	10/30/60 10/30/60	49/55/48 23/17/20	$Ti_7Al_5Si_{12}$ $Ti_7Al_5Si_{12}$
[35]	Ti50–Al50	Ti50–Al50	BAlSi-4 (Al–(11–13) Si–0.3Cu–0.1Mg–0.2Zn–0.15Mn–0.8Fe)	75	Vacuum 8×10^{-5}mbar	900****	3	86.2	$Ti_5Al_{12}Si_3$ + ($TiAl_2$ for *longer brazing time*)
[84]	Ti Grade 2/Grade 12	Tit Grade 2/Grade 12	TiBraze Al-642 (Al–5.3Si–0.8Fe–0.2Ti)	<25	EB + vacuum	>melting point	0.5–0.75		
[39]	Ti Grade 5	Al1060	Al–12Si		Ultrasound-assisted brazing in air	620	4 sec USV + holding of 15 or 5 min + 4 or 0 sec USV	~68	$Ti_7Al_5Si_{12}$, Ti_9Al_{23}

*Transient liquid phase bonding. **For Al–0.1Si/3 min, only Al_3Ti was observed. ***Very weak diffraction line was observed for Ti_9Al_{23}. ****Infrared brazing 900°C/min.

TABLE 7: Ti/Ti–Ti/Al-brazed joints using Al–Cu-type filler alloys.

References	Materials				Conditions	Bonding		Shear strength (MPa)	Intermetallic compounds at Ti/filler interface
	Parent 1	Parent 2	Filler						
			Type	Thickness/gap width (μm)		Temperature (°C)	Time (min)		
[29]	Pure Ti	Pure Ti	Al–1Cu	100	Vacuum 2.7×10^{-3} Pa	680	3		
[28]	Pure Ti	Al (3003)	Al–10Cu–8Sn	100	Vacuum	615	5	~38	$TiAl_3$, Ti_8CuAl_{23}*
[29]	Pure Ti	Al (3003)	Al–33Cu	100	Vacuum 2.7×10^{-3} Pa	630	3		$TiAl_3$
[30]	Ti Grade 5, 2	Ti Grade 5, 2	TiBraze Al-635 (Al–4.2Cu–1.5Mg–0.5Si)	300	Vacuum/torch with flux	>638 liquidus temperature		140 in vacuum and 88 by the torch	$TiAl_3$, *larger on Grade 5*
[3]	Ti Grade 5, 2	Ti Grade 5, 2	TiBraze Al-635 (Al–4.2Cu–1.5Mg–0.5Si)	200	Vacuum/torch with flux	660–700		140.7 G5 100.8 G2	$TiAl_3$, *larger on Grade 5*
			Al–25Cu–9Si	100		580–620		88.6 G5 34.1 G2	
	Ti Grade 2	Ti Grade 2	Al–24Cu–8.6Si–3.5Sn	150		580–620		47.7	
(Cited in [2])	Ti6Al4V	Ti6Al4V	Al–33Cu	6–20	Air	550		25	
[55]	Ti Grade 5	Al (6061)	Al–9.4Si–10Zn–10Cu–2Ni	~200	Ultrasound-assisted induction brazing in air	560	4 sec USV + 60 min + 4 sec USV	~43	$Ti_7Al_5Si_{12}$
			Al–15Sn–8.2Si–8.5Zn–8.5Cu–1.7Ni				4 sec USV + 15 min + 4 sec USV	~76	$Ti_7Al_5Si_{12}$
[40]	Ti-5 (95%)	Ti-5 (95%)	Al–8.54Cu–1.35Ni–0.08Ti	100	By applying pressure/in air	647	13	74.78	
			Al–6.44Cu–1.96Ni–0.92Ti			648	9	73.66	

*Very weak diffraction line was observed for Ti_8CuAl_{23}.

TABLE 8: Ti/Ti–Ti/Al-brazed joints using Al–Ag-type filler alloys.

References	Materials				Conditions	Bonding		Shear strength (MPa)	Intermetallic compounds at Ti/filler interface
	Parent 1	Parent 2	Filler			Temperature (°C)	Time (min)		
			Type	Thickness/gap width (μm)					
[28]	Pure Ti	Al (3003)	Al–20Ag–10Cu	100	Vacuum	600	5	~38	$TiAl_3$
[29]	Pure Ti	Al (3003)	A1–33Ag	100	Vacuum 2.7×10^{-3} Pa	630	3		$TiAl_3$
[30]	Ti Grade 5, 2	Ti Grade 5, 2	Al–38Ag–7.4Si–0.6Fe	260	Vacuum/torch with flux	>580 liquidus temperature		Poor result	$Ti_7Al_5Si_{12}$**
[3]	Ti Grade 2	Ti Grade 2	Al–28Ag	Powder +325/−100 mesh	Vacuum/torch with flux	600–620		70	
			Al–28.5Ag–21Cu–1.3Mg	260		650–710		67.6	

**Another undetermined very thin one.

TABLE 9: Ti/Ti-brazed joints using Al–Ni- and Al–Mn-type filler alloys.

References	Materials				Conditions	Bonding		Shear strength (MPa)	Intermetallic compounds at Ti/filler interface
	Parent 1	Parent 2	Filler			Temperature (°C)	Time (min)		
			Type	Thickness/gap width (μm)					
[29]	Pure Ti	Pure Ti	Al–1Mn, Al–1Ni	100	Vacuum 2.7×10^{-3} Pa	680	3		
[30]	Ti Grade 5, 2	Ti Grade 5, 2	TiBraze Al-700 (Al–21.5Ni–9Cu–9.6Si–1.3Fe–0.6Mn)	300	Vacuum/torch with flux	≥607		93 in vacuum and poor result by torch	Ti_9Al_{23}, $Ti_7Al_5Si_{12}$
[3]	Ti Grade 5, 2	Ti Grade 5, 2	TiBraze Al-700 (Al–21.5Ni–9Cu–9.6Si)	290	Vacuum/torch with flux	700–720		127.3 G5 81.6 G2	Ti_9Al_{23}, $Ti_7Al_5Si_{12}$
Cited in [2]	Ti Grade 5	Ti Grade 5	Al–1.2Mn–0.6Si–0.7Fe–0.2Cu			>660		89	

as a continuous strip at the brazing interface. Table 6 shows a summary of Ti/Ti–Ti/Al brazed joints using Al–Si-type filler alloys.

2.3.4. Al-Cu- and Al–Ag-Type Filler Alloys. Al–Cu and Al–Ag promoted mainly $TiAl_3$ formation at the brazed interface of titanium. Adding Sn did not change the intermetallic type since Al–10Cu–8Sn filler metal mainly promoted $TiAl_3$ formation in Al(A1100)/Ti joints and very weak diffraction line corresponding to Ti_8CuAl_{23}. However, high peak of Sn at the interface was showed by EDX line analysis, and the fracture occurred within the intermetallic compound [29]. In a different manner, the existence of Si in the filler changes the intermetallic type; Chen et al. [55] reported the formation of the intermetallic $Ti_7Al_5Si_{12}$ at the interface between the Grade 5 Ti and each of Al–9.4Si–10Zn–10Cu–2Ni and Al–15Sn–8.2Si–8.5Zn–8.5Cu–1.7Ni fillers; it was concluded that the addition of Sn improved the strength through reducing the residual thermal stress. Eckardt et al. [30] found that the filler alloy TiBraze Al-635 (Al–4.2Cu–1.5Mg–0.5Mn–0.5Si) had the lowest contact angle on Grade 5 Ti and a good one (second-lowest one) on Grade 2 Ti with the highest strength among the other alloys of Al–Ag, Al–Mg, and Ag–Cu–Zn–(Cd–Sn) using vacuum brazing. The interlayer between the parent metal and the filler TiBraze Al-635 consisted of variable and discontinuous layer of $TiAl_3$. According to Wells [41], the addition of Mg or the small addition of Li to the filler metal improved the flow characteristics of Al–Cu–(5–50)Ag-type alloys; on the other hand, alloying with Mg significantly decreased the corrosion resistance of the joints. Si and/or Sn-containing filler metals formed weaker joints or exhibited flow temperature of ~587°C (860 K) or higher. It was found that the flow temperature for the Al–Cu–Ag alloys varied with the employed atmosphere; flow temperatures were 111°C higher in the vacuum than in 6.6×10^4 Pa of argon. In brazing damage-tolerant titanium structures, Al–Cu–Ag filler materials prevent the cracks from spreading by delamination in the crack arrest orientation. The behavior in the crack divider orientation was similar to the behavior of the monolithic material [41]. Tables 7 and 8 show a summary of Ti/Ti–Ti/Al-brazed joints using Al–Cu and Al–Ag-type filler alloys, respectively.

2.3.5. Al–Ni, Al–Mn, and Other Types of Filler Alloys. Nickel addition to the pure aluminum filler had a little effect on the suppression of the intermetallic layer at the Ti/filler interface, whereas Mn addition showed a better result [29]. However, nickel addition to Al–Cu–Ti filler together with a decrease in copper ratio increased the shear strength [40]. Ge improved the corrosion resistance when added to the Al–Cu- and Al–Cu–Ag-type alloys but formed very fragile joints [41]. Table 9 shows a summary of Ti/Ti-brazed joints using Al–Ni- and Al–Mn-type filler alloys.

3. Conclusions and Future Work

In this review, the aluminum-based filler alloys for brazing titanium and the formation of the intermetallic compounds at the interface of Ti/filler have been analyzed. The proper design of time-temperature cycle and the controlled alloying additions in the aluminum filler are crucial factors for the intermetallic compound formation at the interface. The thickness and type of the intermetallic phases will largely determine the mechanical properties of the brazed joint. In the case of brazing Ti/Al joints, further decrease in brazing temperature, adequate time-temperature cycle, and the compatibility of the brazing filler on the both Ti/Al sides should be considered thoroughly.

Several technological parameters have been examined for ultrasound-assisted brazing of titanium by Al-based filler alloys, including the acoustic power, the USV period, the holding time, the filler thickness, and the brazing temperature; accordingly, the following points should be highlighted:

(i) The holding time after USV is necessary for the diffusion process proceeding at the interface between the filler and the titanium parent metal.

(ii) Varying the power of the ultrasound affects the oxide removal at the filler/parent metal interface directly.

(iii) The thickness of the filler should be chosen according to USV parameters and time-temperature brazing cycle; considering the accelerated interaction between the liquid filler and the parent metal by the effect of USV.

(iv) Brazing and ultrasound parameters/conditions should be considered together in ultrasound-assisted brazing process, for instance, the ultrasonic power/intensity, the coupling method of the ultrasound with the workpiece which insures the excellent delivery of the USV to the joint area, the USV activation time, and other traditional brazing parameters.

(v) Studying the USV effects on the microstructure of the brazed joint should be given more attention.

Brazing process has many factors that should be considered to achieve the best possible mechanical properties of the produced joint; the quality of the filler material is one of these factors. Different processes are available for aluminum filler production; the casting process is used as a favorite way to produce new filler aluminum alloys or known ones of lab-scale size for testing purposes. However, the properties and the quality of the cast affect the mechanical properties and the quality of the final brazing joint. The impact of the cast defects and the imperfections varied according to the form of the brazing filler alloy and the brazing process. The filler alloy, as a primary consideration, should have homogeneous chemical composition, with controlled or free of inclusions, and free of porosity. Using USV in the preparation and developing specific, low-melting temperature filler alloys for brazing process from their pure elements would be a great tool in enhancing the filler alloy properties, accelerating the process, and thus, exploring more compatible brazing fillers. Due to the limited works in ultrasound-assisted brazing of Ti by Al-based fillers and the mechanical properties of the produced joints, it is tempting to investigate and seeking to develop further compositions of the Al-based fillers which can be more adequate for ultrasound-assisted brazing of Ti and to conduct deeper studies considering the integration between the USV and the traditional brazing parameters.

Conflicts of Interest

The authors declare that they have no conflicts of interest.

Acknowledgments

This work is supported by Erasmus Mundus through PEACE II Lot 2 Project 2013–2443/001-001-EMA2 and FCT with the reference project UID/EEA/04436/2013, COMPETE 2020, with the code POCI-01-0145-FEDER-006941.

References

[1] T. Onzawa, A. Suzumura, and M. Ko, "Brazing of titanium using low-melting-point Ti-based filler metals," *Welding Journal*, vol. 462, 1990.

[2] A. E. Shapiro and Y. A. Flom, "Brazing of titanium at temperatures below 800°C: review and prospective applications," *DVS Berichte*, vol. 243, p. 254, 2007.

[3] A. E. Shapiro and Y. A. Flom, "New aluminum-based cold-rolled and composite braze foils for brazing titanium below 700°C," in *Proceeding of the 5th International Brazing and Soldering Conference*, pp. 427–432, Las Vegas, NV, USA, April 2012.

[4] M. J. Donachie, *Titanium: A Technical Guide*, ASM International, Materials Park, OH, USA, 2000.

[5] E. Gemelli and N. H. A. Camargo, "Oxidation kinetics of commercially pure titanium," *Matéria (Rio de Janeiro)*, vol. 12, no. 3, pp. 525–531, 2007.

[6] F. H. Froes and I. L. Caplan, *Titanium'92, Science and Technology: Proceedings of a Symposium Sponsored by the Titanium Committee of the Minerals Metals & Materials, Structural Metals Division, Minerals*, Metals and Materials Society, San Diego, CA, USA, 1993.

[7] F. C. Campbell, *Manufacturing Technology for Aerospace Structural Materials*, pp. 119–174, Elsevier, Amsterdam, Netherlands, 2011.

[8] V. V. Voronov, "Development of the technology for brazing of titanium alloys using filler alloys based on the Al-Mg system," *Paton Welding Journal*, vol. 2013, no. 2, pp. 56–58, 2013.

[9] X. Huang and N. L. Richards, "Activated diffusion brazing technology for manufacture of titanium honeycomb structures-a statistical study," *Welding Journal*, vol. 83, no. 3, pp. 73–81, 2004.

[10] A. B. Kolomenskii, V. A. Salikov, A. N. Roshchupkin, and A. V. Degtyarev, "Efficiency of titanium honeycomb packets produced by diffusion bonding," *Welding International*, vol. 9, no. 9, pp. 742–744, 1995.

[11] D. P. Sekulić, *Advances in Brazing: Science, Technology and Applications*, Elsevier Science, Amsterdam, Netherlands, 2013.

[12] C. E. Smeltzer, J. F. Nachman, and A. N. Hammer, *Titanium Braze System for High Temperature Applications*, AFML-TR.76–145, Solar Division of International Harvester, San Diego, CA, USA, 1976.

[13] R. R. Boyer, "An overview on the use of titanium in the aerospace industry," *Materials Science and Engineering: A*, vol. 213, no. 1-2, pp. 103–114, 1996.

[14] K. Matsu, Y. Miyazawa, Y. Nishi, and T. Ariga, "Brazing and interfacial reaction of commercially pure titanium with Ti-Zr-based filler metals," *Materials Transactions*, vol. 48, no. 5, pp. 1055–1059, 2007.

[15] E. Ganjeh and H. Sarkhosh, "Microstructural, mechanical and fractographical study of titanium-CP and Ti–6Al–4V similar brazing with Ti-based filler," *Materials Science and Engineering: A*, vol. 559, pp. 119–129, 2013.

[16] E. Ganjeh, H. Sarkhosh, M. E. Bajgholi, H. Khorsand, and M. Ghaffari, "Increasing Ti–6Al–4V brazed joint strength equal to the base metal by Ti and Zr amorphous filler alloys," *Materials Characterization*, vol. 71, pp. 31–40, 2012.

[17] C. T. Chang, Y. C. Du, R. K. Shiue, and C. S. Chang, "Infrared brazing of high-strength titanium alloys by Ti–15Cu–15Ni and Ti–15Cu–25Ni filler foils," *Materials Science and Engineering: A*, vol. 420, no. 1-2, pp. 155–164, 2006.

[18] T. Onzawa and T. Iiyama, "Development of titanium-based brazing filler metals with low-melting-point," *International Journal of Korean Welding Society*, vol. 2, no. 2, pp. 14–18, 2002.

[19] Titanium Brazing, Inc., Products and Services, November 2017, http://www.titanium-brazing.com/products.html#3.

[20] C. T. Chang, R. K. Shiue, and C. S. Chang, "Microstructural evolution of infrared brazed Ti-15-3 alloy using Ti–15Cu–15Ni and Ti–15Cu–25Ni fillers," *Scripta Materialia*, vol. 54, no. 5, pp. 853–858, 2006.

[21] R. K. Shiue, S. K. Wu, Y. T. Chen, and C. Y. Shiue, "Infrared brazing of Ti50Al50 and Ti–6Al–4V using two Ti-based filler metals," *Intermetallics*, vol. 16, no. 9, pp. 1083–1089, 2008.

[22] M. K. Lee and J. G. Lee, "Mechanical and corrosion properties of Ti–6Al–4V alloy joints brazed with a low-melting-point 62.7Zr–11.0Ti–13.2Cu–9.8Ni–3.3Be amorphous filler metal," *Materials Characterization*, vol. 81, pp. 19–27, 2013.

[23] M. K. Lee, K. H. Kim, J. G. Lee, and C. K. Rhee, "Growth of isothermally-solidified titanium joints using a multi-component Zr–Ti–Cu–Ni–Be amorphous alloy as a brazing filler," *Materials Characterization*, vol. 80, pp. 98–104, 2013.

[24] Ti-Flo Filler Metal for Titanium-Johnson Matthey Metal Joining Company, July 2017, http://www.jm-metaljoining.com/products/ti-flo.

[25] K. B. Pressly, J. A. Brunk, K. Gentry, E. Jimenez, E. M. Lundgren, and G. E. Parkhurst, *Brazing Titanium Structures*, BDX-613–1666(Rev.), National Technical Information Service, Springfield, VA, USA, 1977.

[26] Y. C. Du and R. K. Shiue, "Infrared brazing of Ti–6Al–4V using two silver-based braze alloys," *Journal of Materials Processing Technology*, vol. 209, no. 11, pp. 5161–5166, 2009.

[27] E. Ganjeh, H. Sarkhosh, H. Khorsand, H. Sabet, E. H. Dehkordi, and M. Ghaffari, "Evaluate of braze joint strength and microstructure characterize of titanium-CP with Ag-based filler alloy," *Materials & Design*, vol. 39, pp. 33–41, 2012.

[28] T. Takemoto, H. Nakamura, and I. Okamoto, "Strength of titanium joints brazed with aluminum filler metals," *Transactions of JWRI*, vol. 19, no. 1, pp. 45–49, 1990.

[29] T. Takemoto and I. Okamoto, "Intermetallic compounds formed during brazing of titanium with aluminium filler metals," *Journal of Materials Science*, vol. 23, no. 4, pp. 1301–1308, 1988.

[30] T. Eckardt, B. Hanhold, D. Petrasek, S. Sattler, A. Benatar, and A. Shapiro, "Evaluating low-temperature brazing filler metals for joining titanium," *Welding Journal*, vol. 91, no. 2, pp. 45–50, 2012.

[31] V. F. Khorunov, V. V. Voronov, and S. V. Maksymova, "Brazing of titanium alloys by using aluminium-base filler alloys," *Paton Welding Journal*, vol. 11, pp. 2–5, 2012.

[32] A. E. Shapiro and E. Y. Ivanov, "Heat-resistant brazing filler metals for joining titanium aluminide and titanium alloys," in *Proceedings of the 3rd International Brazing and Soldering Conference (IBSC)*, pp. 12–17, San Antonio, TX, USA, 2006.

[33] W. J. Lewis, P. S. Rieippel, and C. B. Voldrich, *Brazing Titanium to Titanium and to Mild and Stainless Steels, Part 1*, Tech. Rep. 52–313, Battelle Memorial Institute, Columbus, OH, USA, 1952.

[34] W. H. Sohn, H. H. Bong, and S. H. Hong, "Microstructure and bonding mechanism of Al/Ti bonded joint using Al–10Si–1Mg filler metal," *Materials Science and Engineering: A*, vol. 355, no. 1-2, pp. 231–240, 2003.

[35] R. K. Shiue, S. K. Wu, and S. Y. Chen, "Infrared brazing of TiAl using Al-based braze alloys," *Intermetallics*, vol. 11, no. 7, pp. 661–671, 2003.

[36] A. F. Nesterov, "Formation of brazed joints in titanium brazed with an aluminium brazing alloy," *Welding International*, vol. 4, no. 3, pp. 213–215, 1990.

[37] W. Tillmann, M. Zimpel, N. F. L. Dias, J. Pfeiffer, L. Wojarski, and Z. Xu, "Mechanical and microstructural analysis of ultrasonically assisted induction-brazed TiAl6V4 joints," *Welding in the World*, vol. 59, no. 6, pp. 901–909, 2015.

[38] A. Elrefaey, L. Wojarski, J. Pfeiffer, and W. Tillmann, "Preliminary investigation on ultrasonic-assisted brazing of titanium and titanium/stainless steel joints," *Welding Journal*, vol. 92, pp. 148–153, 2013.

[39] X. Chen, J. Yan, S. Ren, J. Wei, and Q. Wang, "Microstructure and mechanical properties of Ti–6Al–4V/Al1060 joints by ultrasonic-assisted brazing in air," *Materials Letters*, vol. 95, pp. 197–200, 2013.

[40] M. K. Karfoul and A. S. Muhrat, "Low Melting Brazing Al-Cu-Ni-Ti Filler Alloys Designed for Joining Titanium under Normal Ambient Air Conditions," in *Proceedings of 10th International Conference Brazing, High Temperature Brazing and Diffusion Welding*, vol. 293, pp. 17–21, Aachen, Germany, June 2013.

[41] R. Wells, "Low temperature large-area brazing of damage tolerant titanium structures," *Welding Journal*, vol. 54, no. 10, pp. 348–356, 1975.

[42] C. T. Chang and R. K. Shiue, "Infrared brazing of Ti-6Al-4V using the Ti-15Cu-15Ni braze alloy," *Journal of Materials Science*, vol. 41, no. 7, pp. 2145–2150, 2006.

[43] American Welding Society (AWS) C3 Committee on Brazing and Soldering, *Brazing Handbook*, American Welding Society, Miami, FL, USA, 2007.

[44] H. Puga, J. Barbosa, J. C. Teixeira, and M. Prokic, "A new approach to ultrasonic degassing to improve the mechanical properties of aluminum alloys," *Journal of Materials Engineering and Performance*, vol. 23, no. 10, pp. 3736–3744, 2014.

[45] H. Puga, J. C. Teixeira, J. Barbosa, E. Seabra, S. Ribeiro, and M. Prokic, "The combined effect of melt stirring and ultrasonic agitation on the degassing efficiency of AlSi9Cu3 alloy," *Materials Letters*, vol. 63, no. 24-25, pp. 2089–2092, 2009.

[46] H. Puga, J. Barbosa, E. Seabra, S. Ribeiro, and M. Prokic, "The influence of processing parameters on the ultrasonic degassing of molten AlSi9Cu3 aluminium alloy," *Materials Letters*, vol. 63, no. 9-10, pp. 806–808, 2009.

[47] H. Xu, X. Jian, T. T. Meek, and Q. Han, "Degassing of molten aluminum A356 alloy using ultrasonic vibration," *Materials Letters*, vol. 58, no. 29, pp. 3669–3673, 2004.

[48] J. Barbosa and H. Puga, "Ultrasonic melt processing in the low pressure investment casting of Al alloys," *Journal of Materials Processing Technology*, vol. 244, pp. 150–156, 2017.

[49] H. Puga, J. Barbosa, S. Costa, S. Ribeiro, A. M. P. Pinto, and M. Prokic, "Influence of indirect ultrasonic vibration on the microstructure and mechanical behavior of Al–Si–Cu alloy," *Materials Science and Engineering: A*, vol. 560, pp. 589–595, 2013.

[50] L. Qingmei, Z. Yong, S. Yaoling, Q. Feipeng, and Z. Qijie, "Influence of ultrasonic vibration on mechanical properties and microstructure of 1Cr18Ni9Ti stainless steel," *Materials & Design*, vol. 28, no. 6, pp. 1949–1952, 2007.

[51] D. Gao, Z. Li, Q. Han, and Q. Zhai, "Effect of ultrasonic power on microstructure and mechanical properties of AZ91 alloy," *Materials Science and Engineering: A*, vol. 502, no. 1-2, pp. 2–5, 2009.

[52] X. Chen, J. Yan, F. Gao, J. Wei, Z. Xu, and G. Fan, "Interaction behaviors at the interface between liquid Al-Si and solid Ti-6Al-4V in ultrasonic-assisted brazing in air," *Ultrasonics Sonochemistry*, vol. 20, no. 1, pp. 144–154, 2013.

[53] W. Cui, C. Wang, J. Yan, Z. Wang, and D. Wei, "Wetting and reaction promoted by ultrasound between sapphire and liquid Al-12Si alloy," *Ultrasonics Sonochemistry*, vol. 20, no. 1, pp. 196–201, 2013.

[54] H. Ji, X. Cheng, and M. Li, "Ultrafast ultrasonic-assisted joining of bare α-alumina ceramics through reaction wetting by aluminum filler in air," *Journal of the European Ceramic Society*, vol. 36, no. 16, pp. 4339–4344, 2016.

[55] X. Chen, R. Xie, Z. Lai, L. Liu, G. Zou, and J. Yan, "Ultrasonic-assisted brazing of Al–Ti dissimilar alloy by a filler metal with a large semi-solid temperature range," *Materials & Design*, vol. 95, pp. 296–305, 2016.

[56] Y. Xiao, H. Ji, M. Li, and J. Kim, "Ultrasound-assisted brazing of Cu/Al dissimilar metals using a Zn–3Al filler metal," *Materials & Design*, vol. 52, pp. 740–747, 2013.

[57] W. Guo, T. Luan, J. He, and J. Yan, "Ultrasonic-assisted soldering of fine-grained 7034 aluminum alloys using ZnAl filler metals," *Materials & Design*, vol. 125, pp. 85–93, 2017.

[58] W. Guo, X. Leng, T. Luan, J. Yan, and J. He, "Ultrasonic-promoted rapid TLP bonding of fine-grained 7034 high strength aluminum alloys," *Ultrasonics Sonochemistry*, vol. 36, pp. 354–361, 2017.

[59] W. L. Cotton, "report Effects of service environments on aluminum-brazed titanium (ABTi)," NASA Contractor Report 2943, Boeing Commercial Airplane Company, Seattle, Washington, USA, 1978.

[60] K. S. Suslick, D. A. Hammerton, and R. E. Cline, "Sonochemical hot spot," *Journal of the American Chemical Society*, vol. 108, no. 18, pp. 5641-5642, 1986.

[61] J. B. Jones, A. L. Fuchs, and W. R. Bailey, *Ultrasonic Brazing Unit*, US2833238, Serial No. 600198, USA, 1958.

[62] Z. Xu, J. Yan, G. Wu, X. Kong, and S. Yang, "Interface structure and strength of ultrasonic vibration liquid phase bonded joints of Al_2O_{3p}/6061Al composites," *Scripta Materialia*, vol. 53, no. 7, pp. 835–839, 2005.

[63] G. I. Eskin and D. G. Eskin, *Ultrasonic Treatment of Light Alloy Melts*, CRC Press, Boca Raton, FL, USA, 2014.

[64] M. A. Margulis and I. M. Margulis, "Mechanism of sonochemical reactions and sonoluminescence," *High Energy Chemistry*, vol. 38, no. 5, pp. 285–294, 2004.

[65] D. G. Shchukin, E. Skorb, V. Belova, and H. Möhwald, "Ultrasonic cavitation at solid surfaces," *Advanced Materials*, vol. 23, no. 17, pp. 1922–1934, 2011.

[66] S. Tamura and M. Hatakeyama, "The role of acoustic cavitation in liquid pressurization in narrow tubes," *Journal of Applied Physics*, vol. 113, no. 14, p. 144905, 2013.

[67] I. Hansson and K. A. Morch, "The dynamics of cavity clusters in ultrasonic (vibratory) cavitation erosion," *Journal of Applied Physics*, vol. 51, no. 9, pp. 4651–4658, 1980.

[68] Y. Li, X. Leng, S. Cheng, and J. Yan, "Microstructure design and dissolution behavior between 2024 Al/Sn with the ultrasonic-associated soldering," *Materials & Design*, vol. 40, pp. 427–432, 2012.

[69] L. Ma, Z. Xu, K. Zheng, J. Yan, and S. Yang, "Vibration characteristics of aluminum surface subjected to ultrasonic waves and their effect on wetting behavior of solder droplets," *Ultrasonics*, vol. 54, no. 3, pp. 929–937, 2014.

[70] F. Zhang, S. L. Chen, Y. A. Chang, and U. R. Kattner, "A thermodynamic description of the Ti-Al system," *Intermetallics*, vol. 5, no. 6, pp. 471–482, 1997.

[71] C. Leyens and M. Peters, *Titanium and Titanium Alloys: Fundamentals and Applications*, Wiley-VCH, Weinheim, Germany, 2003.

[72] R. Pretorius, A. M. Vredenberg, F. W. Saris, and R. de Reus, "Prediction of phase formation sequence and phase stability in binary metal-aluminum thin-film systems using the effective heat of formation rule," *Journal of Applied Physics*, vol. 70, no. 7, pp. 3636–3646, 1991.

[73] M. Sujata, S. Bhargava, and S. Sangal, "On the formation of $TiAl_3$ during reaction between solid Ti and liquid Al," *Journal of Materials Science Letter*, vol. 16, no. 13, pp. 1175–1178, 1997.

[74] S. Wöhlert and R. Bormann, "Phase selection governed by different growth velocities in the early stages of the Ti/Al phase reaction," *Journal of Applied Physics*, vol. 85, no. 2, pp. 825–832, 1999.

[75] C. Michaelsen, S. Wöhlert, and R. Bormann, "Phase formation and microstructural development during solid-state reactions in Ti-Al multilayer films," *MRS Proceedings*, vol. 343, pp. 205–210, 1994.

[76] Z. G. Zhang, Y. J. Wang, L. J. Xiao, L. Q. Zhang, Y. Su, and J. S. Lin, "High-temperature oxidation of hot-dip aluminizing coatings on a Ti_3Al–Nb alloy and the effects of element additions," *Corrosion Science*, vol. 64, pp. 137–144, 2012.

[77] Z. G. Zhang, Y. P. Peng, Y. L. Mao, C. J. Pang, and L. Y. Lu, "Effect of hot-dip aluminizing on the oxidation resistance of

Ti–6Al–4V alloy at high temperatures," *Corrosion Science*, vol. 55, pp. 187–193, 2012.

[78] L. Xu, Y. Y. Cui, Y. L. Hao, and R. Yang, "Growth of intermetallic layer in multi-laminated Ti/Al diffusion couples," *Materials Science and Engineering: A*, vol. 435-436, pp. 638–647, 2006.

[79] J. G. Luo and V. L. Acoff, "Interfacial reactions of titanium and aluminum during diffusion welding," *Welding Journal*, vol. 79, no. 9, pp. 239s–243s, 2000.

[80] F. J. J. Loovan and G. D. Rieck, "Diffusion in the titanium-aluminium system—II. interdiffusion in the composition range between 25 and 100 at.% Ti," *Acta Metallurgica*, vol. 21, no. 1, pp. 73–84, 1973.

[81] Y. Mishin and C. Herzig, "Diffusion in the Ti–Al system," *Acta Materialia*, vol. 48, no. 3, pp. 589–623, 2000.

[82] C. Buque and F. Appel, "Diffusion bonding of gamma-based titanium aluminides," *Zeitschrift für Metallkunde*, vol. 93, no. 8, pp. 784–789, 2002.

[83] D. Herrmann and F. Appel, "Diffusion bonding of γ(TiAl) alloys: influence of composition, microstructure, and mechanical properties," *Metallurgical and Materials Transactions A*, vol. 40, no. 8, pp. 1881–1902, 2009.

[84] Y. Flom, "Electron Beam Brazing of Titanium for Construction in Space," in *Proceedings of The Third International Brazing and Soldering Conference (Brazing and Soldering)*, pp. 1–5, ASM International, San Antonio, TX, USA, April 2006.

[85] S. W. Kim, D. H. Kim, B. D. You, J. W. Han, and M. S. Kim, "Evaporation behavior of magnesium under reduced pressure," *Materials Science Forum*, vol. 439, pp. 238–243, 2003.

[86] W. L. Winterbottom and G. A. Gilmour, "Vacuum brazing of aluminum: auger studies of wetting and flow characteristics," *Journal of Vacuum Science and Technology*, vol. 13, no. 2, pp. 634–643, 1976.

[87] O. Ohashi and K. Sasabe, "Effect of alloying elements on the behaviour of oxide films at diffusion-welded interfaces of aluminium alloys—study of surface films at diffusion-welded interfaces (report 4)," *Welding International*, vol. 4, no. 10, pp. 775–780, 1990.

[88] F. Cverna, *ASM Ready Reference: Thermal Properties of Metals*, ASM International, Materials Park, OH, USA, 2002.

[89] S. Y. Chang, L. C. Tsao, Y. H. Lei, S. M. Mao, and C. H. Huang, "Brazing of 6061 aluminum alloy/Ti–6Al–4V using Al–Si–Cu–Ge filler metals," *Journal of Materials Processing Technology*, vol. 212, no. 1, pp. 8–14, 2012.

14

Experimental Investigation and FE Analysis on Constitutive Relationship of High Strength Aluminum Alloy under Cyclic Loading

Yuanqing Wang and Zhongxing Wang

Department of Civil Engineering, Tsinghua University, Beijing 10084, China

Correspondence should be addressed to Zhongxing Wang; 1534307298@qq.com

Academic Editor: Hiroshi Noguchi

Experiments of 17 high strength aluminum alloy (7A04) specimens were conducted to investigate the constitutive relationship under cyclic loading. The monotonic behavior and hysteretic behavior were focused on and the fracture surface was observed by scanning electron microscope (SEM) to investigate the microfailure modes. Based on Ramberg-Osgood model, stress-strain skeleton curves under cyclic loading were fitted. Parameters of combined hardening model including isotropic hardening and kinematic hardening were calibrated from test data according to Chaboche model. The cyclic tests were simulated in finite element software ABAQUS. The test results show that 7A04 aluminum alloy has obvious nonlinearity and ultra-high strength which is over 600 MPa, however, with relatively poor ductility. In the cyclic loading tests, 7A04 aluminum alloy showed cyclic hardening behavior and when the compressive strain was larger than 1%, the stiffness degradation and strength degradation occurred. The simulated curves derived by FE model fitted well with experimental curves which indicates that the parameters of this combined model can be used in accurate calculation of 7A04 high strength aluminum structures under cyclic loading.

1. Introduction

With the development of aluminum alloy structures, the research on constitutive relationship of aluminum alloy and aluminum alloy members is gradually deepening. Much recent research activity about aluminum alloy has centred on static research including stability behavior research [1–3], research on joints [4, 5], research on fatigue crack [6, 7], and research on overall aluminum alloy structures [8]. However, the research work of aluminum alloy members or structures under dynamic loading [9] is scant. With the rapid development of metallurgical industry, high strength aluminum alloy (HSAA) starts to be applied as main supporting members in engineering structures. The research on seismic performance of aluminum alloy structures or members will be a big trend. As the premise of the seismic performance analysis, the constitutive relationship of HSAA under cyclic loading is the important basis.

The research on the constitutive relationship of aluminum alloy was focused on its monotonic behavior. Ramberg-Osgood presented a nonlinear constitutive model of aluminum alloy in 1943 [10]:

$$\varepsilon = \frac{\sigma}{E} + \left(\frac{\sigma}{f_{0.2}}\right)^n. \tag{1}$$

In 1971, Steinhardt presented a method which can simplify the calculation of "n" in Ramberg-Osgood model and has been applied widely [11], as is shown in

$$10n = f_{0.2}\ (\text{MPa}). \tag{2}$$

Mazzolani also presented his constitutive model of aluminum alloy [12], but this model is seldom used.

Chaboche plastic constitutive model [13] is usually applied to describe the constitutive relationship of metal material under cyclic loading. Chaboche model is a combined hardening model including isotropic hardening component and nonlinear kinetic hardening component. However, there is nearly no test of aluminum alloy conducted to calibrate the

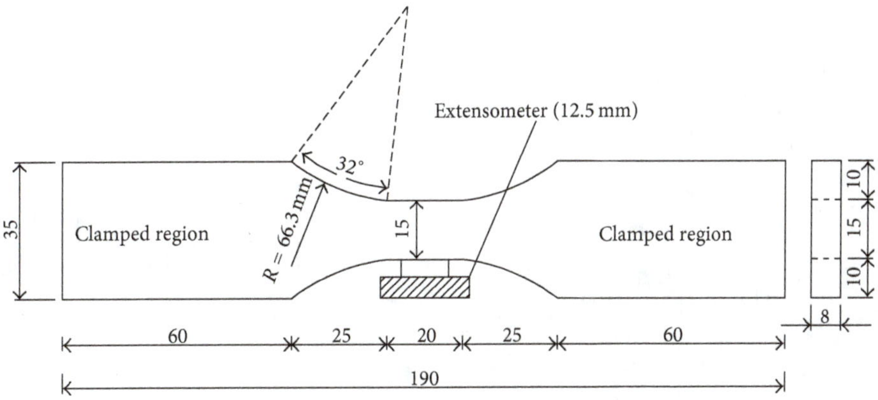

FIGURE 1: Dimension of the specimen.

TABLE 1: Chemical composition (%) of 7A04 aluminum alloy.

Material	Mass fraction/%				
	Si	Fe	Cu	Mn	Zn
7A04	0.6	0.7	0.05	1.0	0.1

parameters in Chaboche model which limits the research of aluminum alloy structures under cyclic or seismic loading.

7A04 HSAA is Al-Zn-Mg-Cu series aluminum alloy, with nominal yield stress over 500 MPa, which has been applied in aerospace field. Recent years, several transmission tower structures and military engineering structures intend to use the 7A04 HSAA as the main structural material. Zhang et al. conducted a research on static constitutive relationship on 7A04 HSAA in 2011 [14]. However, there is no research of 7A04 HSAA on the constitutive relationship under cyclic loading.

In the current paper, 2 monotonic loading tests and 15 cyclic loading tests were conducted. Monotonic behavior and hysteretic behavior of the material were analyzed. SEM was used to observe the crack surface in order to study the microfailure modes of the material. Stress-strain skeleton curves under cyclic loading were fitted based on Ramberg-Osgood model. After calibration of the parameters in Chaboche model, FE software ABAQUS was used to simulate the corresponding tests. The purpose of this paper is to supply a basis for calculation of 7A04 HSAA structures under cyclic loading.

2. Experimental Investigation

2.1. Test Specimens. Plates are usually used in structure engineering; therefore, 17 specimens were cut from HSAA plates. The dimension of the specimen is shown in Figure 1. The chemical composition of 7A04 HSAA is shown in Table 1.

2.2. Test Set-Up. Monotonic loading and cyclic loading were carried out on INSTRON 8801 testing machine, as is shown in Figure 2. The strain of specimens was measured by extensometer connected with the testing machine with gauge length of 12.5 mm and measuring range of ±35%.

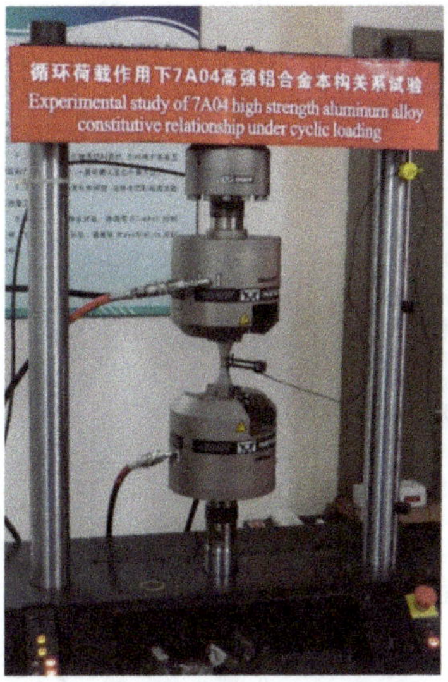

FIGURE 2: Test set-up.

The layout of the extensometer is shown in Figure 1. Strain controlled loading was adopted in the test. Stress, strain, and displacement were collected by computer in real time, and the acquisition frequency is 20 Hz.

Specimens M1 and M2 were tested under monotonic loading. Specimens H1~H15 were tested under different kinds of cyclic loading to investigate the hysteretic behavior and constitutive characteristics. After the cyclic loading, the specimens were loaded until tension failure. All of the cyclic loading system is shown in Figure 3.

3. Test Results and Analysis

3.1. Test Results. Test results of monotonic loading and cyclic loading are summarized in Tables 2 and 3. Specimen H1 failed in the 6th loading cycle, not reaching the ultimate strength.

3.2. Monotonic Behavior. Stress-strain curves of monotonic loading are shown in Figure 4. It can be found from Figure 4 and Table 2 that the nominal yield strength of 7A04 HSAA is about 540 MPa, and the ultimate strength is about 600 MPa.

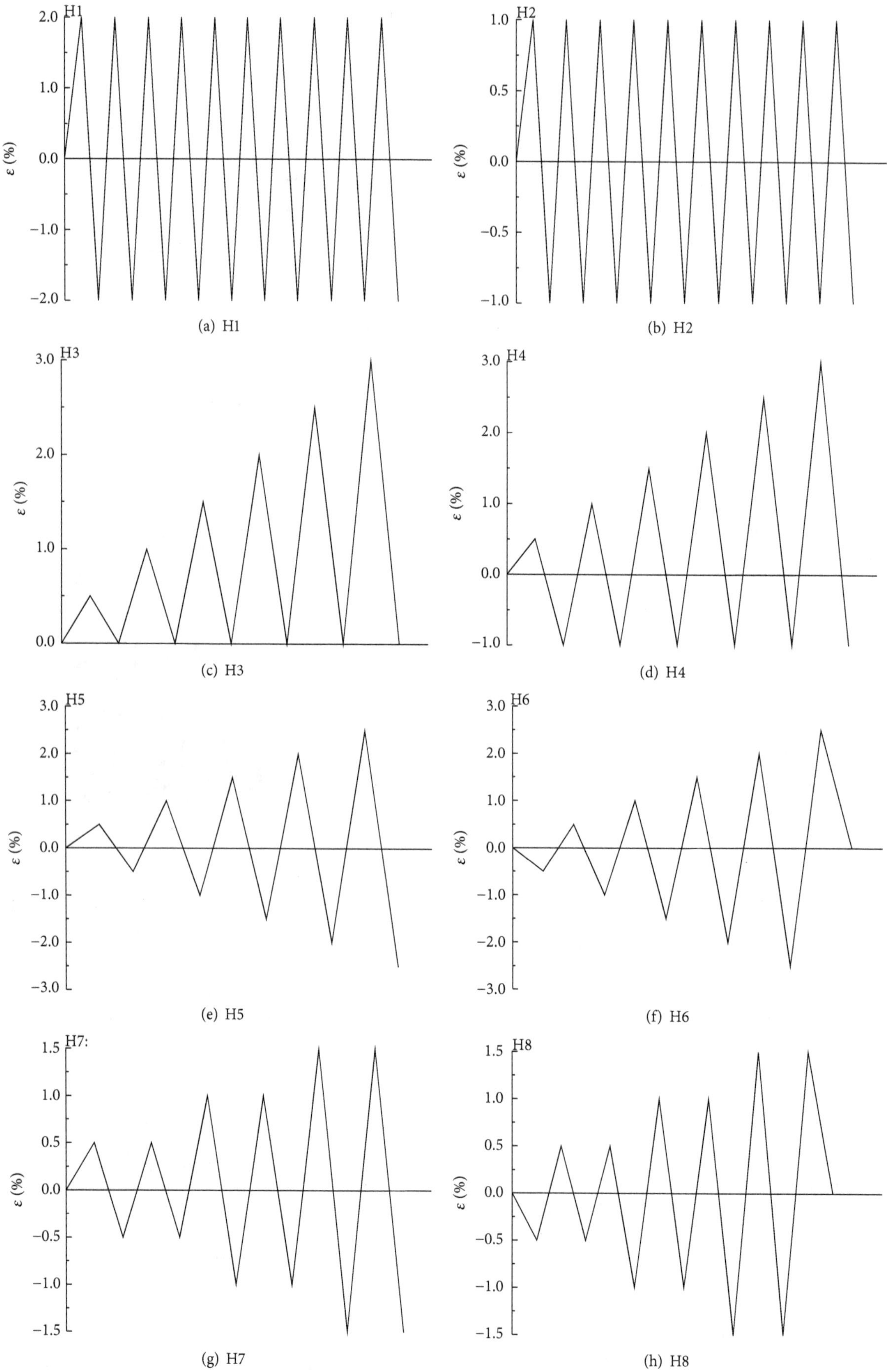

(a) H1
(b) H2
(c) H3
(d) H4
(e) H5
(f) H6
(g) H7
(h) H8

FIGURE 3: Continued.

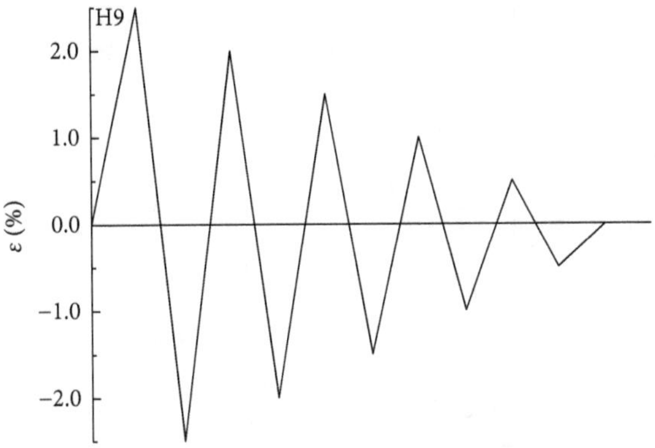

(i) H9

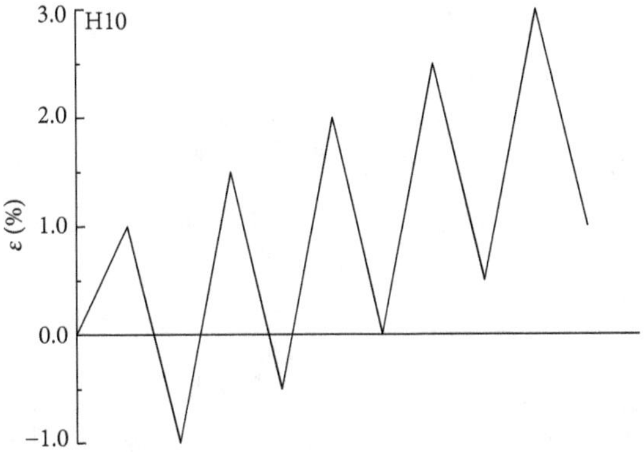

(j) H10

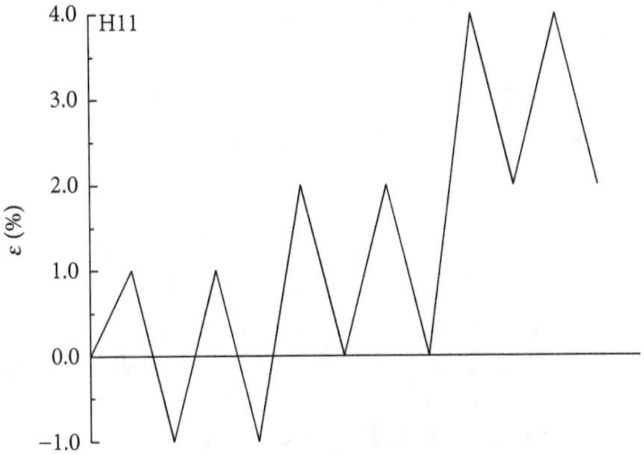

(k) H11

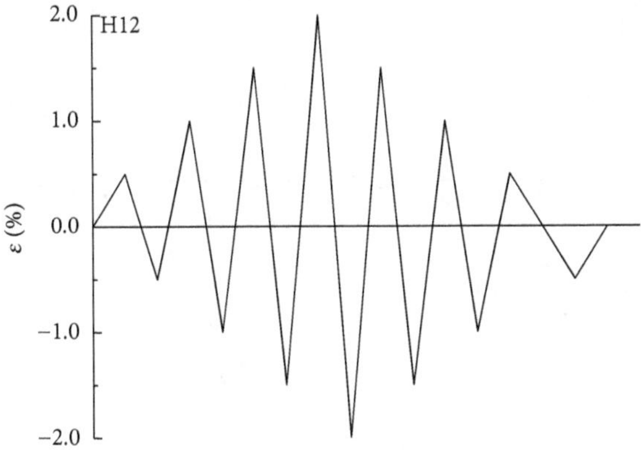

(l) H12

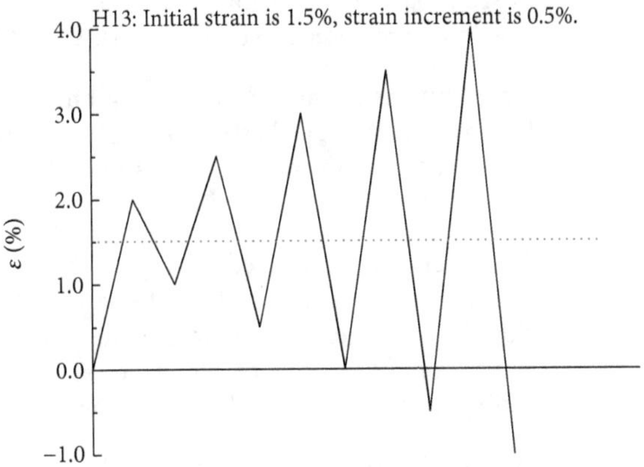

(m) H13

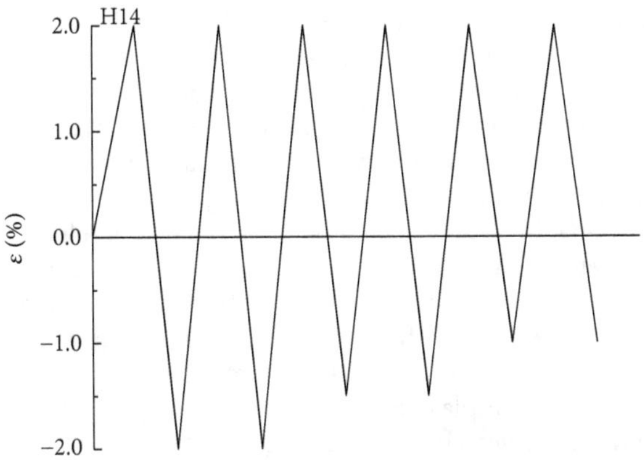

(n) H14

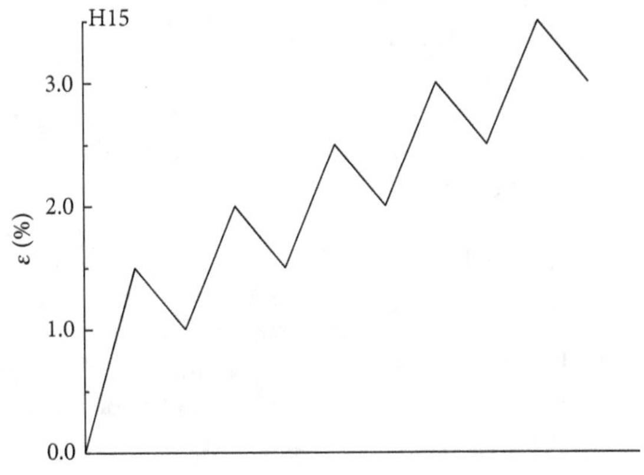

(o) H15

FIGURE 3: Cyclic loading system.

TABLE 2: Test results of monotonic loading.

Specimen	$f_{0.2}$/MPa	f_u/MPa	f_{u2}/MPa	ε_1/%	ε_u/%	ε_2%
M1	552.95	605.04	593.43	0.92	8.77	11.71
M2	537.36	593.76	589.84	0.89	8.01	9.98

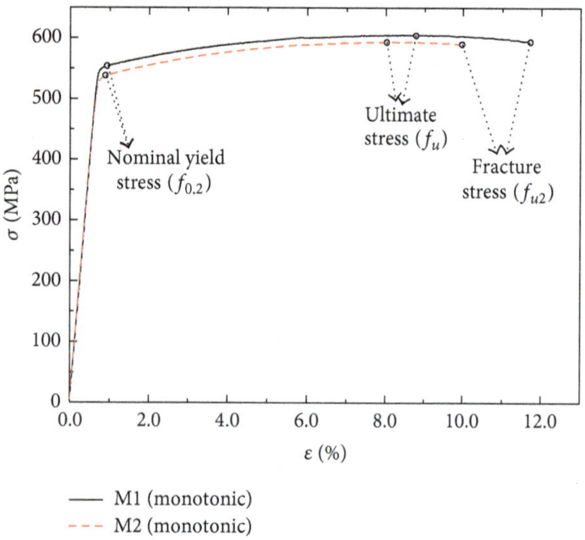

FIGURE 4: Stress-strain curves of monotonic loading.

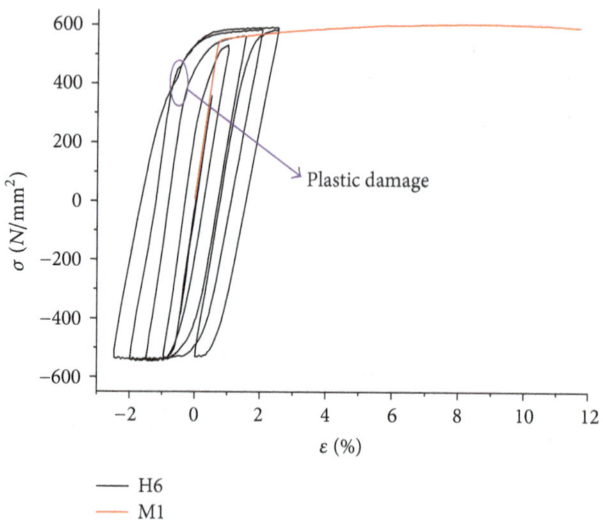

FIGURE 5: Comparison between monotonic and cyclic loading curves.

The strength of 7A04 HSAA is close to Q460 high strength steel. There are no obvious yield plateau and descending stage in the stress-strain curve. The fracture strength is decreased by 1.3% compared with the ultimate strength. The failure process is very sudden without visible necking phenomenon. The ductility is relatively poor with 10% ultimate strain.

Figure 5 is the comparison between monotonic curves and cyclic loading curves. Because of Bauschinger effect, cyclic hardening effect, and cyclic softening effect, there is difference between the monotonic curve and cyclic loading curves. The strength in strain-hardening ranges of cyclic loading curves is higher than that of monotonic curves. However, because of plastic damage, the ultimate strengths of monotonic curves and cyclic loading curves are nearly the same. The plastic damage has been marked in Figure 5 which makes a sudden drop of the curve when there is a relatively large stress. Plastic damage phenomenon is found in the loading process of H1, H3~H6, H9~H10, and H12~H15.

3.3. Hysteretic Behavior. To study the hysteretic behavior of material is to study the constitutive response of the material under cyclic loading. The stress-strain curves of H1~H15 under cyclic loading are shown in Figure 6; the hysteretic behavior of 7A04 HSAA is shown as follows:

(1) By constant-amplitude cyclic loading, the stress of specimens H1 and H2 increases at the same strain with the increase of hysteretic cycles. The material shows typical cyclic hardening effect.

(2) For most specimens, the stiffness degradation and strength degradation occurred after 1% compressive strain, which is largely due to two reasons. The first one is that local buckling led to a strength decrease; the second one is that the strain measured by extensometer included buckling deformation leading to the stiffness degradation. However, both the stiffness degradation and strength degradation are caused by specimen buckling damage rather than the constitutive relationship of 7A04 HSAA itself [15].

(3) For specimens H6 and H8 (first press and then pull), the tensile nominal yield stress is nearly the same with the one derived in monotonic loading. Bauschinger effect of the material is not significant.

(4) For specimens subjected to cyclic loading, the decrease range of the fracture strength is the same with the specimens under monotonic loading. The ductility of 7A04 HSAA did not decrease due to cyclic loading, which is different from the properties of steel.

(5) 7A04 HSAA showed combined hardening behavior including isotropic hardening and nonlinear kinetic hardening.

(6) Under many kinds of cyclic loading systems, the curves in Figure 6 are plump which show good hysteretic behavior of 7A04 HSAA.

Stress-strain skeleton curves were used to investigate the difference between the constitutive relationship of 7A04 HSAA under cyclic loading and the one under monotonic loading. The point on the skeleton curve is the point with the maximum stress and maximum strain in each loading cycle. Ramberg-Osgood model was adopted to fit the stress-strain skeleton curves. The formula is as follows:

$$\frac{\Delta\varepsilon}{2} = \frac{\Delta\varepsilon_e}{2} + \frac{\Delta\varepsilon_p}{2} = \frac{\Delta\sigma}{2E} + \left(\frac{\Delta\sigma}{2K'}\right)^{1/n'}. \quad (3)$$

TABLE 3: Test results of cyclic loading.

Specimen	$f_{0.2}$/MPa	f_u/MPa	f_{u2}/MPa	ε_1/%	N_c	E_n/kN·mm
H1	545.14	\	\	0.94	6	185.68
H2	543.56	615.99	594.53	0.90	10	77.75
H3	544.18	606.12	598.96	0.94	6	53.80
H4	542.66	627.07	597.42	0.95	6	117.26
H5	542.08	631.12	602.66	0.88	5	113.75
H6	548.83	604.14	604.03	0.89	5	116.57
H7	543.27	619.57	593.52	0.95	6	55.68
H8	530.48	613.02	587.04	0.87	6	55.57
H9	541.29	605.20	601.27	0.90	5	112.44
H10	544.15	618.60	592.09	0.84	5	59.52
H11	532.39	599.03	580.14	0.98	6	61.24
H12	542.88	615.72	590.75	0.87	7	95.67
H13	542.89	623.28	595.65	0.91	5	121.96
H14	550.99	628.30	613.66	0.93	6	173.97
H15	554.52	612.95	599.64	1.00	5	23.17

TABLE 4: Cyclic hardening parameters.

Specimen	E/MPa	K'/MPa	n'
H3	73970	714.24	0.052
H4	72463	798.89	0.071
H5	79870	822.35	0.078
H6	79820	761.11	0.063
H10	84678	734.98	0.056
H12	81132	839.89	0.087
H13	76688	734.01	0.062
H15	69647	723.16	0.053

In order to facilitate the fitting process, (3) is simplified as follows:

$$\varepsilon = \frac{\sigma}{E} + \left(\frac{\sigma}{K'}\right)^{1/n'}. \quad (4)$$

The initial elastic modulus E of 7A04 HSAA and the parameters of R-O model (K' and n') are shown in Table 4. Figure 7 shows the comparison between the typical fitting curves and monotonic curves. From Figure 7, the Ramberg-Osgood model can fit the stress-strain skeleton curves of 7A04 HSAA fairly well. With the increase of strain amplitude, the cyclic hardening effect is more and more obvious. Combined with the data in Tables 2 and 3, the cyclic hardening effect could increase the ultimate strength by 3%. The strength increasing degree by cyclic hardening effect of 7A04 HSAA is less than steel.

3.4. Test Phenomenon and Failure Modes

3.4.1. Macroscopic Failure Modes. In the loading process, there are no visible deformation and necking of the specimens before its failure. In the cyclic loading, when the compressive strain was larger than 2.0%, the obvious local buckling could be observed as shown in Figure 8. The failure process of the specimen is very sudden with relatively small fracture strain. The failure mode is shown in Figure 9. The fracture surface forms a 45-degree angle with the direction of tensile stress. The fracture section is very even without deformation and shrinkage.

3.4.2. Microfailure Modes. Fracture surface was observed by scanning electron microscope (SEM) with magnifications of 2,000 times. The microfailure modes are shown in Figure 10. The specimens in Figures 10(a), 10(b), and 10(c) have many round or oval dimples which have better ductility than the specimen in Figure 10(d) (H1). There is obvious fish scale cleavage observed in the fracture surface of H1 which shows typical characteristic of fatigue failure.

4. Numerical Simulation of the Cyclic Loading Experiments

In order to provide a foundation for the analysis of high strength aluminum alloy structures in seismic design, finite element analysis software ABAQUS [16] was used to simulate the cyclic experiments. To describe the elastoplastic response completely, von Mises rule was adopted for flow rule and a combined model including isotropic hardening and kinematic hardening was adopted for hardening criterion according to the research work [13]. Figure 11 shows the schematic diagram of the isotropic hardening and kinematic hardening.

4.1. Isotropic Hardening. Isotropic hardening describes the variation of the size of yield surface σ^0, which is a function of equivalent plastic strain ε^p, and is shown in the following equation:

$$\sigma^0 = \sigma|_0 + Q_\infty\left(1 - e^{-b_{iso}\varepsilon^p}\right), \quad (5)$$

where $\sigma|_0$ is the yield stress when the equivalent plastic strain equals zero; according to the research work [17], take

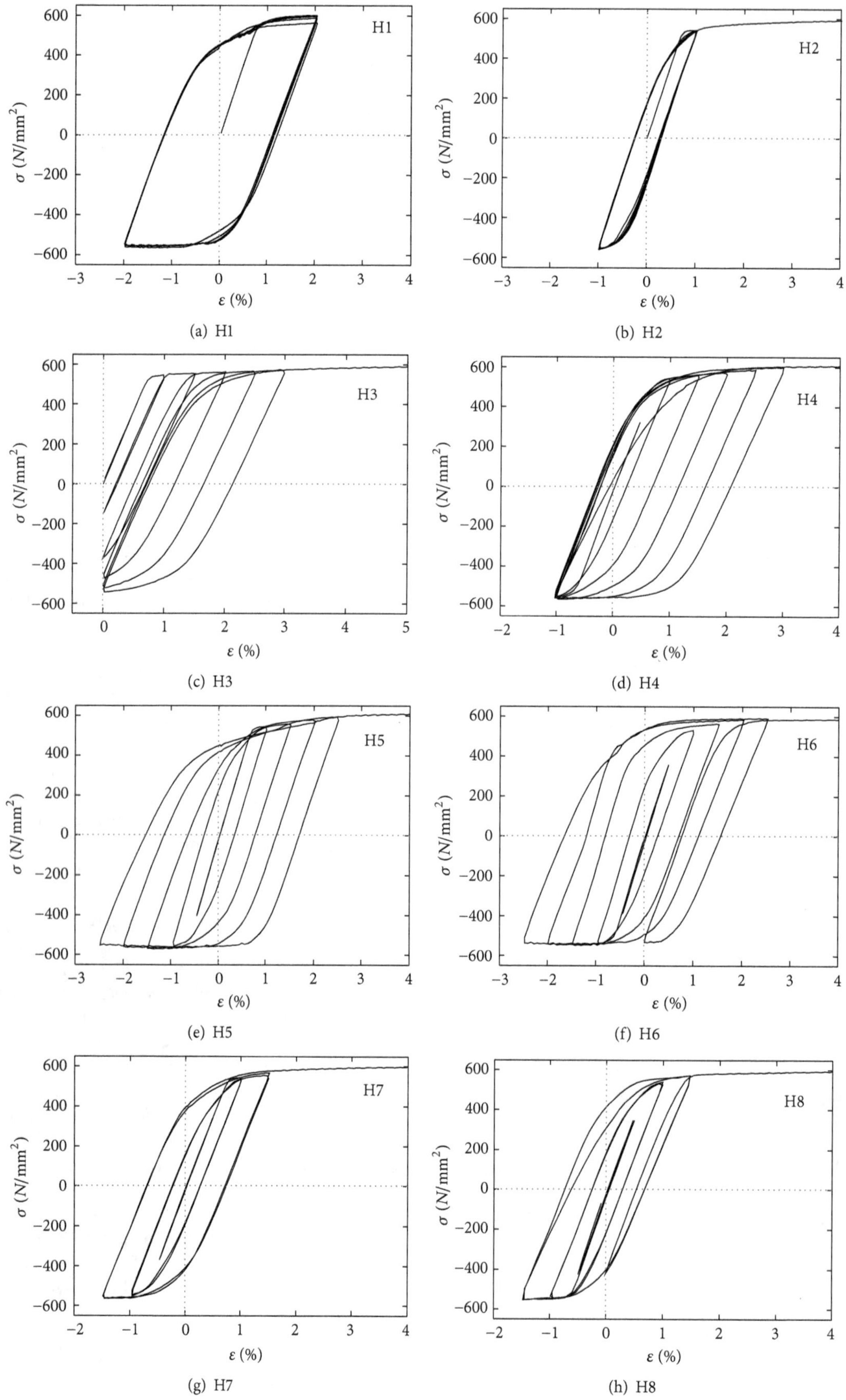

(a) H1 (b) H2 (c) H3 (d) H4 (e) H5 (f) H6 (g) H7 (h) H8

FIGURE 6: Continued.

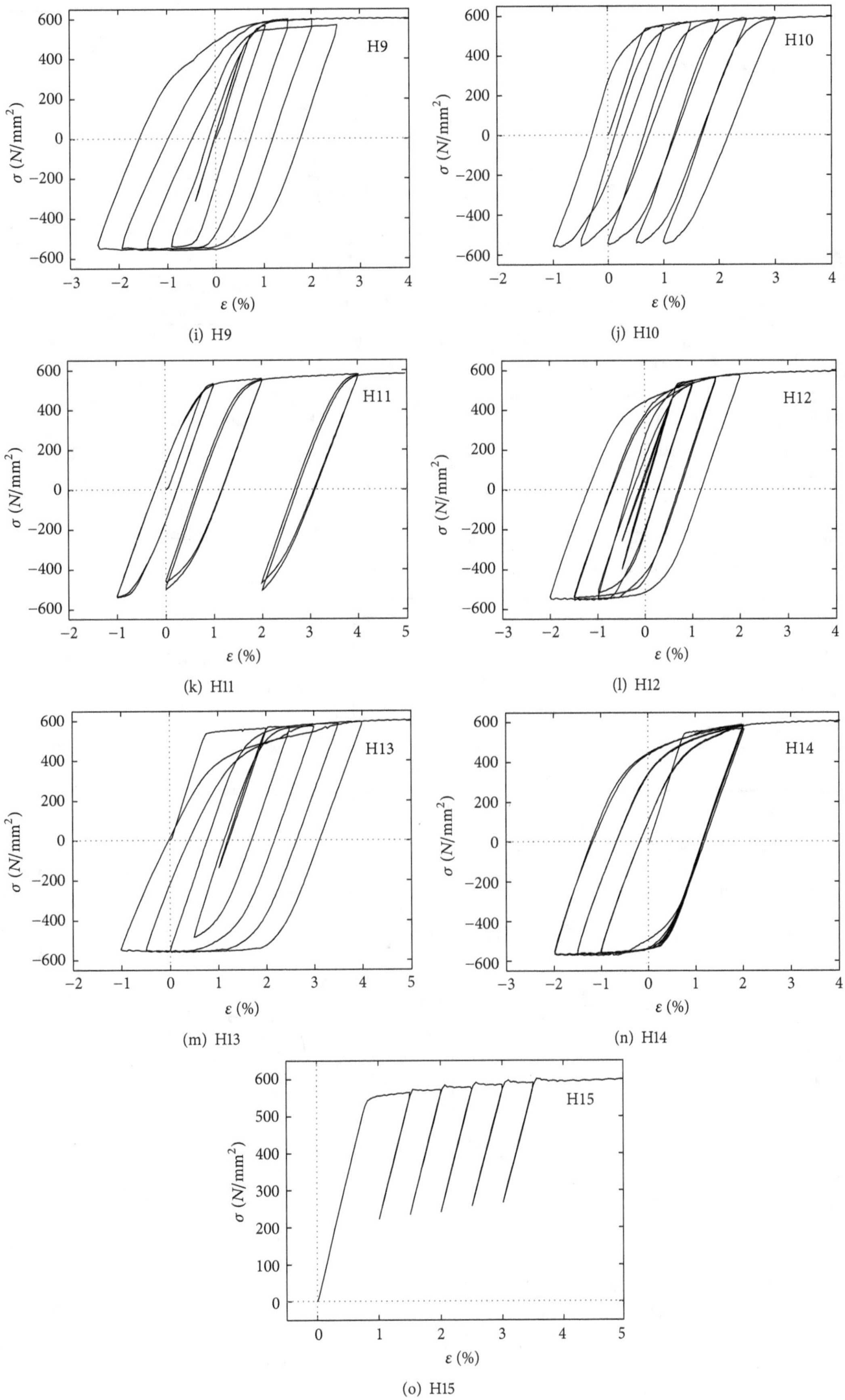

FIGURE 6: Stress-strain curves of 7A04 HSAA under cyclic loading.

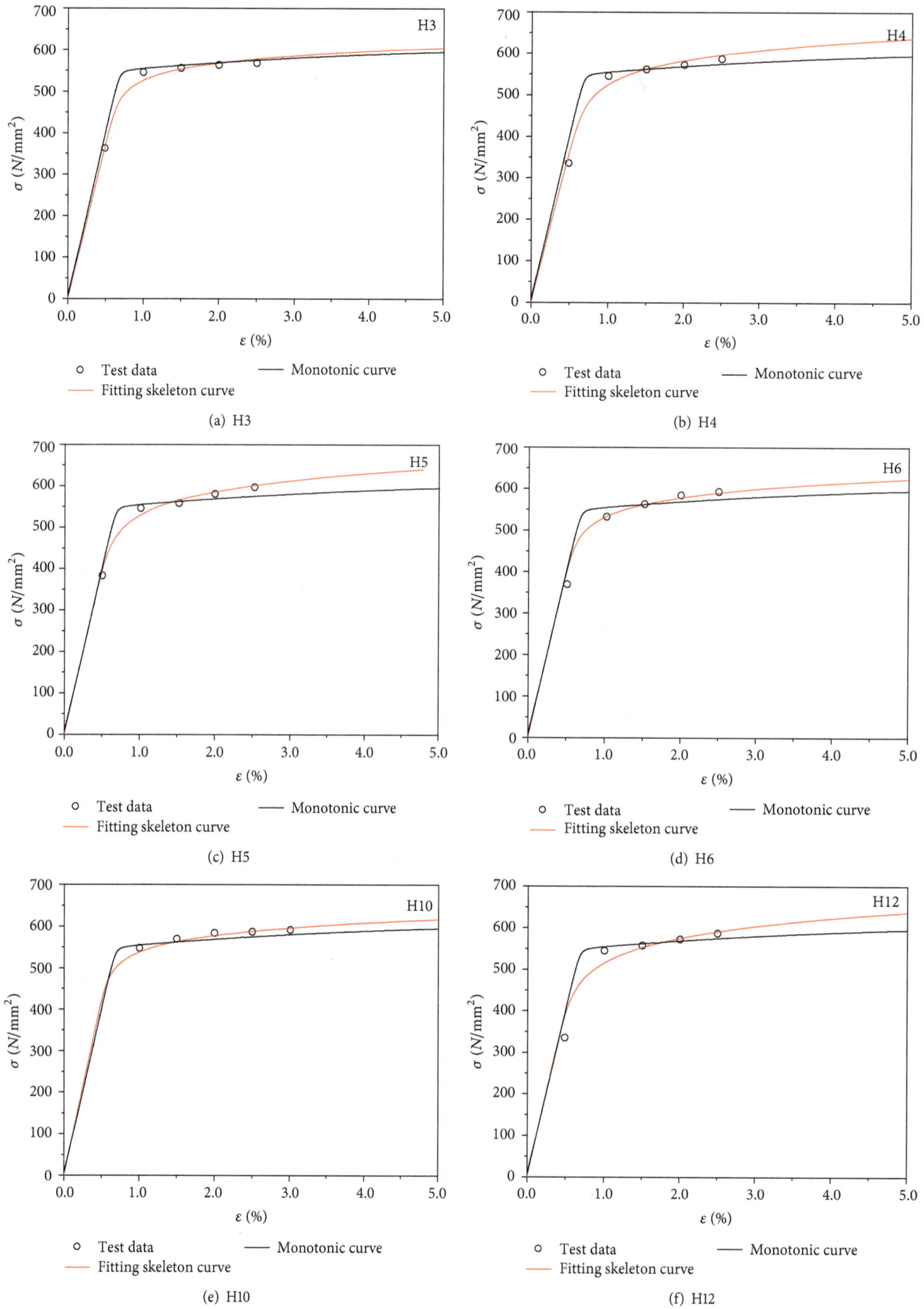

Figure 7: Continued.

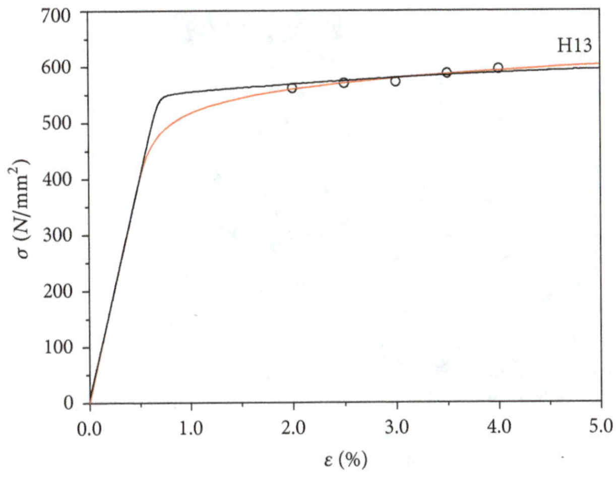

(g) H13

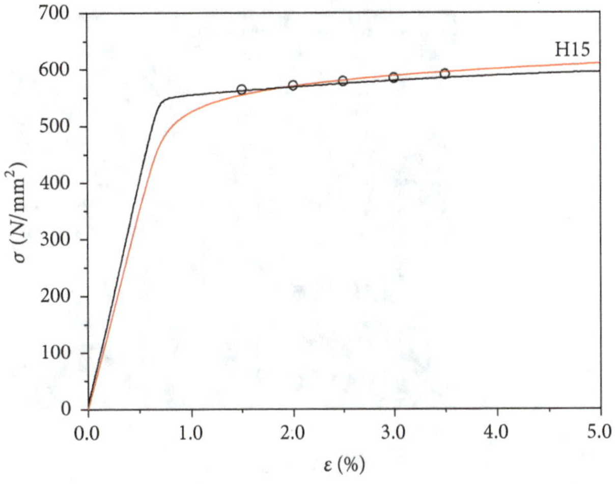

(h) H15

FIGURE 7: The comparison between cyclic skeleton curves and monotonic loading curves.

TABLE 5: Parameters of 7A04 aluminum alloy in the combined model.

Parameter	$\sigma\|_0$/MPa	Q_∞	b_{iso}	$C_{kin,1}$/MPa	γ_1	$C_{kin,2}$/MPa	γ_2	$C_{kin,3}$/MPa	γ_3
Value	521.11	60.20	26.63	83.92	0.78	33.33	0.80	12.74	0.88

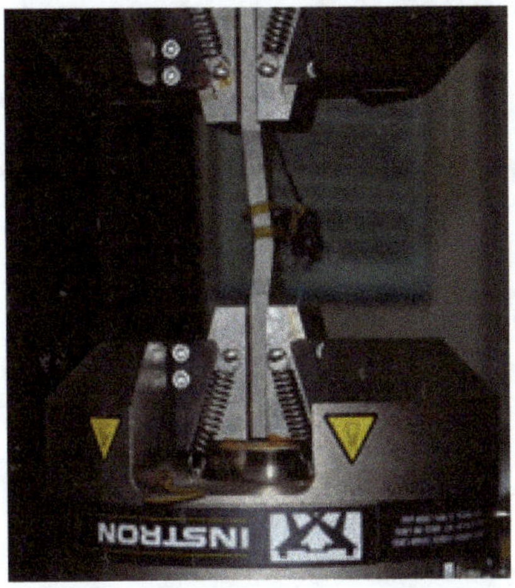

FIGURE 8: Buckling by compression.

FIGURE 9: Macroscopic failure modes of the specimen.

$\sigma_{0.01}$ as $\sigma|_0$ for nonlinear metal material; Q_∞ and b_{iso} are undetermined parameters which mean the maximum change in the size of the yield surface and rate at which the size of the yield surface changes as plastic strain increases, respectively; σ_i^0 is the size of the yield surface in ith cycle which is defined as the following equation:

$$\sigma_i^0 = \frac{\sigma_i^t - \sigma_i^c}{2}, \tag{6}$$

where σ_i^t is the maximum tensile stress and σ_i^c is the maximum compressive stress (Figure 11(a)).

The equivalent plastic strain ε_i^p is defined as

$$\varepsilon_i^p = \frac{1}{2}(4i-3)\,\Delta\varepsilon_p, \qquad \Delta\varepsilon_p \approx \Delta\varepsilon - \frac{2\sigma_1^t}{E}, \tag{7}$$

where $\Delta\varepsilon_p$ is the plastic strain range. Q_∞ and b_{iso} in (5) can be calibrated through the data pair $(\sigma_i^0, \varepsilon_i^p)$ including the data pair $(\sigma|_0, 0)$. The parameters of isotropic hardening component are listed in Table 5.

4.2. Nonlinear Kinematic Hardening. Kinematic hardening describes the variation of backstress α_k which is the function of plastic strain ε^p in a stabilized cycle, as is shown in (8):

$$\alpha_k = \frac{C_{kin,k}}{\gamma_k}\left(1 - e^{-\gamma_k \varepsilon^p}\right) + \alpha_{k,1} e^{-\gamma_k \varepsilon^p}, \tag{8}$$

$$\alpha = \sum_{k=1}^{N} \alpha_k, \tag{9}$$

where $C_{kin,k}$ and γ_k are undetermined parameters which mean initial kinematic hardening modulus and rate at which

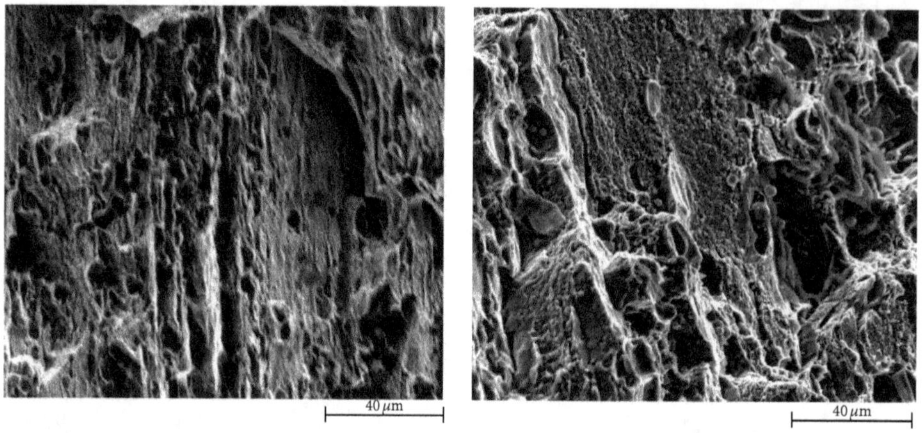

(a) Monotonic loading (M1) (b) Constant-amplitude cyclic loading (H2)

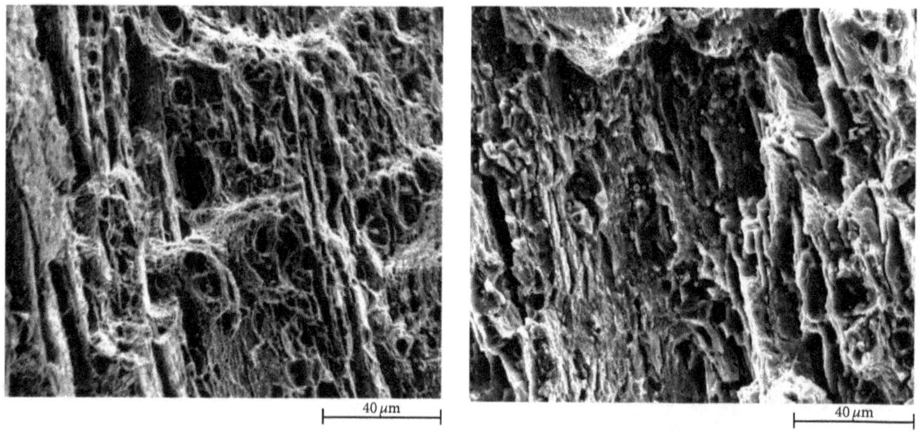

(c) Gradual cyclic loading (H5) (d) Fatigue failure (H1)

FIGURE 10: Microfailure modes of fracture surface.

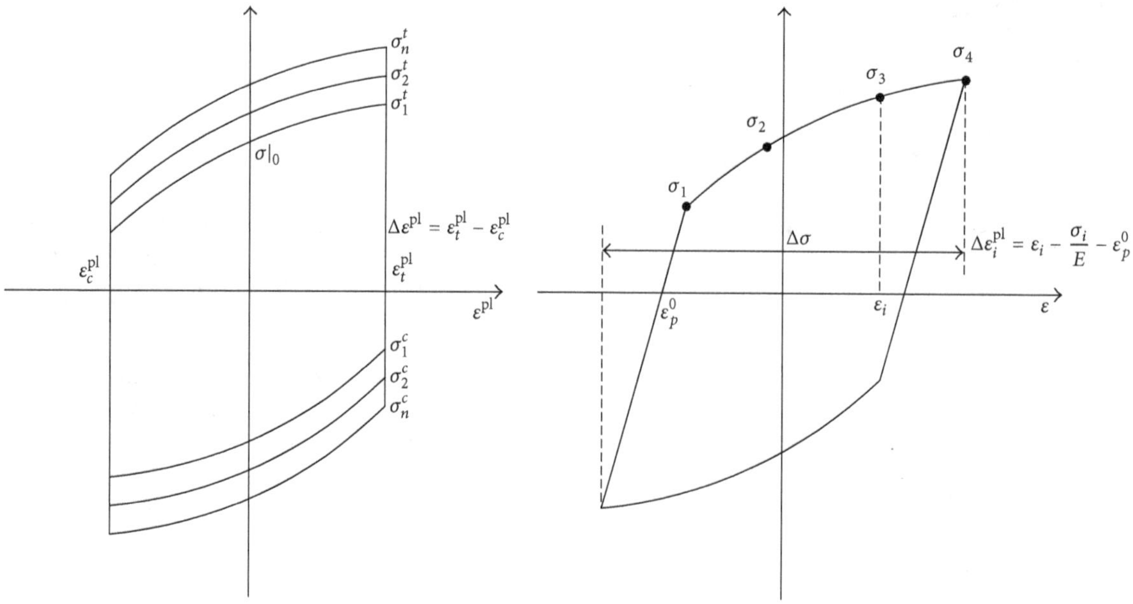

(a) Isotropic hardening (b) Kinematic hardening

FIGURE 11: Schematic diagram of the isotropic hardening and kinematic hardening.

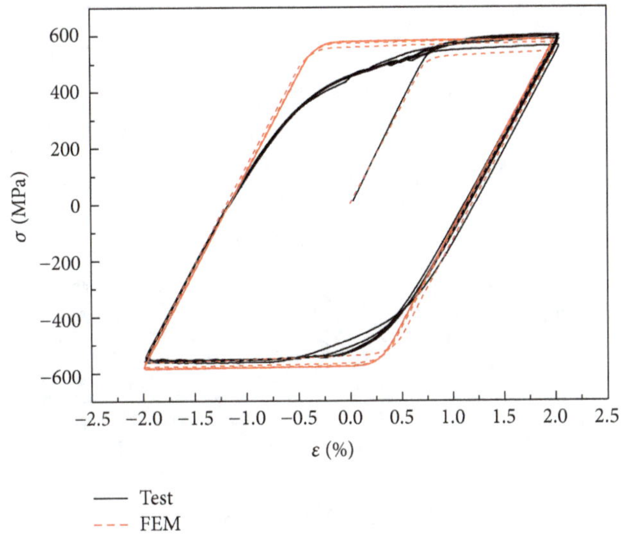

(a) H1

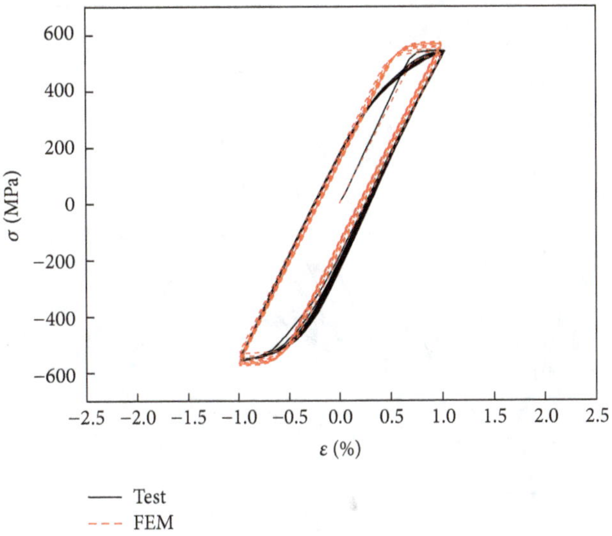

(b) H2

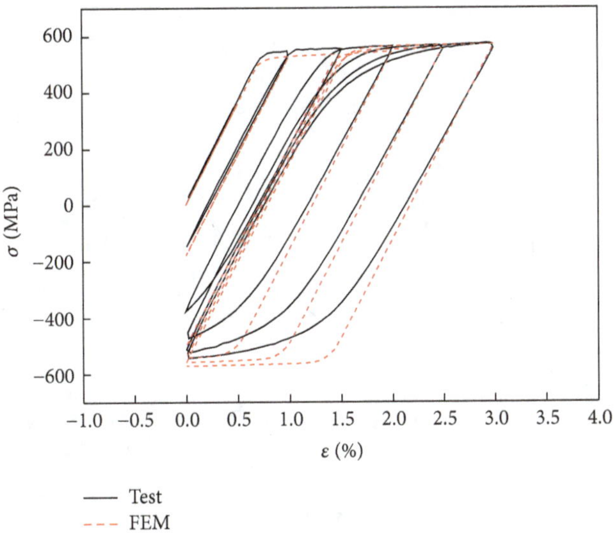

(c) H3

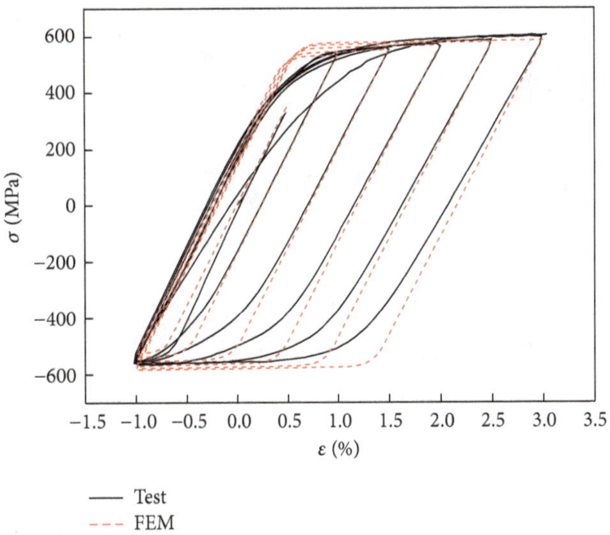

(d) H4

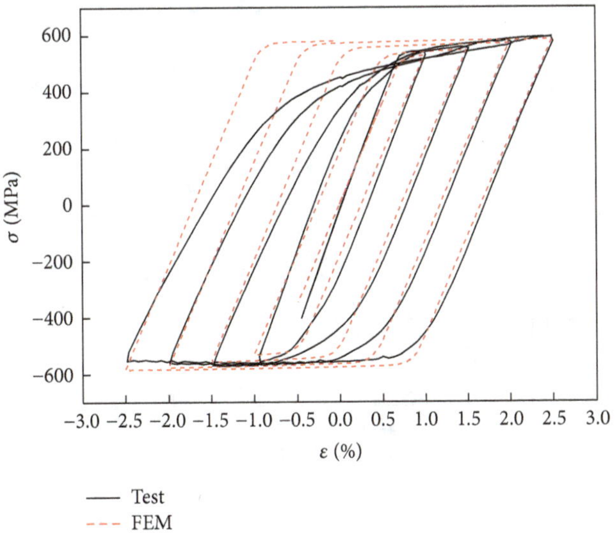

(e) H5

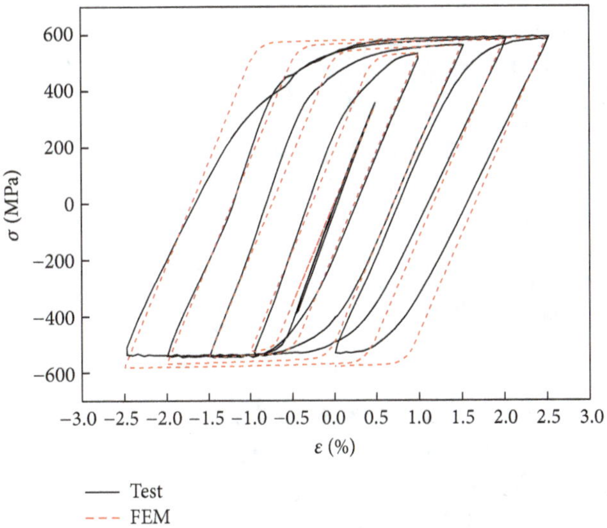

(f) H6

FIGURE 12: Continued.

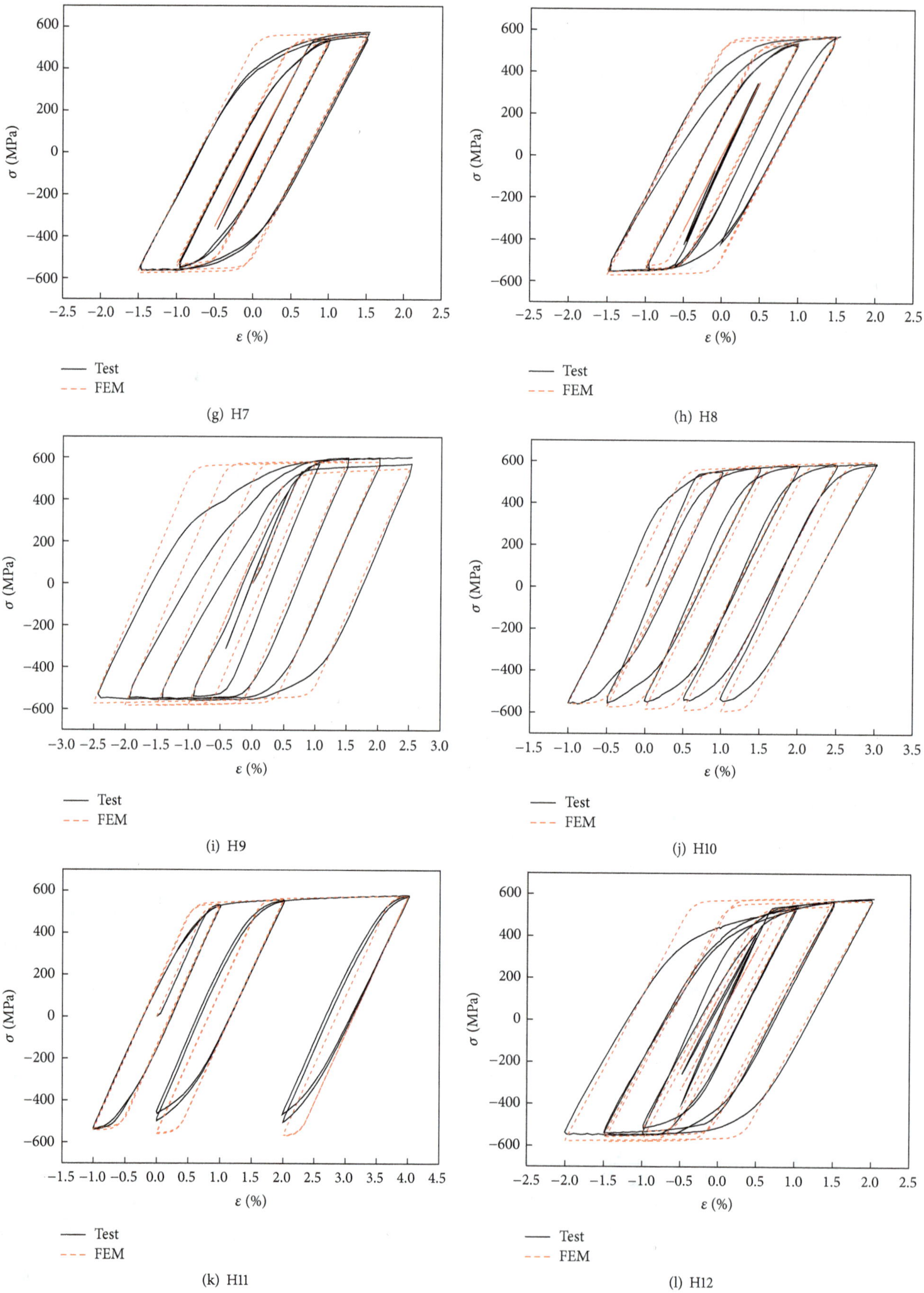

(g) H7 (h) H8

(i) H9 (j) H10

(k) H11 (l) H12

FIGURE 12: Continued.

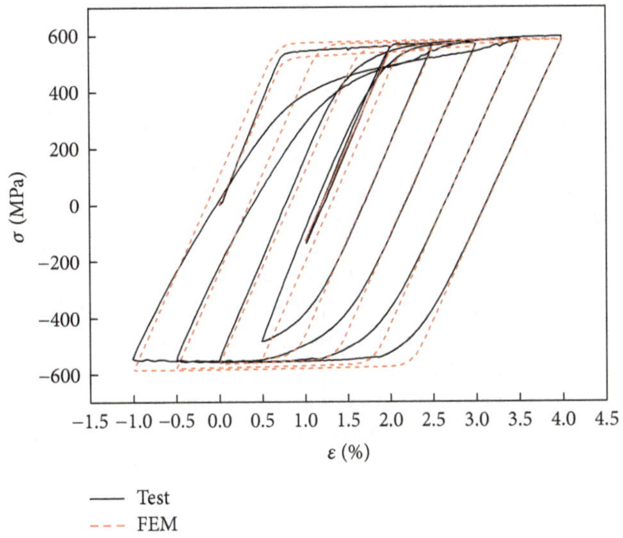

(m) H13

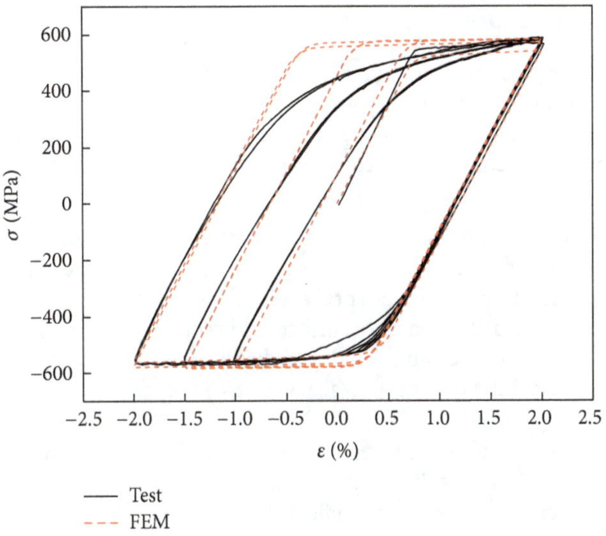

(n) H14

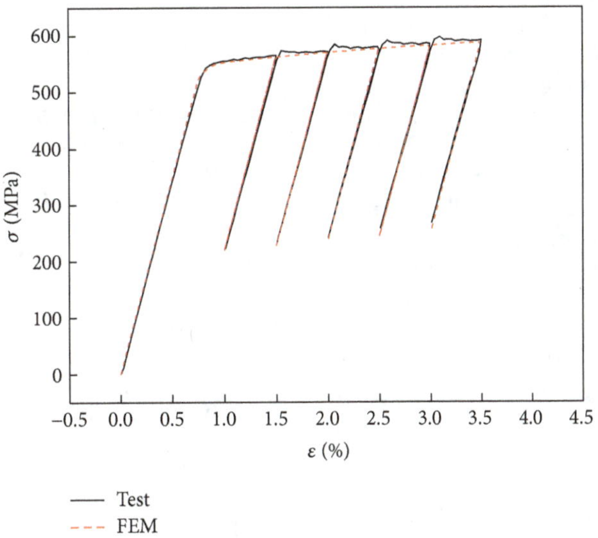

(o) H15

FIGURE 12: Comparison analysis between test curves and FE curves.

the kinematic hardening modulus varies with increasing plastic deformation, respectively. α_k is kth backstress and $\alpha_{k,1}$ is kth backstress of the first data point, $k = 1, 2, 3, \ldots, N$.

The multiple backstress superposition method (9) is adopted for fitting the curve more accurately. In the current paper, k equals 3 in (9).

Plastic strain ε_i^{pl} in data pairs $(\sigma_i, \varepsilon_i^{pl})$ is shifted as

$$\varepsilon_i^{pl} = \varepsilon_i - \frac{\sigma_i}{E} - \varepsilon_p^0, \quad (10)$$

where ε_i^{pl} is the plastic strain value when the curve intercepts the strain axis and ε_1^{pl} equals zero. For each data pair $(\sigma_i, \varepsilon_i^{pl})$, backstress could be obtained from (11):

$$\alpha_i = \sigma_i - \sigma^s, \quad (11)$$

where $\sigma^s = (\sigma_1 + \sigma_n)/2$; that is, σ^s is average value of the first data σ_1 and last data σ_n. Lots of data pairs $(\alpha_i, \varepsilon_i^{pl})$ were used to calibrate the parameters $C_{\text{kin},k}$ and γ_k which are obtained in Table 5.

4.3. FE Analysis. The comparison between the FE results and test results is shown in Figure 12. The simulated curves fit well with the experimental curves. However, because of unknown imperfections in the test specimens, the FE results overestimate the stress in certain strain region. Because of the model assumption in ABAQUS, in the upper left corner area, the simulated curves have little difference with the test curves which is also found in research work [18, 19].

In general, the parameters of 7A04 HSAA in the combined model can be used in the analysis of 7A04 HSAA structures or members under cyclic loading.

5. Conclusion

Experimental investigation of 7A04 HSAA under monotonic loading and cyclic loading was conducted and a corresponding FE model was established in the current paper. From all above research, main conclusions can be obtained as follows:

(1) There are no visible deformation and necking of the specimens under monotonic loading and cyclic loading. The failure process is very sudden with a loud sound. The fracture surface was very even and forms a 45-degree angle with the direction of tensile stress. And the fracture surface was observed by SEM.

(2) By hysteretic curves, the hysteretic behavior of 7A04 HSAA was investigated. The material shows typical cyclic hardening effect; however, its Bauschinger effect is not significant. For most specimens, the stiffness degradation and strength degradation occurred after 1% compressive strain. The ductility of 7A04 HSAA did not decrease due to cyclic loading. And it shows combined hardening behavior including isotropic hardening and nonlinear kinetic hardening.

(3) The stress-strain skeleton curves were fitted according to Ramberg-Osgood model. And the cyclic hardening parameters of the skeleton curves were determined. The comparison was conducted between cyclic loading skeleton curves and monotonic loading curves.

(4) The curves simulated by ABAQUS fit well with the experimental curves and the cyclic hardening parameters based on Chaboche model derived in the current paper can be used in engineering practice.

(5) This paper supplies a basis for accurate calculation of 7A04 HSAA structures under cyclic loading.

Nomenclature

$f_{0.2}$: Nominal yield stress (stress at 0.2% plastic strain)
f_u: Ultimate strength
f_{u2}: Fracture strength
ε_1: Strain when stress is $f_{0.2}$
ε_u: Strain when stress is f_u
ε_2: Fracture strain when stress is f_{u2}
$\Delta\varepsilon$: Total strain amplitude
$\Delta\varepsilon_e$: Elastic strain amplitude
$\Delta\varepsilon_p$: Plastic strain amplitude
$\Delta\sigma$: Stable stress amplitude
E: Initial elastic modulus of aluminum alloy
K': Cyclic strength coefficient
n': Cyclic strain-hardening exponent
E_n: Hysteretic energy
N_c: Hysteretic loops
n: Strain-hardening coefficient in R-O model
σ^0: Size of yield surface
ε^p: Equivalent plastic strain
$\sigma|_0$: Yield stress when the equivalent plastic strain equals zero
$\sigma_{0.01}$: Stress at 0.01% plastic strain
Q_∞: Maximum change in the size of the yield surface
b_{iso}: Rate at which the size of the yield surface changes as plastic strain increases
σ_i^0: Size of the yield surface in ith cycle
σ_i^t: Maximum tensile stress
σ_i^c: Maximum compressive stress
$C_{\text{kin},k}$: Initial kinematic hardening modulus
γ_k: Rate at which the kinematic hardening modulus varies with increasing plastic deformation
α_k: kth backstress
$\alpha_{k,1}$: kth backstress of the first data point
σ^s: Average value of the first data σ_1 and last data σ_n
$\varepsilon_i^{\text{pl}}$: The plastic strain value when the curve intercepts the strain axis.

Competing Interests

The authors declare that they have no competing interests.

Acknowledgments

The paper is supported by the National Natural Science Foundation of China (Grant no. 51038006) and the Special Research Fund for the Doctoral Program of Higher Education (Grant no. 20110002130002).

References

[1] Y. Q. Wang, Z. X. Wang, F. X. Yin et al., "Experimental study and finite element analysis on the local buckling behavior of aluminum alloy beams under concentrated loads," *Thin-Walled Structures*, vol. 105, pp. 44–56, 2016.

[2] H. X. Yuan, Y. Q. Wang, T. Chang, X. X. Du, Y. D. Bu, and Y. J. Shi, "Local buckling and postbuckling strength of extruded aluminium alloy stub columns with slender I-sections," *Thin-Walled Structures*, vol. 90, pp. 140–149, 2015.

[3] G. O. Adeoti, F. Fan, Y. Wang, and X. Zhai, "Stability of 6082-T6 aluminium alloy columns with H-section and rectangular hollow sections," *Thin-Walled Structures*, vol. 89, pp. 1–16, 2015.

[4] W. Yuanqing, L. Xiaochen, S. Yongjiu et al., "Finite element analysis on mechanical performance of TEMCOR joints in aluminum alloy shell structures," *Journal of Tianjin University (Science and Technology)*, vol. 7, pp. 1–8, 2015.

[5] G. D. Matteis, A. Mandara, and F. M. Mazzolani, "T-stub aluminum joints: influence of behavioural parameters," *Computers and Structures*, vol. 78, no. 1–3, pp. 311–327, 2000.

[6] B. Atzori and P. Lazzarin, "Notch sensitivity and defect sensitivity under fatigue loading: two sides of the same medal," *International Journal of Fracture*, vol. 107, no. 1, pp. 1–8, 2000.

[7] A. De-Andrés, J. L. Pérez, and M. Ortiz, "Elastoplastic finite element analysis of three-dimensional fatigue crack growth in aluminum shafts subjected to axial loading," *International Journal of Solids & Structures*, vol. 36, no. 15, pp. 2231–2258, 1999.

[8] Y. Hiyama, K. Ishikawa, S. Kato, and S. Okubo, "Experiments and analysis of the post-buckling behaviors of aluminum alloy double layer space grids applying ball joints," *Structural Engineering and Mechanics*, vol. 9, no. 3, pp. 289–304, 2000.

[9] J. Bin, Z. Qilin, L. Xiaoqun et al., "Study on hysteretic behavior of aluminum alloy energy dissipation braces," *Journal of Building Structures*, vol. 36, no. 8, pp. 49–57, 2015.

[10] W. Ramberg and W. R. Osgood, "Description of stress-strain curves by three parameters," Tech. Rep. 902, Archive and Image Library, 1943.

[11] O. Steinhardt, "Aluminum construction in the field of civil engineering," *Aluminium*, vol. 47, no. 2, pp. 131–139, 1971.

[12] F. M. Mazzolani, *Proposal to Classify the Aluminum Alloy on the Basis of Mechanical Behavior*, vol. 16 of *ECCS Committee*, 1974.

[13] J. L. Chaboche, "Time-independent constitutive theories for cyclic plasticity," *International Journal of Plasticity*, vol. 2, no. 2, pp. 149–188, 1986.

[14] W. Zhang, X.-K. Xiao, and G. Wei, "Constitutive relation and fracture model of 7A04 aluminum alloy," *Explosion and Shock Waves*, vol. 31, no. 1, pp. 81–87, 2011.

[15] H. Krawinkler and M. Zohrei, "Cumulative damage in steel structures subjected to earthquake ground motions," *Computers & Structures*, vol. 16, no. 1–4, pp. 531–541, 1983.

[16] H. Hibbitt, B. Karlsson, and P. Sorensen, *ABAQUS Analysis User's Manual Version 6.10*, Dassault Systemes Simulia Corp, Providence, RI, USA, 2011.

[17] K. H. Nip, L. Gardner, C. M. Davies, and A. Y. Elghazouli, "Extremely low cycle fatigue tests on structural carbon steel and stainless steel," *Journal of Constructional Steel Research*, vol. 66, no. 1, pp. 96–110, 2010.

[18] Y. Shi, M. Wang, and Y. Wang, "Experimental and constitutive model study of structural steel under cyclic loading," *Journal of Constructional Steel Research*, vol. 67, no. 8, pp. 1185–1197, 2011.

[19] Y. Q. Wang, T. Chang, Y. J. Shi, H. X. Yuan, L. Yang, and D. F. Liao, "Experimental study on the constitutive relation of austenitic stainless steel S31608 under monotonic and cyclic loading," *Thin-Walled Structures*, vol. 83, pp. 19–27, 2014.

15

Numerical and Experimental Research on Cold Compression Deformation Method for Reducing Quenching Residual Stress of 7A85 Aluminum Alloy Thick Block Forging

Cui Jin-dong, Yi You-ping, and Luo Guo-yun

School of Materials Science and Engineering, Central South University, Changsha 410083, China

Correspondence should be addressed to Yi You-ping; yyp@mail.csu.ed.cn

Academic Editor: Sutasn Thipprakmas

In aeronautical machining industry, the most difficult problem to deal with is the distortion of aviation integral component, one main cause of which is the existence of quenching residual stress of forgings, especially for large-sized ones. Therefore, it is important to study the methods that can reduce the quenching residual stress. In this work, the distribution of quenching residual stress of 7A85 aluminum alloy thick block forging, as well as the effect of cold compression deformation method on reducing quenching residual stress, has been investigated by simulation. The results show that, in length direction of 7A85 aluminum alloy thick block with a large size of 260 (H) × 1150 (W) × 5300 (L) mm, quenching residual stress can be significantly reduced by about 2.5% cold compression deformation along the direction of highness, with residual stress in length direction ranging from −65 MPa to 60 MPa, compared with its counterpart after quenching from −170 MPa to 140 MPa. Then a cold compression experiment was carried out, in which the forging residual stress on the surface was measured by X-ray diffraction device. The experimental results indicate that the optimal compression deformation value is 1%-2%, reducing 70% residual stress for 7A85 aluminum alloy specimens in size of 100 (L) × 60 (W) × 40 (H) mm.

1. Introduction

Aluminum alloys are the primary material for structural components of military aircrafts, helicopters, amphibious warships, and so forth for several decades. This is because of the low specific gravity of aluminum which favors the selection of aluminum alloys in weight-critical applications [1]. 7A85 aluminum alloy developed in China in recent years is a new type of high-strength aluminum alloys, whose performance is similar to 7085 aluminum alloy that has been developed by Alcoa company since 2002. In order to gain high-strength performance, a series of heat treatment processes must be carried out including a severe quenching operation which introduces a very high level of residual stress in these aluminum alloys forgings. Quenching residual stress not only results in unexpected machining distortion, but also increases the possibility of unpredicted fatigue failures and stress corrosion cracks (SCC) [2, 3]. Therefore, it is of great significance to investigate quenching residual stress prediction and reduction methods in aircraft aluminum parts.

A lot of research on quenching residual stress has been carried out by researchers. Particularly with development of computer technology in recent years, numerical simulation technology has become an effective way to simulate the generation and distribution of residual stress during quenching process. Koç et al. used the finite element method to simulate the quenching and prestretching process to predict residual stress value of quenched and prestretched aluminum blocks [4–6]. Wu et al. studied the residual stress in aluminum blocks with different quenching processes [7–9]. Prime and Hill used the crack compliance method to measure the residual stress value in products of Alcoa and obtained residual stress distribution of quenched aluminum block and the corresponding prestretched block [10, 11]. Muammer et al. studied the residual stresses in quenched aluminum blocks

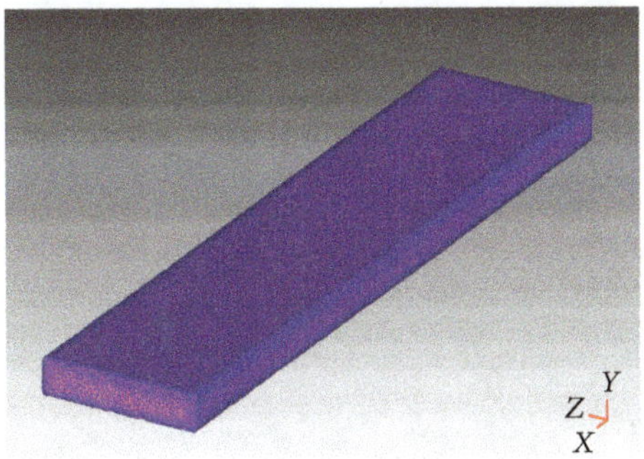

FIGURE 1: Finite element model.

TABLE 1: Heat transfer coefficient of 7A85 aluminum alloy.

Temperature (°C)	20	100	200	300	400	500
Transfer coefficient (W/m^2·C)	2500	5000	13857	10000	3000	700

and their reduction through cold working processes [12, 13]. In this work, the effect of cold compression deformation on reducing quenching residual stress was studied through numerical simulation and experiment.

2. Simulation

2.1. Finite Element Model and Simulation Parameters. A thick block forging of 7A85 aluminum alloy with a large size of 260 (H) × 1150 (W) × 5300 (L) mm was chosen as the test component for the numerical analysis. The quenching and cold compression processes were subjected to numerical simulation in this study. The block was quenched in water of 20°C from initial 470°C and then 1%–5% cold compression deformation was conducted on the block along the highness direction. Finite Element Code Deform-3D was used to simulate the two processes. The shape and finite element model are shown in Figure 1.

Heat transfer coefficient between 7A85 aluminum alloy block and water which is the most important parameter in quenching process is given in Table 1 [14]. The heat capacity of 7A85 aluminum alloy obtained through DSC method by using the instrument NETZSCH STA 449 C is shown in Table 2. Flash method was used to get the thermal conductivity coefficient of 7A85 aluminum alloy by JR-3 machine. The measurement of thermal expansion coefficient was completed through NETZSCH DIL402C device. Both of them are given in Table 2.

TABLE 2: Thermal conductivity and thermal expansion coefficient of 7A85 aluminum alloy.

Temperature (°C)	15	100	200	300	400	550
Heat capacity (J/g·°C)	1.02	0.95	1.1	1.07	1.24	1.07
Conductivity (W/m°C)	185.3	182.0	217.3	218.2	228.1	187.2
Expansion (10^{-6}/°C)	*	24.3	25.9	26.7	28.2	*

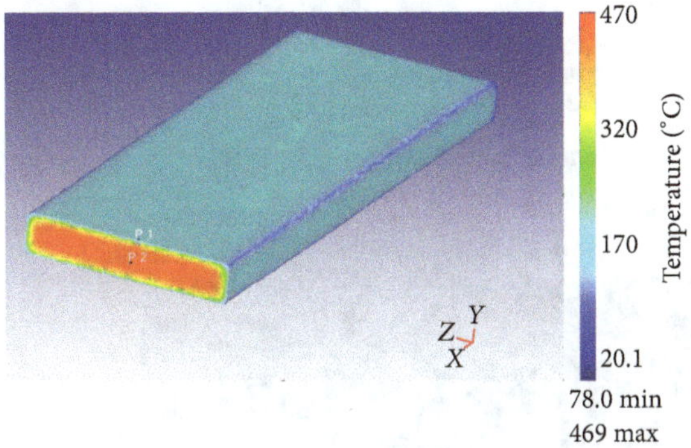

FIGURE 2: Results of temperature after cooling time 10 s.

2.2. Quenching Simulation Results and Discussion. Figure 2 shows the simulated results of temperature profile for the cooling time 10 s. Obviously, the temperature of edge at the block is the lowest and the temperature difference between the edge and the highest temperature of the block is about 400°C. This is due severely to heat convection at the edge along any two axis directions. The temperature gradient and temperature difference gradually decrease with prolonging cooling time. Two characteristic points 1 (surface) and 2 (core) of the block shown in Figure 2 were employed for observing the temperature (T) change with cooling time (t) from 470°C to 20. Figure 3 gives the relative calculated results. The largest temperature difference between points 1 and 2 occurs in range of cooling time 10 to 15 s. The temperature difference produces the thermal stress in the internal, which leads to the deformation of the block. At the same time, the release of phase transition latent heat results in a decrease in the slope of T-t curves. Because the quenching process is very severe, the effect of the phase transition latent heat release on the temperature could be ignored. After quenching for about 350 seconds, the temperature distribution in the block evolves to be uniform and the magnitude is close to the water temperature. With increasing time, the temperature of the two points gradually reduces; it takes about 650 s for them to be cooled to the target temperature 20°C.

The final residual stress distribution of quenched block in length direction is shown in Figure 4. Figure 5 gives stress (X)-time curves for points 1 (surface) and 2 (core) of the block. In the initial stage of quenching, the normal stress was tensile at the surfaces and compressive in the core. Because of the sudden contraction of the surfaces and the great temperature gradient in the vicinity of the surface, the normal stress at surface point 1 increases sharply to about 70 MPa and then decreases slowly with the decreasing temperature gradient. At the same time, the compressive normal stress at the center point 2 gradually increases. During this process, the block undergoes plastic deformation. The plastic deformation finally results in the fact that the surface tensile residual stress shifts to compressive and the core

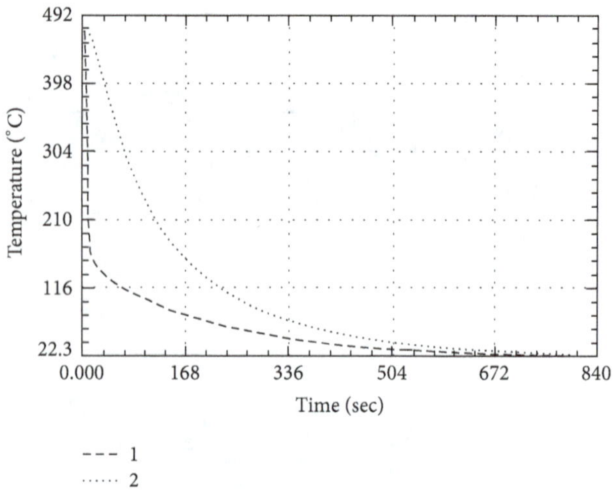

FIGURE 3: Temperature-time curves for points 1 and 2 in Figure 2.

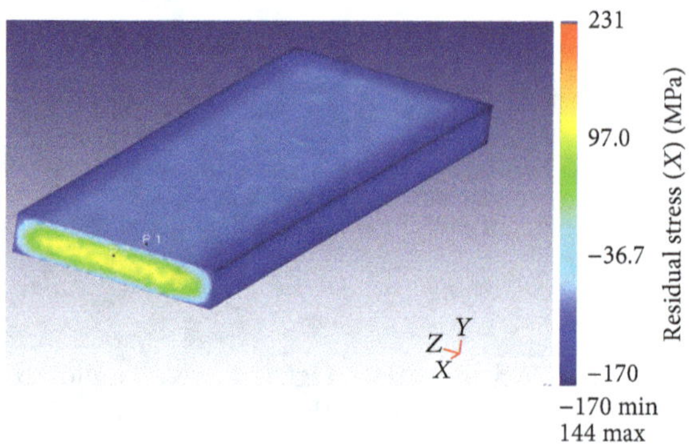

FIGURE 4: Results of residual stress (X) after quenching.

compressive stress switches into tensile stress at the later stage of quenching.

2.3. Compression Simulation Results and Discussion. Figure 6 shows the simulated results of residual stress distribution of the block for different cold compression values 0%, 1%, 2.5%, 4%, and 5%. The x- and y-axes stand for residual stress value and percentage of the block forging, respectively. It can be seen that the residual stress in length direction ranged from −170 MPa to 140 MPa when the quenching process was completed (Figure 6(a)). After conducting 1% cold compression deformation on the block along the direction of highness, the distribution range of residual stress in length direction significantly reduced that ranged from −110 MPa to 120 MPa (Figure 6(b)). This is due to the plastic deformation that redistributed the residual stress of the block. The range of residual stress in length direction of the block sharply reduces to the range from −65 MPa to 60 MPa as the compression deformation value runs up to 2.5%, and the uniformity was better than before (Figure 6(c)). However, when the deformation continued to come up to 4% and 5%, the ranges of residual stress in length direction of the block sharply widened which are from −90 MPa to 100 MPa and from-110 MPa to 120 MPa, respectively (Figures 6(d) and 6(e)).

TABLE 3: Cold compression deformation scheme.

Percentage	1%	2%	3%	4%	5%
Plastic deformation (mm)	0.4	0.8	1.2	1.6	2.0
Experimental deformation (mm)	0.9	1.2	1.6	2.0	2.4

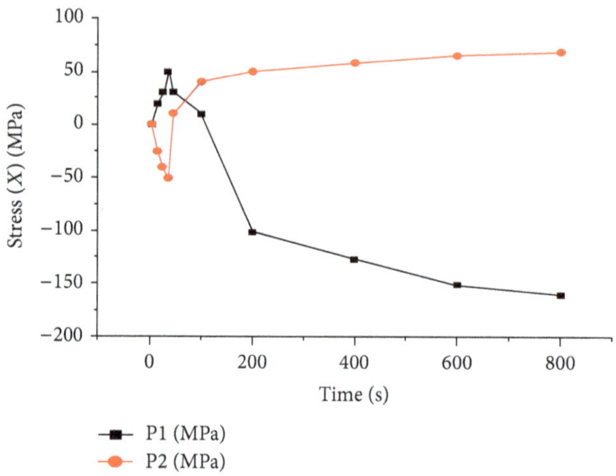

FIGURE 5: Stress (X)-time curves for points 1 and 2 in Figure 4.

This phenomenon can be explained by the fact that if the cold compression deformation continues to increase, it will produce much more elastic deformation in the block which is the main cause of the increase of residual stress. Therefore, for this simulation, it can be concluded that the optimal compression deformation value is about 2.5% for this large-sized block forging.

3. Cold Compression Experiment

Figure 7 shows the cold compression experimental specimens of 7A85 aluminum alloy forging with the size of 100 (L) × 60 (W) × 40 (H) mm provided by Southwest Aluminum Company and equipment of Central South University including X-ray diffraction device, and 4000t CNC hydraulic press machine. The entire experimental course was subdivided into four operations, which are quenching, measuring residual stress for the first time, cold compression deformation along the highness direction of specimens, and measuring residual stress for the second time. 1%–5% cold compression deformation was conducted on the quenched specimens along highness direction, respectively. The data of plastic deformation is shown in Table 3. Compression speed was 0.1 mm/s, which can be accurately controlled by the hydraulic control system. The residual stress of the same points on the surface of compressed specimens was measured by X-ray diffraction device in X, Y, and Z directions.

The results of cold compression deformation experiment are shown in Table 4 and Figure 8. The average values of residual stress for quenched specimens are −174.9 MPa,

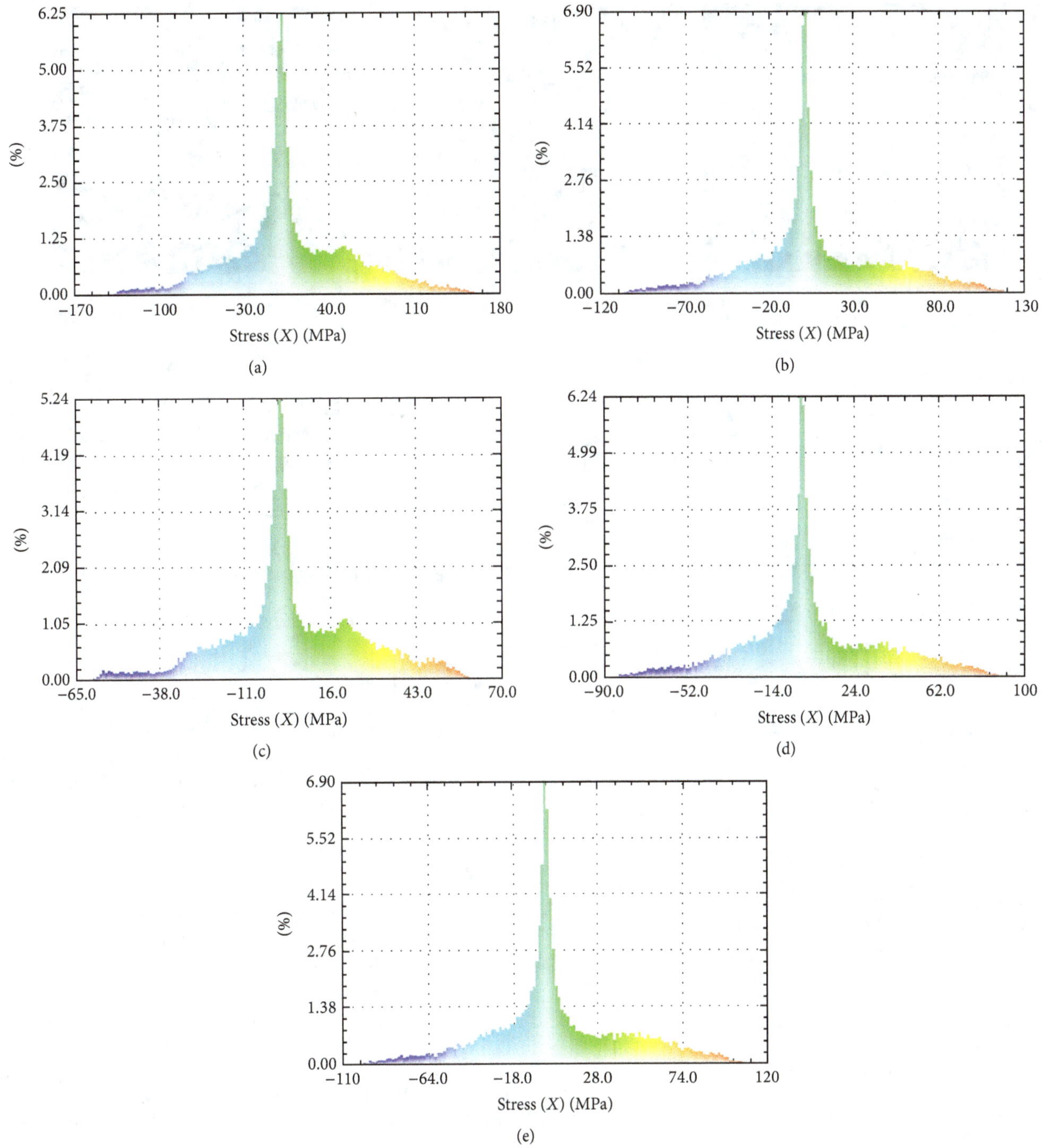

FIGURE 6: Results of residual stress (X) distribution for different compression value: (a) quenched; (b) 1%; (c) 2.5%; (d) 4%; (e) 5%.

−156.7 MPa, and −150.6 MPa, in X, Y, and Z directions, respectively. After conducting 1% compression deformation, the values of them reduced to 7 MPa, 53.4 MPa, and −5.9 MPa, respectively. When compression deformation exceeded 2%, the level of residual stress in three directions both obviously increased. It can be concluded that 1%-2% is the optimal compression deformation value which can reduce 70% residual stress for these 7A85 aluminum alloy specimens sized 100 (L) × 60 (W) × 40 (H) mm.

4. Conclusions

(1) Physical property parameters of 7A85 aluminum alloy such as heat capacity, thermal conductivity,

FIGURE 7: Experimental specimens and equipment.

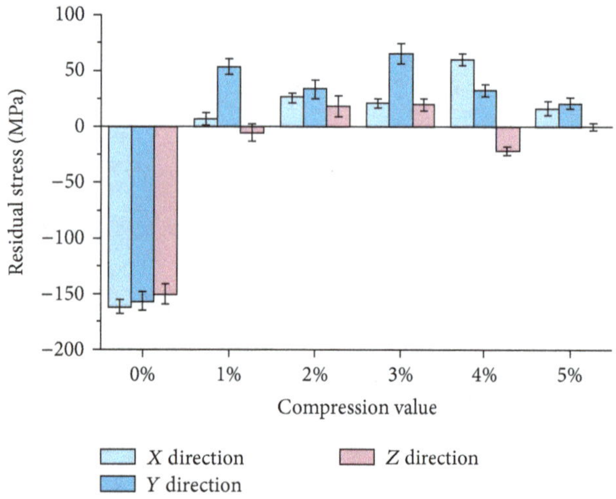

FIGURE 8: The error for measurement data.

TABLE 4: Residual stress of quenched and compressed specimens.

Compression value	X direction (MPa)	Y direction (MPa)	Z direction (MPa)
0%	−174.9	−156.7	−150.6
1%	7.0	53.4	−5.9
2%	28.2	33.7	18.3
3%	22.4	65.3	19.1
4%	64.9	32.4	−22
5%	17.8	20.8	−0.1

and expansion coefficient were obtained through experiments, which are the foundation of subsequent simulative and experimental research.

(2) Cold compression deformation method has a significant effect on reducing quenching residual stress if an appropriate amount of compression deformation is chosen. The simulation results show that, in length direction of 7A85 aluminum alloy thick block with a large size of 260 × 1150 × 5300 mm, quenching residual stress can be significantly reduced by about 2.5% cold compression deformation along the direction of highness, with residual stress in length direction ranging from −65 MPa to 60 MPa, compared with that from −170 MPa to 140 MPa after quenching. The experimental results indicate that the optimal compression deformation value is about 1%-2%, reducing 70% residual stress for 7A85 aluminum alloy specimens sized 100 × 60 × 40 mm.

Competing Interests

The authors declare that they have no competing interests.

Acknowledgments

The authors would like to thank The State Key Laboratory of High Performance and Complex Manufacturing (zzyjkt2014-02) and the National Basic Research Program of China (Grant 2012CB619504) for their financial support.

References

[1] E. A. Starke Jr. and J. T. Staley, "Application of modern aluminum alloys to aircraft," *Progress in Aerospace Sciences*, vol. 32, no. 2-3, pp. 131–172, 1996.

[2] P. Jeanmart and J. Bouvaist, "Finite element calculation and measurement of thermal stress in quenched plates of high strength 7075 aluminum alloy EJ," *Materials Science and Technology*, vol. 1, no. 3, p. 765, 1985.

[3] D. J. Smith, C. G. C. Poussard, and M. J. Pavier, "An assessment of the Sachs method for measuring residual stresses in cold worked fastener holes," *Journal of Strain Analysis for Engineering Design*, vol. 33, no. 4, pp. 263–274, 1998.

[4] M. Koç, J. Culp, and T. Altan, "Prediction of residual stresses in quenched aluminum blocks and their reduction through cold working processes," *Journal of Materials Processing Technology*, vol. 174, no. 1–3, pp. 342–354, 2006.

[5] H. F. Wang, D. W. Zuo, and L. T. Wang, "Numerical analysis of surface residual stress of NC milling 7075-T7451 aluminum alloy," *Key Engineering Materials*, vol. 40, no. 7, pp. 718–722, 2009.

[6] A. Arcari, R. De Vita, and N. E. Dowling, "Mean stress relaxation during cyclic straining of high strength aluminum alloys," *International Journal of Fatigue*, vol. 31, no. 11-12, pp. 1742–1750, 2009.

[7] H. Gong, Y.-X. Wu, and K. Liao, "Influence of pre-stretching onresidual stress distribution in 7075 aluminum alloy thick-plate," *Transactions of Materials and Heat Treatment*, vol. 30, no. 6, pp. 201–205, 2009.

[8] C.-Y. Yao, *Numerical Simulation of Quench Temperature Field and Internal Stress Field of Aluminum Alloy 7050 Thick Plate*, Central South University, Changsha, China, 2007.

[9] K. Liao, Y.-X. Wu, H. Gong, and S.-Y. Zhang, "Application of integral method on residual stress calculation along depth in aluminium alloy thick plate," *The Chinese Journal of Nonferrous Metals*, vol. 19, no. 6, pp. 1006–1011, 2009.

[10] M. B. Prime and M. R. Hill, "Residual stress, stress relief, and inhomogeneity in aluminum plate," *Scripta Materialia*, vol. 46, no. 1, pp. 77–82, 2002.

[11] M. B. Prime, "Residual stress measurement by successive extension of a slot: the crack compliance method," *Applied Mechanics Reviews*, vol. 52, no. 2, pp. 75–96, 1999.

[12] K. Muammer, "Prediction of residual stresses in quenched aluminum blocks and their reduction through cold working processes," *Journal of Materials Processing Technology*, vol. 174, pp. 342–354, 2006.

[13] D. A. Tanner and J. S. Robinson, "Modelling stress reduction techniques of cold compression and stretching in wrought aluminium alloy products," *Finite Elements in Analysis and Design*, vol. 39, no. 5-6, pp. 369–386, 2003.

[14] S. Wang, *Analysis of Quenching Residual Stress and Its Reduction Process of 7A85 Aluminum Alloy Component*, Central South University, ChangSha, China, 2011.

16

Design of Laser Welding Parameters for Joining Ti Grade 2 and AW 5754 Aluminium Alloys using Numerical Simulation

Mária Behúlová, Eva Babalová, and Miroslav Sahul

Faculty of Materials Science and Technology in Trnava, Slovak University of Technology in Bratislava, Ulica Jána Bottu č. 2781/25, 917 24 Trnava, Slovakia

Correspondence should be addressed to Mária Behúlová; maria.behulova@stuba.sk

Academic Editor: Ying Li

Joining of dissimilar Al-Ti alloys is very interesting from the point of view of weight reduction of components and structures in automotive or aerospace industries. In the dependence on cooling rate and chemical composition, rapid solidification of Al-Ti alloys during laser welding can lead to the formation of metastable phases and brittle intermetallic compounds that generally reduce the quality of produced weld joints. The paper deals with design and testing of welding parameters for preparation of weld joints of two sheets with different thicknesses from titanium Grade 2 and AW 5754 aluminium alloy. Temperature fields developed during the formation of Al-Ti butt joints were investigated by numerical simulation in ANSYS software. The influence of laser welding parameters including the laser power and laser beam offset on the temperature distribution and weld joint formation was studied. The results of numerical simulation were verified by experimental temperature measurement during laser beam welding applying the TruDisk 4002 disk laser. The microstructure of produced weld joints was assessed by light microscopy and scanning electron microscopy. EDX analysis was applied to determine the change in chemical composition across weld joints. Mechanical properties of weld joints were evaluated using tensile tests and Vickers microhardness measurements.

1. Introduction

The specific properties of different materials are jointly utilized to achieve the required product performance. This trend is reported for several industries including automotive, aircraft, aeronautic, or marine applications. For example, the light-weight construction of the Boeing 787 Dreamliner consists of 50% composite materials (the primary material of the airframe structure), 20% aluminium, 15% titanium, 10% steel, and 5% other materials [1]. Aluminium is used for the wing and tail leading edges; titanium is applied mainly on engines and fasteners.

Joining two dissimilar metals is often difficult due to the distinct material properties and chemical reactions between different materials which can lead to the formation of intermetallic compounds (IMC) at the metals interface during the welding process. The presence of IMC in weld joints generally degrades the mechanical properties and quality of the welds. Therefore, the effort during the design of technology for welding of dissimilar material is aimed at elimination of IMC formation. The thickness of the IMC layer depends on the temperature profile generated by the heat source exploited for welding and the heat flux distribution across the interface. However, the thermal fields are also responsible for geometry of the fusion zone and thus the dimension of the bonding area [2].

The successful joining of aluminium to titanium requires the reduction in thickness of potentially brittle intermetallic zones such as Ti_3Al, TiAl, and $TiAl_3$. Therefore, it is difficult to join the Al-Ti compound structures by means of conventional fusion welding [3]. Laser welding with its high energy density alleviates these problems and provides additional benefits [4]. The small size and accuracy of the laser beam provide the possibility to control the microstructure development. Rapid solidification results in a narrow heat-affected zone (HAZ) and less grain growth. Formation of a brittle layer of IMC can occur, although its thickness is smaller than that in conventional arc welding, for example [5]. Moreover, laser welding allows obtaining thin and high precision welds and reducing thermal distortion. Small interaction zone

TABLE 1: Chemical composition of titanium Grade 2.

Element	Fe	C	O	H	N	Ti
wt.%	0.3	0.1	0.25	0.015	0.03	Balance

TABLE 2: Chemical composition of AW 5754 aluminium alloy.

Element	Mg	Mn	Fe	Si	Al
wt.%	2.6 to 3.2	0.5	0.4	0.4	Balance

and high welding speed promote high thermal gradients, which are useful for local phase content optimization. So, the importance of mixing and diffusion phenomena can be reduced. In addition, the effect of mismatch on thermo-physical properties of welded dissimilar materials can be eliminated by the offset of laser beam to one of the welded component. Laser offset welding was demonstrated to be a promising technique to join dissimilar metals, being capable of producing effective bonding with good tensile strength [6]. The use of higher welding speeds (>4 m·min^{-1}) would minimize the interdiffusion between the molten materials and reduce the development of brittle intermetallic zones [7].

Modeling and numerical simulation of welding processes have been exploited with many advantages for investigation of physical principles and complex phenomena connected with joining processes [8–13]. Together with testing and inspection, numerical simulations are mutually important to understand and predict the performance of the welding processes. Moreover, the application of numerical methods and computer simulation represents one of the ways to ensure design and optimization of welding parameters. Results of computer simulation can be applied for the relatively fast and cheap evaluation of the influence of welding parameters on the quality of the weld joints and the definition of optimal process parameters for weld manufacturing [14–17]. In general, numerical simulation is regarded as a valid tool and is widely applied in welding process designing [18].

In connection with numerical simulation of welding processes, the features of distribution and variation of the temperature fields are highly important. Heat propagation during the welding process essentially depends on the geometry of welded components and thermophysical properties of the materials used to make them. Another important factor is the application and modeling of a welding heat source. In case of simple shaped sources like point, line, or plane sources, it is possible to find an analytical solution for [19]. If the heat source is of more complex shape, the solution may be obtained only by numerical integration or using, for example, the FEM method [20]. Extensive literature is available on the modeling and exploitation of heat sources for different welding technologies including surface or volumetric heat source models [12, 20–27]. For numerical simulation of laser beam welding, various volumetric heat source models have been tested and applied, ranging from Gaussian ellipsoidal heat source, Goldak's double-ellipsoidal heat source, cylindrical heat source, and cone-shaped source to some of their modifications and combinations [22, 28–33].

The paper deals with the numerical simulation of temperature fields by laser beam welding of titanium Grade 2 and AW 5754 aluminium alloy using the finite-element code ANSYS. Based on the obtained results, the appropriate welding parameters were designed to realize the real experiments of the laser welding of dissimilar Ti-Al alloy sheets by the TruDisk 4002 disk laser. The quality of the prepared weld joints was evaluated by light microscopy, EDX analysis, X-ray diffraction analysis, tensile testing, and microhardness measurements.

2. Problem Description

Laser beam welding is applied for joining two sheets from dissimilar materials and with different thicknesses. The sheet from Ti Grade 2 with the chemical composition given in Table 1 has the thickness of 2 mm. The thickness of the sheet from AW 5754 aluminium alloy with the chemical composition given in Table 2 is 1.5 mm. Butt joints of the samples with dimensions of 30 mm × 75 mm are assumed to be produced with different laser parameters (laser power and laser offset) using the required welding speed of 10 mm·s^{-1} and no filler material. Argon gas with the maximum flow rate of 30 l·min^{-1} can be used as a shielding gas to protect both weld bead and root.

In order to save time and financial costs, numerical simulation of laser beam welding of dissimilar Al-Ti alloys is implemented to study the influence of chosen welding parameters on the temperature distribution in welded sheets and to set appropriate welding parameters for experimental preparation of weld joints by the TruDisk 4002 disk laser. The welding parameters considered in numerical experiments are summarized in Table 3.

3. Theoretical Background and Simulation Model

Generally, numerical simulation of laser welding processes requires solution of coupled thermal-fluid and stress-strain problems involving phase transformations [32–36]. Using this approach, it is possible to evaluate the temperature fields taking into account the melt convection in the weld pool and also the deformations of welded components. In addition, the stresses during the welding process and residual stresses remaining in the welded material after the cooling of a weld to the room temperature can be predicted in order to eliminate

Table 3: Welding parameters applied for numerical experiments.

Experimental sample number	Laser power [W]	Laser beam offset [μm]
1	750	250
2	800	250
3	850	250
4	850	150
5	850	100
6	850	50
7	900	250
8	900	150
9	900	100
10	950	250

the possible cracking and failure of the weld joint. In our case, the initial numerical simulation of the laser welding was focused particularly on the analysis of temperature field and possibility to join materials with very distinct thermal properties.

Transient temperature fields developed during welding processes can be described by the heat diffusion partial differential equation in the following form [37]:

$$\rho c_p \frac{DT}{Dt} = \left[\frac{\partial T}{\partial x}\left(\lambda_x \frac{\partial T}{\partial x}\right) + \frac{\partial T}{\partial y}\left(\lambda_y \frac{\partial T}{\partial y}\right) + \frac{\partial T}{\partial z}\left(\lambda_z \frac{\partial T}{\partial z}\right)\right] + q_v, \quad (1)$$

where $T(x, y, z, t)$ is the temperature [°C, K], t is the time [s], ρ is the density [kg·m^{-3}], c_p is the specific heat [J·kg^{-1}·K^{-1}], λ_x, λ_y, and λ_z are the thermal conductivities [W·m^{-1}·K^{-1}] in the x, y, and z directions of Cartesian coordinate system, respectively, and q_v is the volumetric density of internal heat sources [W·m^{-3}], that is, the heat generated in the unit volume of material per unit time. The expression DT/Dt represents the substantial derivative of temperature. To solve the heat diffusion equation (1), it is necessary to specify geometrical, physical, initial, and boundary conditions, that is, geometrical shapes and dimensions of welded components, thermophysical properties of welded materials, intensity of internal heat sources, and initial temperature distribution at the beginning of the welding process, and also the conditions and thermal effects at the interface of welded structures and environment.

Considering the problem definition, the computational domains representing the welded plates had dimensions of 30 mm × 30 mm × 1.5 mm (Al plate) and 30 mm × 30 mm × 2 mm (Ti plate). The 3D finite-element mesh was generated using the ANSYS finite-element software [38] according to supposed temperature gradients; that is, the mesh density was the highest in the region of the weld pool and the heat-affected zone. The eight-node three-dimensional solid elements of the type of SOLID70 were applied for the thermal

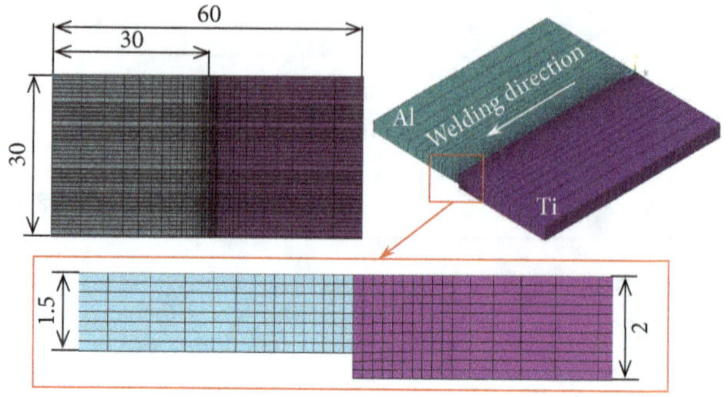

Figure 1: FEM model of welded sample with the detail of generated mesh.

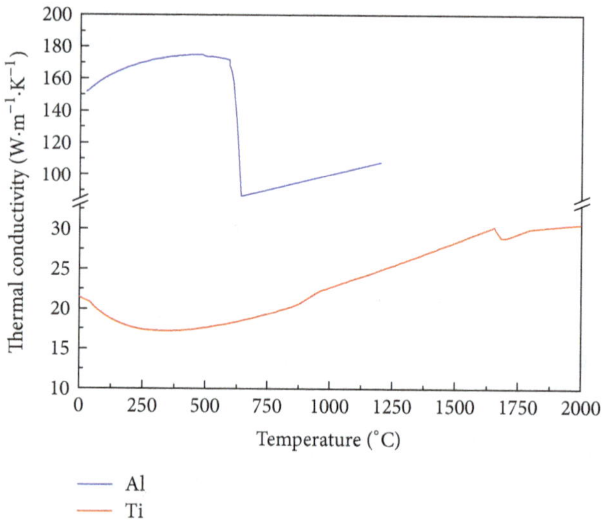

Figure 2: Thermal conductivity of the Ti (Grade 2) and Al (AW 5754) alloy.

analysis. The finest mesh density along the weld line is 0.2 mm. FEM model of a welded sample with the detail of generated mesh is shown in Figure 1. It consists of a total number of 79858 nodes and 68400 3D elements.

Thermophysical material properties of the Ti (Grade 2) and Al (AW 5754) alloys in the dependence on temperature computed using the JMatPro software [39] are plotted in Figures 2–4. Titanium Grade 2 represents commercially pure titanium (99.2%) with the liquidus temperature of T_L = 1660°C. The equilibrium solidus and liquidus temperatures for the AW 5754 alloy were computed to be 600°C and 660°C, respectively. The enthalpy of fusion was taken into account applying the method of modified specific heat [40].

The initial condition can be written in the following form:

$$T(x, y, z, t = 0) = T_0(x, y, z), \quad (2)$$

where T_0 is the initial temperature of welded components, mostly equal to the ambient (room) temperature. The initial temperature of plates was supposed to be 20°C.

The boundary condition of the 3rd kind is possible to exploit for the definition of welded plates cooling by

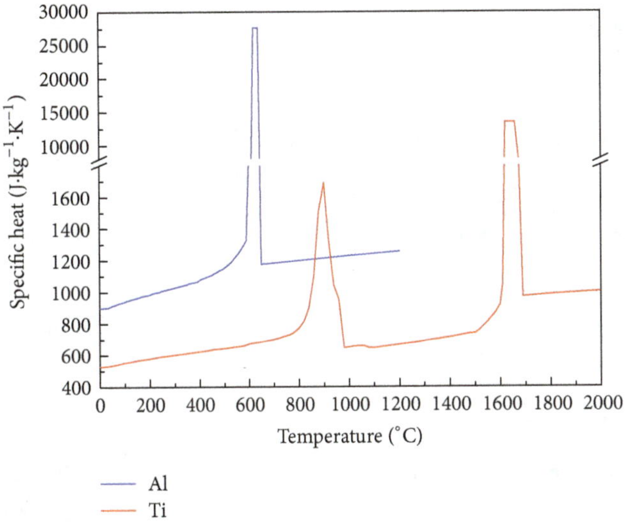

FIGURE 3: Specific heat of the Ti (Grade 2) and Al (AW 5754) alloy.

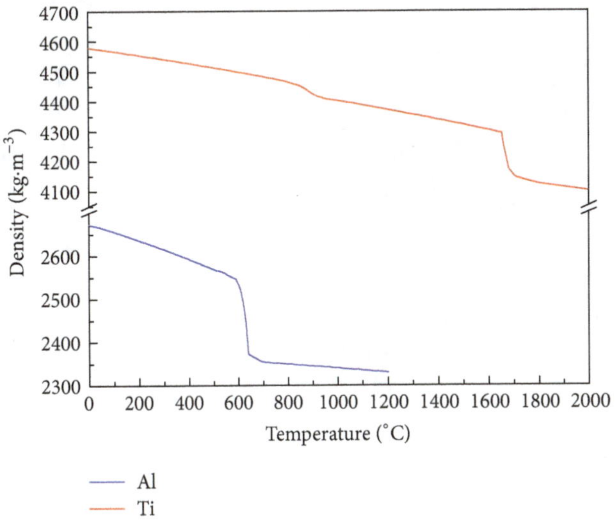

FIGURE 4: Density of the Ti (Grade 2) and Al (AW 5754) alloy.

convection and radiation to the argon shielding gas and to the surrounding air:

$$-\lambda\frac{\partial T}{\partial \mathbf{n}} = h\left(T_w - T_f\right), \tag{3}$$

where T_w is the sample surface temperature, T_f is the surrounding fluid temperature, and h represents the combine heat transfer coefficient given by the sum of convection heat transfer coefficient h_C and radiation heat transfer coefficient h_R:

$$h = h_C + h_R = h_C + \frac{\varepsilon\sigma_0\left(T_w^4 - T_f^4\right)}{T_w - T_f}, \tag{4}$$

where ε is the emissivity and σ_0 is the Stefan-Boltzmann constant.

As it was mentioned, the heat input in fusion welding can be modeled by different methods including moving point or line heat sources and surface or volumetric heat sources. For the presented numerical simulation of laser welding of Ti-Al sheets, the Gaussian ellipsoidal volumetric heat source model (Figure 5) was applied [17, 41]. Using this model, it is possible to model sufficiently accurate various types of welding processes with different shape and depth of the weld pool [12]. The effective thermal power of a heat source is calculated according to the supposed volume of the weld pool. The Gaussian ellipsoidal volumetric heat source is defined by three parameters: the ellipsoid half-axes a, b, and c. The volumetric density of internal heat source q_v [W·m^{-3}] is given by the following function [41]:

$$\begin{aligned} q_v(x,y,z) &= \frac{\Phi f(x,y,z)}{V} \\ &= \Phi\frac{6\sqrt{3}}{abc\pi^{3/2}}e^{-3x^2/a^2}e^{-3y^2/b^2}e^{-3z^2/c^2}, \end{aligned} \tag{5}$$

where Φ represents the effective laser power, $V = a\cdot b\cdot c\cdot \pi^{3/2}/(6\sqrt{3})$ is the volume of one half of the ellipsoid, and $f(x,y,z) = \exp(-3x^2/a^2)\cdot\exp(-3y^2/b^2)\cdot\exp(-3z^2/c^2)$ is the Gaussian distribution function. Taking into account the welding speed w [m·s^{-1}] and the distance X [m] from the initial position of the source centre in the time $t = 0$ s (Figure 5(a)), it is possible to calculate the distribution of internal heat sources in an arbitrary time as illustrated in Figure 5(b).

For nonlinear transient thermal analyses in this study, the following parameters of the Gaussian volumetric heat source were applied: $a = 1$ mm, $b = 1$ mm, and $c = 2.5$ mm. The laser power varied from 750 W to 950 W supposing that the efficiency is 63.5%. The welding speed was set to the required value of 10 mm·s^{-1}. In order to eliminate intermixing of both materials and thus the formation of hard and brittle intermetallic compounds, it was suggested to offset the laser beam on the titanium sheet side. The offset of laser beam to titanium side ranged from 50 μm to 250 μm.

4. Equipment for Experimental Laser Beam Welding

For the preparation of experimental butt weld joints, the TruDisk 4002 disk laser [42] with the maximum power of 2 kW, the wavelength of 1.03 μm, and the beam quality (BPP) of 8 mm·mrad was available. The laser beam can be transported to the BEO D70 focusing optics via laser light cable with the core diameter of Ø 200 μm or Ø 400 μm. In case of application of the laser light cable with the core diameter of Ø 400 μm and the laser focusing on the top surface of components to be welded, the spot size of the laser beam is also 400 μm. The focusing optics are mounted on a 6-axis robot FANUC M-710iC/50 (Figure 6), enabling very accurate setting and control of welding parameters. Argon with the maximum flow rate of 30 l·min^{-1} can be used for shielding the weld pool and/or the weld root.

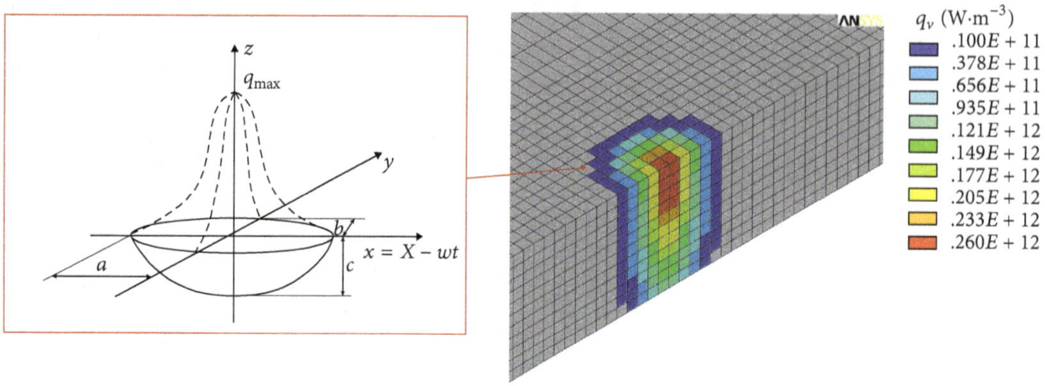

(a) (b)

FIGURE 5: Gaussian ellipsoidal volumetric heat source: (a) the scheme with definition of parameters and (b) application of volumetric heat source to the FE model.

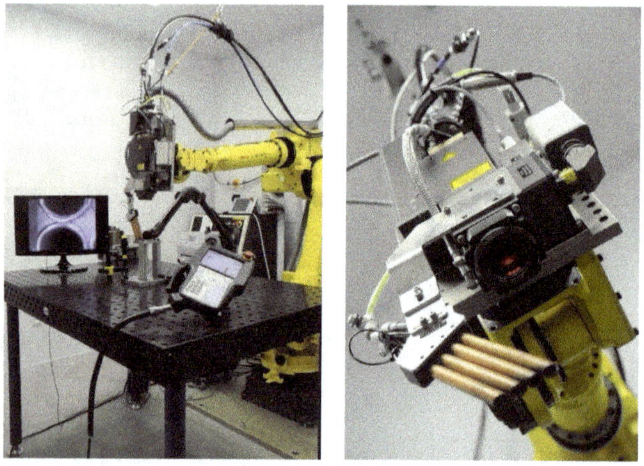

FIGURE 6: Laboratory of laser beam welding at the Faculty of Materials Science and Technology of STU in Trnava: (a) laser workplace with 6-axis robot FANUC M-710iC/50 and (b) detail of BEO D70 focusing optics.

5. Results and Discussion

In the first step, numerical simulation of laser beam welding of dissimilar Al-Ti sheets using the described simulation model was performed for the laser beam powers and the laser offsets given in Table 3. Based on the results of numerical simulation, parameters for the preparation of experimental weld joints were suggested. During experimental laser welding, the temperatures in selected points were measured and consequently applied for the verification of developed simulation model. Finally, the microstructure and the quality of produced weld joints were evaluated using light and scanning electron microscopy including EDX analysis, X-ray diffraction analysis, Vickers microhardness measurements, and tensile tests.

5.1. Results of Numerical Simulation. Numerical simulation of laser beam welding of dissimilar Ti-Al sheets using the described simulation model was performed for the laser beam powers from 750 W to 950 W and the laser offsets to the Ti sheet from 50 μm to 250 μm. For illustration, the computed temperature fields in chosen times for sample 3 (the laser power of 850 W and the laser offset of 250 μm) are shown in Figure 7. The influence of laser beam offset as well as the effect of differences in the material properties of Ti and Al alloys on the temperature distribution in welded sheets can be seen from this figure.

The time history of temperatures in the node P located in the centre of a weld pool on the top surface of Ti plate (Figure 7) for different laser beam powers and the laser offset of 250 μm is plotted in Figure 8. For the laser power of 750 W, the maximum temperature in the node P reaches the value of 1668°C which is only a few degrees above the titanium liquidus temperature. This result indicates that the laser power of 750 W will be insufficient for the formation of a weld joint with the desired quality. The maximum computed temperature during laser welding with the maximum considered laser power of 950 W is 1944°C.

The influence of the laser offset on the maximum temperatures of the weld pool during laser welding is illustrated in Figure 9. The larger the laser offset, the higher the maximum

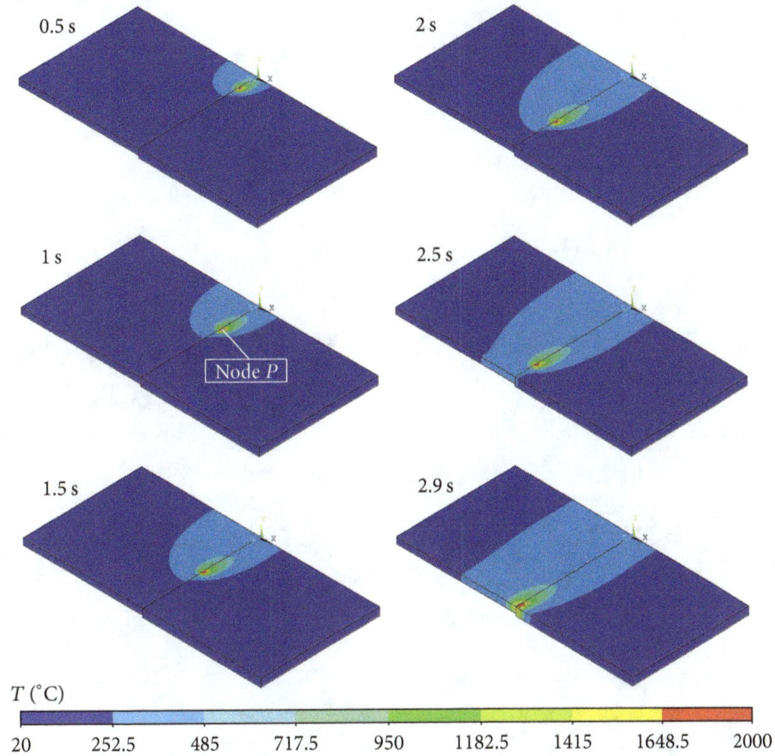

FIGURE 7: Temperature distribution in welded sheets in chosen times for sample 3 (the laser power of 850 W and the laser offset of 250 μm).

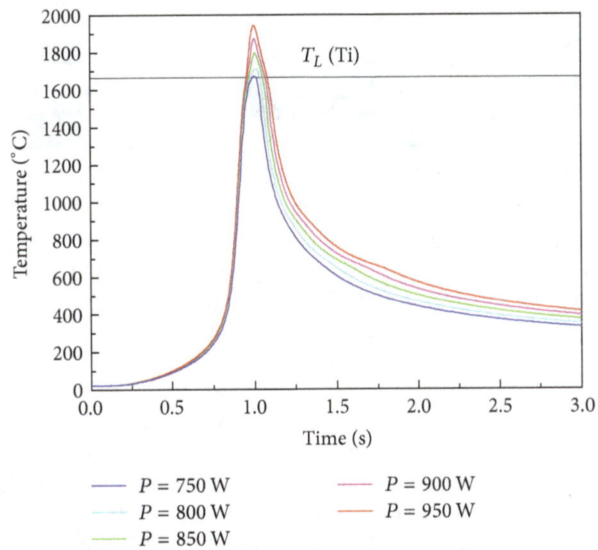

FIGURE 8: Time history of temperatures in the node P located on the top surface in the centre of weld pool of Ti plate for different laser beam powers and the offset of 250 μm.

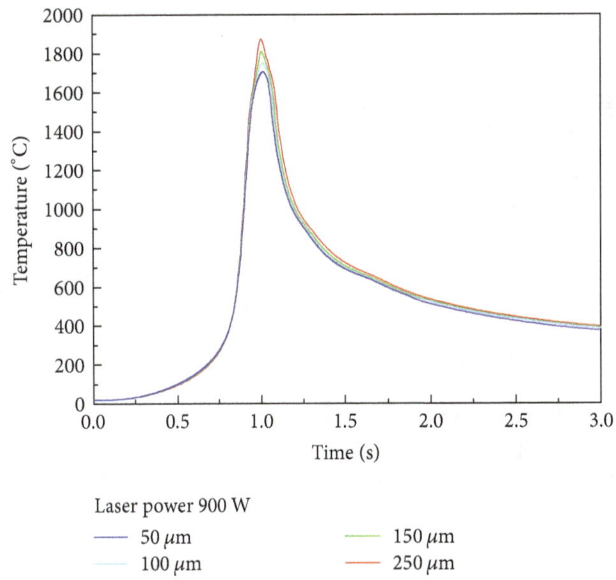

FIGURE 9: Influence of the laser offset on the time history of temperatures in the node P for the laser power of 900 W.

temperature of the weld pool, a consequence of the smaller thermal diffusivity of titanium in comparison to aluminium alloy. In case of smaller laser offset, the heat from the weld pool is dissipated faster to the AW 5754 alloy plate. This results in a smaller maximum temperature of the weld pool but on the other hand results in the higher temperature at the interface of welded Al-Ti plates. The dependence of temperatures on the distance from the weld centre for different laser offsets and the laser power of 850 W and 900 W, respectively, is shown in Figure 10. The influence of laser power on the temperature distribution along the line perpendicular to the weld centre is documented in Figure 11.

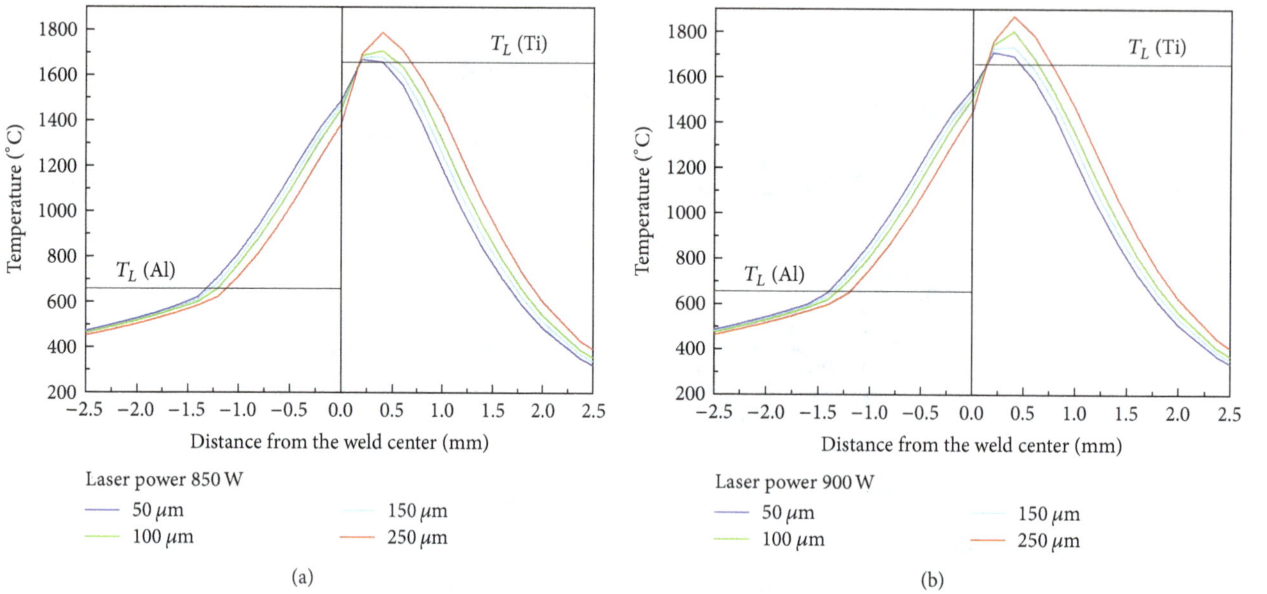

FIGURE 10: The dependence of computed temperatures on the distance from the weld centre for different laser offsets and the laser power of (a) 850 W and (b) 900 W.

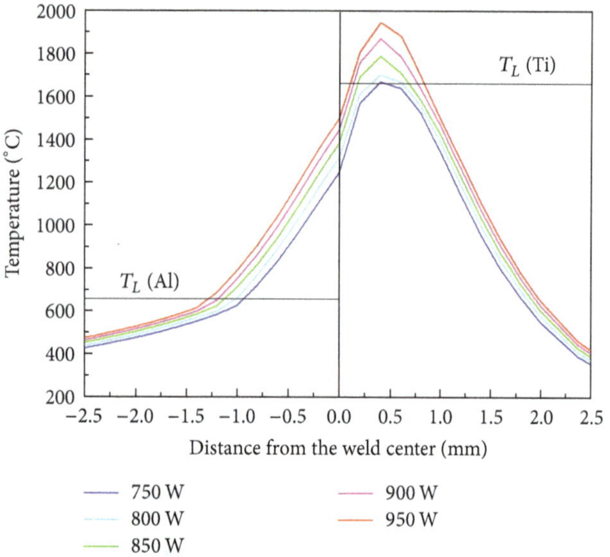

FIGURE 11: Dependence of computed temperatures on the distance from the weld centerline for different laser powers and the laser beam offset of 250 μm.

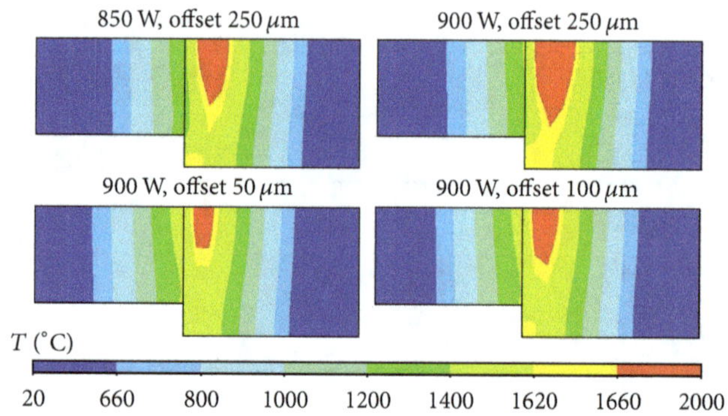

FIGURE 12: Temperature fields in the cross sections perpendicular to the Al-Ti weld centre through the node P in the time of 1 second illustrating the effect of laser offset and the laser power.

The increase in laser power leads to the enhancement of the maximum weld pool temperature and also leads to the increase of the interface temperature. A better idea about the temperature distribution in the welded sheets near the weld centre can be provided by Figures 12 and 13. In these figures, the temperature fields in perpendicular and longitudinal weld cross sections in the time of 1 second for chosen laser powers and considered laser offsets are illustrated. Figure 12 confirms the fact that the larger laser offset leads to the smaller interface temperature of the weld joint in the moment of laser beam passing. However, the maximum weld pool temperatures are higher for greater laser offsets. Thus, due to the mechanism of heat conduction and also convection in the weld pool, the maximum interface temperatures will achieve with some delay the higher values for the larger laser offsets, as can be seen from Figure 13. This figure also clearly demonstrates that the utilization of small laser offsets (50 μm and 100 μm) cannot provide suitable thermal conditions for the sound weld joint development as the temperatures of Ti plate are below the liquidus temperature for Ti Grade 2.

Based on the results of numerical simulation of laser beam welding of Al-Ti sheets, it can be supposed that, for the laser beam offsets of 150 μm and 250 μm, the Ti Grade 2 sheet will be melted across the whole thickness when the minimum laser power of 800 W is applied. In this reason, the following welding parameters were recommended for the experimental verification of simulation model and production of butt weld joints of Al-Ti sheets:

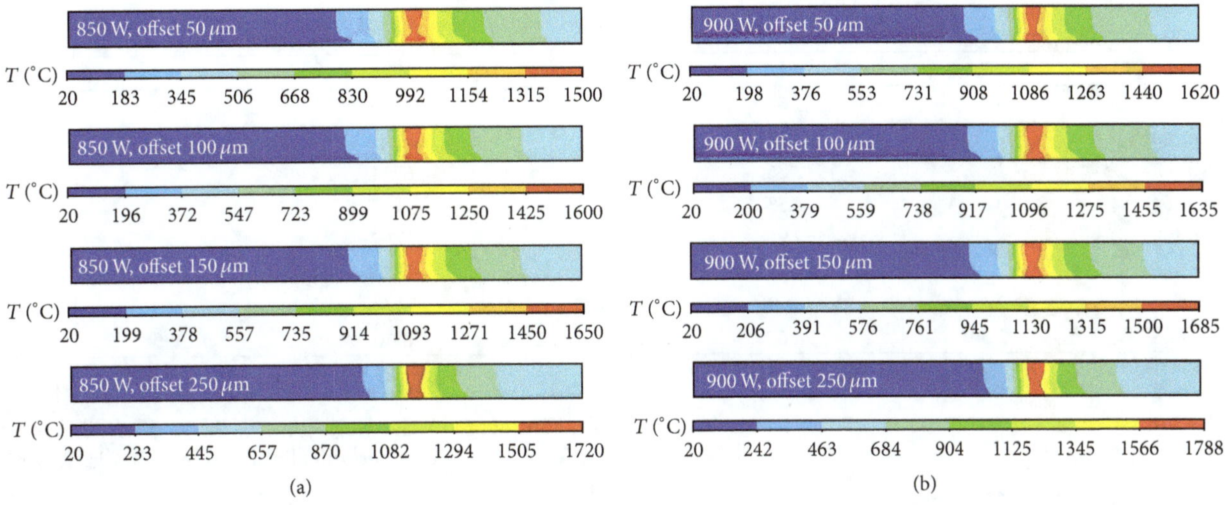

FIGURE 13: Temperature fields in the cross section along the plane of weld centre in the time of 1 second from the beginning of laser welding. (a) Laser power 850 W; (b) laser power 900 W.

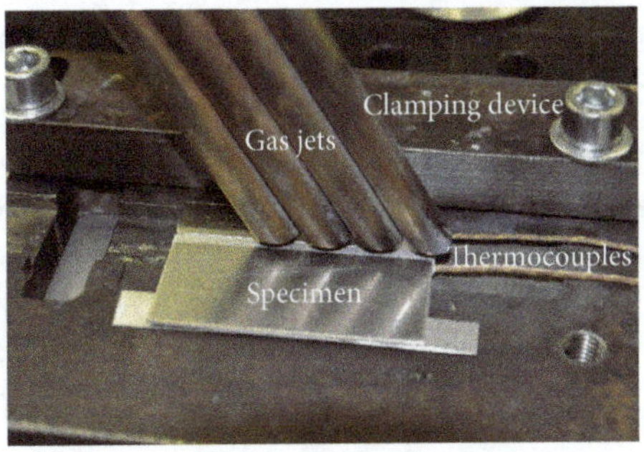

FIGURE 14: Specimen with thermocouples prepared for welding.

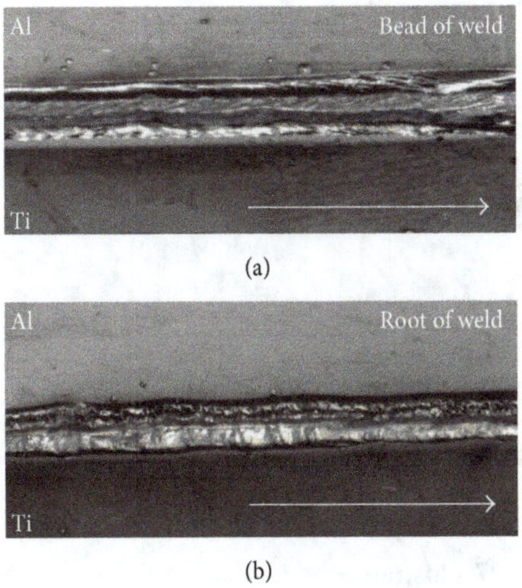

FIGURE 15: Appearance of the butt weld joint prepared with the laser power of 800 W and the laser offset of 250 μm: (a) bead of weld and (b) root of weld.

(i) The laser power from 800 W to 900 W

(ii) The laser offset from 150 μm to 250 μm

5.2. Verification of Simulation Model by Experimental Temperature Measurements. Experimental weld joints were prepared by the continuous wave TruDisk 4002 disk laser described in Section 3 using the laser light cable with the core diameter of 400 μm. Argon with the flow rate of 30 l·min^{-1} was used as shielding gas. The laser beam was focused on the top surface of welded samples. During welding experiments, temperatures were measured by two thermocouples of the K-type located in the depth of 0.5 mm from the bottom side of samples and 3 mm from the weld centerline on the aluminium sheet and 4 mm on the titanium sheet. The specimen with thermocouples prepared for welding is shown in Figure 14. Figure 15 illustrates the bead and the root of butt weld joint prepared using the laser power of 800 W and the laser offset of 250 μm (sample 2).

The time dependence of temperatures measured by thermocouples and computed in the nodes corresponding to the location of thermocouples is plotted in Figure 16. The maximum temperature measured by thermocouples in experiment 2 (sample 2) was 300.9°C on the Ti side and 365.9°C on the Al side. The computed temperatures are slightly higher in comparison with experimentally measured temperatures. The temperature deviations can be caused by not pretty accurate setting of the heat source parameters or weld cooling by the shielding gas from the bottom side. Another reason can be also the fact that the thermal contact between Al and Ti plates was considered to be ideal in numerical simulations. If the plates are in solid state, the contact resistance definitely influences the temperatures at the interface of welded sheets. On the other hand, in the weld pool, the convection and intermixing of both welded materials contribute to the decrease of thermal gradients in molten zone. Nevertheless, it can be concluded that the results of numerical simulation are in a good agreement with the experimentally measured temperatures.

Table 4: Results of point EDX analysis.

Spectrum	Mg	Al	Si	Ti	Mn	Fe
Spectrum 1		22.07		77.93		
Spectrum 2		25.57		74.43		
Spectrum 3		32.86		67.14		
Spectrum 4	0.30	34.53		64.67		0.50
Spectrum 5		62.98		37.02		
Spectrum 6		34.72		65.28		
Spectrum 7	2.43	73.59	0.33	23.66		
Spectrum 8	2.50	94.67		0.60	0.23	

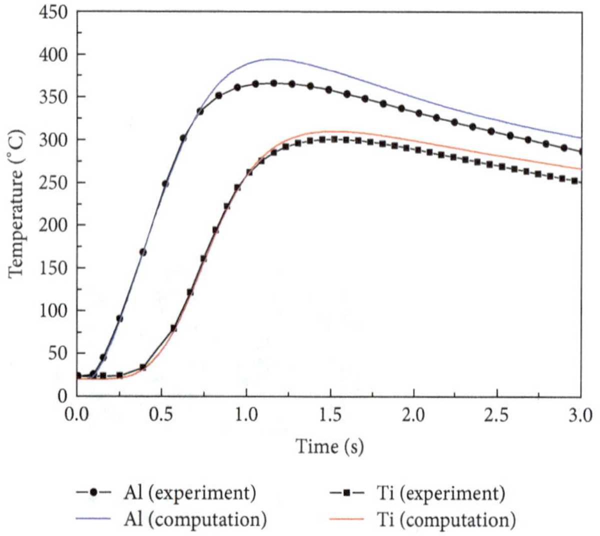

Figure 16: Time dependence of temperatures measured by thermocouples and computed in the nodes corresponding to the location of thermocouples (laser power of 800 W; laser offset of 250 μm).

Figure 17 illustrates the influence of laser power on the computed temperatures in the nodes corresponding to the thermocouple location on the Al and Ti sides. The temperatures computed for the laser power of 750 W are below the experimentally recorded temperatures during the laser welding of Al-Ti plates with the laser power of 800 W. This result also confirms that the simulation model is developed correctly.

5.3. Evaluation of the Quality of Produced Joints. After the laser welding, the samples were cut for the preparation of metallographic specimens. The macrostructure and microstructure of weld joints and the interface characteristic were investigated on the cross section of weld joints perpendicular to the direction of welding. After metallographic preparation of cross sections of welded joints, the specimens were etched with Keller's reagent with the chemical composition of 1.0 ml HF + 1.5 ml HCl + 2.5 ml HNO_3 + 95 ml H_2O.

The light microscopy (NEOPHOT 32 light microscope) was used for the analysis of dissimilar Al-Ti weld joints. Figure 18(a) documents the macrostructure of the weld joint for sample 2. In the weld joint of Al-Ti dissimilar metals, different zones can be typically identified, that is, AW 5754 aluminium alloy, aluminium based weld metal, interface layer, titanium based weld metal, titanium heat-affected zone (HAZ), and titanium Grade 2 base metal.

Weld root concavity was observed on the aluminium alloy side. The microstructures at the interfaces of titanium Grade 2 and the heat-affected zone are shown in Figure 18(b). The microstructure of titanium Grade 2 consists of fine equiaxed α grains. Grain coarsening was observed in HAZ due to higher heat input used. The width of HAZ on titanium side reached the value of 1.085 mm. This is also associated with the lower thermal conductivity of titanium. The microstructure of aluminium AW 5754 alloy and weld metal (WM) interface is illustrated in Figure 18(c). The microstructure of weld metal is quite different from the microstructure of original material. Columnar grains present in the weld metal are oriented towards the weld centre.

Figure 19 exhibits the microstructure and heterogeneous character of the weld metal for samples 3 and 4. Zones with different chemical compositions can be seen in the microstructure. It is evident that intermixing of both metals (titanium Grade 2 and AW 5754 aluminium alloy) occurred resulting in the development of complex structures.

Electron microscopy was carried out in order to study the microstructure of produced weld joints in more detail. Scanning electron microscope JEOL 7600F fitted with EDX analyzer X-max 50 mm^2 from Oxford Instruments was applied. The microstructure of aluminium alloy and titanium weld metal is depicted in Figure 20. The needle-shaped particles represent zones with the presence of titanium. This statement can be proven by the results of EDX analysis summarized in Table 4. The locations of point EDX analysis are designated in Figure 21. The large bright zones (Spectrums 1 to 3) are characterized by higher content of titanium. The results of EDX analysis indicate that the titanium content decreases in the direction to the aluminium side of the weld metal. Higher content of Al compared to Ti was documented in Spectrum 5. On the contrary, the dark zones are characterized by high content of aluminium. Aluminium content in Spectrum 8 is almost 95 wt.%. Small amount of Mg was also detected.

Vickers microhardness measurements were performed in the weld interface zone on Buehler IndentaMet 1100 microhardness tester. The load of 100 g acting during 10 sec

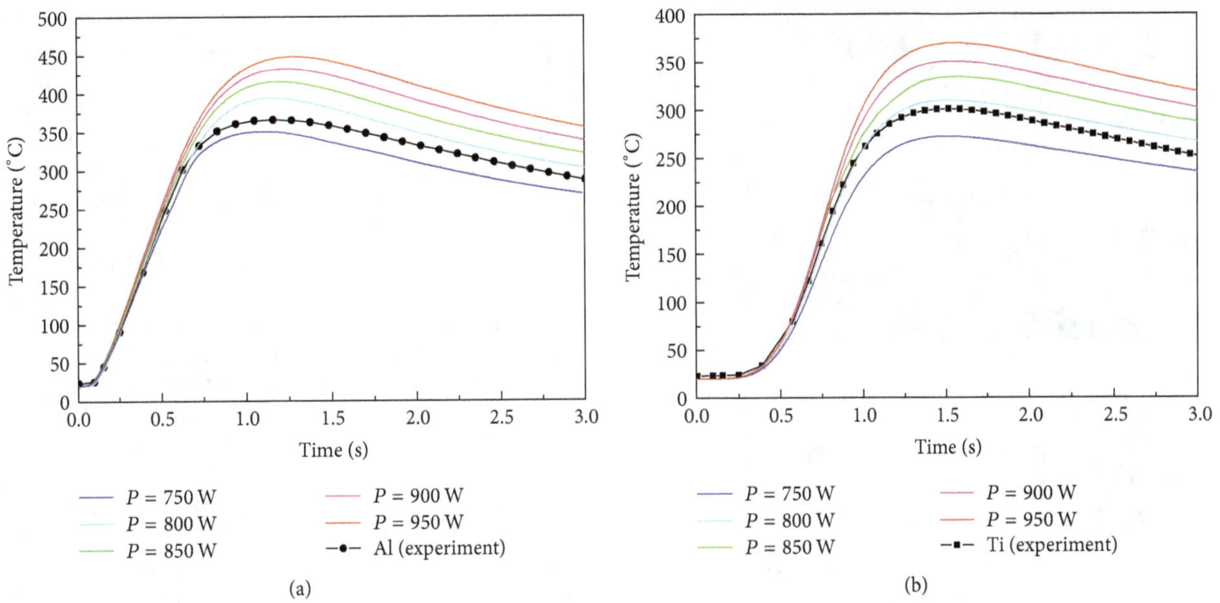

FIGURE 17: Influence of laser power on the computed temperatures in the nodes corresponding to the thermocouple location on the (a) Al side and (b) Ti side (laser offset of 250 μm).

was applied. The distance between individual indents was 200 μm.

Figure 22 shows the microhardness distribution across dissimilar Al-Ti metals weld joints for sample 3 and sample 4. The microhardness of titanium ranged from approximately 150 to 190 HV0.1. The microhardness increase was recorded towards weld metal. The maximum microhardness was measured in weld metal, reaching the values of 512.9 HV0.1 and 481.1 HV0.1 for sample 3 and sample 4, respectively. From the microhardness values measured, it can be concluded that the maximum microhardness found in weld metal is more than 2.5 times higher than that measured in titanium Grade 2. On other hand, the average microhardness of the weld metal at the aluminium side is similar to the microhardness of AW 5754 aluminium alloy (approximately 60 HV0.1).

The higher values of microhardness in the weld metal (in the intermixing zones) could be associated with high cooling rates during laser beam welding and the development of hard intermetallic compounds during rapid solidification of weld pool. The chemical composition measured in the zone close to the indent (Figure 23) is given in Table 5. Based on the titanium and aluminium binary diagram [43], microhardness measurements, and results of EDX analysis, the presence of hard and brittle intermetallic compounds of TiAl, Ti_3Al, and Al_3Ti types can be supposed. This assumption was confirmed by the X-ray diffraction analysis (Figure 24). In addition to intermetallic compounds of Al_3Ti and $AlTi_3$, $Al_{18}Mg_3Ti_2$ was recorded by X-ray diffraction as well.

The tensile testing of prepared dissimilar Al-Ti welds was carried out at the room temperature using the LabTest 5.250 Spl tensile testing machine (max. loading force 250 kN) of LABORTECH Company. The influence of the laser power and laser offset on the tensile strength of weld joints was

TABLE 5: Chemical composition measured in Spectrums 1 to 3.

Spectrum	Mg	Al	Ti
Spectrum 1	0.68	51.66	47.67
Spectrum 2		30.90	69.10
Spectrum 3		52.69	47.31

investigated. In all tensile tests, the fractures were observed in the weld metal.

The dependence of tensile strength (UTS) of weld joints on the laser power is given in Figure 25 (remaining welding parameters were constant: the welding speed of 10 $mm \cdot s^{-1}$, focal position was 0 mm, and laser offset to titanium Grade 2 sheet was 250 μm). The maximum tensile strength was reached for the weld joint produced with the laser power of 800 W. The measured tensile strength was at the level of 80 MPa. This highest tensile strength can be associated with the low heat input used for welding of dissimilar metals. The calculated linear heat input in this case was 80 J/mm. It resulted in the development of a narrow molten zone leading to elimination of the intermixing of both metals and thus the formation of hard and brittle intermetallic compounds. On the contrary, the increase in laser power to 850 W resulted in the decrease of UTS of weld joints to about 64 MPa. The calculated linear heat input was 85 J/mm in this case. The further increase of laser power to 900 W and thus the increase of linear heat input to 90 J/mm led to the higher production of IMC and reduction of tensile strength of welds.

The dependence of UTS of joints on the laser offset is plotted in Figure 26. The maximum tensile strength of 80 MPa was attained for the weld joint produced with the laser offset of 250 μm. This can be also attributed to the reduced amount of hard and brittle IMC. The reduction of tensile strength was

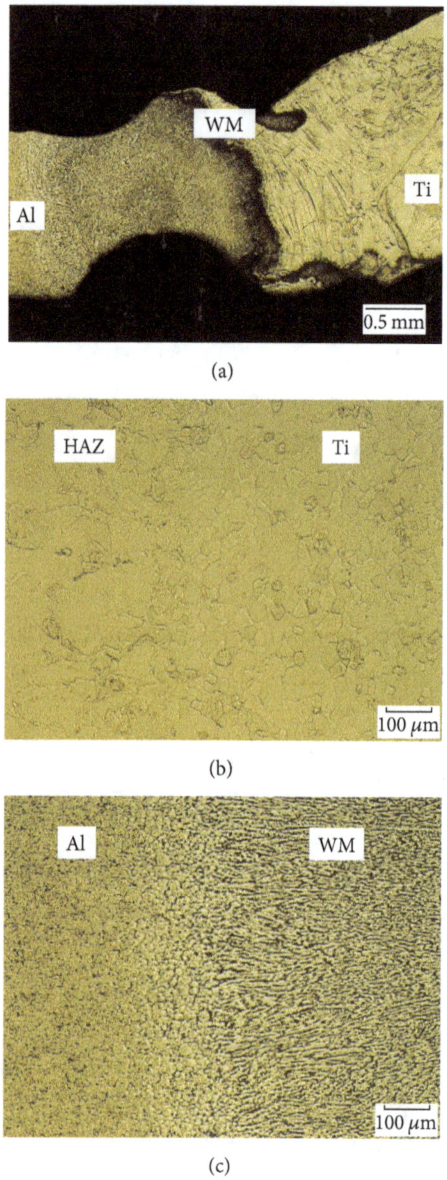

FIGURE 18: Macrostructure and microstructure of the weld joint (sample 2): (a) macrostructure of weld joint; (b) HAZ, titanium Grade 2 interface; (c) aluminium alloy, WM interface.

observed when the laser beam offset dropped to 150 μm. It can be supposed that in this case the intermixing of metals to be welded was higher. Tensile strength decreased to 32 MPa. The minimum tensile strength was recorded for the weld joints when the beam offset was only 100 μm.

The obtained results of tensile testing of Al-Ti dissimilar metals weld joints correspond to some published research in this field. Tomashchuk et al. [7] found that the linear tensile force measured during tensile testing of weld produced with beam offset towards Ti6Al4V titanium alloy led to the value of 120 N/mm. The division of this value by the original thickness of welded sheets (2 mm) provides the value of UTS at the level of 60 MPa. This value is approximate because it does not take into account the variation of thickness of the weld joint compared to the thickness of base materials. On the other hand, the authors observed the tensile strength of 120 MPa when the laser beam was offset to the aluminium alloy [7]. This tendency was confirmed also by laser brazing of Ti6Al4V titanium alloy to A6061 aluminium alloy [44] or laser weld brazing of titanium Grade 2 to AW 5083 aluminium alloy using 5087 aluminium filler wire [45]. However, Casalino et al. [46] obtained the UTS of the dissimilar metals butt joints of AA5754 and T40 alloys as high as 191 MPa using the laser power of 1.5 kW, the welding speed of 1.80 $m \cdot s^{-1}$ (corresponding to linear heat input of 50 J/mm), and the beam offset of 0.75 mm. In accordance with our findings, they proved significant influence of the linear energy input and also the laser beam offset on the mechanical properties of dissimilar metals weld joints. In case of the possibility to produce another shape of the weld joint, its UTS can be improved using the V-groove as reported in [47] for welding-brazing of Ti-6Al-4V titanium alloy to 5A06 aluminium alloy or in [48] for joining of AA5754 aluminum alloy to T40 titanium with the use of Al-Si filler wires in both studies.

For this research, the butt joint of Al-Ti alloys together with the welding speed of 10 $mm \cdot s^{-1}$ was required. According to our investigation and published research, the mechanical properties of dissimilar Al-Ti weld joints can be improved by the enhancement of laser beam offset to the titanium side or by the use of laser weld brazing process applying Al based filler wire and setting the beam offset to the aluminium alloy side.

6. Conclusions

The possibility of producing butt joints of sheets from dissimilar metals (AW 5754 aluminium alloy and commercially pure titanium Grade 2) with different thicknesses using the disk laser welding was investigated.

Applying numerical simulation of the laser welding process in the program code ANSYS, the influence of the welding parameters on the temperature fields developed during the laser welding of dissimilar Al-Ti alloys was studied. For the fixed welding speed of 10 $mm \cdot s^{-1}$, the laser power was changed from 750 W to 950 W. In order to eliminate intermixing of both welded materials and formation of brittle intermetallic compounds, the setting of laser beam offset towards the titanium side from 50 μm to 250 μm was tested. Based on the obtained results of numerical simulations, the following welding parameters were recommended for production of butt weld joints of Al-Ti sheets:

(i) The welding speed of 10 $mm \cdot s^{-1}$ (required)

(ii) The laser beam offset of 250 μm (or more)

(iii) The laser power from 800 W to 850 W

The quality of weld joints prepared using the TruDisk 4002 disk laser was evaluated by light and scanning electron microscopy, EDX analysis, X-ray diffraction, microhardness measurements, and the tensile tests.

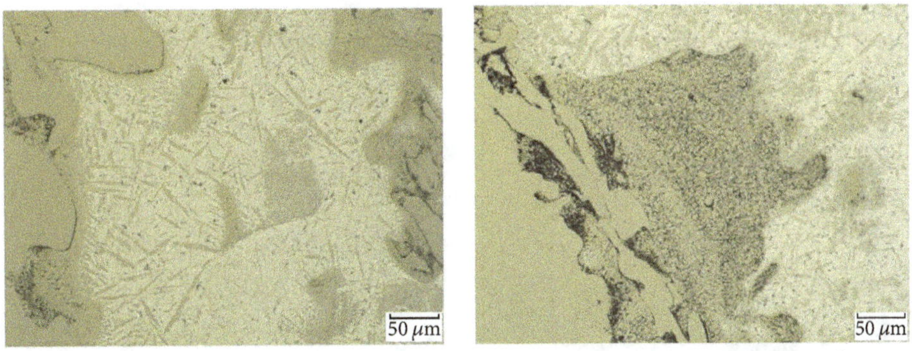

(a) (b)

FIGURE 19: Microstructure of the weld metal for (a) sample 3 and (b) sample 4.

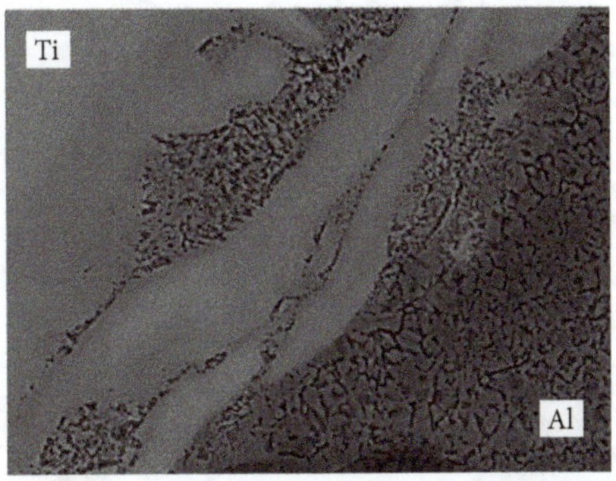

FIGURE 20: Microstructure of aluminium AW 5754 alloy and Ti Grade 2 weld metal.

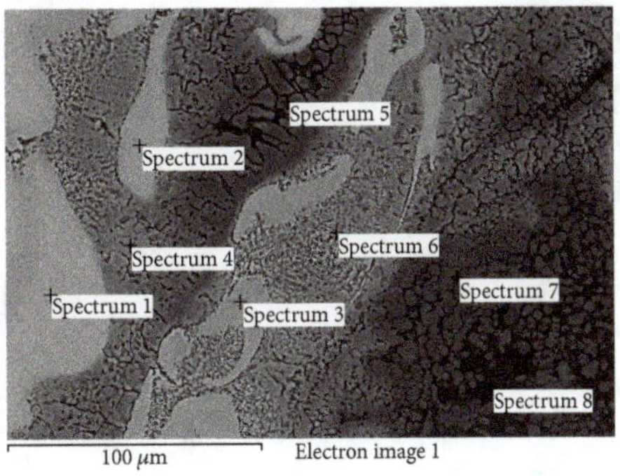

FIGURE 21: Location of point EDX analysis carried out in different zones of weld metal.

The following can be stated based on the attained results:

(i) The produced weld joints consist of different areas, namely, AW 5754 aluminium alloy, aluminium based weld metal, interface layer, titanium based weld metal, titanium HAZ, and titanium Grade 2 base metal.

(ii) Intermixing of titanium Grade 2 and aluminium alloy AW 5754 occurred at the interface of welded sheets, resulting in the development of hard and brittle intermetallic compounds of Al_3Ti, Ti_3Al, and $Al_{18}Mg_3Ti_2$ types (confirmed by XRD).

(iii) The formation of hard and brittle IMC reduced with the increase of laser beam offset to the titanium side.

(iv) The microhardness increased in the weld metal in comparison to the base metals (due to intermixing of both metals).

(v) The highest tensile strength of 80 MPa was recorded for the weld joint prepared using the laser power of 800 W and laser beam offset of 250 μm to the titanium Grade 2 sheet.

(vi) The tensile strength increased with decreasing the linear heat input and with enhanced laser beam offset to the titanium side as a result of elimination of the intermixing of welded metals and thus the development of hard and brittle intermetallic

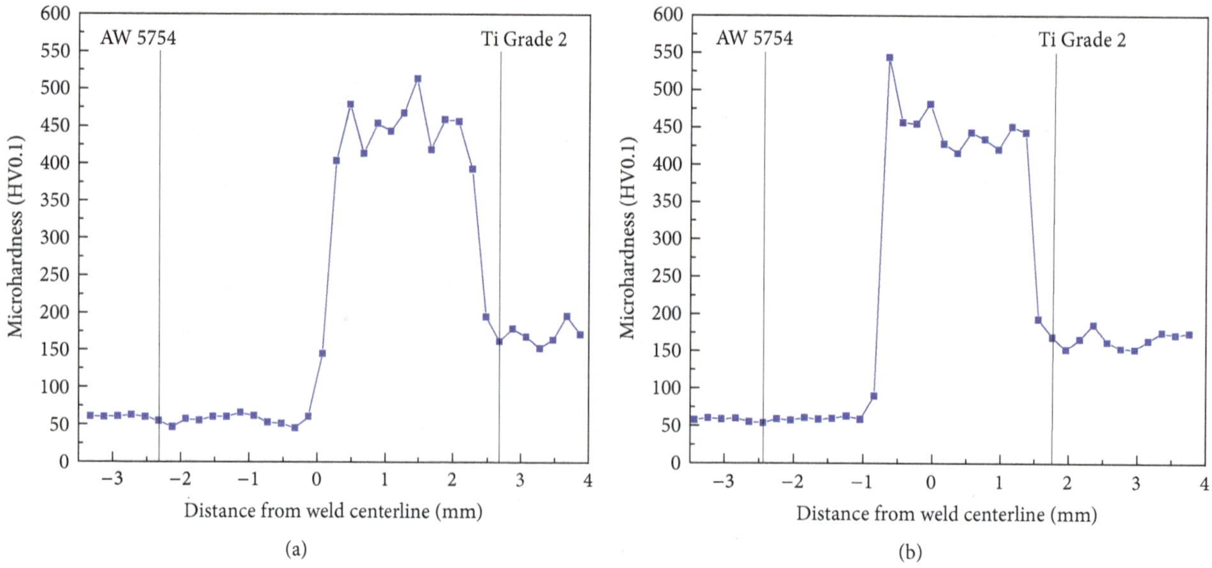

Figure 22: Results of microhardness measurements: the course of microhardness across AW 5754 aluminium alloy, weld metal, and titanium Grade 2 for (a) sample 3 and (b) sample 4.

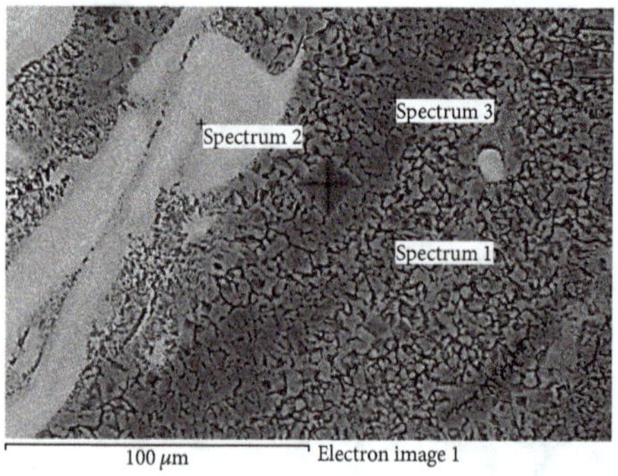

Figure 23: Indentation present in Al-Ti weld metal (sample 4).

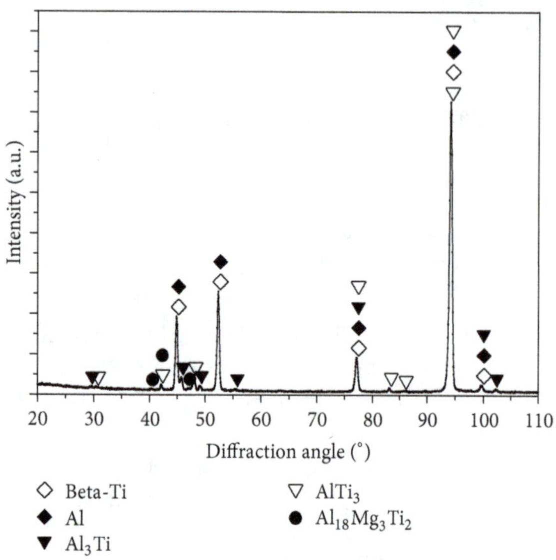

Figure 24: X-ray diffraction pattern of the weld metal.

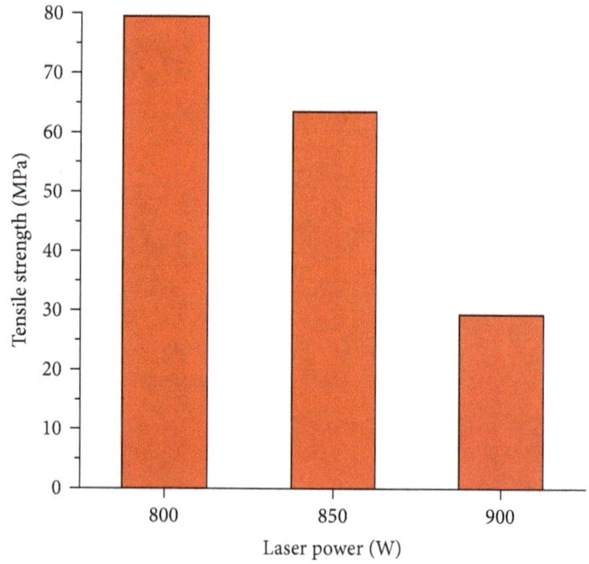

Figure 25: Dependence of tensile strength on laser power (laser beam offset of 250 μm).

compounds mostly responsible for the fracture of dissimilar metals weld joints.

(vii) Dissimilar Al-Ti butt weld joints with higher tensile strength prepared according to required condition (welding speed of 10 mm$\cdot$s^{-1}) could be in the next research produced with the laser offset larger than 250 μm to the titanium sheet or by the laser weld brazing process using an aluminium based filler wire and setting the beam offset to the aluminium alloy side.

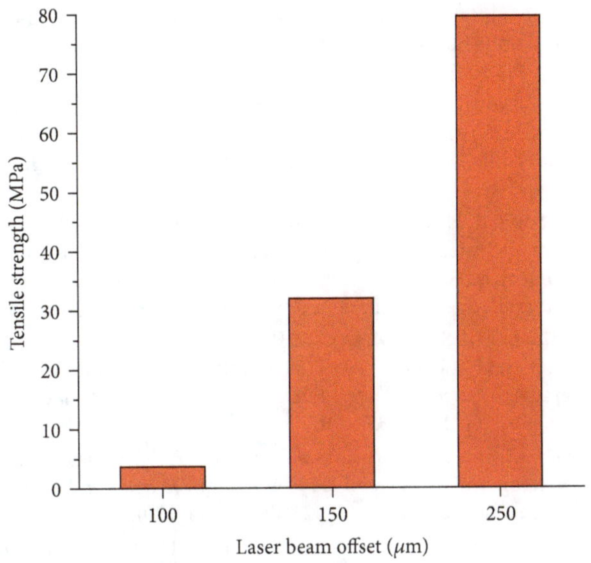

FIGURE 26: The dependence of tensile strength of joints on the laser offset (laser power of 800 W).

Conflicts of Interest

The authors declare that they have no conflicts of interest.

Acknowledgments

The research has been supported by the Scientific Grant Agency of the Slovak Republic (VEGA) within the Project no. 1/1010/16, the program to support excellent young research teams of the STU Bratislava (1347/PRO-MAT-ZVAR), and the Project ITMS 26220120048 under the Research & Development Operational Programme funded by the ERDF.

References

[1] K. Martinsen, S. J. Hu, and B. E. Carlson, "Joining of dissimilar materials," *CIRP Annals—Manufacturing Technology*, vol. 64, no. 2, pp. 679–699, 2015.

[2] G. Pardal, S. Meco, A. Dunn et al., "Laser spot welding of laser textured steel to aluminium," *Journal of Materials Processing Technology*, vol. 241, pp. 24–35, 2017.

[3] R. Jiangwei, L. Yajiang, and F. Tao, "Microstructure characteristics in the interface zone of Ti/Al diffusion bonding," *Materials Letters*, vol. 56, no. 5, pp. 647–652, 2002.

[4] K.-M. Hong and Y. C. Shin, "Analysis of microstructure and mechanical properties change in laser welding of Ti6Al4V with a multiphysics prediction model," *Journal of Materials Processing Technology*, vol. 237, pp. 420–429, 2016.

[5] Z. Sun and J. C. Ion, "Laser welding of dissimilar metal combinations," *Journal of Materials Science*, vol. 30, no. 17, pp. 4205–4214, 1995.

[6] G. Casalino, P. Guglielmi, V. D. Lorusso, M. Mortello, P. Peyre, and D. Sorgente, "Laser offset welding of AZ31B magnesium alloy to 316 stainless steel," *Journal of Materials Processing Technology*, vol. 242, pp. 49–59, 2017.

[7] I. Tomashchuk, P. Sallamand, E. Cicala, P. Peyre, and D. Grevey, "Direct keyhole laser welding of aluminum alloy AA5754 to titanium alloy Ti6Al4V," *Journal of Materials Processing Technology*, vol. 217, pp. 96–104, 2015.

[8] Z. Ma, W. Zhao, J. Yan, and D. Li, "Interfacial reaction of intermetallic compounds of ultrasonic-assisted brazed joints between dissimilar alloys of Ti-6Al-4V and Al-4Cu-1Mg," *Ultrasonics Sonochemistry*, vol. 18, no. 5, pp. 1062–1067, 2011.

[9] X. Chen, R. Xie, Z. Lai, L. Liu, G. Zou, and J. Yan, "Ultrasonic-assisted brazing of Al-Ti dissimilar alloy by a filler metal with a large semi-solid temperature range," *Materials and Design*, vol. 95, pp. 296–305, 2016.

[10] S. A. Tsirkas, P. Papanikos, and T. Kermanidis, "Numerical simulation of the laser welding process in butt-joint specimens," *Journal of Materials Processing Technology*, vol. 134, no. 1, pp. 59–69, 2003.

[11] F. Boitout, D. Dry, and Y. Gooroochurn, *Transient Simulation of Welding Processes*, ESI Group, 2004.

[12] A. P. Mackwood and R. C. Craferb, "Thermal modelling of laser welding and related processes: a literature review," *Optics & Laser Technology*, vol. 37, pp. 99–115, 2005.

[13] L.-E. Lindgren, "Numerical modelling of welding," *Computer Methods in Applied Mechanics and Engineering*, vol. 195, no. 48-49, pp. 6710–6736, 2006.

[14] A. G. Olabi and G. Casalino, "Mathematical modeling of weld phenomena, part 1: finite-element modeling," *Comprehensive Materials Processing*, vol. 6, pp. 101–109, 2014.

[15] G. Casalino and M. Mortello, "Modeling and experimental analysis of fiber laser offset welding of Al-Ti butt joints," *International Journal of Advanced Manufacturing Technology*, vol. 83, no. 1-4, pp. 89–98, 2016.

[16] G. Casalino, M. Mortello, and P. Peyre, "FEM Analysis of Fiber Laser Welding of Titanium and Aluminum," in *Proceedings of the 48th CIRP International Conference on Manufacturing Systems (CIRP CMS '15)*, pp. 992–997, Isle of Ischia, Italy, June 2015.

[17] E. Babalová and M. Behúlová, "Numerical simulation of temperature fields by welding of Ti-Al alloys applying volumetric heat source," *Advanced Materials Research*, vol. 887-888, pp. 1280–1283, 2014.

[18] Y. Ai, P. Jiang, X. Shao et al., "The prediction of the whole weld in fiber laser keyhole welding based on numerical simulation," *Applied Thermal Engineering*, vol. 113, pp. 980–993, 2017.

[19] D. Radaj, *Heat Effects of Welding*, Springer, Berlin, Germany, 1992.

[20] Z. Saternus, "Computer methods for determination of deformations in welded closed profiles," *Procedia Engineering*, vol. 177, pp. 188–195, 2017.

[21] A. Franco, L. Romoli, and A. Musacchio, "Modelling for predicting seam geometry in laser beam welding of stainless steel," *International Journal of Thermal Sciences*, vol. 79, pp. 194–205, 2014.

[22] J. Rahman Chukkan, M. Vasudevan, S. Muthukumaran, R. Ravi Kumar, and N. Chandrasekhar, "Simulation of laser butt welding of AISI 316L stainless steel sheet using various heat sources and experimental validation," *Journal of Materials Processing Technology*, vol. 219, pp. 48–59, 2015.

[23] R. Spina, L. Tricarico, G. Basile, and T. Sibillano, "Thermo-mechanical modeling of laser welding of AA5083 sheets," *Journal of Materials Processing Technology*, vol. 191, no. 1-3, pp. 215–219, 2007.

[24] N. S. Shanmugam, G. Buvanashekaran, K. Sankaranarayanasamy, and S. Ramesh Kumar, "A transient finite element

simulation of the temperature and bead profiles of T-joint laser welds," *Materials and Design*, vol. 31, no. 9, pp. 4528–4542, 2010.

[25] L. J. Zhang, J. X. Zhang, A. Gumenyuk, M. Rethmeier, and S. J. Na, "Numerical simulation of full penetration laser welding of thick steel plate with high power high brightness laser," *Journal of Materials Processing Technology*, vol. 214, no. 8, pp. 1710–1720, 2014.

[26] J. Goldak, A. Chakravarti, and M. Bibby, "A new finite element model for welding heat sources," *Metallurgical Transactions B*, vol. 15, no. 2, pp. 299–305, 1984.

[27] C. S. Wu, G. Wang, and Y. M. Zhang, "A new heat source model for keyhole plasma arc welding in FEM analysis of the temperature profile," *Welding Journal*, vol. 85, pp. 284–291, 2006.

[28] N. S. Shanmugam, G. Buvanashekaran, K. Sankaranarayanasamy, and K. Manonmani, "Some studies on temperature profiles in AISI 304 stainless steel sheet during laser beam welding using FE simulation," *International Journal of Advanced Manufacturing Technology*, vol. 43, no. 1-2, pp. 78–94, 2009.

[29] N. Siva Shanmugam, G. Buvanashekaran, and K. Sankaranarayanasamy, "Some studies on weld bead geometries for laser spot welding process using finite element analysis," *Materials and Design*, vol. 34, pp. 412–426, 2012.

[30] S.-J. Na and W.-I. Cho, "Developments in modelling and simulation of laser and hybrid laser welding," in *Handbook of Laser Welding Technologies*, S. Katayama, Ed., Electronic and Optical Materials, pp. 522–560, Woodhead Publishing, Cambridge, UK, 2013.

[31] M. Zain-Ul-Abdein, D. Nélias, J. F. Jullien, and D. Deloison, "Thermo-mechanical analysis of laser beam welding of thin plate with complex boundary conditions," *International Journal of Material Forming*, vol. 1, supplement 1, pp. 1063–1066, 2008.

[32] M. Dal and R. Fabbro, "An overview of the state of art in laser welding simulation," *Optics and Laser Technology*, vol. 78, pp. 2–14, 2016.

[33] M. Zain-ul-Abdein, D. Nelias, J.-F. Jullien, and D. Deloison, "Prediction of laser beam welding-induced distortions and residual stresses by numerical simulation for aeronautic application," *Journal of Materials Processing Technology*, vol. 209, no. 6, pp. 2907–2917, 2009.

[34] L.-E. Lindgren, "Finite element modeling and simulation of welding part 1: increased complexity," *Journal of Thermal Stresses*, vol. 24, no. 2, pp. 141–192, 2001.

[35] B. Souloumiac, F. Boitout, and J.-M. Bergheau, "A new local global approach for the modelling of welded steel component distortions," *Mathematical Modelling of Weld Phenomena*, vol. 6, pp. 573–590, 2002.

[36] D. Deng, "FEM prediction of welding residual stress and distortion in carbon steel considering phase transformation effects," *Materials and Design*, vol. 30, no. 2, pp. 359–366, 2009.

[37] P. F. Incropera and D. P. De Witt, *Fundamentals of Heat and Mass Transfer*, Wiley and Sons, New York, NY, USA, 1996.

[38] ANSYS Mechanical ADPL Theory Reference. ANSYS, Inc. Release 15.0. SAS IP, Inc., 2015.

[39] JMatPro Help, release 6.1, Sente Software Ltd., 2012.

[40] H. Hu and S. A. Argyropoulos, "Mathematical modelling of solidification and melting: a review," *Modelling and Simulation in Materials Science and Engineering*, vol. 4, no. 4, pp. 371–396, 1996.

[41] E. Friedman, "Thermomechanical analysis of the welding process using the finite element method," *Journal of Pressure Vessel Technology*, vol. 97, no. 3, pp. 206–213, 1975.

[42] 2017, http://www.mtf.stuba.sk/english/development-and-eu-structural-funds/centers-of-excellence/center-of-excellence-5-axis-machining.html?page_id=10353.

[43] E. Illeková, P. Švec, and D. Janičkovič, "Influence of the processing on the ordering process in the Al-Ti binary system with composition close to Al_3Ti," *Journal of Physics: Conference Series*, vol. 144, no. 1, 2009.

[44] Z. Song, K. Nakata, A. Wu, and J. Liao, "Interfacial microstructure and mechanical property of Ti6Al4V/A6061 dissimilar joint by direct laser brazing without filler metal and groove," *Materials Science and Engineering A*, vol. 560, pp. 111–120, 2013.

[45] M. Sahul, M. Sahul, M. Vyskoč, L'. Čaplovič, and M. Pašák, "Disk laser weld brazing of AW5083 aluminum alloy with titanium grade 2," *Journal of Materials Engineering and Performance*, vol. 26, no. 3, pp. 1346–1357, 2017.

[46] G. Casalino, M. Mortello, and P. Peyre, "Yb-YAG laser offset welding of AA5754 and T40 butt joint," *Journal of Materials Processing Technology*, vol. 223, pp. 139–149, 2015.

[47] S. Chen, L. Li, Y. Chen, J. Dai, and J. Huang, "Improving interfacial reaction nonhomogeneity during laser welding-brazing aluminum to titanium," *Materials and Design*, vol. 32, no. 8-9, pp. 4408–4416, 2011.

[48] I. Tomashchuk, P. Sallamand, A. Méasson, E. Cicala, M. Duband, and P. Peyre, "Aluminum to titanium laser welding-brazing in V-shaped groove," *Journal of Materials Processing Technology*, vol. 245, pp. 24–36, 2017.

17

Atmospheric Corrosion Behavior of 2A12 Aluminum Alloy in a Tropical Marine Environment

Zhongyu Cui,[1,2] Xiaogang Li,[1] Huan Zhang,[1] Kui Xiao,[1] Chaofang Dong,[1] Zhiyong Liu,[1] and Liwei Wang[3]

[1]*Corrosion and Protection Center, University of Science and Technology Beijing, Beijing 100083, China*
[2]*Institute of Materials Science and Engineering, Ocean University of China, Qingdao 266100, China*
[3]*Pipeline Research Institute of China National Petroleum Corporation, Langfang 065000, China*

Correspondence should be addressed to Xiaogang Li; lixiaogang@ustb.edu.cn

Academic Editor: Somchai Thongtem

Atmospheric corrosion behavior of 2A12 aluminum alloy exposed to a tropical marine environment for 4 years was investigated. Weight loss of 2A12 alloy in the log-log coordinates can be well fitted with two linear segments, attributing to the evolution of the corrosion products. EIS results indicate that the corrosion product layer formed on the specimens exposed for 12 months or longer presents a good barrier effect. Corrosion morphology changes from pitting corrosion to severe intergranular corrosion with the extension of exposure time, resulting in the reduction of the mechanical properties.

1. Introduction

Due to their high strength and low density, 2xxx series aluminum alloys have been extensively used as structural materials in the aircraft and aerospace industries [1, 2]. However, they are prone to different forms of corrosion, such as pitting corrosion, intergranular corrosion, exfoliation corrosion, and even stress corrosion cracking when exposed in outdoor atmosphere especially in marine atmosphere.

Atmospheric corrosion studies have been conducted on aluminum alloys in laboratory controlled environments [3–5]. The main focus was on the effects of environmental factors, such as relative humidity (RH), chloride contents, CO_2, and SO_2. Sun et al. [4] investigated the exfoliation corrosion of extruded 2024-T4 by laboratory-accelerated test and found that exfoliation could occur when the RH was higher than 75%, and the higher the temperature or the Cl^- amount was, the more serious exfoliation the specimen suffered. Blücher et al. [5] investigated the effect of CO_2 on the NaCl-induced atmospheric corrosion of AA 1070 aluminum and demonstrated that corrosion is about 10–20 times faster in CO_2-free humid air compared with air containing ambient levels of CO_2. However, the laboratory simulation tests are not sufficient for predicting actual service performance of aluminum alloys because of the differences in environment, and field exposure tests are needed.

Some field exposure tests have been conducted on aluminum alloys in various atmospheric environments and pitting corrosion [6–9], intergranular corrosion [7], exfoliation corrosion [4, 9], and stress corrosion cracking [7] have been detected. It has been illustrated that atmospheric corrosion is more serious for high strength aluminum alloys because of the addition of some alloy elements which can form second phases and accelerate the localized corrosion. For instance, Sun et al. [9] investigated atmospheric corrosion behavior of 2024 and 7075 aluminum alloys in urban, coastal, and industrial environments for 20 years. The results showed that exfoliation corrosion occurred on extruded 2024 and 7075 aluminum alloys in coastal and industrial atmospheres, and mechanical properties decreased greatly with the occurrence of exfoliation. Wang et al. [7] investigated the atmospheric corrosion of 2024-T3 alloy exposed to salt lake environment and found that weight loss of the alloy monotonically increased as exposure time prolonged in the first 2 years and then tended to be stable over time. Generally, the stabilization of weight loss during long-term exposure was attributed

Table 1: Chemical composition of 2A12 aluminum alloy used in this work.

Alloy	Chemical composition (wt.%)							
	Si	Fe	Cu	Mn	Mg	Zn	Ti	Al
2A12	0.13	0.35	4.68	0.53	1.58	0.10	0.023	Bal.

to the formation of a layer of oxide film on the surface which could control the diffusion of the reaction species and inhibit the corrosion process [10]. The relationship between the protectiveness of corrosion products and the weight loss has been investigated by using electrochemical techniques [8, 11]. Liu et al. [8] investigated the influence of native oxide and corrosion products on atmospheric corrosion of pure Al through electrochemical impedance spectroscopy (EIS). They revealed that the structure and thickness of the native oxide and corrosion products were the main factors that influenced the atmospheric corrosion of pure Al. Vilche et al. [11] indicated that a bilayer structure formed on the surface of pure Al exposed to aggressive sites for 2 years by using EIS.

In this work, corrosion behavior of 2A12 aluminum alloy exposed in a tropical marine atmosphere for 4 years was investigated. The corrosion rate was determined by the weight loss method. The composition, structure, and protectiveness of the corrosion products were characterized with morphology analysis and EIS. The corrosion kinetics, the protectiveness of the corrosion products, and the corrosion mechanisms of this alloy in tropical marine atmosphere were discussed.

2. Experimental Details

2.1. Materials Preparation. A sheet of 2A12 aluminum alloy was used in this work. Chemical composition of the 2A12 alloy was listed in Table 1. Samples were wet-ground through successive grades of silicon carbide abrasive papers from P120 to P2000 followed by diamond finishing to 0.1 mm. Then the specimen was subjected to a scanning electron microscope (SEM, Quanta 250) to observe the distribution and compositions of the second phases. For the metallographic characterization, the polished surface was further etched using Keller's reagent with the following composition: 1 mL hydrofluoric acid (HF), 1.5 mL hydrochloric acid (HCl), 2.5 mL nitric acid (HNO_3), and 95 mL water (H_2O). After ultrasonic cleaning in ethanol for 1 min and drying in air, the samples were taken immediately to a KEYENCE VHX2000 stereology microscope for microstructure observation. The microstructure of the 2A12 alloy was further investigated by transmission electron microscopy (TEM; Tecnai G^2 F20), with the work function of 300 kV. Thin foils for TEM were cut to about 500 μm by an electron discharge cutter and later ground and polished on both sides to a thickness of about 70 μm. Then the thin foils were ion-milled at room temperature in a Gatan PIPS with a small incident angle until perforation.

Figure 1 showed the optical micrograph (a) of the 2A12 aluminum alloy and the distribution (b) and composition ((c) and (d)) of the second phases. The grain size of this alloy varied between 20 and 200 μm (Figure 1(a)) with many intermetallic (IM) particles locating in the substrate (white particles in Figure 1(b)). EDS results indicated that these particles were mainly composed of Al_2CuMg (Figure 1(c)) and AlCuFeMn(Si) (Figure 1(d)) which was consistent with the results of Chen et al. [12].

Figure 2 showed the TEM images of the IM particles and the grain boundary zones. Figure 2(a) was a typical rod-like particle which distributed in the grain. According to the selected area electron diffraction result and the previous studies [13, 14], this phase was the Mn-rich T ($Al_{20}Cu_2Mn_3$) dispersoid which usually formed in the ingot homogenization process and could refine the grain through the way of pinning grain boundary migration during recrystallization annealing process [14]. As shown in Figure 2(b), this phase was also found along the grain boundaries. Figure 2(c) showed an IM particle at the grain boundary, which was identified as S phase (Al_2CuMg) based on the selected area diffraction pattern. This kind of phase played an important role in the pitting corrosion and intergranular corrosion of aluminum alloys [15]. In addition, it is worthwhile to note that precipitate-free zone (PFZ) was not found in this alloy. Instead, there are some clean grain boundaries without precipitation particles as shown in Figure 2(d).

2.2. Field Exposure Test. Samples with the dimension of 100 mm × 50 mm × 2.5 mm were used for field exposure test. Prior to the test, all the specimens were ground down to 800 grit paper and then degreased by acetone followed by cleaning in ethanol. All the specimens were weighed (original weight) and the surface area (S) was measured before test. Then all the specimens were installed on a test rack with an inclination angle of 45° horizontal to the sky and facing south in Xisha islands (112°20′E, 16°50′N), 100 meters from the coastline. The test duration was 4 years. Four replicate metal samples were retrieved from the exposure site after 1, 3, 6, 9, 12, 24, and 48 months. Three replicas were employed to determine the weight loss, and the remaining one was used to analyze the corrosion morphology and the corrosion products.

Table 2 lists the annual average environmental parameters and atmospheric pollutants measured at Xisha islands during the exposure, which is consistent with the description in our previous work [16]. The annual average temperature in Xisha marine atmosphere was 27°C with the maximum temperature increasing up to 33.3°C and the minimum temperature reducing to 20.1°C which was also a relatively high temperature. The extreme relative humidity (RH) values oscillated between 61% and 94% with a mean value of 77%. Time of wetness, which was defined as the period during which a metallic surface was covered by adsorptive and/or liquid film of electrolyte, was another important factor that affects the corrosion behavior of metals in atmosphere. Schindelholz et al. [17] summarized that the TOW determination methods represented in the literature and practice could be categorized by the way in which TOW was treated either as an environmental parameter or as a surface parameter. However, both of the two methods including ISO 9223 and other indirect electrode sensors had their limitations [18, 19]. In this work, the TOW was treated as an environmental parameter

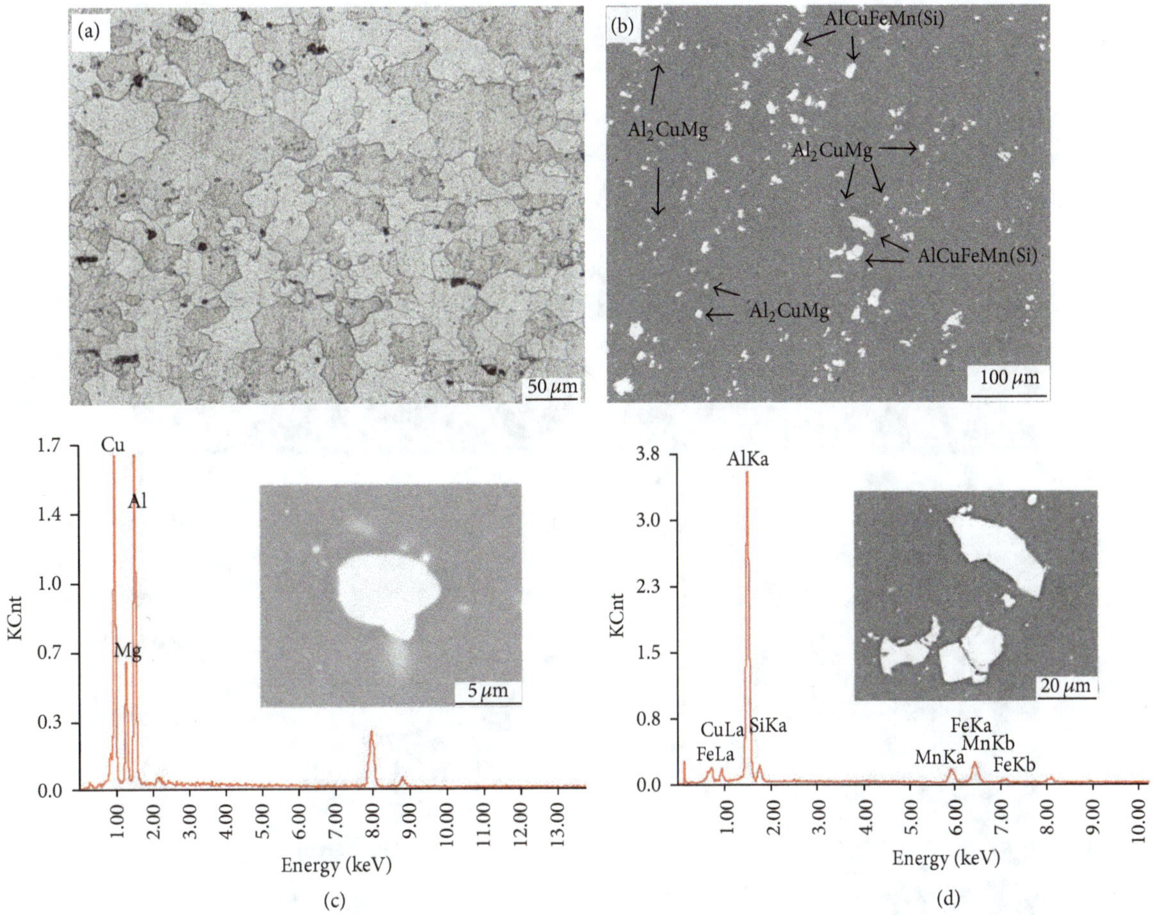

FIGURE 1: Microstructure of 2A12 aluminum alloy used in this work: (a) optical micrograph, (b) SEM images of the distribution of second phases, (c) EDS of AlCuMg particles, and (d) EDS of AlCuFeMnSi particles.

TABLE 2: Climatic parameters and atmospheric pollutants of Xisha islands during 4 years of exposure.

Exposure site	Xisha	TOW	2562 h/year (τ_4)
Location	112°20′E, 16°50′N	Rainfall	1526 mm/year
Max. temperature	33.3°C	Cl^- deposition rate	112.68 mg/m^2 d (S_2)
Min. temperature	20.1°C	SO_2 deposition rate	<0.1 mg/m^2 d (P_0)
Average temperature	27°C	Sunshine	2675 h/year
Max. RH	94%	pH of rain	6.5
Min. RH	61%	Distance to sea	100 m
Average RH	77%		

and was calculated according to ISO 9223, which defines TOW as the length of time when the RH is greater than 80% at a temperature above 0°C. With this method, the approximate TOW in Xisha islands was 2562 h/year which belonged to the classification of τ_4 [20]. As for the corrosive species precipitated on the samples, the Cl^- and SO_2 deposition rates were considered and measured using the method given by ISO 9225 [21]. The Cl^- deposition rate was determined by the wet candle method and the SO_2 deposition rate was determined using alkaline surfaces of porous filter plates saturated by a solution of sodium carbonate. According to the results, the annual average deposition rate of chloride was 112.68 mg/m^2 d, which was classified under S_2 [20]. Meanwhile, the precipitation of SO_2 was less than 0.1 mg/m^2·d and the rain had a near-neutral pH of 6.5 because of the absence of heavy industry at the exposure site.

Figure 3 showed the monthly average temperature, RH, TOW, and Cl^- deposition rate during the exposure test. The temperature presented a seasonal characteristic with a higher value in summer and lower value in winter (Figure 3(a)). As for the RH and TOW, the highest value all appeared at October, while the gap between other months was not so great (Figures 3(b) and 3(c)). In addition, variation of the Cl^- deposition rate exhibited no trend here. It fluctuated with

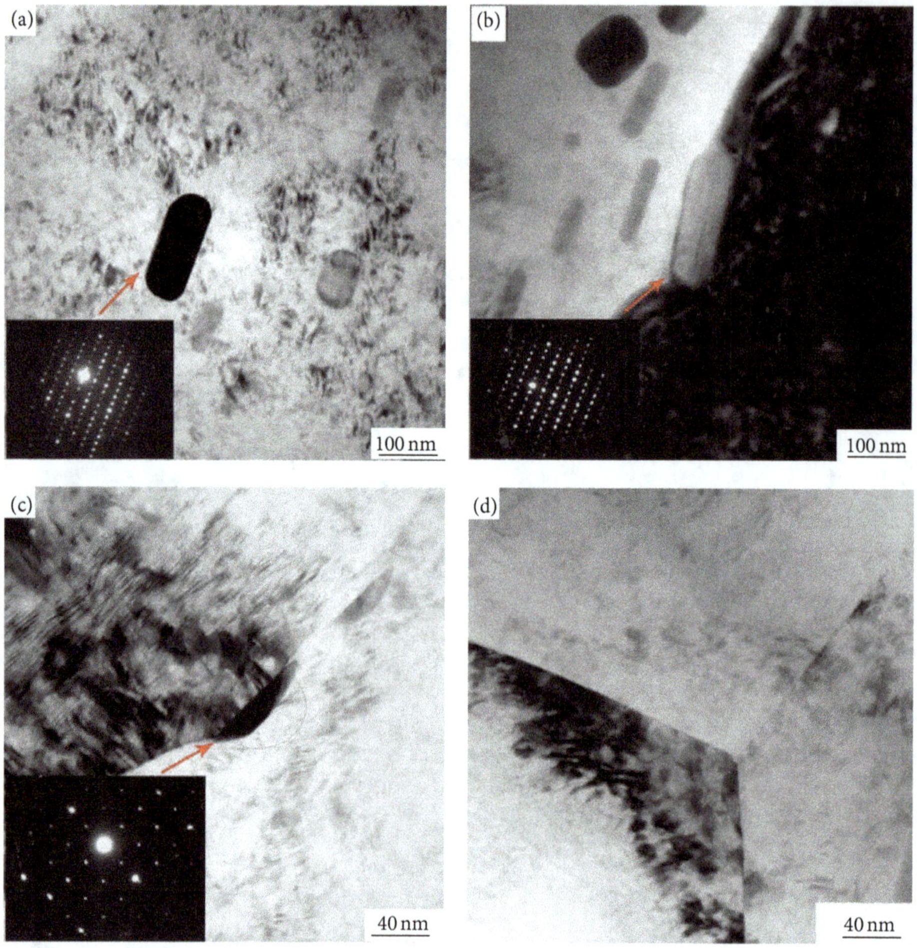

FIGURE 2: TEM images of the intermetallic precipitates in the matrix (a), along the grain boundaries (b, c) and the clean grain boundaries (d).

the exposure time with a higher value in the second half of one year (Figure 3(d)).

2.3. Weight Loss Measurement. Corrosion products of the withdrawn triplicate specimens were chemically removed by pickling in the solution (50 mL H_3PO_4 + 20 g CrO_3 + 1 L H_2O) for 10 min at 80–100°C. After that, the specimens were rinsed with distilled water, dried in warm air, and then weighed to obtain their final weights (w_1). Meanwhile, the metal loss caused by picking was determined by using uncorroded sample. Then, the weight loss was calculated as follows:

$$W = \frac{w_0 - w_1}{S} - W_{un}, \quad (1)$$

where W was the weight loss of the metal due to corrosion (g/m^2), w_0 was the original weight (g), w_1 was the final weight (g), W_{un} was the weight loss of a fresh specimen due to the picking, and S was the surface area (m^2). The average corrosion rate of aluminum after exposure for different periods was calculated by using the equation in [22]:

$$V_n = \frac{12\left(w_n - w_{n-1}\right)}{t_n - t_{n-1}}, \quad (2)$$

where V_n was the corrosion rate (g/m^2·year), w_n was the weight loss (g/m^2), t was the exposure time (month), and n was the period of exposure (n = 1, 2, 3, 4, 5, 6, and 7 referred to the sample exposed for 1, 3, 6, 9, 12, 24, and 48 months, resp.).

2.4. Mechanical Properties Measurement. After removing the corrosion products, the corroded tensile specimens were machined according to the ASTM E8M specification with 25 mm gauge length and 6 mm gauge width, and the thickness remained unchanged. Then, the tensile tests were carried out according to ASTM E8M with a constant deformation rate of 1 mm/min. A data logger was used to store the data in a digital file. To get representative average values of the tensile properties, three tensile tests have been carried out per each test series.

2.5. Characterization of Corrosion Morphology and Corrosion Products. After the exposure test, the macroscopic appearances of the specimens with corrosion products were taken by a D200 digital camera. Then, the microscopic corrosion morphologies with and without corrosion products were observed by scanning electron microscope (SEM,

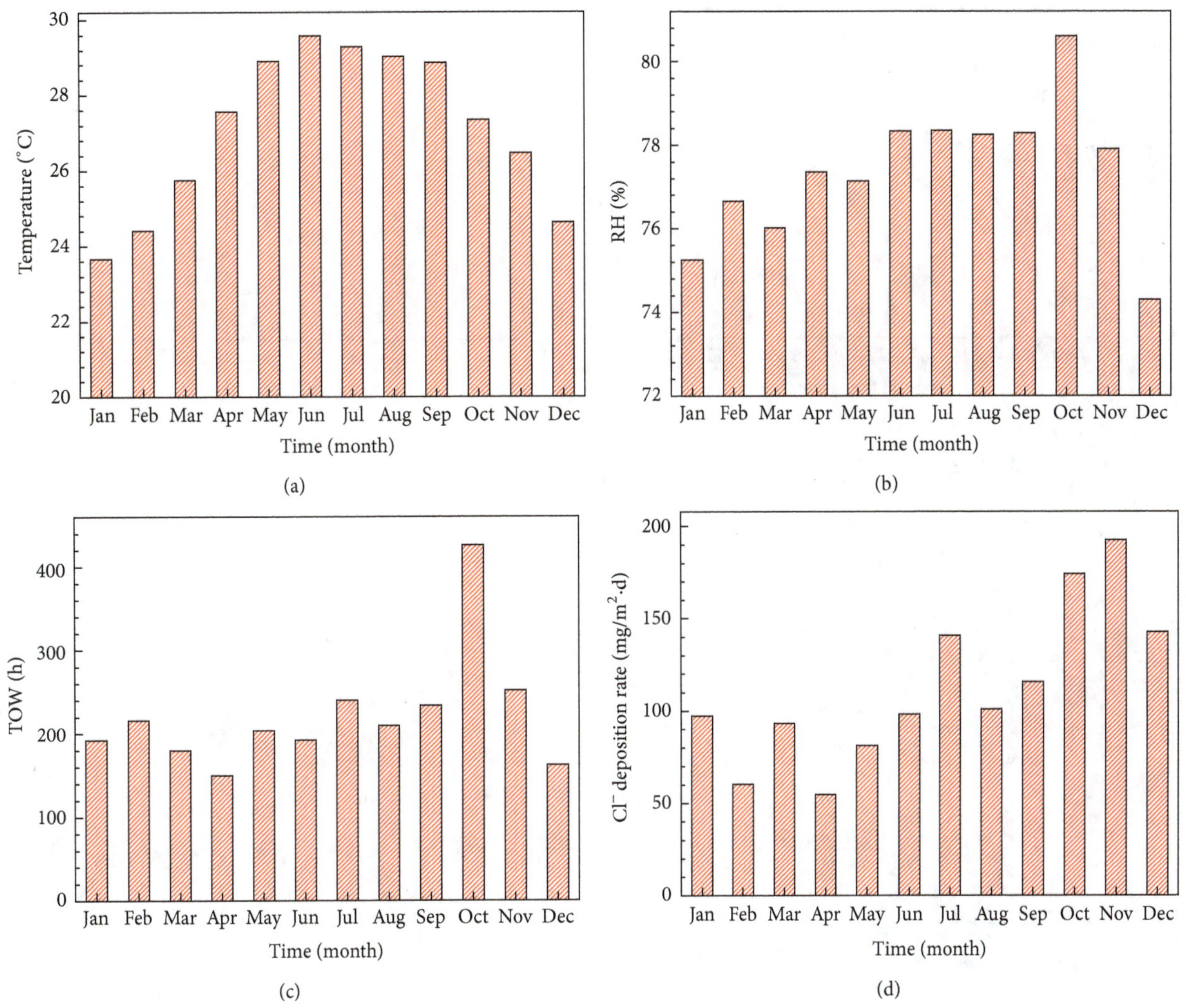

FIGURE 3: Monthly average temperature (a), RH (b), TOW (c), and Cl^- deposition rate (d) during the exposure test.

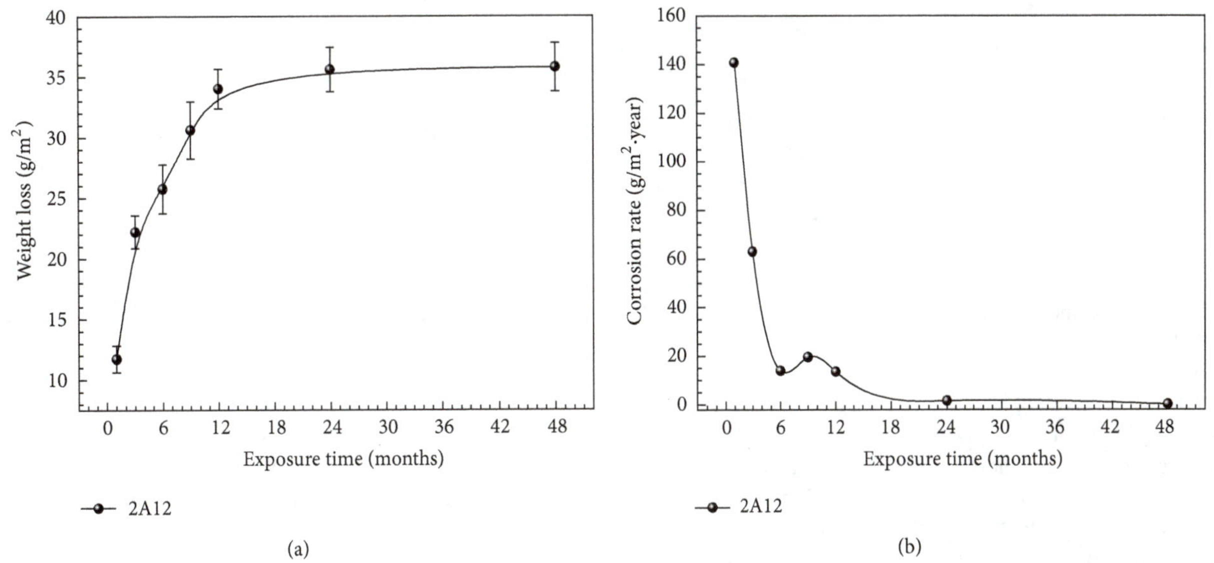

FIGURE 4: Weight loss (a) and corrosion rate (b) of 2A12 aluminum alloy during exposure in Xisha marine atmosphere for 4 years.

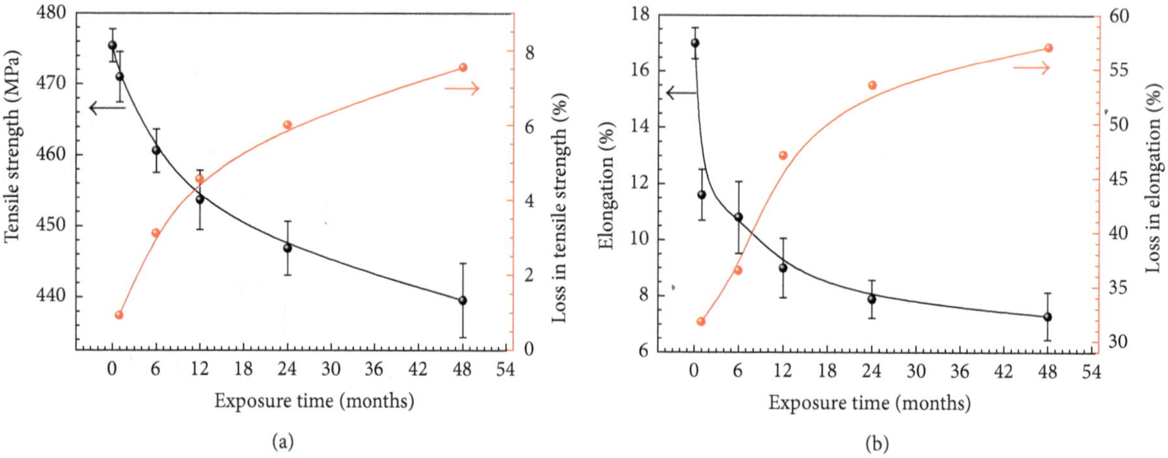

FIGURE 5: Changes of the tensile strength (a) and elongation (b) of 2A12 alloy exposed for different time.

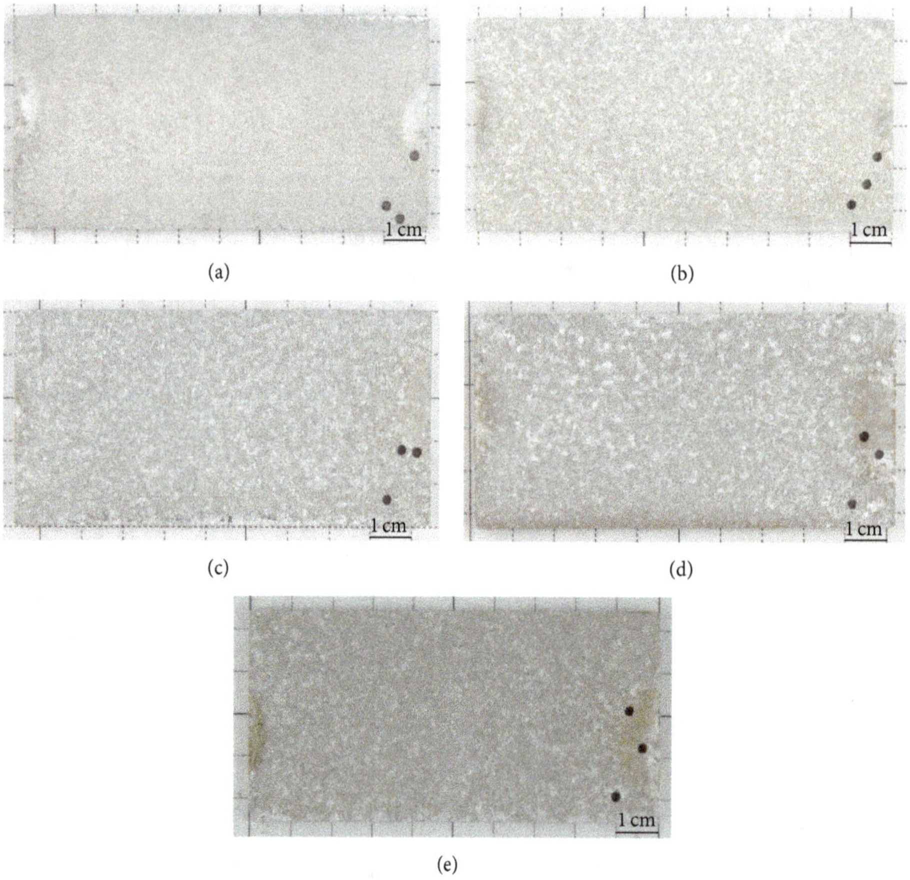

FIGURE 6: Surface appearances of the 2A12 aluminum alloy exposed for 1 month (a), 6 months (b), 12 months (c), 24 months (d), and 48 months (e).

Quanta 250). Element compositions of the corrosion products were analyzed by energy dispersive X-ray spectroscopy (EDS). Elemental distributions in the corrosion product layer were examined by electron probe microanalyzer (EPMA, JXA8230) with accelerating voltage of 15 kV and sample current of 0.02 μA. Before the EPMA experiment, the specimen with corrosion products was embedded in the epoxy resin. Then the specimen was mechanically ground down to 2000 grit SiC paper and polished with 1.5 μm diamond paste. The X-ray diffraction (XRD, Rigaku Dmax-rc) investigation was carried out in situ on the exposed specimens and conducted in a step scanning mode from 10° to 90° with Cu Kα radiation. After that, the data was analyzed with Jade 5.0 software.

FIGURE 7: Surface morphology of the corrosion products formed on 2A12 aluminum alloy after exposure for 1 month (a), 12 months (b), and 48 months (c).

2.6. Electrochemical Impedance Spectroscopy (EIS) Measurement. EIS measurements were performed with VMP3 electrochemical workstation in 0.1 M NaCl solution in a conventional three-electrode cell, using a platinum sheet as counter electrode and a saturated calomel electrode (SCE) as reference electrode. The test system was always in a steady state with no stirring. The whole metal sample was manually sectioned into coupons of dimensions 15 mm × 15 mm × 2.5 mm which served as working electrode. The working electrode surface was covered with silicone rubber leaving an exposed area of 1.0 cm^2. The impedance spectra were recorded in the frequency range between 100 kHz and 10 mHz with 10 mV amplitude peak to peak perturbing signal. The experimental data were analyzed by the commercial software ZSimpWin. All the measurements were performed at ambient temperature (25 ± 2°C) and repeated at least three times to maintain the reproducibility.

3. Results

3.1. Weight Loss and Corrosion Rate. Figure 4 shows the weight loss (a) and the corrosion rate (b) of 2A12 aluminum alloy during the 4-year exposure in Xisha marine atmosphere. An increasing trend in weight loss consisting of two different segments is observed in Figure 4(a). The weight loss increases rapidly during the initial 12 months of exposure and then the increasing rate decreases to a much lower value as the exposure time prolonged. Corrosion rate of the alloy decreases remarkably at the initial six months of exposure and then presents a slight augment from 6 months to 9 months. With further increasing of the weathering test, corrosion rate decreases gradually and stabilizes to a certain value. Corrosivity of aluminum alloys in atmospheric environment has been classified into five grades in some literatures [3, 9, 23] according to ISO 9223. The corrosion rate of 2A12 alloy after exposure in Xisha islands for 1 year is 34.49 g/m^2·year, so it belongs to the classification of C_5 (very high) based on ISO 9223 [20].

3.2. Loss in Mechanical Properties. Changes of the tensile strength and elongation for 2A12 alloy versus exposure time in Xisha marine atmosphere are shown in Figure 5. The tensile strength decreases gradually as the exposure test extends (Figure 5(a)). The biggest loss is about 40 MPa and 7.6% in percentage. As for the elongation, it reduces from 17% for the uncorroded specimen to 7.3% for the specimen exposed for 48 months. That is, the toughness loss reaches 57% in the end of the exposure test. The reduction in mechanical properties has been reported by many literatures [24, 25]. This reduction can be attributed either to hydrogen embrittlement or to the stress concentration of the corrosion-induced intergranular cracks [24].

3.3. Macroscopic Appearances. Figure 6 shows the macroscopic appearances of the 2A12 specimens exposed for different time. The surfaces are covered with corrosion products so that the original appearances of the samples cannot be seen. It is seen that the specimens become grey with increasing of the exposure time and the corrosion product layers are relatively uniform.

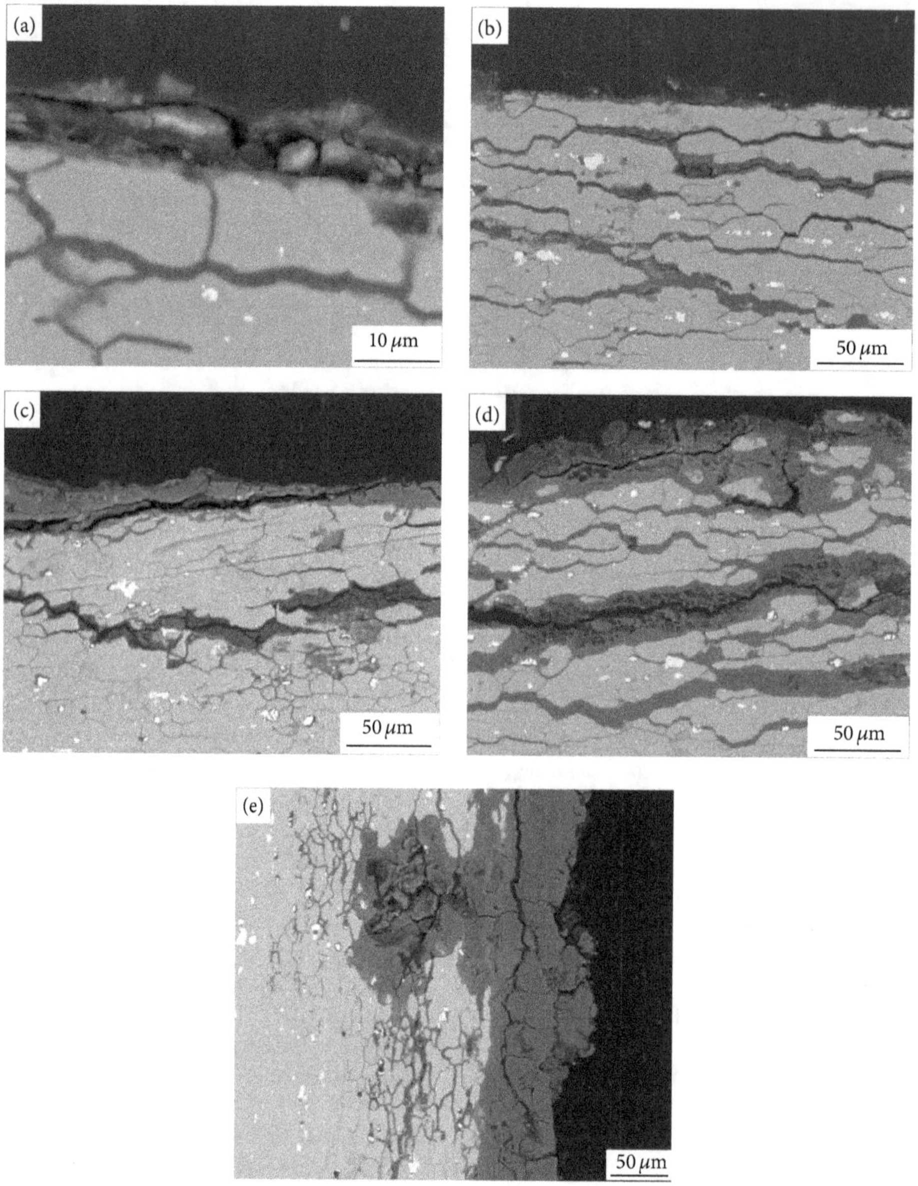

FIGURE 8: Cross-sectional morphology of 2A12 aluminum alloy after exposure for 1 month (a), 6 months (b), 12 months (c), 24 months (d), and 48 months (e).

3.4. Corrosion Product Analysis

3.4.1. Surface Morphology. Figure 7 shows the surface morphologies of the corrosion products formed on 2A12 aluminum alloy after exposure for different time. The images demonstrate that many microcracks exist in the corrosion products which are mainly attributed to the dehydration process during the wet-dry cycles [22]. The chemical compositions of the points a, b, c, and d marked in Figure 7 are listed in Table 3. The primary elements included in the corrosion products are Al and O, with a trace amount of Mg, Na, Cl, and Ca, which mainly come from the salt particles that deposited on the surface of the metal.

TABLE 3: Element contents of corrosion products marked in Figure 4 determined by EDS.

Point	Chemical composition (wt.%)					
	O	Al	Mg	Na	Cl	Ca
a	67.21	27.84	1.21	0.98	1.16	1.60
b	22.73	75.35	1.92	—	—	—
c	65.49	34.51	—	—	—	—
d	64.23	34.71	1.07	—	—	—

3.4.2. Cross-Sectional Morphology. Cross-sectional morphologies of the corrosion product layer on 2A12 aluminum alloy after exposure for different time are shown in Figure 8. It can be seen that intergranular corrosion occurs on the specimen after only 1 month of exposure (Figure 8(a)). As the weathering time prolonged, the intergranular corrosion

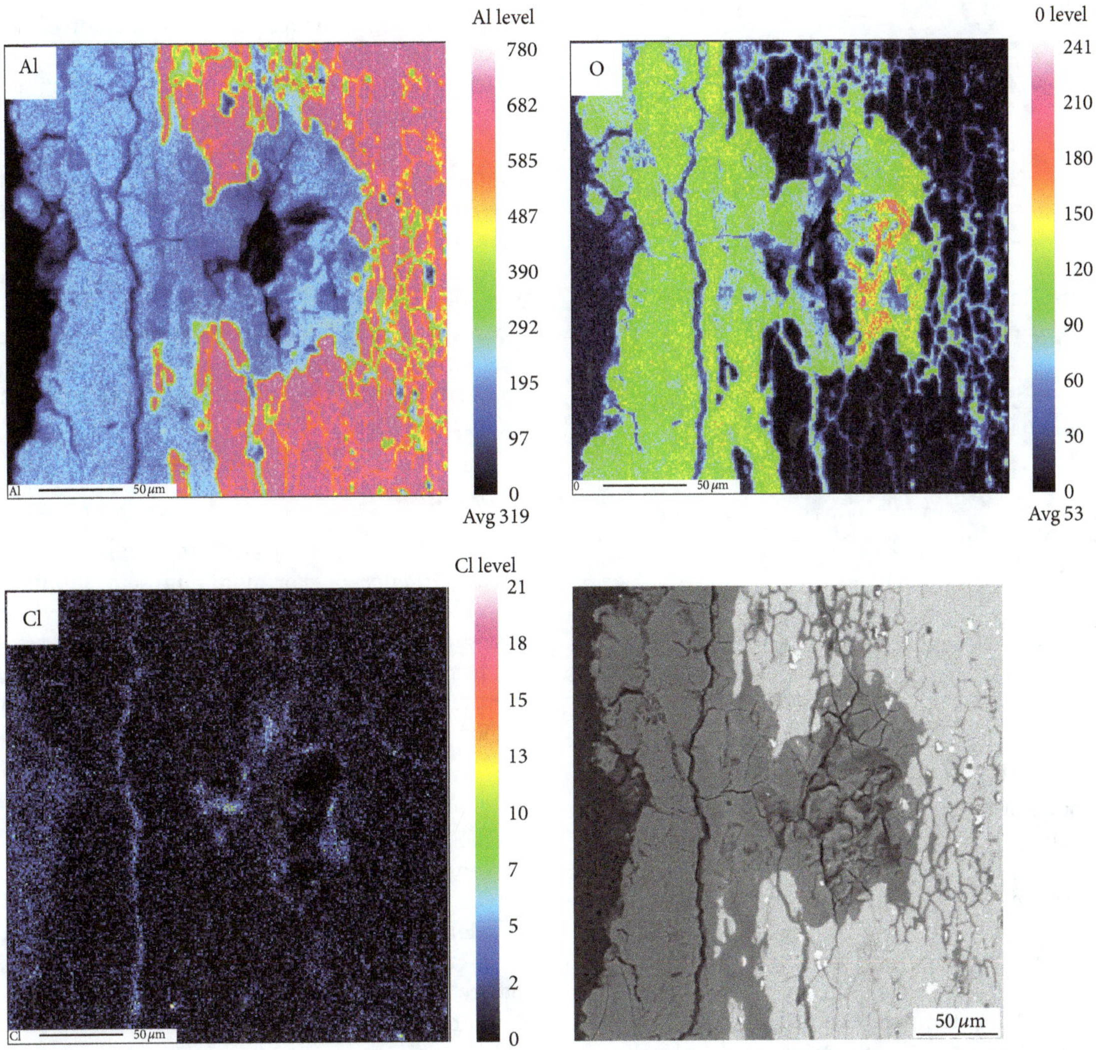

FIGURE 9: EPMA mapping of the main elements distributing on the cross section of the specimen after exposure for 48 months.

crack width increases. After exposure for 48 months, the specimen surface is covered by a compact and thick corrosion product layer under which there exist some isolated pits. To determine the distribution of the elements on the corrosion products of 2A12 aluminum alloy, the EPMA map scanning analysis is performed and shown in Figure 9. It can be seen that corrosion products are mainly composed of Al and O and the chloride is mainly present along the cracks. It can be attributed that the surface $AlCl_3$ has dissolved away during the water polishing, while some $AlCl_3$ remains inside the cracks.

3.4.3. XRD Analysis. Figure 10 shows the XRD patterns of the corrosion products on the surface of 2A12 alloy after exposure for different time. It is seen that only AlO(OH) is detected in the corrosion products during the initial three months of exposure. As the exposure test increases up to 6 months, two emerging diffraction peaks around $2\theta = 18^\circ$ and $2\theta = 20^\circ$, which are identified to be $Al(OH)_3$, are detected.

3.5. EIS Measurements. EIS is an efficient method to study the surface condition of metals and also a powerful tool for evaluating the protectiveness of the corrosion product layer formed on metals [8, 22, 26]. Figure 11 shows the time variation in EIS diagrams of the 2A12 aluminum alloy before and after exposure for 1, 6, 12, and 24 months. It is seen that all impedance spectra of the specimens consist of two capacitive loops except for that of the sample exposed for 1 month as shown in Figure 11(b). It shows that a low-frequency diffusion tail appears after immersing in the 0.1 M NaCl solution for 24 h and 48 h. This is because the surface corrosion product layer on the specimen exposed for 1 month will be destroyed after immersing in the NaCl solution for a long time. Meanwhile, the intergranular corrosion crack fronts will be activated due to the permeation of the reactant species like H_2O and O_2. Once the corrosion fronts are reactivated, there is need for abundant reactant species to support the reactions. Therefore, the diffusion tail appears because the reactant species are not sufficient to sustain the corrosion reactions under the corrosion products. For the

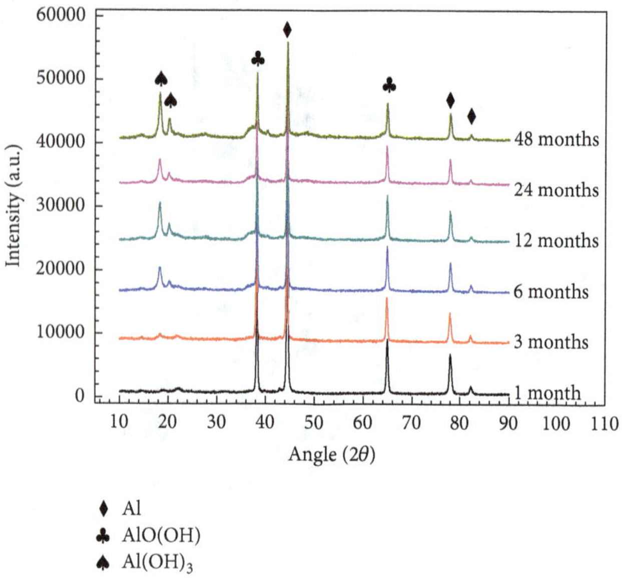

Figure 10: XRD patterns of the corrosion products formed on 2A12 aluminum alloy after different exposure time.

specimens exposed for 6 months or longer, the corrosion reactions under the surface layers are slow due to the thick corrosion product films. Therefore, the diffusion is not the rate controlling step and the diffusion tail does not appear.

The equivalent circuit models in Figure 12 are used to fit the EIS results in Figure 11. Among them, Model A is used to fit the spectra with two capacitive loops and Model B is used for modeling the spectra with a diffusion tail in Figure 11(b). In Model A, R_s is solution resistance; Q_f and R_f are the constant phase element of the corrosion product layer and its resistance for the corroded 2A12 and the constant phase element of the oxide film and its resistance for the freshly prepared 2A12 samples. CPE_{dl} and R_t are the double layer capacitance and the charge transfer resistance, respectively. In Model B, the redundant W is the Warburg impedance which is associated with the diffusion process through the corrosion product layer. The fitting results are also shown as solid lines in Figure 11, which match the experimental data well, indicating that the equivalent circuits in Figure 12 are suitable for fitting the EIS results.

The polarization resistance, R_p, the reciprocal of which is proportional to the corrosion rate [22], can be calculated from the two equivalent circuits as follows: $R_p = R_f + R_t$. The R_p values of 2A12 alloy before and after exposure varying with immersion time are shown in Figure 13. It is seen that the R_p values of all the corroded specimens are higher than the 2A12 substrate after immersion in the NaCl solution for only 0.5 h. This is the reason why the corrosion rate decreases in the initial exposure periods. As the immersion test extends, the R_p values of the specimen exposed for 6 months fluctuate around that of the freshly prepared sample. Meanwhile, the R_p of the samples exposed for only 1 month exhibits lower values than the specimen without exposure, while for the specimens exposed for 12 months or longer, the R_p present significantly higher values as compared with the 2A12 substrate. This may explain why the weight loss increments become ignorable in the later stage of the exposure test (Figure 4(a)).

3.6. Corrosion Morphology without Corrosion Products. The surface corrosion morphologies of 2A12 alloy after removing corrosion products are shown in Figure 14. It is observed that some pits form after exposure for 1 month (Figure 14(a)). With the increase of exposure time (Figure 14(b)), deeper and wider corrosion spots are observed and the corrosion zone almost covers the whole surface. More and more grains are corroded away, leaving a rugged surface as shown in Figure 14(c).

4. Discussion

4.1. Corrosion Kinetics of 2A12 Aluminum Alloy in Tropical Marine Atmosphere. For aluminum and aluminum alloys, a number of studies have reported the development of corrosion loss with time, both for laboratory investigation [5] and field exposure test [6, 7, 9]. Most of these studies have interpreted the data in terms of the well-known and widely applied power function [27]:

$$W = At^n, \tag{3}$$

where A and W are the corrosion losses after one and t years of exposure, respectively, and n is a constant. This equation is a simplification of the mathematical result for oxygen diffusion through increasing thickness of the corrosion films. Townsend and Zoccola [28] have proposed a method to analyze corrosion data using linear regression analysis to fit a straight line on log-log plot by taking logarithm of (3):

$$\log W = \log A + n \log t. \tag{4}$$

In this work, however, weight loss of 2A12 alloy is found to deviate from (3). This phenomenon can be easily observed in Figure 15, which reproduces the result in Figure 4(a) by plotting the weight loss against the exposure time in log-log coordinates. It is seen that the log-log plot of weight loss versus time presents two straight lines taking the 12 months of exposure as the turning point. This behavior has been reported by some researchers [29] who described this irregular corrosion behavior as in the following representative equation:

$$W = At_1^{n_1-n_2}t^{n_2} \quad (t \geq t_1), \tag{5}$$

where t_1 is the length in months of the first period of slope n_1 and n_2 is the slope in the second period. The fitting results of the two linear segments are shown in Figure 15. It is found that the slope of the second segment (n_2) is smaller than that of the initial segment (n_1) and both of them are less than 0.5. The constant n is a significant parameter that can be considered as an indicator for the physicochemical behavior of the corrosion product layer [30]. Value of n lower than 0.5 suggests a parabolic growth of the layer by formation

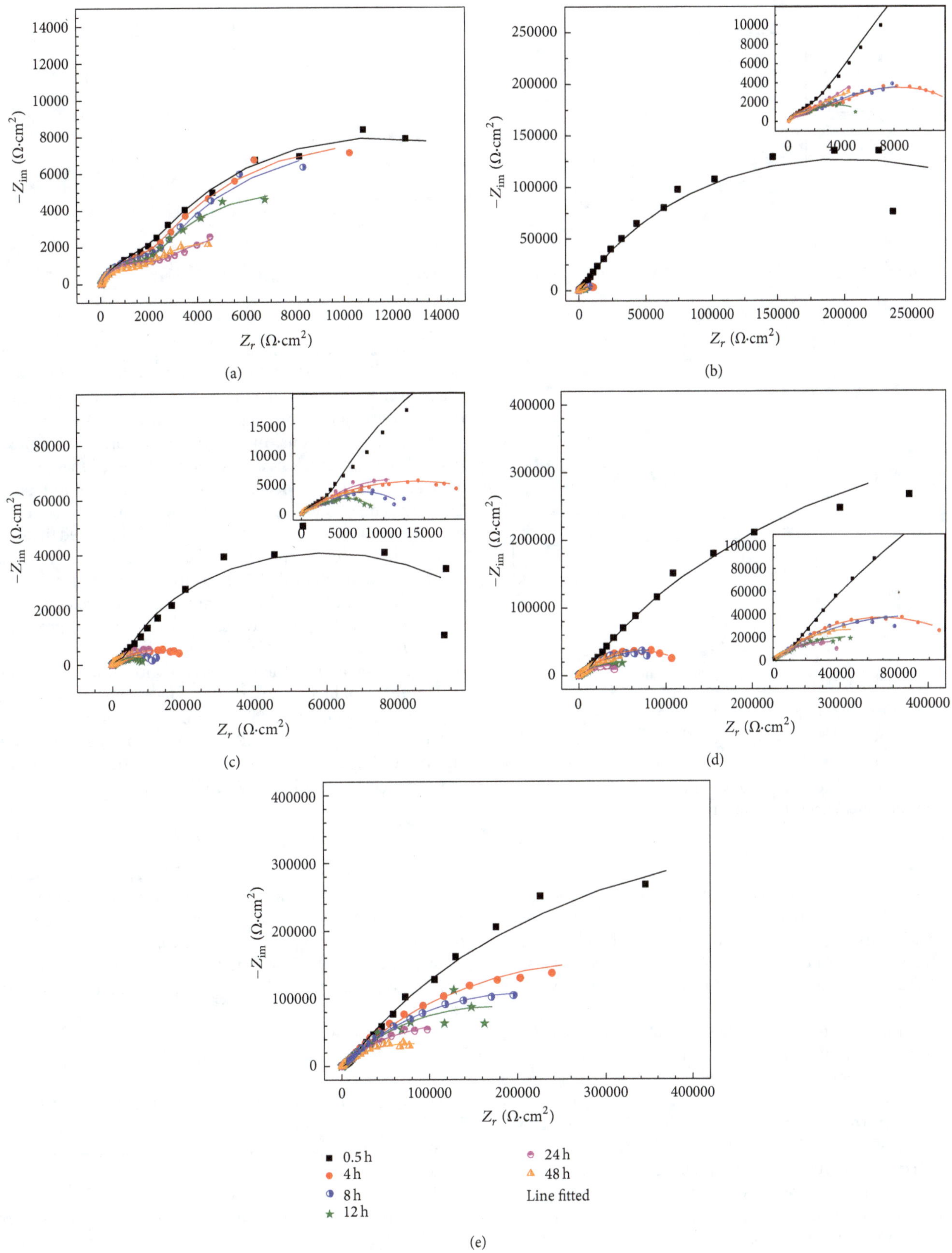

FIGURE 11: Nyquist diagrams for 2A12 aluminum alloy substrate (a) and specimens after exposure for 1 month (b), 6 months (c), 12 months (d), and 24 months (e) during the immersion test.

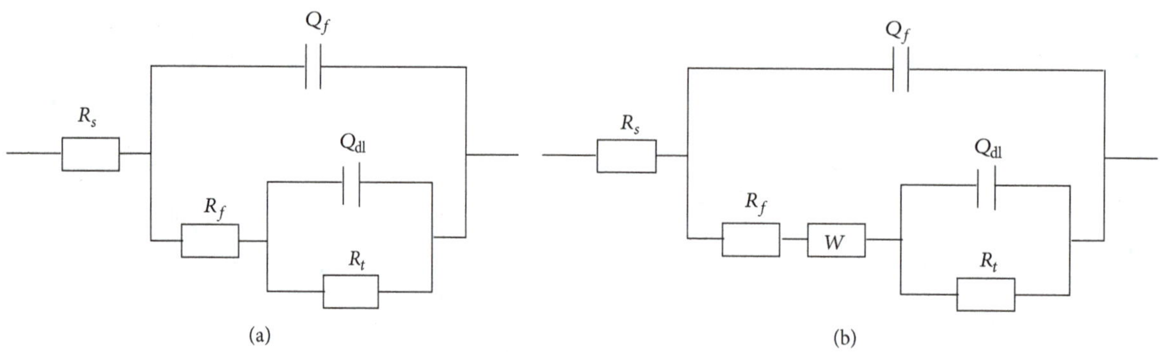

FIGURE 12: Equivalent circuits for 2A12 aluminum alloy with different exposure time in 0.1 M NaCl: (a) Model A and (b) Model B.

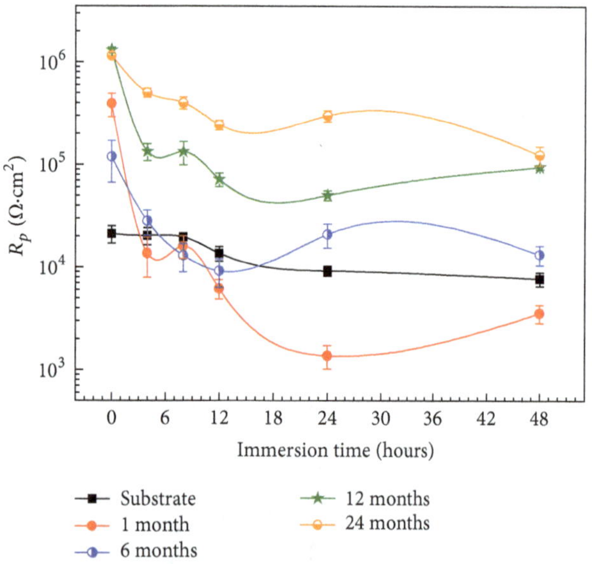

FIGURE 13: Calculated polarization resistance (R_p) of 2A12 aluminum alloy after exposure for different time.

of protective corrosion products that decrease the corrosion rate by acting as a barrier to the transportation of aggressive agents. In this work, both n_1 and n_2 are less than 0.5, which indicates that a protective corrosion product layer forms on 2A12 aluminum during the exposure test. The relatively lower value of n_2 demonstrates that corrosion products formed on 2A12 aluminum after exposure for 12 months or longer exhibit a better protectiveness, which is certified by the EIS results in the following section.

Corrosion rate of the 2A12 aluminum alloy presents a deceleration trend with some fluctuations during the exposure test. Ma et al. [29] revealed that this special behavior only occurred in marine environment because of the effect of chloride. In our previous work, the same phenomenon was also demonstrated for AZ31 magnesium in Xisha marine atmosphere [22]. It can be speculated that, in this work, the fluctuation of the corrosion rate is mainly attributed to the competition between the deterioration effect caused by chloride ions and the stabilisation process of the corrosion product layer.

4.2. Electrochemical Behavior of Corroded 2A12 Aluminum and Its Implication on the Protectiveness of Corrosion Products. EIS is an effective method to investigate the corrosion mechanism and evaluate the protectiveness of the corrosion product layer formed on metals. In our previous work [22], protectiveness of the corrosion film generated on AZ31 magnesium after atmospheric exposure was investigated by immersing the corrosion product-covered specimens in NaCl solution for different time and measuring the EIS. It illustrated that EIS data of the corroded specimens from long time immersion is more accurate than that only from immersion for a short time. In this work, the EIS measurements are also employed to characterize the protectiveness of the corrosion products based on the calculated R_f and CPE$_f$ values.

The depression of the semicircles with immersion time in Figure 8 might be caused by surface roughness and/or heterogeneities in the metal/electrolyte interface. Therefore, constant phase element (CPE) is used for more accurate analysis of impedance behavior of the electric double layer. The impedance of CPE is expressed as [8]

$$Z_{\mathrm{CPE}} = Y_0^{-1} (j\omega)^{-n}, \quad (6)$$

where Z_{CPE} is the impedance of a CPE, Y_0 is the modulus, j is the imaginary component, ω is the angular frequency, and n is the power index number ($0 < n \leq 1$). CPE is commonly used to replace capacitance, but CPE is different from capacitance [8]. The film capacitance can be calculated from CPE parameters values (Y_0 and n) using the following expression [31]:

$$C_f = Y_0^{1/n} R_f^{(1-n)/n}, \quad (7)$$

where C_f is the film capacitance and R_f is the film resistance. The R_f, Y_0, and n values fitted from the EIS data by the equivalent circuits and the calculated C_f values are listed in Table 4.

Figure 16 shows the variation of C_f and R_f with the immersion time. The film resistance R_f represents the barrier against migration of ions in association with the corrosion reaction and is useful for estimating the protective properties of the corrosion scales [32]. A higher R_f value means higher protectiveness of the corrosion scales. So it can be speculated from Figure 13 that the corrosion scales formed on the surface

FIGURE 14: Corrosion morphologies of specimens without corrosion products after exposure for 1 month (a), 12 months (b), and 48 months (c).

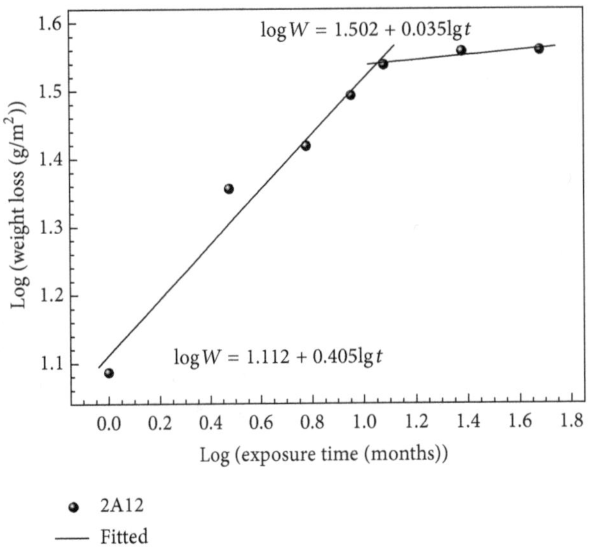

FIGURE 15: Bilogarithmic plots of data points from Figure 3(a) as well as the fitting lines.

of the 2A12 aluminum alloy after exposure in the tropical marine atmosphere for 12 months and 24 months have better protectiveness than the others based on the fact that they have higher R_f values. This explains why the weight loss versus exposure time plot (Figure 12) exhibits a turning point at 12 months.

The C_f value also has been used to evaluate the protective effect of the corrosion products using the parameter [8, 32]

$$C = \varepsilon\varepsilon_0 \frac{A}{d}, \tag{8}$$

where C is the capacitance, ε is the relative permittivity of the oxide, ε_0 is the vacuum permittivity, d is the thickness of the product layer, and A is the effective area which is highly dependent on the structure of the corrosion product layer. Loose and porous structure corresponds to a large A value while dense and packed structure corresponds to a small value [26]. The relative permittivity of the film, ε, is recognized as a constant because the composition change of the corrosion products can be ignored. Figure 17 compares the C_f values of the 2A12 substrate and the exposed specimens after immersion in 0.1 M NaCl solution for 0.5 h. A nonmonotonous trend of C_f value is observed, which indicates that both A and d vary during the exposure test. For the freshly prepared sample, the C_f is about one order greater in magnitude than the corrosion product covered samples, ascribed to the thin film formed during the short-term immersion. After exposure for 1 month, pitting corrosion occurs and corrosion products form on the specimen, which results in the decrease of the effective area A and the increase of the thickness d. Therefore, C_f drops sharply to a low value. With increasing of the exposure time, corrosion product layer covers the whole surface that gives rise to an increase of the effective area A. This is responsible for the large C_f of the specimen exposed for 6 months. As the exposure continues, corrosion products become denser and compact

TABLE 4: Values of the parameters Y_0, R_f, and n as well as the calculated C_f.

Specimens	Immersion time	Y_0 ($F{\cdot}cm^{-2} \cdot s^{n-1}$)	n	R_f ($\Omega{\cdot}cm^2$)	C_f ($F{\cdot}cm^{-2}$)
Substrate	0.5 h	1.43×10^{-4}	0.846	4142	1.30×10^{-4}
	4 h	1.62×10^{-4}	0.853	3425	1.46×10^{-4}
	8 h	1.58×10^{-4}	0.850	3489	1.42×10^{-4}
	12 h	1.79×10^{-4}	0.880	2953	1.64×10^{-4}
	24 h	2.18×10^{-4}	0.825	2956	1.99×10^{-4}
	48 h	2.19×10^{-4}	0.827	2216	1.88×10^{-4}
1 month	0.5 h	3.1×10^{-6}	0.772	3600	8.22×10^{-7}
	4 h	2.9×10^{-5}	0.748	4248	1.43×10^{-5}
	8 h	4.76×10^{-5}	0.816	2239	2.87×10^{-5}
	12 h	1.05×10^{-4}	0.784	1486	6.29×10^{-5}
	24 h	9.76×10^{-5}	0.838	1070	6.31×10^{-5}
	48 h	1.59×10^{-4}	0.826	1867	1.23×10^{-4}
6 months	0.5 h	1.17×10^{-5}	0.689	7369	3.87×10^{-6}
	4 h	2.6×10^{-5}	0.691	3979	9.42×10^{-6}
	8 h	4.19×10^{-5}	0.717	3800	2.03×10^{-5}
	12 h	5.12×10^{-5}	0.742	3388	2.79×10^{-5}
	24 h	5.9×10^{-5}	0.796	2120	3.46×10^{-5}
	48 h	1.07×10^{-4}	0.823	1801	7.51×10^{-5}
12 months	0.5 h	6.9×10^{-6}	0.579	29280	2.15×10^{-6}
	4 h	9.6×10^{-6}	0.625	17890	3.33×10^{-6}
	8 h	1.31×10^{-5}	0.651	11240	4.68×10^{-6}
	12 h	1.33×10^{-5}	0.749	8530	6.40×10^{-6}
	24 h	1.96×10^{-5}	0.781	7677	1.15×10^{-5}
	48 h	2.05×10^{-5}	0.801	9099	1.35×10^{-5}
24 months	0.5 h	3.7×10^{-6}	0.571	13400	3.89×10^{-6}
	4 h	7.9×10^{-6}	0.606	10590	1.58×10^{-6}
	8 h	1.27×10^{-5}	0.601	44720	8.72×10^{-6}
	12 h	1.3×10^{-5}	0.626	65310	1.18×10^{-5}
	24 h	1.82×10^{-5}	0.646	25430	1.19×10^{-5}
	48 h	1.43×10^{-5}	0.704	28600	9.81×10^{-6}

accompanied with the growth of the thickness, resulting in the decrease of A and d and consequently the decrease of C_f.

As the immersion test extends, the corrosion product layer is destroyed by the aggressive chloride ions and thus the C_f value varies (Figure 16(b)). The C_f values present an increase trend with the prolonged immersion time, indicating the reduction of the layer thickness and the enlargement of the surface area. Moreover, the C_f of the specimen after 12- and 24-month exposure always keep lower values throughout the immersion test. These results once again certify the better protectiveness of the product layer generated after exposure for 12 and 24 months.

4.3. Localized Corrosion Behavior of 2A12 Aluminum Alloy in Tropical Marine Atmosphere

4.3.1. Pitting Corrosion of 2A12 Aluminum Alloy. Due to the high content of chloride ions existing in Xisha marine atmosphere, pitting of aluminum alloys is thus highly expected. When exposed to the Xisha marine atmosphere, an oxide layer composed of alumina (Al_2O_3) and boehmite (γ-AlOOH) or bayerite ($Al(OH)_3$) [6] forms on the surface of the 2A12 aluminum alloy. However, there exist many intermetallic (IM) particles, such as S-phase, θ-phase, and AlCuFeMn phase in the alloy (Figures 1 and 2). Because of the presence of the IM particles, some flaws occur in the oxide surrounding the IM particles which are preferentially attacked by chloride ions and leading to the initiation of pitting corrosion. The galvanic coupling effect between the IM particles and the surrounding matrix promotes the growth of the pits.

4.3.2. Intergranular Corrosion of 2A12 Aluminum Alloy. IGC (intergranular corrosion) is detected on the 2A12 aluminum alloy during exposure in Xisha atmosphere. It is recognized that IGC is caused by the formation of a microgalvanic cell between the grain matrix and the grain boundary region, which includes the grain boundary precipitates and the precipitate-free zone. In this work, no obvious PFZ is detected in 2A12 aluminum alloy by TEM. Therefore, the grain boundary precipitates shown in Figure 2 are mainly responsible for the IGC propagation. According to Buchheit

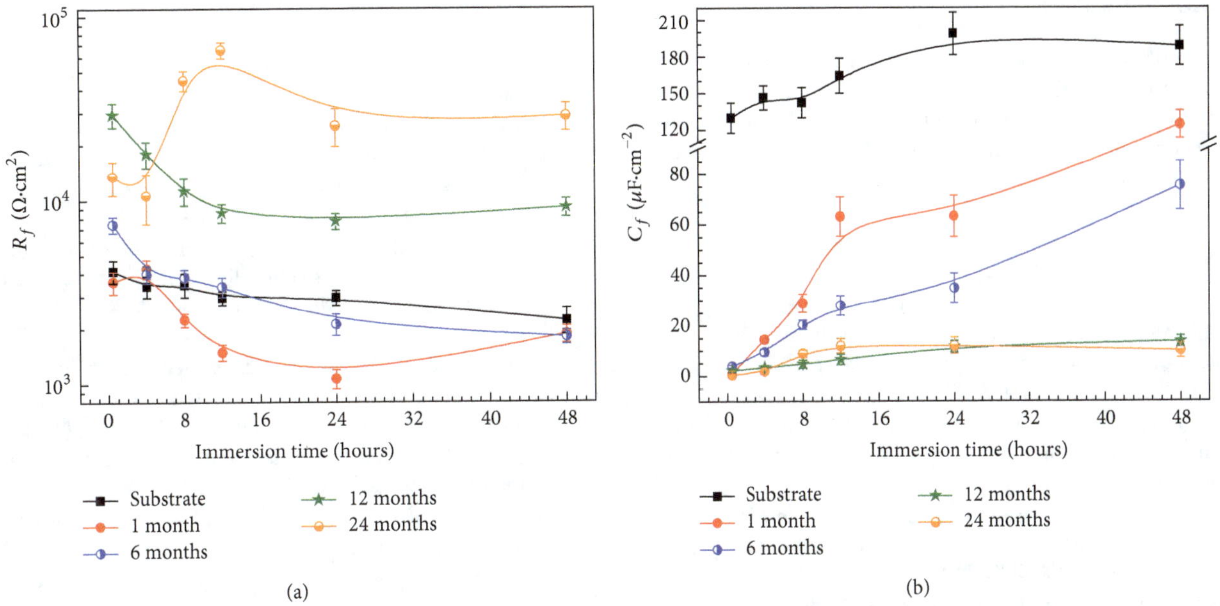

FIGURE 16: Fitting results of film resistance R_f (a) and the calculated capacity C_f (b) of the corrosion product layer during immersion test.

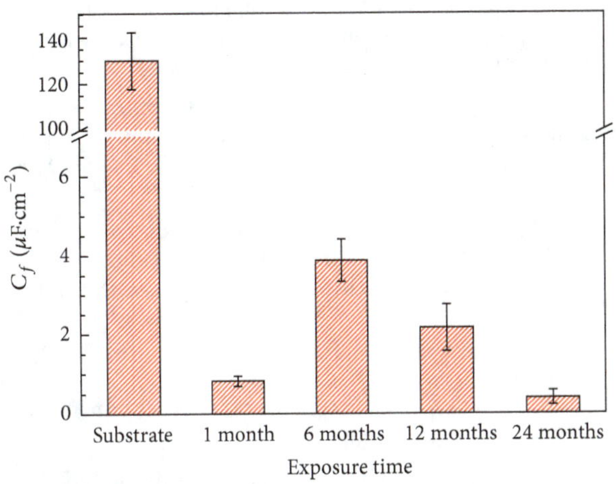

FIGURE 17: The calculated capacity C_f of the corrosion products after immersion in 0.1 M NaCl for 0.5 h.

[33, 34] and other researchers [35, 36], the S-phase particles (Al_2CuMg) are anodic with respect to the adjacent grains and can be expected to undergo preferential dissolution, leading to acidification of the anolyte solution through hydrolysis of aluminum ions in the intergranular crevice and also to copper enrichment on the surface of the grain crevice behind the active anodic head at the point of attack. In comparison, the T-phase particles ($Al_{20}Cu_2Mn_3$) are also anodic with respect to the matrix [36] and might preferentially dissolve. Due to the dissolution of the grain boundary precipitates, an anodic path along the grain boundaries forms, resulting in the propagation of IGC. Besides, IGC is thought to have a more detrimental effect on fatigue life than pits as it can act as a more severe stress raiser [37]. Therefore, the mechanical properties reduce heavily as shown in Figure 5.

5. Conclusion

Corrosion behavior of 2A12 aluminum alloy in a tropical marine atmosphere was investigated in this work. The main conclusions were made as follows:

(1) The weight loss of 2A12 aluminum alloy exposed to the tropical marine environment deviates from the well-known bilogarithmic equation $W = At^n$.

(2) The weight loss versus exposure time plot presents two segments, taking the 12 months as turning point. The slope of the second segment is significantly lower than that of the first segment, which is attributed to the good protectiveness of the corrosion product layer.

(3) Mechanical properties of the 2A12 alloy decrease heavily as the exposure time extends.

(4) The protectiveness of corrosion products formed on 2A12 aluminum after atmospheric exposure varies with the exposure time. Corrosion products formed after exposing for 12 and 24 months present better barrier effect against further attacks.

(5) Pitting corrosion and intergranular corrosion occur on 2A12 aluminum alloy in this tropical marine atmosphere, resulting in the deterioration of the mechanical properties during the exposure.

Conflict of Interests

The authors declare that there is no conflict of interests regarding the publication of this paper.

Acknowledgments

The authors wish to acknowledgement the financial support of the National Basic Research Program of China (973 Program Project no. 2014CB643300) and the National Science and Technology Basic Project of the Ministry of Science and Technology of China (no. 2012FY113000).

References

[1] A. Heinz, A. Haszler, C. Keidel, S. Moldenhauer, R. Benedictus, and W. S. Miller, "Recent development in aluminium alloys for aerospace applications," *Materials Science and Engineering A*, vol. 280, no. 1, pp. 102–107, 2000.

[2] T. Dursun and C. Soutis, "Recent developments in advanced aircraft aluminium alloys," *Materials and Design*, vol. 56, pp. 862–871, 2014.

[3] T. E. Graedel, "Corrosion mechanisms for aluminum exposed to the atmosphere," *Journal of the Electrochemical Society*, vol. 136, no. 4, pp. 204C–212C, 1989.

[4] S. Sun, Q. Zheng, D. Li, S. Hu, and J. Wen, "Exfoliation corrosion of extruded 2024-T4 in the coastal environments in China," *Corrosion Science*, vol. 53, no. 8, pp. 2527–2538, 2011.

[5] D. B. Blücher, R. Lindström, J.-E. Svensson, and L.-G. Johansson, "The effect of CO_2 on the NaCl-induced atmospheric corrosion of aluminum," *Journal of the Electrochemical Society*, vol. 148, no. 4, pp. B127–B131, 2001.

[6] D. de la Fuente, E. Otero-Huerta, and M. Morcillo, "Studies of long-term weathering of aluminium in the atmosphere," *Corrosion Science*, vol. 49, no. 7, pp. 3134–3148, 2007.

[7] B. B. Wang, Z. Y. Wang, W. Han, and W. Ke, "Atmospheric corrosion of aluminium alloy 2024-T3 exposed to salt lake environment in Western China," *Corrosion Science*, vol. 59, pp. 63–70, 2012.

[8] Y. Liu, Z. Wang, and W. Ke, "Study on influence of native oxide and corrosion products on atmospheric corrosion of pure Al," *Corrosion Science*, vol. 80, pp. 169–176, 2014.

[9] S. Sun, Q. Zheng, D. Li, and J. Wen, "Long-term atmospheric corrosion behaviour of aluminium alloys 2024 and 7075 in urban, coastal and industrial environments," *Corrosion Science*, vol. 51, no. 4, pp. 719–727, 2009.

[10] Z. Szklarska-Smialowska, "Pitting corrosion of aluminum," *Corrosion Science*, vol. 41, no. 9, pp. 1743–1767, 1999.

[11] J. R. Vilche, F. E. Varela, G. Acuña et al., "A survey of Argentinean atmospheric corrosion: I—aluminium and zinc samples," *Corrosion Science*, vol. 37, no. 6, pp. 941–961, 1995.

[12] G. S. Chen, M. Gao, and R. P. Wei, "Microconstituent-induced pitting corrosion in aluminum alloy 2024-T3," *Corrosion*, vol. 52, no. 1, pp. 8–15, 1996.

[13] S. C. Wang and M. J. Starink, "Two types of S phase precipitates in Al–Cu–Mg alloys," *Acta Materialia*, vol. 55, no. 3, pp. 933–941, 2007.

[14] Z. Q. Feng, Y. Q. Yang, B. Huang, M. H. Li, Y. X. Chen, and J. G. Ru, "Crystal substructures of the rotation-twinned T ($Al_{20}Cu_2Mn_3$) phase in 2024 aluminum alloy," *Journal of Alloys and Compounds*, vol. 583, pp. 445–451, 2014.

[15] R. Buchheit, R. Grant, P. Hlava, B. McKenzie, and G. Zender, "Local dissolution phenomena associated with S Phase (Al_2CuMg) particles in aluminum alloy 2024–T3," *Journal of The Electrochemical Society*, vol. 144, pp. 2621–2628, 1997.

[16] Z. Y. Cui, X. G. Li, K. Xiao, C. F. Dong, Z. Y. Liu, and L. W. Wang, "Corrosion behavior of field-exposed zinc in a tropical marine atmosphere," *Corrosion*, vol. 70, no. 7, pp. 731–748, 2014.

[17] E. Schindelholz, R. G. Kelly, I. S. Cole, W. D. Ganther, and T. H. Muster, "Comparability and accuracy of time of wetness sensing methods relevant for atmospheric corrosion," *Corrosion Science*, vol. 67, pp. 233–241, 2013.

[18] I. S. Cole, W. D. Ganther, J. D. Sinclair, D. Lau, and D. A. Paterson, "A study of the wetting of metal surfaces in order to understand the processes controlling atmospheric corrosion," *Journal of the Electrochemical Society*, vol. 151, no. 12, pp. B627–B635, 2004.

[19] I. S. Cole and W. D. Ganther, "Experimental determination of duration of wetness on metal surfaces," *Corrosion Engineering Science and Technology*, vol. 43, no. 2, pp. 156–162, 2008.

[20] ISO 9223, "Corrosion of metals and alloys—classification of corrosivity of atmospheres," 1992.

[21] ISO 9225, Corrosion of metals and alloys—corrosivity of atmospheres—measurement of pollution, 1992.

[22] Z. Y. Cui, X. G. Li, K. Xiao, and C. F. Dong, "Atmospheric corrosion of field-exposed AZ31 magnesium in a tropical marine environment," *Corrosion Science*, vol. 76, pp. 243–256, 2013.

[23] Z. Dan, I. Muto, and N. Hara, "Effects of environmental factors on atmospheric corrosion of aluminium and its alloys under constant dew point conditions," *Corrosion Science*, vol. 57, pp. 22–29, 2012.

[24] N. D. Alexopoulos and P. Papanikos, "Experimental and theoretical studies of corrosion-induced mechanical properties degradation of aircraft 2024 aluminum alloy," *Materials Science and Engineering A*, vol. 498, no. 1-2, pp. 248–257, 2008.

[25] N. D. Alexopoulos, "On the corrosion-induced mechanical degradation for different artificial aging conditions of 2024 aluminum alloy," *Materials Science and Engineering A*, vol. 520, no. 1-2, pp. 40–48, 2009.

[26] J. Wang, Z. Y. Wang, and W. Ke, "A study of the evolution of rust on weathering steel submitted to the Qinghai salt lake atmospheric corrosion," *Materials Chemistry and Physics*, vol. 139, no. 1, pp. 225–232, 2013.

[27] M. Benarie and F. L. Lipfert, "A general corrosion function in terms of atmospheric pollutant concentrations and rain pH," *Atmospheric Environment*, vol. 20, no. 10, pp. 1947–1958, 1986.

[28] H. E. Townsend and J. C. Zoccola, "Eight-year atmospheric corrosion performance of weathering steel in industrial, rural and marine environments," in *Atmospheric Corrosion*, ASTM STP 767, p. 45, ASTM, Philadelphia, Pa, USA, 1982.

[29] Y. Ma, Y. Li, and F. Wang, "Corrosion of low carbon steel in atmospheric environments of different chloride content," *Corrosion Science*, vol. 51, no. 5, pp. 997–1006, 2009.

[30] D. de la Fuente, I. Díaz, J. Simancas, B. Chico, and M. Morcillo, "Long-term atmospheric corrosion of mild steel," *Corrosion Science*, vol. 53, no. 2, pp. 604–617, 2011.

[31] B. Hirschorn, M. E. Orazem, B. Tribollet, V. Vivier, I. Frateur, and M. Musiani, "Determination of effective capacitance and film thickness from constant-phase-element parameters," *Electrochimica Acta*, vol. 55, no. 21, pp. 6218–6227, 2010.

[32] Y. Ma, Y. Li, and F. Wang, "The effect of β-FeOOH on the corrosion behavior of low carbon steel exposed in tropic marine environment," *Materials Chemistry and Physics*, vol. 112, no. 3, pp. 844–852, 2008.

[33] R. G. Buchheit, "A compilation of corrosion potentials reported for intermetallic phases in aluminum alloys," *Journal of the Electrochemical Society*, vol. 142, no. 11, pp. 3994–3996, 1995.

[34] R. G. Buchheit, L. P. Montes, M. A. Martinez, J. Michael, and P. F. Hlava, "The electrochemical characteristics of bulk–synthesized Al_2CuMg," *Journal of the Electrochemical Society*, vol. 146, no. 12, pp. 4424–4428, 1999.

[35] C. Blanc, A. Freulon, M.-C. Lafont, Y. Kihn, and G. Mankowski, "Modelling the corrosion behaviour of Al_2CuMg coarse particles in copper-rich aluminium alloys," *Corrosion Science*, vol. 48, no. 11, pp. 3838–3851, 2006.

[36] S. M. Skolianos, T. Z. Kattamis, and O. F. Devereux, "Microstructure and corrosion behavior of as-cast and heat-treated Al-4.5 Wt pct Cu-2.0 wt pct Mn alloys," *Metallurgical Transactions A*, vol. 20, no. 11, pp. 2499–2516, 1989.

[37] J. R. Davis, *Corrosion of Aluminum and Aluminum Alloys*, ASM International, 1999.

18

Deposition of Cerium-Based Conversion Coatings on Aluminum Alloy 380

Ci Lin, Surender Maddela, William G. Fahrenholtz, and Matthew J. O'Keefe

Materials Research Center and Department of Materials Science and Engineering, Missouri University of Science & Technology, Rolla, MO 65409, USA

Correspondence should be addressed to William G. Fahrenholtz, billf@mst.edu

Academic Editor: Ingrid Milošev

Cerium-based conversion coatings were deposited on as-cast aluminum alloy 380 substrates by a spontaneous immersion process. In this study, the effects of rinsing temperature prior to immersion in the coating deposition solution were studied with respect to the surface morphology, electrochemical response, and corrosion resistance of the coatings. Panels rinsed at 25°C prior to coating had large cracks and holes in the coating. In contrast, panels rinsed at 100°C prior to coating had a uniform coating morphology with fewer, smaller cracks. Electrochemical testing revealed that coatings deposited on substrates rinsed at 100°C had higher impedance ($\sim$80 k$\Omega\cdot$cm^2) and lower corrosion current ($\sim$0.34 μA/cm^2) compared to coatings deposited on substrates rinsed at 25°C, which had 10 k$\Omega\cdot$cm^2 impedance and 2.7 μA/cm^2 corrosion current. Finally, ASTM B117 salt spray testing showed that rinsing at 100°C prior to coating resulted in cerium-based conversion coatings that could resist the formation of salt tails for at least 8 days.

1. Introduction

Aluminum cast alloys are widely used in the automotive and aerospace industries where innovative, lightweight materials and product forms are needed to improve performance [1]. Common aluminum casting alloys contain alloying additives that affect the mechanical properties, fluidity, and corrosion resistance [2]. For example, the 3xx.x series contain noble alloying elements such as Si and Cu that can promote pitting corrosion and lead to the failure of aluminum alloy components [3].

Chromate conversion coatings are used for corrosion protection for a wide variety of aluminum alloy components [4]. However, the toxicity and carcinogenic properties of hexavalent chromium (Cr^{6+}) have caused severe restrictions to be imposed on the use of chromates [5–7]. As a result, environmentally benign alternatives to chromates have been extensively investigated [8]. Potential replacements for chromate conversion coatings include anodized coatings [9], rare-earth-based inhibitors in conversion coatings [10], and sol-gel coatings [11]. Among potential chromate replacements, rare-earth inhibitors have attracted significant attention. Hinton et al. were the first to investigate cerium-based conversion coatings (CeCCs) as an environmentally benign alternative to chromate conversion coatings [12, 13].

The corrosion resistance of CeCCs is thought to arise from a combination of barrier properties and active response to the environment [14, 15]. The barrier protection properties of CeCCs have been studied by changing processing parameters, specifically surface preparation [16, 17], post-treatment [18], and the use of gelatin [19]. The corrosion resistance of CeCCs on high-strength aluminum alloys has been optimized through screening studies that have examined variables such as cerium concentration, pH of the deposition solution, and hydrogen peroxide concentration [20, 21]. The composition of the spray deposition solution used in the present research is based on the work of Pinc et al. [19] that exhibited the best corrosion resistance on Al 2024-T3.

Previous studies have focused on the deposition mechanisms of CeCCs from cerium chloride solutions onto aluminum alloy substrates. Hinton and Wilson [22] speculated that many electrochemical cells would arise due to the different activities of intermetallic particles on the surface of an aluminum alloy when the alloy was immersed into the solution. When immersed, anodic dissolution as expressed

TABLE 1: Nominal compositions of Al alloy 380 [1].

Element	Si	Fe	Cu	Mn	Mg	Ni	Zn	Sn	Other	Al
wt%	7.5–9.5	2.0	3.0–4.0	0.50	0.10	0.50	3.0	0.35	0.50	Balance

by Reaction (1) could occur at different locations according to the inhomogeneous surface activity.

$$Al \longrightarrow Al^{3+} + 3e^-. \quad (1)$$

As Al dissolves, the corresponding cathodic reaction could be hydrogen evolution (Reaction (2)), the reduction of peroxide (Reaction (3)), and/or reduction of oxygen dissolved in solution (Reaction (4)):

$$2H^+ + 2e^- \longrightarrow H_2, \quad (2)$$

$$H_2O_2 + 2H^+ + 2e^- \longrightarrow 2H_2O, \quad (3)$$

$$O_2 + 4H^+ + 4e^- \longrightarrow 2H_2O. \quad (4)$$

The pH in the electrolyte near local cathodic sites increases as the electrode reactions proceed. At pH values lower than 2.5, H_2O_2 complexes with Ce(III) species in solution as $Ce(H_2O_2)^{3+}$, and with increasing pH values (2.5 to 3.3) through deprotonating steps peroxo species form $Ce(O_2)^{2+}$ and when the pH value reaches 6.5 $Ce(O_2)^{2+}$ transforms to $Ce(O_2)(OH)_2$ [23, 24]. In the deposition process, H_2O_2 acts as a complexing agent, oxidant, crystallization inhibitor, and OH^- generator [24].

The purpose of this paper is to investigate the deposition of cerium-based conversion coatings on cast aluminum alloy 380 (AA 380). The effects of pretreatment and the temperature of the panel before immersion on coating morphology, electrochemical response, and corrosion resistance were investigated.

2. Experimental Procedure

Aluminum alloy 380 sheets 0.3 cm thick were cut into test panels 2.5 cm by 7.6 cm in size. The nominal composition of AA 380 is summarized in Table 1 [1]. Pretreatment of the panels prior to coating started with an isopropyl alcohol wipe followed by degreasing in an aqueous solution of a commercial alkaline cleaner (5 wt% Turco 4215 NC-LT in deionized water) for 5 minutes at 55°C. After degreasing, the panels were activated by immersion for 10 min at 50°C in an aqueous solution containing 1 wt% sulfuric acid. Following cleaning and activation, the panels were rinsed in deionized water that was either at room temperature (nominally 25°C) or heated to 100°C for about 1 minute. After rinsing, the panels were immersed in the deposition solution for different time intervals of up to 8 min.

The CeCC deposition solution was prepared from a stock solution consisting of 40 g $CeCl_3 \cdot xH_2O$ (Alfa Aesar, 99.9%) and 780 g of deionized water. The pH of the solution was adjusted to 2.07 with HCl after dissolution of the cerium chloride. For the deposition solution, 205 g of the stock solution was mixed with 0.8 g of a water soluble gelatin (DSF, Rousselot) that was dissolved in 25 g of deionized water. Just before deposition, 15 mL of H_2O_2 (Fisher Chemical, 30 wt%) was added.

Coated panels were posttreated by immersion for 5 minutes in a water solution containing 2.5 wt% Na_3PO_4 (pH adjusted to 4.5 with phosphoric acid) that was heated to 85°C. Posttreatment converts the as-deposited hydrated cerium oxide phase to hydrated cerium phosphate, which has superior corrosion protection [14, 18]. The corrosion resistances of the coated panels were evaluated using salt spray testing (Q-Fog, Q-Panel Lab Products) per ASTM standard B117. Panels with CeCCs were stored at room temperature in the laboratory for at least 24 h before characterization or salt spray testing.

Scanning electron microscopy (SEM; Hitachi S-4700) with energy dispersive X-ray spectroscopy (EDS; Phoenix System) was used to characterize the surface morphologies and compositions of CeCCs. Coating thicknesses were determined from cross-sections of coated panels that were produced by focus ion beam milling in a dual column instrument (FIB; FEI Helios NanoLab 600).

Electrochemical impedance spectroscopy (EIS) was carried out at open-circuit potential with amplitude of 10 mV in the frequency range from 10^5 to 10^{-2} Hz. Measurements were made after stabilization in the test electrolyte for 1500 s. All experiments were conducted with a frequency response analyzer (Schlumberger SI 1255 HF) coupled with a potentiostat/galvanostat (EG&G Princeton Applied Research Model 273A). Spectra were fit to equivalent circuit models to determine coating characteristics including coating impedance (R_P) according to previously reported methods [25, 26]. Potentiodynamic polarization measurements were carried out after EIS, the initial potential was $-400\ mV_{SCE}$ and final potential is $800\ mV_{SCE}$ with respect to the open-circuit potential, and the scan rate was 1 mV/sec. A standard flat cell (EG&G Princeton Applied Research) was used for the electrochemical polarization measurements. The cell electrolyte was a modified prohesion solution, which consisted of 0.70 wt% $(NH_4)_2SO_4$ and 0.35 wt% NaCl in deionized water. Specimens were tested at the natural pH of the solution, which was about 5.5. The exposed area of the working electrode was 1 cm^2. A saturated calomel electrode (SCE) was used as the reference electrode and Pt mesh was used as the counter electrode. Open-circuit potential (E_{corr}) and corrosion current (I_{corr}) were determined by the Tafel fit method using CorrView and ZView software (Scribner Associates) as previously reported [18].

3. Results and Discussion

3.1. Effect of Pretreatment. Aluminum alloy 380 is composed of several different phases including the Al matrix phase, a major second phase of Si, and trace intermetallic phases

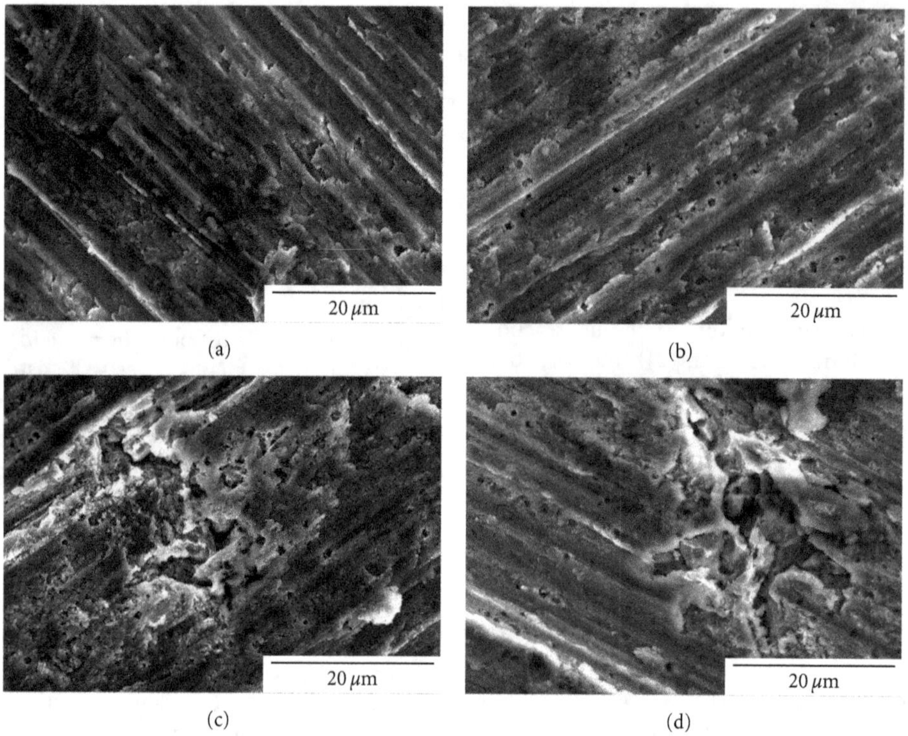

FIGURE 1: SEM surface morphology of AA 380 panels after alkaline cleaning and acid activation and (a) polishing with 180 grit abrasive paper, (b) water rinsing at 25°C, (c) water rinsing at 25°C showing an area with removal of material, and (d) water rinsing at 100°C.

such as Al_2Cu, $FeSi_2$, and $Al_{4.5}FeSi$. Analysis of surface morphology (Figures 1(a) and 1(b)) and chemistry of the substrate after pretreatment revealed that intermetallic particles containing Fe and Si were dissolved by the alkaline or acid pretreatments. Dissolution sometimes left holes in the substrates that were up to about 20 μm in diameter (Figures 1(c) and 1(d)). No discernible differences in surface morphology were observed for panels that had been rinsed with water at room temperature or 100°C.

Pretreatment altered the electrochemical response of the substrates. From the potentiodynamic curves shown in Figure 2(a), pretreatment increased the open-circuit potential from about −600 mV_{SCE} for the starting panel to about −500 mV_{SCE} after alkaline cleaning, acid activation, and rinsing. Figure 2(a) also shows that the pretreatment increased the corrosion current from 0.3 $\mu A/cm^2$ before pretreatment to 2.0 $\mu A/cm^2$ after pretreatment. However, the rinsing temperature, 25°C or 100°C, did not produce significant differences in either the open-circuit potential or the corrosion current. The increase in corrosion current after pretreatment indicated that the alloy surface was more electrochemically active, which made the panels easier to coat. From the corresponding electrochemical impendence spectra (Figure 2(b)), the impedance of the starting panel was around 40 $k\Omega \cdot cm^2$, which was much larger than the impedance after pretreatment (~7.5 $k\Omega \cdot cm^2$ for both rinsing temperatures). Based on comparisons to the results of previous studies [27], the pretreatment process likely reduced the thickness of the native oxide layer, which decreased the impedance of the substrate. Previous analysis showed that the reduction in impedance and increase in corrosion current were necessary to promote deposition of CeCCs [27].

TABLE 2: Electrochemical parameters derived from potentiodynamic polarization measurements and impedance spectra.

Immersion time (min)	Corrosion parameters					
	R_p ($k\Omega \cdot cm^2$)		I_{corr} ($\mu A/cm^2$)		E_{corr} (mV_{SCE})	
	25°C	100°C	25°C	100°C	25°C	100°C
2	32	36	0.82	0.73	−540	−480
5	16	49	1.60	0.54	−550	−480
8	10	76	2.70	0.34	−510	−540

3.2. Electrochemical Behavior of CeCCs. For substrates rinsed at 25°C prior to coating deposition, increasing the immersion time in the coating solution made the surface more electrochemically active (values summarized in Table 2). From the potentiodynamic curves in Figure 3(a), the corrosion current increased from 0.82 $\mu A/cm^2$ after two minutes in the coating solution to 1.6 $\mu A/cm^2$ after 5 minutes and 2.7 $\mu A/cm^2$ after eight minutes. In contrast, the open-circuit potential did not vary significantly among the three panels, with all having values around −530 mV_{SCE}. The increasing corrosion current with the increasing immersion time suggests that the surface of the AA 380 is more susceptible to chloride attack and dissolution. Therefore, for substrates rinsed at 25°C, increasing the immersion time appears to increase attack of the substrate by chloride ions in the coating solution.

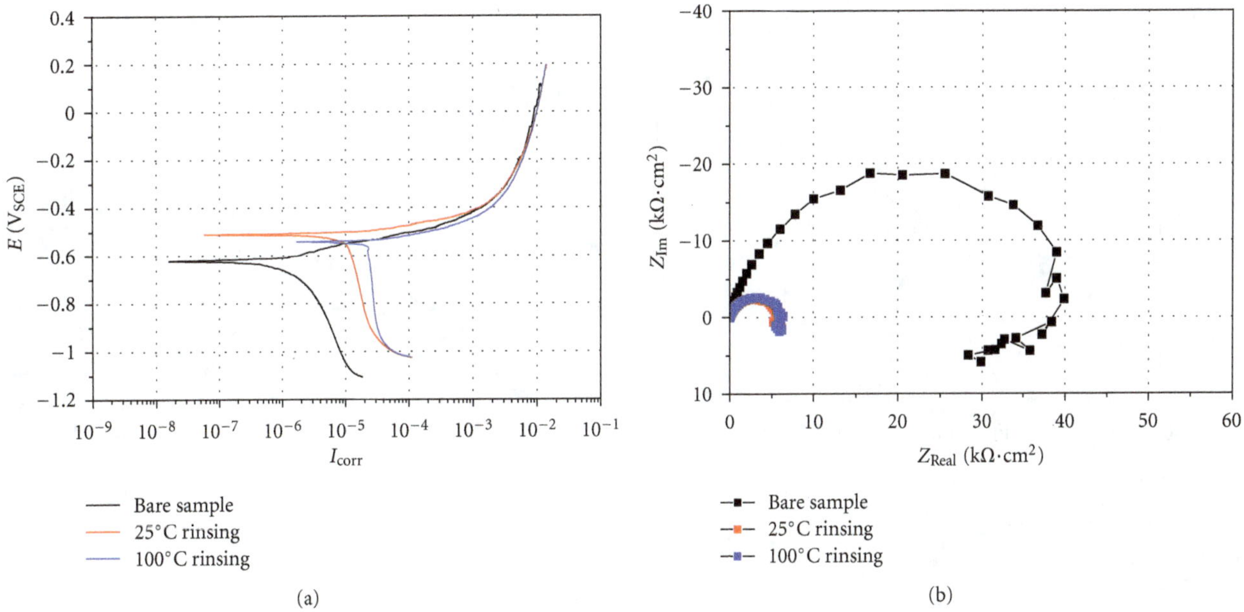

FIGURE 2: (a) Potentiodynamic polarization curves and (b) impedance spectroscopy of AA 380 panels before pretreatment (bare) and after rinsing at 25°C and 100°C.

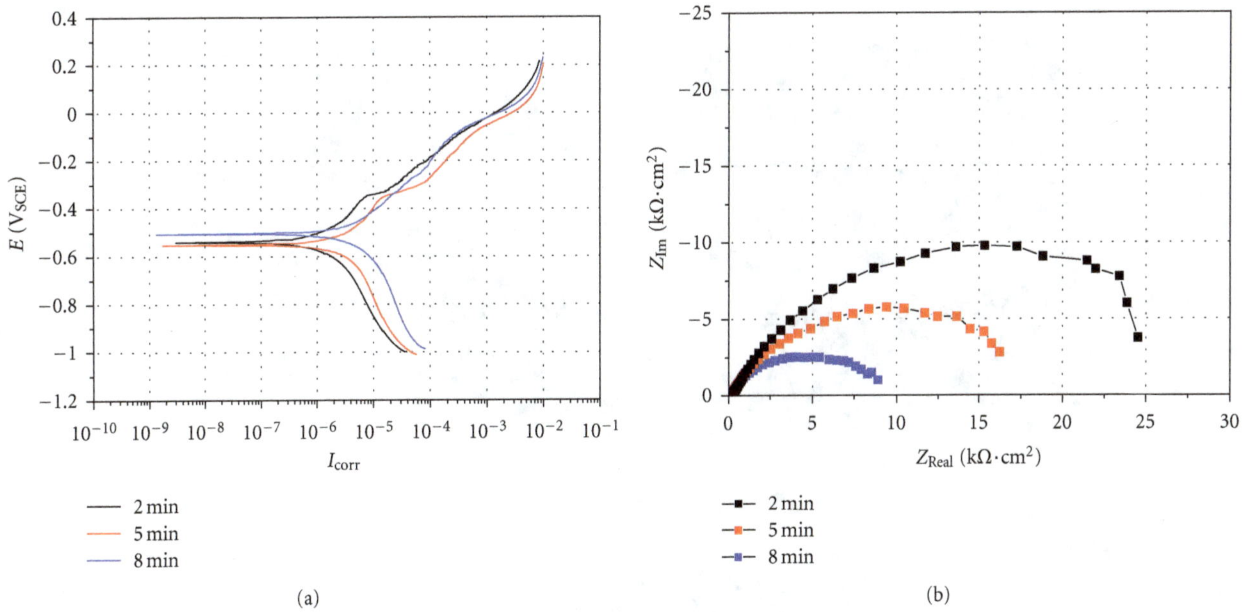

FIGURE 3: (a) Potentiodynamic polarization curves and (b) impedance spectra for CeCCs deposited AA 380 substrates rinsed at 25°C as a function of coating deposition time.

For substrates rinsed at 100°C prior to coating deposition, increasing the immersion time in the coating solution increased the corrosion resistance (values summarized in Table 2). The potentiodynamic curves shown in Figure 4(a) reveal that the corrosion current decreased from 0.73 μA/cm^2 after immersion for 2 minutes to 0.54 μA/cm^2 after five minutes and to 0.34 μA/cm^2 after eight minutes. The open circuit potential was about the same after immersion in the deposition solution for 2 or 5 minutes (-480 mV$_{SCE}$), but decreased to about -540 mV$_{SCE}$ after immersion for eight minutes. The decreasing corrosion current indicated that the coatings had better corrosion resistance as the immersion time increased. Compared to panels rinsing at 25°C (0.82 μA/cm^2), rinsing at 100°C for 2 minutes resulted in a coating with better corrosion resistance (0.73 μA/cm^2) and the corrosion resistance, as measured by corrosion current, continued to increase as deposition time increased. Therefore, rinsing at 100°C prior to coating deposition

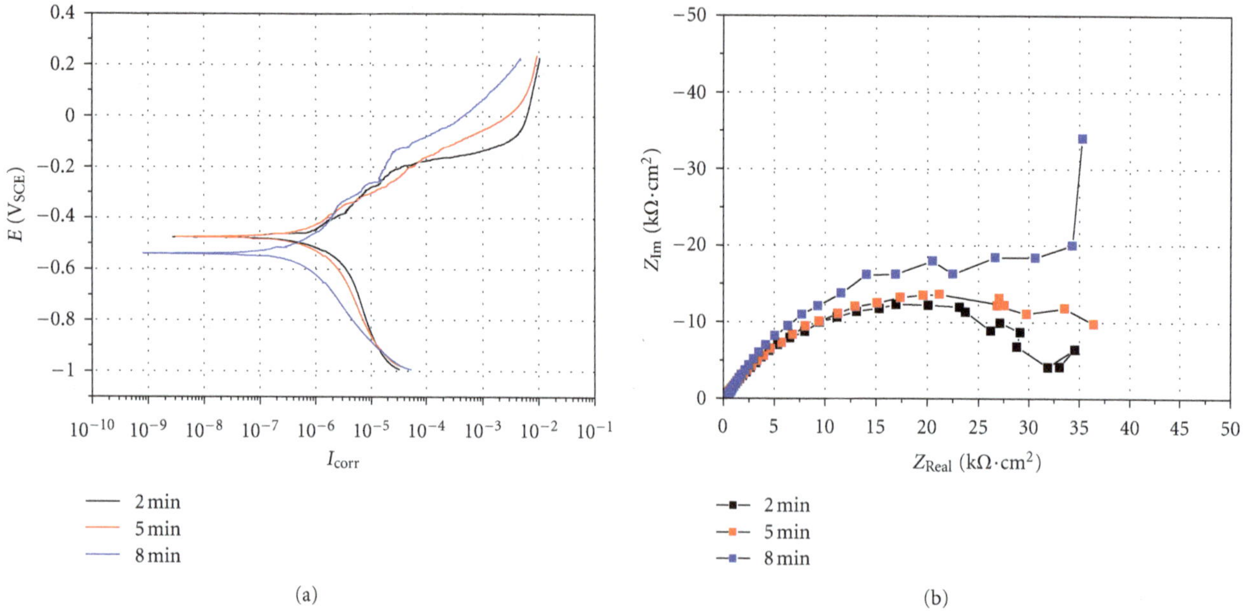

FIGURE 4: (a) Potentiodynamic polarization curves and (b) impedance spectra for CeCCs deposited AA 380 substrates rinsed at 100°C as a function of coating deposition time.

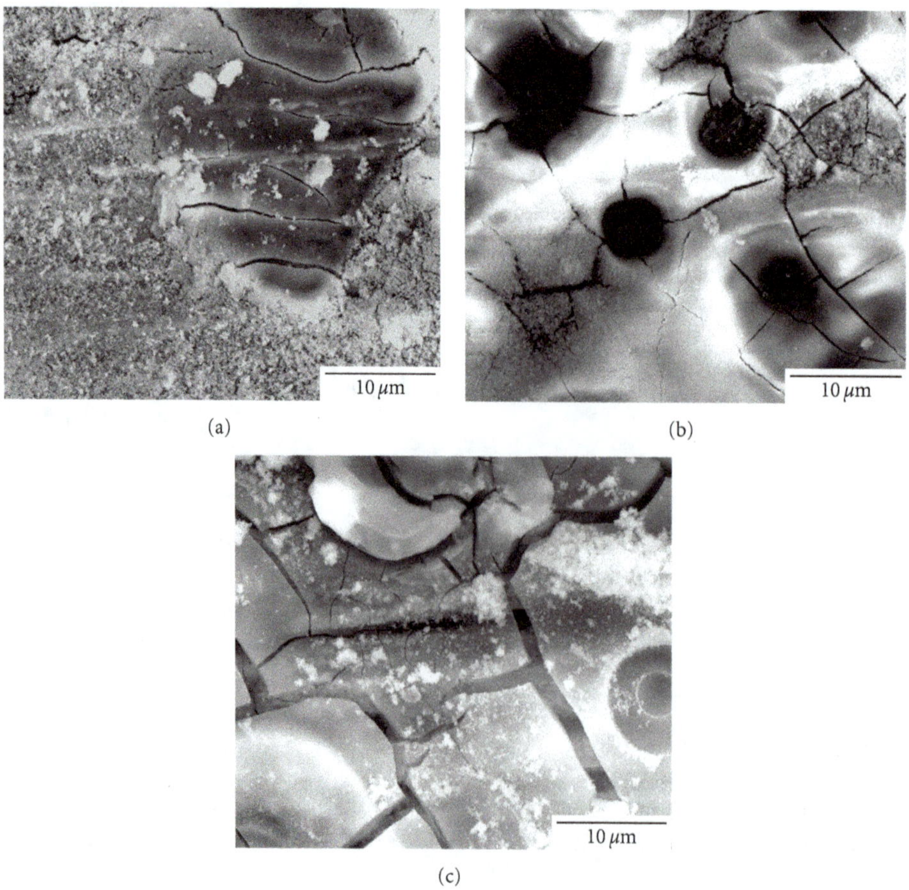

FIGURE 5: Surface morphology of coatings deposited on panels rinsed at 25°C after immersion in the coating solution for (a) 2 min, (b) 5 min, and (c) 8 min.

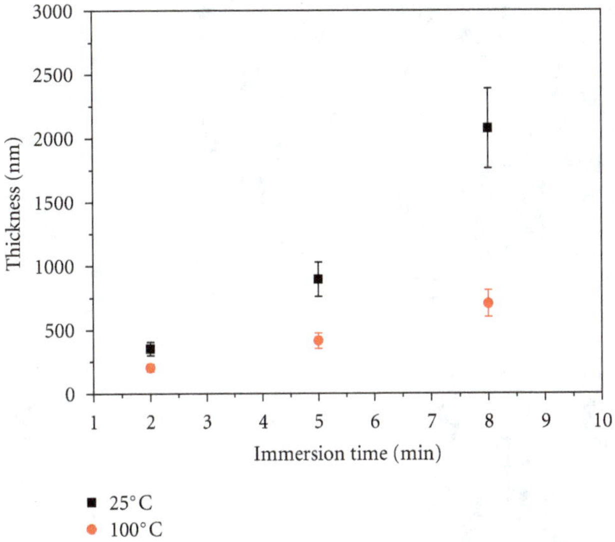

FIGURE 6: Thickness as a function of immersion time for CeCCs deposited on panels rinsed at 25°C and 100°C.

appears to produce CeCCs on AA 380 with better corrosion resistance.

The EIS results as a function of immersion time were consistent with potentiodynamic analysis (Table 2). Rinsing at 25°C prior to deposition resulted in coatings with impedance values that decreased as immersion time increased (Figure 3(b)). For panels rinsed at 25°C prior to coating, the impedance was 32 kΩ·cm^2 after 2 minutes of immersion in the coating solution and the value decreased to 16 kΩ·cm^2 after 5 minutes and 10 kΩ·cm^2 after eight minutes. However, for panels rinsed at 100°C prior to coating deposition, the impedance was 36 kΩ·cm^2 after immersion in the coating solution for 2 minutes, and it increased to 49 kΩ·cm^2 after 5 minutes and 79 kΩ·cm^2 after 8 minutes (Figure 4(b)). For panels rinsed at 100°C prior to coating deposition, the impedance of 79 kΩ·cm^2 after 8 minutes of immersion in the coating solution was more than double the highest value for coatings deposited on panels rinsed at 25°C, which was 32 kΩ·cm^2 after 2 minutes of immersion in the coating solution. These results indicate that rinsing at 25°C prior to deposition led to corrosion resistance that decreased with increasing immersion time in the coating solution whereas rinsing at 100°C prior to coating deposition led to corrosion resistance that increased with the increasing immersion time.

3.3. Morphology and Thickness of the CeCCs. Coatings that were deposited on panels rinsed at 25°C were cracked and had large holes. As shown in Figure 5, the cracks became larger as coating time increased. In addition to the cracks, large holes (dark areas), ~5 μm in diameter, were also observed on the panel surfaces. Even though the coatings were cracked, the average thickness (Figure 6) in areas with coating increased as immersion time increased. For example, the coating thickness was about 350 nm after 2 min, but increased to nearly 2 μm after 8 min of immersion in the coating solution. However, the thickness was not uniform and varied across the panel. The surface morphology supports the electrochemical testing results that showed that the coating impedance decreased as immersion time increased for coatings on substrates rinsed at 25°C. As coating time increased, the size of cracks grew, which exposed more of the substrate to the coating solution and allowed for attack of the substrate. The impedance approached a value of 10 kΩ·cm^2 for deposition times of 8 min, which was about one-third of the value (32 kΩ·cm^2) after immersion for 2 minutes. Hence, SEM analysis was consistent with electrochemical results that showed coatings deposited on substrates rinsed at 25°C did not serve as effective barriers to corrosion due to nonuniform coverage of the AA 380 surface.

Coatings that were deposited on substrates that were rinsed at 100°C had a nodular appearance (Figure 7) for all deposition times, which is similar to what has been reported in previous work [13]. The coatings covered the substrates and only a few, small cracks were observed. As with coatings on substrates rinsed at 25°C, the thickness of CeCCs deposited on substrates that were rinsed at 100°C increased with increasing deposition time. After 2 minutes of immersion, the coating thickness was just over 200 nm and it increased to ~700 nm after immersion for 8 min (Figure 6). Despite being approximately twice as thick as CeCCs on high-strength aluminum alloys, such as Al 2024-T3 and Al 7075-T6 [28, 29], the coatings had only a few small cracks. As shown by the electrochemical analysis in Figure 4(b), coating impedance increased as immersion time increased, which is consistent with the formation of a thicker, continuous coating. Therefore, the surface morphology and thickness of the coatings are consistent with the electrochemical results, which showed that coatings on substrates that were rinsed at 100°C increased the impedance compared to uncoated substrates.

3.4. Corrosion Protection. Cerium-based conversion coatings had a yellow-gold appearance (Figures 8(a) and 9(a)) after posttreatment. After salt spray testing, coatings deposited on substrates rinsed at 25°C prior to immersion showed a significant amount of white corrosion produce (i.e., salting). Corrosion pits and salt tails were visible after just 24 hours in salt spray testing as shown in Figure 8(b). Coatings continued to degrade with further time in salt spray testing (Figures 8(c) and 8(d)). In contrast, coatings deposited on substrates rinsed at 100°C prior to deposition showed better corrosion resistance. After 24 hours, a few corrosion pits were present (Figure 9(b)), but significant salting was not observed until after 192 hours of salt spray testing. Therefore the results of salt spray testing were consistent with electrochemical characterization and surface morphology. Coatings deposited on substrates that were rinsed at 25°C prior to immersion had higher corrosion currents, lower impedance values, and were cracked, which led to more severe corrosion in salt spray testing. However, coatings that were deposited on substrates that were rinsed at 100°C prior to deposition had lower corrosion currents, higher impedance values, and were free of large cracks. As a result, coatings deposited on substrates rinsed at 100°C provided improved corrosion protection for AA 380 alloy substrates.

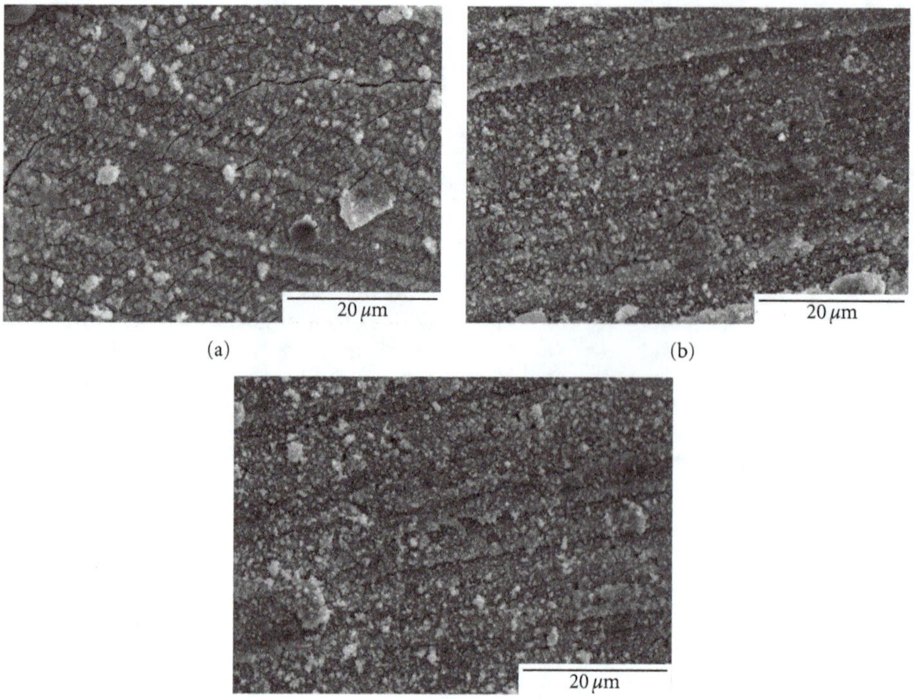

(a) (b) (c)

FIGURE 7: Surface morphology of coatings deposited on panels rinsed at 100°C after immersion in the coating solution for (a) 2 min, (b) 5 min, (c) 8 min.

(a) (b) (c) (d)

FIGURE 8: Optical images of coatings deposited on substrates rinsed at 25°C (a) as deposited, and after salt spray testing for (b) 24 hours, (c) 96 hours, and (d) 192 hours.

4. Summary

Cerium-based conversion coatings were deposited on AA 380 alloy substrates. This study examined the effect of changing the rinsing temperature prior immersion in the coating solution on the electrochemical response, coating morphology, and corrosion resistance of CeCCs. From the results described above, the conclusions that can be drawn include the following.

(1) Differences in the rinsing temperature did not produce any noticeable differences in the morphology of the substrates prior to coating or the electrochemical response of the panels prior to coating. Based on

(a) (b) (c) (d)

FIGURE 9: Optical images of coatings deposited on substrates rinsed at 100°C (a) as deposited, and after salt spray testing for (b) 24 hours, (c) 96 hours, and (d) 192 hours.

SEM/EDS analysis, pretreatment led to the removal of Fe and Si-rich intermetallic compounds from the substrate surfaces, but the open-circuit potential ($-500\,mV_{SCE}$), corrosion current ($2\,\mu A/cm^2$), and impedance ($7.5\,k\Omega\cdot cm^2$) of the pretreated panels were nearly identical regardless of the final rinsing temperature.

(2) Coatings deposited on panels rinsed at 25°C had cracks and large holes in the CeCCs. The cracks and holes became larger and deeper with increasing immersion time. The surface morphology was consistent with results of the electrochemical tests. For example, immersion for two minutes resulted in a corrosion current of $0.82\,\mu A/cm^2$ and an impedance of $32\,k\Omega\cdot cm^2$. As the coating deposition time increased to 5 minutes, the cracks become bigger and large holes appeared. This resulted in a corrosion current that increased to $1.6\,\mu A/cm^2$ and an impedance that decreased to $16\,k\Omega\cdot cm^2$. After 8 minutes of immersion in the coating solution, the coating had even lower values of corrosion current ($2.7\,\mu A/cm^2$) and impedance ($10\,k\Omega\cdot cm^2$).

(3) Rinsing at 100°C prior to coating deposition resulted in coatings that had uniform appearance and fewer cracks. With increasing immersion time, the thickness of the coatings increased uniformly, which enhanced the corrosion resistance. From electrochemical testing, immersion for 2 minutes in the deposition solution produced a corrosion current of $0.73\,\mu A/cm^2$ and an impedance of $36\,k\Omega\cdot cm^2$ with a coating thickness of ~200 nm. Increasing the coating deposition time to 5 minutes increased the thickness of the coating to ~400 nm, decreased the corrosion current to $0.54\,\mu A/cm^2$, and increased the impedance to $49\,k\Omega\cdot cm^2$. Further increasing the immersion time to 8 minutes increased the coating thickness to ~750 nm, decreased the corrosion current to $0.34\,\mu A/cm^2$, and increased the impedance to $76\,k\Omega\cdot cm^2$. Apparently, the higher rinsing temperature led to more uniform nucleation and growth of the resulting CeCC, which decreased the corrosion current and increased the impedance compared to similar coatings deposited on substrates rinsed at 25°C.

(4) Cerium-based conversion coatings deposited on AA 380 panels that were rinsed in water heated to 100°C after pretreatment, but before CeCC deposition provided significant corrosion resistance. Whereas coatings deposited on substrates rinsed at 25°C had both pits and salt tails after only 24 hours in salt spray testing, coatings deposited on substrates rinsed at 100°C were able to inhibit formation of salt tails for at least 96 hours. The improved corrosion performance of CeCCs on substrates that were rinsed at 100°C was attributed to the more uniform coating that was deposited, which led to higher impedance and lower corrosion currents.

Acknowledgments

The authors acknowledge the technical guidance and support of Bruce Sartwell at the Strategic Environmental Research and Development Program (SERDP). This work was funded through SERDP under Contract W912HQ-08-C-0008 as Project WP-1618. The assistance of Clarissa Wisner (SEM), Dr. Elizabeth Kulp (FIB), Dr. Kai Song (FIB), and Carlos Castano (FIB) from the Graduate Center for Materials Research at Missouri S&T is acknowledged.

References

[1] J. G. Kaufman and E. L. Rooy, *Aluminum Alloy Castings: Properties, Processes, and Applications*, American Society for Materials International, Materials Park, Ohio, USA, 2004.

[2] J. E. Hatch and Aluminum:, *Properties and Physical Metallurgy*, American Society for Materials International, Materials Park, Ohio, USA, 1984.

[3] J. R. Davis, *Corrosion of Aluminum and Aluminum Alloys*, American Society for Materials International, Materials Park, Ohio, USA, 1999.

[4] E. H. Hollingsworth and H. Y. Hunsicker, *Corrosion*, vol. 13 of *ASM Handbook*, American Society for Materials International, Materials Park, Ohio, USA, 1990.

[5] S. S. Wise, J. H. C. Schuler, A. L. Holmes et al., "Comparison of two particulate hexavalent chromium compounds: barium chromate is more genotoxic than lead chromate in human lung cells," *Environmental and Molecular Mutagenesis*, vol. 44, no. 2, pp. 156–162, 2004.

[6] http://www.osha.gov/SLTC/hexavalentchromium/index.html.

[7] H. J. Gibb, P. S. J. Lees, P. F. Pinsky, and B. C. Rooney, "Lung cancer among workers in chromium chemical production," *American Journal of Industrial Medicine*, vol. 38, no. 2, pp. 115–126, 2000.

[8] R. G. Buchheit and A. E. Hughes, "Chromate and chromate-free conversion coatings," in *Corrosion: Fundamentals, Testing and Protection*, vol. 13A of *ASM Handbook*, pp. 720–735, American Society for Materials International, Materials Park, Ohio, USA, 2003.

[9] B. Meyers and S. Lynn, *Surface Engineering*, vol. 5 of *ASM Handbook*, American Society for Materials International, Materials Park, Ohio, USA, 1990.

[10] D. Zhao, J. Sun, L. Zhang, Y. Tan, and J. Li, "Corrosion behavior of rare earth cerium based conversion coating on aluminum alloy," *Journal of Rare Earths*, vol. 28, no. 1, pp. 371–374, 2010.

[11] G. R. Salazar-Banda, S. R. Moraes, A. J. Motheo, and S. A. S. MacHado, "Anticorrosive cerium-based coatings prepared by the sol-gel method," *Journal of Sol-Gel Science and Technology*, vol. 52, no. 3, pp. 415–423, 2009.

[12] B. R. W. Hinton, D. R. Arnott, and N. E. Ryan, "Cerium conversion coatings for the corrosion protection of aluminum," *Materials Forum*, vol. 9, no. 3, pp. 162–173, 1986.

[13] B. R. W. Hinton, "Corrosion inhibition with rare earth metal salts," *Journal of Alloys and Compounds*, vol. 180, no. 1-2, pp. 15–25, 1992.

[14] D. K. Heller, *Phosphate post-treatment of cerium-based conversion coatings on Al 2024-T3 [Ph.D. dissertation]*, Missouri University of Science and Technology, 2010.

[15] S. Joshi, E. A. Kulp, W. G. Fahrenholtz, and M. J. O'Keefe, "Dissolution of cerium from cerium-based conversion coatings on Al 7075-T6 in 0.1M NaCl solutions," *Corrosion Science*, vol. 60, pp. 290–295, 2012.

[16] S. Geng, W. R. Pinc, P. Yu et al., "Influence of cleaning on the deposition rate of cerium based conversion coatings on Al alloy 2024-T3," *Journal of Applied Surface Finishing*, vol. 2, no. 4, pp. 276–282, 2007.

[17] W. Pinc, S. Geng, M. O'Keefe, W. Fahrenholtz, and T. O'Keefe, "Effects of acid and alkaline based surface preparations on spray deposited cerium based conversion coatings on Al 2024-T3," *Applied Surface Science*, vol. 255, no. 7, pp. 4061–4065, 2009.

[18] D. K. Heller, W. G. Fahrenholtz, and M. J. O'Keefe, "The effect of post-treatment time and temperature on cerium-based conversion coatings on Al 2024-T3," *Corrosion Science*, vol. 52, no. 2, pp. 360–368, 2010.

[19] W. Pinc, P. Yu, M. O'Keefe, and W. Fahrenholtz, "Effect of gelatin additions on the corrosion resistance of cerium based conversion coatings spray deposited on Al 2024-T3," *Surface and Coatings Technology*, vol. 203, no. 23, pp. 3533–3540, 2009.

[20] B. F. Rivera, B. Y. Johnson, M. J. O'Keefe, and W. G. Fahrenholtz, "Deposition and characterization of cerium oxide conversion coatings on aluminum alloy 7075-T6," *Surface and Coatings Technology*, vol. 176, no. 3, pp. 349–356, 2003.

[21] S. Geng, P. Yu, M. J. O'Keefe, W. G. Fahrenholtz, and T. J. O'Keefe, "Screening study of spray solution parameters for depositing cerium-based conversion coatings on Al alloy 2024-T3," *Journal of Applied Electrochemistry*, vol. 40, no. 3, pp. 551–559, 2010.

[22] B. R. W. Hinton and L. Wilson, "The corrosion inhibition of zinc with cerous chloride," *Corrosion Science*, vol. 29, no. 8, pp. 967–985, 1989.

[23] S. A. Hayes, P. Yu, T. J. O'Keefe, M. J. O'Keefe, and J. O. Stoffer, "The phase stability of cerium species in aqueous systems. I. E-pH diagram for the Ce-$HClO_4$-H_2O system," *Journal of the Electrochemical Society*, vol. 149, no. 12, pp. C623–C630, 2002.

[24] F. H. Scholes, C. Soste, A. E. Hughes, S. G. Hardin, and P. R. Curtis, "The role of hydrogen peroxide in the deposition of cerium-based conversion coatings," *Applied Surface Science*, vol. 253, no. 4, pp. 1770–1780, 2006.

[25] P. Zoltowski, "The error function for fitting of models to immittance data," *Journal of Electroanalytical Chemistry*, vol. 178, no. 1, pp. 11–19, 1984.

[26] J. R. Dygas and M. W. Breiter, "Variance of errors and elimination of outliers in the least squares analysis of impedance spectra," *Electrochimica Acta*, vol. 44, no. 24, pp. 4163–4174, 1999.

[27] C. S. Lin and S. K. Fang, "Formation of cerium conversion coatings on AZ31 magnesium alloys," *Journal of the Electrochemical Society*, vol. 152, no. 2, pp. B54–B59, 2005.

[28] B. L. Treu, S. Joshi, W. R. Pinc, M. J. O'Keefe, and W. G. Fahrenholtz, "Characterization of localized surface states of Al 7075-T6 during deposition of cerium-based conversion coatings," *Journal of the Electrochemical Society*, vol. 157, no. 8, pp. C282–C287, 2010.

[29] P. S. Jones, P. Yu, W. R. Pinc, M. J. O'Keefe, W. G. Fahrenholtz, and T. J. O'Keefe, "Spray deposition of cerium oxide-based conversion coatings on Al 2024-T3," *International Journal of Applied Ceramic Technology*, vol. 5, no. 1, pp. 63–73, 2008.

19

Corrosion Behavior of Commercial Aluminum Alloy Processed by Equal Channel Angular Pressing

Atef Korchef[1] and Abdelkrim Kahoul[2]

[1] *National Center of Researches in Materials Sciences, Technopark of Borj-Cédria, BP 73, 8027 Soliman, Tunisia*
[2] *Laboratoire d'Energétique et d'Electrochimie des Solides (LEES), Faculté de Technologie, Université Ferhat Abbas, 19000 Sétif, Algeria*

Correspondence should be addressed to Abdelkrim Kahoul; kahoulabdelkrim@yahoo.fr

Academic Editor: W. Ke

A commercial aluminum alloy was subjected to severe plastic deformation through equal channel angular pressing (ECAP). The alloy contains a low volume fraction of α-AlFeSi located essentially at the grain boundaries. The corrosion behavior of the ECAP'ed alloy was investigated in NaCl solution using potentiodynamic polarization and immersion tests. The effects of scan rate and NaCl concentration on the alloy susceptibility to corrosion were also studied. The results obtained were compared with those of the nonpressed alloy. ECAP leads to an intensive grain refinement accompanied by an increased dislocation density. All electrochemical tests confirm that corrosion resistance of the alloy remarkably diminished with increasing the ECAP number of passes. This is presumably due to the breakdown of the α-AlFeSi after ECAP leading to higher number of galvanic cells and enhanced dissolution of the aluminum matrix.

1. Introduction

It is now well known that severe plastic deformation (SPD) leads to enhanced grain refinement to the submicrometer (100–1000 nm) or nanometer (<100 nm) levels. The intensive grain refinement is responsible for the enhancement of physical and mechanical properties of the processed materials, such as high strength and ductility. SPD techniques include, among others, torsion under high pressure [1], accumulative roll bonding (ARB) [2], and equal channel angular pressing (ECAP) [3, 4]. The latter procedure, introduced almost 30 years ago to achieve enhanced plastic strains, is especially attractive for several near superplastic applications. Since the cross-section of the sample remains unchanged during ECAP, repetitive pressings on the same sample are possible to achieve the required strains and to invoke different slip systems by rotating differently the sample between consecutive passes [4]. Thus, the microstructure obtained after ECAP depends on the introduced strains and whether and how the sample is rotated between two successive passes. In the early stages of pressing, reported results in the literature showed that the microstructure was reasonably homogeneous and consisted of parallel bands of subgrains. For further pressings, these subgrains evolved into an array of refined grains separated by high angle boundaries [5]. As a result of the grain refinement, a significant improvement of the mechanical properties was observed after ECAP [6].

In addition, ECAP process alters the bulk of the material leading to changes in material flow, residual stress, texture, size and distribution of a second phase if any exists in the material, and so forth. In addition to grain refinement, these changes significantly affect the corrosion behavior of the material [7]. However, the results reported in the literature are often controversial claiming that the same metric could both decrease and increase susceptibility to corrosion. For example, it was suggested that the decrease of the grain size can either decrease [8, 9] or increase [10] the corrosion resistance of aluminum. In fact, depending on whether grains were equiaxed or columnar, corrosion rate would variably increase or decrease with grain refinement [7].

Table 1: Chemical composition (cc) of the aluminum alloy determined by ICPOES.

Metal	cc (wt.%)
Fe	0.2500
Si	0.1071
Cr	0.0056
Cu	0.0558
Mn	0.0600
Mg	0.2200
Co	0.0002
Ni	0.0036
Pb	<0.0001
Ca	0.0014
Ti	0.0060
Zn	0.1660
Al	balance

In previous works [11–13], we studied the effect of ECAP process on the microstructure evolution and the mechanical properties of a commercial aluminum containing a low volume fraction of α-AlFeSi located essentially at the grain boundaries. X-ray diffraction (XRD) analysis showed an important broadening of the reflections of the alloy attributed mainly to grain refinement and lattice distortions. The grain refinement and the accumulation of dislocations generated by ECA pressing lead to a significant improvement of the mechanical properties in compression and tension tests [13]. The present work aims to investigate the effect of ECAP on corrosion behavior of the same material. The effects of grain refinement and the α-AlFeSi on the susceptibility to corrosion of aluminum alloy were especially discussed. This allows gaining a more detailed insight into the effect of SPD through ECAP on corrosion behavior of aluminum alloys. The results obtained for the ultrafine grained (UFG) aluminum alloy were compared with those of the coarse grained counterpart.

2. Experimental Procedure

2.1. Sample Preparation. The experiments were carried out using low purity aluminum (99.1%). The material was received in the form of cast ingots. It was analysed by Inductively Coupled Plasma Optical Emission Spectrometry (ICPOES). The chemical composition is given in Table 1. The material was annealed for 24 hours at 500°C and quenched in iced water. Samples with dimensions ~$10 \times 10 \times 70\,mm^3$ were cut for ECA pressing. The die used for ECA pressing has two square channels of $10 \times 10\,mm^2$ cross sectional area intersecting at an angle $\Phi = 90°$. The arc of curvature at the outer point of the intersection of the two channels delineated an angle $\Psi = 20°$. According to the equation given by Iwahashi et al. [14], the total equivalent strain, ε_N, accumulated by N passes through the die was equal to ~N.

In the present work, repetitive pressings of the same sample were conducted, at room temperature, to achieve an equivalent strain up to ~4 without rotation of the sample between two successive passes (the so-called route A [15]).

2.2. Electrochemical Measurements. The electrochemical tests were conducted at room temperature in aerated stagnant NaCl solutions used as the corrosive medium at pH = 6.4. NaCl (Aldrich) was added to bidistilled water in required amounts to prepare solutions with concentrations varying between 0.15 and 0.9 M (0.875 and 5.25 wt.%, resp.).

The working electrodes used were cut from the central part of the ECA pressed ($N = 2$ and 4) samples. The non-pressed ($N = 0$) aluminum alloy was also studied for comparison. Before each experiment, the working electrodes of cross-section 1 cm^2 were mechanically polished with grinding paper up to 1000 grit to a mirror finish. The electrodes were then degreased with acetone, finally washed with bidistilled water, and dried in air. In order to avoid a prolonged exposure to atmosphere leading to oxide formation on the Al surface, the samples were tested immediately after drying. A Pt sheet was used as a counter electrode. The potential was measured against a saturated calomel electrode (SCE) as a reference. Potentiodynamic polarization curves, using a computerized potentiostat (Radiometer model VoltaLab 40) and Voltamaster 4 software, were performed by starting scanning electrode potential from an initial potential equal or below the open circuit potential (OCP) to an anodic potential of 0.3 V in 3.5 wt.% NaCl solution at pH = 6.4. Electrochemical impedance spectroscopy (EIS) measurements were carried out at open circuit potential (OCP) using the same Voltamaster 4 software. A 10 mV amplitude was applied to the electrodes under study at sweeping frequencies between 100 kHz and 0.1 Hz. Some experiments were repeated two or three times, and reproducibility was found to be satisfactory.

2.3. Weight Loss Tests. The weight loss tests were carried out at room temperature in 100 mL glass beakers containing 3.5 wt.% NaCl solution at pH = 6.4. The polished and cleaned specimens were weighed and immersed for various time intervals. Once the interval time of immersion was finished, the samples were removed from the solution and cleaned with a 70 wt% HNO_3 solution [16], washed with acetone and weighed again. The above steps were repeated, and the mass loss in every interval was measured to evaluate the corrosion rate ($mg{\cdot}cm^{-2}$) of the samples.

2.4. Surface Analysis. For X-ray diffraction (XRD), optical microscopy (OM), and scanning electron microscopy (SEM) investigations, specimens of ~$5 \times 5 \times 10\,mm^3$ were cut from the central part of the tested ECA pressed ($N = 4$) and unpressed ($N = 0$) samples. Their surfaces were electrolytically polished for 30 min at a temperature below 10°C under an operating voltage of 30 V, using a mixture of 66% $(CH_3CO)_2O$ and 34% $HClO_4$. Added polishing in a 10% H_2SO_4 solution was used for OM under polarized light.

2.5. X-Ray Diffraction Analysis. XRD was carried out with a Philips X'PERT PRO diffractometer in step scanning mode using Co-K_α radiation. The XRD patterns were recorded at room temperature in scan rate of 2×10^{-3} deg/s in the scanning range $2\theta = 20$–155°. The exact position and the full-width at half maximum (FWHM) of each reflection was determined using X'PERT HighScore Plus software. A correction for the contribution from the instrumental effects to the experimental FWHM, β_{exp}, for each peak were obtained using the following expression [12]:

$$\beta = \left(\left[\left(\beta^2\right)_{exp} - \left(\beta^2\right)_{ins}\right]^{1/2} \times \left[\beta_{exp} - \beta_{ins}\right]\right)^{1/2}, \quad (1)$$

where β_{ins} is the instrumental FWHM which was determined using NaCl or the undeformed aluminum as a standard reference material.

The mean coherent crystalline domain size and strain ε for each pass were calculated using the Halder-Wagner (HW) method [17]. In the HW method, which assumes a Voigt peak shape, the crystallite size, D, and the strain, ξ, are related to β^* by the following expression:

$$\left(\frac{\beta^*}{d^*}\right)^2 = \left[\left(\frac{1}{D}\right) \times \frac{\beta}{(d^*)^2}\right] + \left(\frac{\xi}{2}\right)^2. \quad (2)$$

From (2), the intercept of the plot of $(\beta^*/d^*)^2$ versus $\beta/(d^*)^2$ gives the value of the equivalent strain, and the slope gives the mean cell size.

The lattice parameter of the aluminum before and after ECA pressings was obtained from a linear regression analysis of the measured lattice parameter, a_{app}, obtained from each reflection, plotted against the Nilson-Reley function [18]:

$$\text{N-R} = \frac{\cos^2\theta/\sin\theta + \cos^2\theta/\theta}{2}, \quad (3)$$

and extrapolated to N-R = 0; that is, $2\theta = 180^\circ$.

Assuming that the dislocation density before ECAP deformation ρ_i is significantly lower than the remaining dislocation density ρ after ECAP ($\rho_i \ll \rho$), ρ could be calculated using the following equation [19, 20]:

$$\rho = \frac{1}{D(y + b/\varepsilon_N)}, \quad (4)$$

where D is the coherency length (cell size) provided by XRD, ε_N (=1.07 N) is the plastic strain introduced, N is the number of ECAP passes, b (=0.2867 nm) is the Burgers vector of dislocation, and y (=2 nm) is the annihilation distance of dislocation [20].

3. Experimental Results

3.1. Microstructure Investigation

3.1.1. XRD Analysis. The XRD patterns of the unpressed sample ($N = 0$) and sample processed by ECAP ($N = 4$) between $2\theta = 20$ and 155° are given in Figure 1(a). Besides the fundamental reflections of the aluminum matrix, very fine reflections with low intensity, are observed (inset of Figure 1(a)). These reflections correspond to the presence of a second phase. In several AlFeSi commercial alloys, the phase present is usually denoted by α-AlFeSi (α-phase). Possible chemistry of the phase is Al_8Fe_2Si, $Al_{12}Fe_3Si$, or $Al_{15}Fe_3Si_2$. It should be noted, however, that the iron-rich phases form in commercial purity aluminum alloys in two very different forms. First is the coarse 1-2 μm "constituents" formed by eutectic solidification, at a spacing of the dendrite cell size set by the solidification rate. Depending on the particular Fe/Si ratio and the casting rate, different phases may result, that is, the metastable Al_6Fe or the α-phase. Besides, there is a residual Fe content typically about 0.04 wt% left in solution after casting. This can be partially precipitated as the α-phase during homogenization anneals especially at the temperature used in the present work, only 500°C, or during low temperature anneals following deformation. These solid state precipitates (usually called dispersoids) are very fine (0.1 to 0.2 μm) particles. In this work, the X-ray emission spectrometry by SEM showed the presence of Al, Fe, and Si in the precipitates. Therefore, the α-phase is present in the aluminum alloy before deformation and persists during ECAP since reflections, with lower intensities, corresponding to this phase are still observed after ECAP (inset of Figure 1(a)).

After ECAP, a pronounced reflection broadening was observed (Figure 1(b)). This is due to grain refinement and enhanced lattice distortions of the processed samples. In addition, the $K\alpha_2$ reflection, which was well resolved for the nondeformed sample ($N = 0$), became scarcely visible after ECAP because of the reflection broadening. The cell size and the dislocation density evolutions during ECAP are given in Figure 2. An intensive grain refinement was observed after the ECAP process. In fact, the grain size reached ~200 nm after the two first passes and slightly continued to decrease to attain ~192 nm after four passes. The decrease of the cell size during ECAP was accompanied by a significant increase of the dislocation density. It should be noted that during severe plastic deformation such as ECAP, subgrain boundaries evolve in the grain structures [21] developing areas with coherent crystalline domain (cell structure) [22, 23]. The dislocations distribution may affect the size of these areas.

3.1.2. Microscopic Analysis. After annealing of the as-cast aluminum, at 500°C, OM reveals that the microstructure was heterogeneous and consists of equiaxed coarse grains with an average size of ~80 μm and significantly coarser ones with an average grain size of about 500 μm, due to prolonged annealing (Figure 3).

The SEM analysis gives information about the resulting microstructure of the material before and after ECAP. Before ECAP, SEM observations show that the aluminum matrix contains precipitates located essentially at grain boundaries, even though some precipitates were also observed in the grain interior (Figures 4(a) and 4(b)). As demonstrated by XRD and X-ray emission spectrometry by SEM, these precipitates correspond to the α-phase. After ECAP, a heterogeneous microstructure was obtained consisting of ultrafine equiaxed grains and elongated grains (Figures 4(c) and 4(d)). We showed in a previous work [13] that the presence of the α-AlFeSi essentially at grain boundaries allowed to reveal

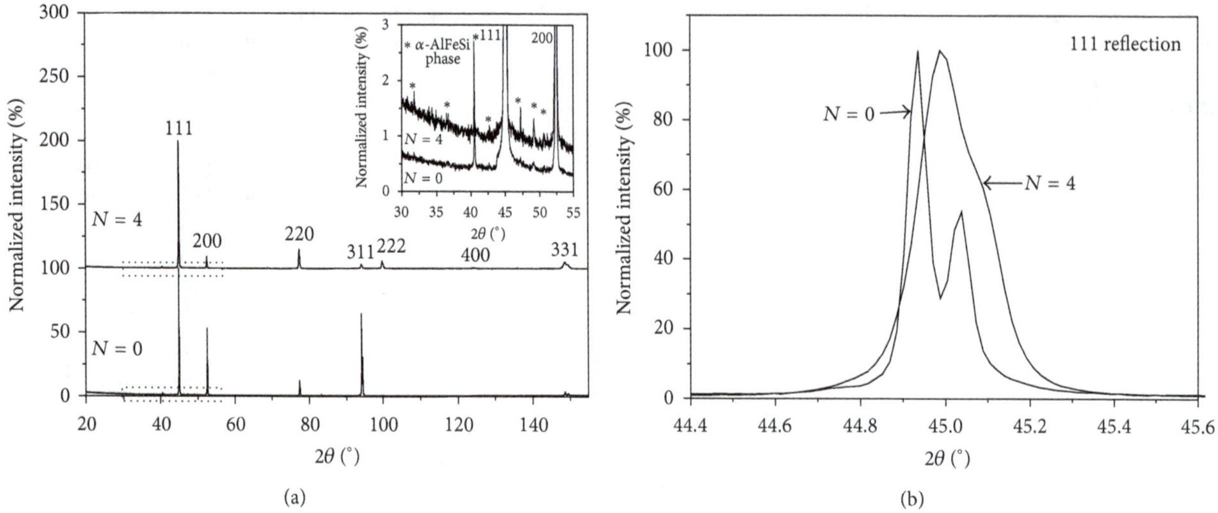

Figure 1: (a) X-ray diffraction pattern and (b) the 111 aluminum reflection of the nondeformed ($N = 0$) and deformed ($N = 4$) aluminium samples. Dashed rectangles are enlarged in the inset showing the α-AlFeSi particles reflections.

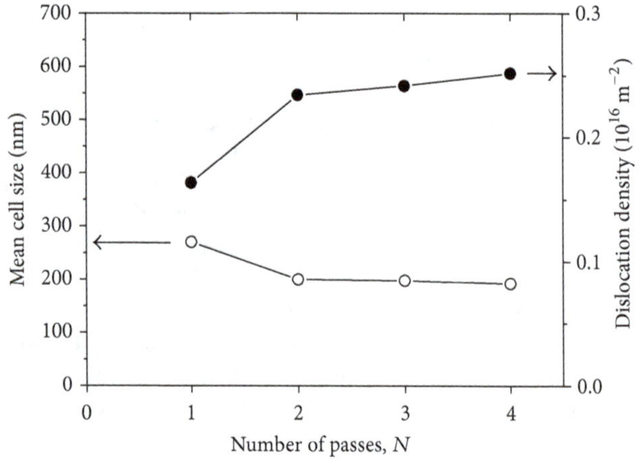

Figure 2: The cell size and the dislocation density evolutions during ECAP.

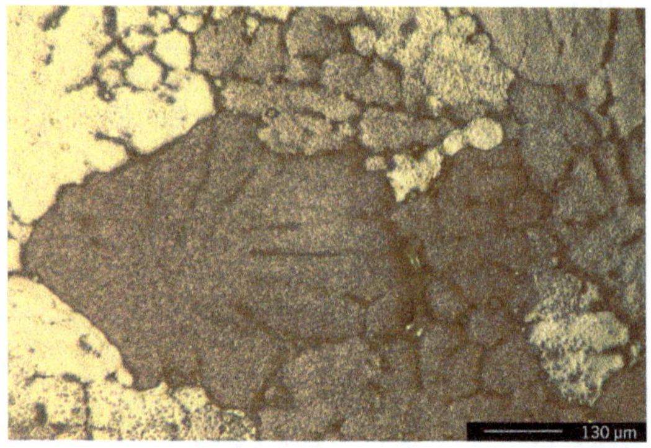

Figure 3: OM micrograph showing the microstructure of the nondeformed ($N = 0$) aluminium.

the initial shape of the grains and to follow their evolution during ECA pressing by OM; that is, the grains appeared to be elongated in the side face of the sample after ECAP. Therefore, ECAP leads to grain refinement of the aluminum alloy to some extent. Similar trend has been reported in microstructure investigation by transmission electron microscopy (TEM) on high purity aluminum processed by ECAP [24, 25].

In addition to grain refinement and enhanced dislocation density, high strain imposed by ECAP leads to the breakdown (Figure 4(e)) and/or partial dissolution of the α-phase particles (Figure 4(f)).

3.2. Corrosion Behavior. When Al contacts with water solution, a rapid hydrated oxide film (AlOOH, also regarded as $Al_2O_3 \cdot H_2O$) will form on the surface. The hydrated passive film formation process is as follows [26]:

$$Al + H_2O \longrightarrow AlOH + H^+ + e^- \quad (5)$$

$$AlOH + H_2O \longrightarrow Al(OH)_2 + H^+ + e^- \quad (6)$$

$$Al(OH)_2 + H_2O \longrightarrow AlOOH + H^+ + e^- \quad (7)$$

Then, the total electrode reaction is

$$Al + 2H_2O \longrightarrow AlOOH + 3H^+ + 3e^- \quad (8)$$

In contact with Cl^- containing solution, the film will be attacked, and the anodic dissolution products (AlOHCl, $AlOHCl_2$) having a good water solubility will promote the anodic dissolution of the passive film [27]. Figure 5 shows typical potentiodynamic polarization curves plotted from −1.2 V to 0 V at 10 mV/s for the ECAP'ed ($N = 2$ and 4) and unpressed ($N = 0$) Al samples. All samples were immersed in a 3.5 wt% solution for about 3 min at pH = 6.4

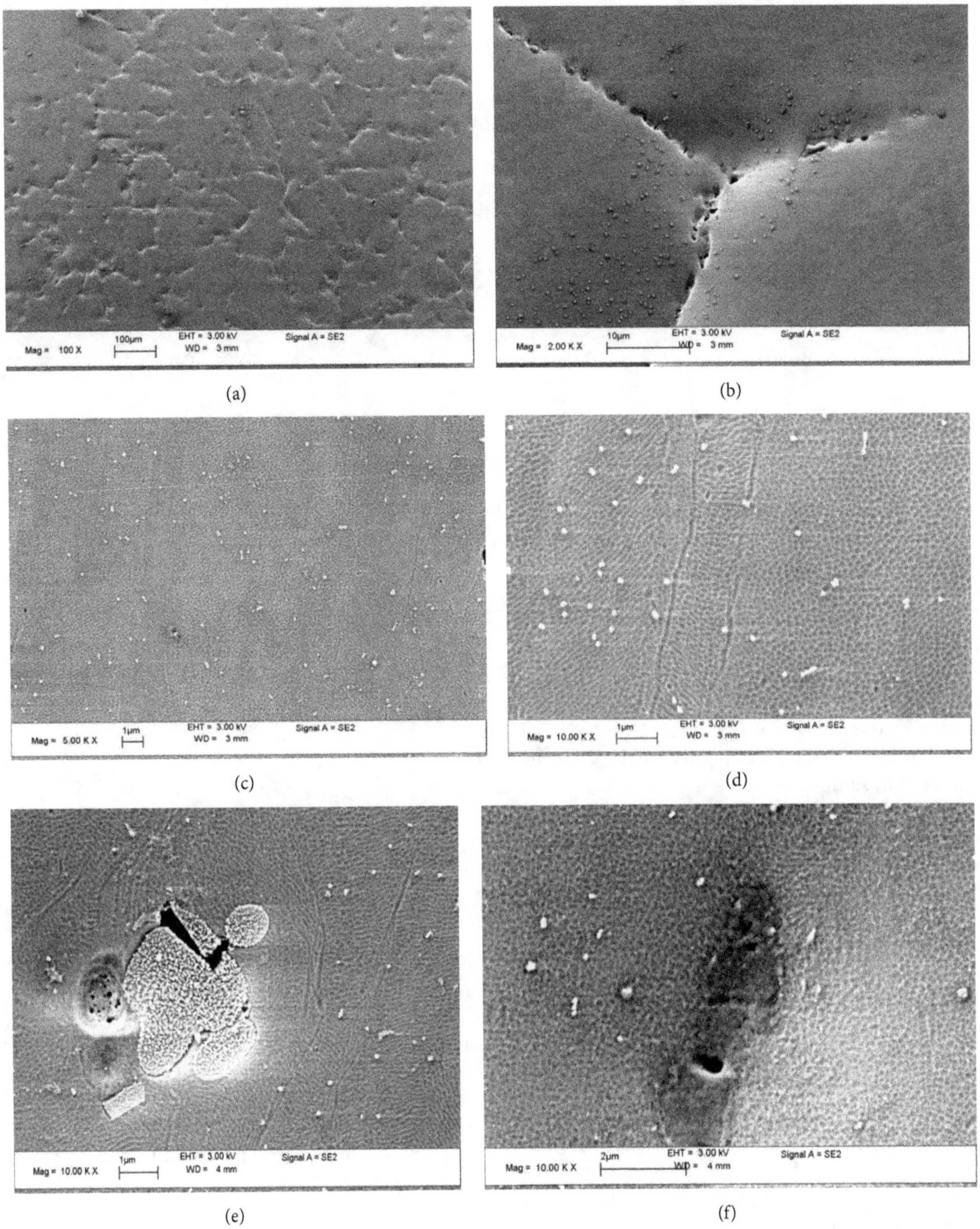

FIGURE 4: SEM micrographs with two magnifications showing the microstructure of (a, b) the nondeformed ($N = 0$) aluminium matrix containing α-phase particles, (c, d) after ECAP ($N = 4$), and the effect of ECAP on the α-phase particles, that is, (e) breakdown and (f) partial dissolution.

before polarization tests to achieve their stable OCP values. A similar shape is observed for the three samples. The more ECAP passes correlate with the nobler corrosion potentials and the higher corrosion current densities (i_{corr}). In fact, corrosion current densities (i_{corr}) of the unpressed ($N = 0$) and ECAP'ed samples ($N = 2$ and 4) are equal to about 8×10^{-6}, 2×10^{-5}, and 4×10^{-5} A/cm^2, respectively. The i_{corr} values were determined, from the polarization curves, by extrapolation of the linear Tafel plots of the cathodic branches. The results show that the ECAP'ed Al sample ($N = 4$) has nobler corrosion potential value (E_{corr}) but higher i_{corr} and less corrosion resistance when compared with nondeformed ($N = 0$) and ECAP'ed ($N = 2$). Therefore, ECAP increased the susceptibility to corrosion of the aluminum alloy. While the value of E_{corr} represents a thermodynamic characteristic of a given metal-electrolyte

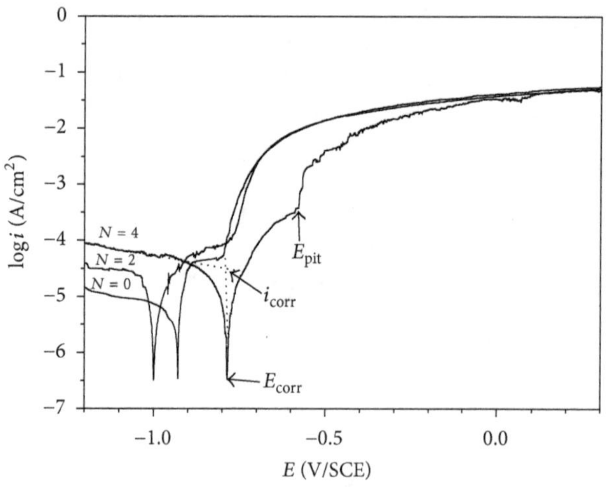

FIGURE 5: Potentiodynamic polarization curves obtained at 25°C and 10 mV/s with the nonpressed ($N = 0$) and ECAP'ed ($N = 2$ and 4) samples in solution of 3.5 wt.% NaCl solution at pH 6.4.

system [28], the i_{corr} value represents the kinetics of the metal corrosion and reflects more accurately the general corrosion rate than the value of E_{corr}. It is important to note that i_{corr} values of the ECAP'ed samples increased with increasing the ECAP number of passes, and the i_{corr} after 4 passes is of several orders of magnitude higher than that obtained for the nondeformed ($N = 0$) sample.

Comparing the cathodic branches controlling the uniform corrosion process for all samples, it can be seen that, while the nondeformed sample shows low cathodic current density i_{cath}, the ECAP'ed sample ($N = 4$) exhibits the highest i_{cath}, which is equivalent to corrosion rate. During the positive potential scan, curves do not exhibit active dissolution region near E_{corr}. The lack of active dissolution could be assigned to the hydrated passive films formed on the Al samples. When the potential moves towards more anodic values, narrow anodic plateaus are observed for all samples, indicating formation of a passive layer on the electrode surface, and then the current rises rapidly, denoting destruction of this layer at a pitting potential E_{pit} [29]. It is important to point out that, besides the higher i_{corr} value, the current density of the passive layer increases, and its potential range decreases with ECAP, as a result of the decrease in pitting corrosion resistance for the ECAP'ed samples ($N = 2$ and 4).

Figure 6(a) shows the impedance spectra on Nyquist plots obtained at open circuit potential (OCP) values from the ECAP'ed ($N = 2$ and 4) and nondeformed ($N = 0$) Al samples. For all samples, the results of EIS measurements showed that the diameter of the arc, giving the polarization resistance (R_p), exceeds $5\,k\Omega{\cdot}cm^2$. High R_p values indicate that the aluminum forms a relatively stable oxide layer and has an excellent corrosion resistance in the naturally aerated 3.5 wt% NaCl solution. Compared with that of the unpressed sample, the R_p value decreased with increasing ECAP passes; that is, the corrosion resistance decreased with increasing ECAP passes. The adopted Randles equivalent circuit, which is very common for these plots, is illustrated in Figure 6(b). It shows

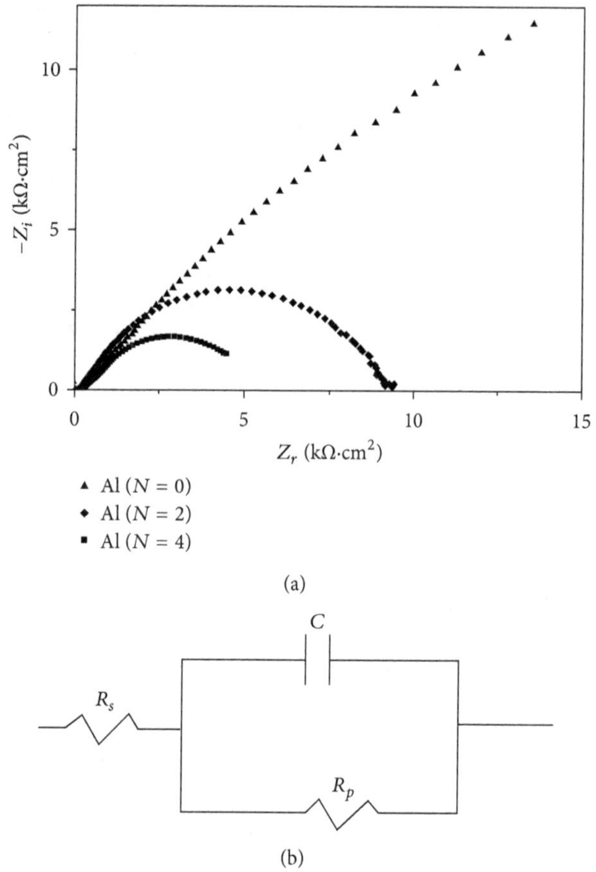

FIGURE 6: Impedance spectra of the ECAP'ed samples and unpressed sample obtained at open circuit (a) and 25°C in solution of 3.5 wt.% NaCl solution at pH 6.4 and the electrical equivalent circuit (b).

the electrochemical system of metal/oxide layer/electrolyte, where R_s, R_p, and C represent the electrolyte resistance, the charge-transfer resistance of corrosion reaction and the interfacial capacitance, respectively. This capacitance contains the contribution of the passive film C_f and the double layer C_{dl}, which are in series. The results of EIS measurements are consistent with those of the Tafel plots since i_{corr} is inversely proportional to R_p and can be obtained from R_p via the Stern-Geary equation [30] given as

$$i_{corr} = \frac{\beta_a \beta_c}{2.3 (\beta_a + \beta_c) R_p}, \quad (9)$$

where i_{corr} is the corrosion current density, β_a and β_c are the anodic and cathodic Tafel coefficients, respectively, and R_p is the polarization resistance. The Tafel coefficients are generated from the anodic and cathodic branches of the Tafel plots.

The influence of scan rate, v (in the range 5–100 $mV{\cdot}s^{-1}$), on the potentiodynamic polarization behavior of the unpressed and ECAP'ed samples has been studied in 3.5 wt.% NaCl solution and pH 6.4 (data obtained are not shown herein). It has been shown that the increase in v shifts E_{pit}, in all cases, towards more noble values. The variation of E_{pit} with $\log v$, for the unpressed and ECAP'ed samples, is

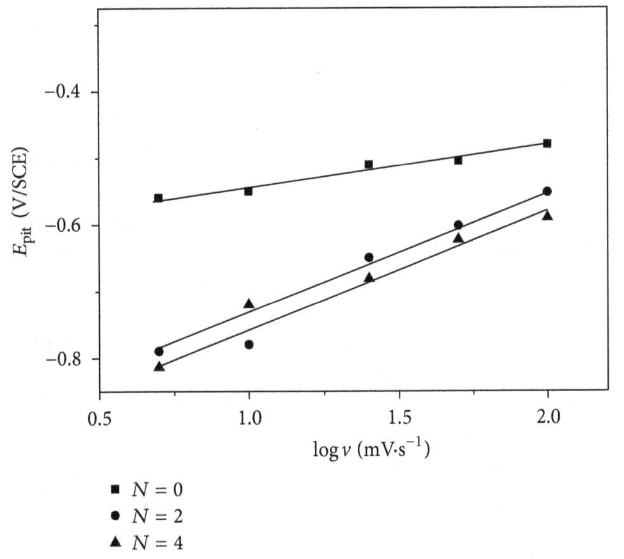

FIGURE 7: Dependence of E_{pit} on the logarithm of scan rate ($\log v$) for the ECAP'ed ($N = 2$ and 4) and unpressed ($N = 0$) samples, obtained at 25°C in 3.5 wt.% NaCl solution at pH 6.4.

depicted in Figure 7. A linear relationship is obtained fitting the relation

$$E_{pit} = a + b \log v, \quad (10)$$

where a and b are constants depending on whether the sample was pressed or not. Similar results were obtained for Zn [31], pure Al, and Al-Cu alloys [32]. The linear trend has been explained in terms of incubation time (the time needed for Cl^- ions to penetrate the passive film and to reach the metal surface) [33, 34]. At high scan rate, initiation of passivity breakdown occurs only at more positive potentials corresponding to a sufficient short pit incubation time. For all samples, the increase of the scan rate up to 100 mV·s^{-1} leads to the increase of the pitting potential. It is easy to note that E_{pit} increases for all samples in the following order: 4 passes sample < 2 passes sample < nondeformed sample; that is, the pitting corrosion resistance decreases with increasing the ECAP passes.

Typical polarization curves of the ECAP'ed sample ($N = 2$) in different NaCl concentrations at pH = 6.4 and scan rate of 10 mV·s^{-1} in the anodic potential domain are shown in Figure 8(a). The curves show that an increase in the chloride concentration produces changes in both the anodic current density and the pitting potential E_{pit} of the sample. Figure 8(b) shows, for the unpressed and ECAP'ed samples, the relation between E_{pit} and logarithm of Cl^- concentration ($\log[Cl^-]$); straight lines are observed, fitting the following equation:

$$E_{pit} = c - d \log [Cl^-], \quad (11)$$

where c and d are constants depending on whether the sample was pressed or not. The linear dependence is well established for pure Al [35] and is entirely due to the variation of the solution ohmic drop IR at the pit mouth. These results show

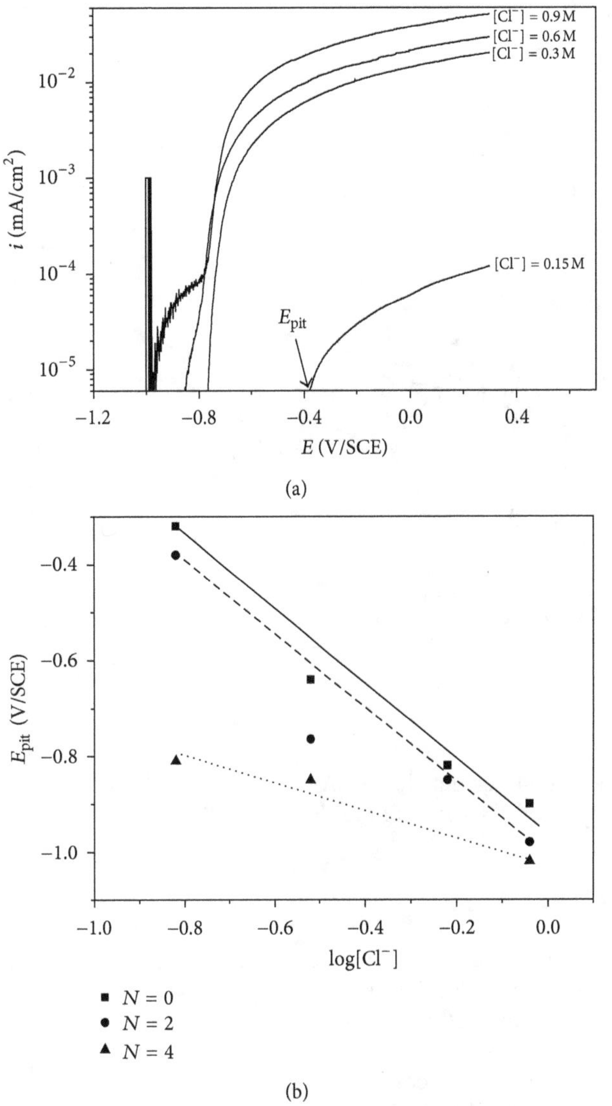

FIGURE 8: Polarization curves of the ECAP'ed sample ($N = 2$) in different NaCl concentrations at pH = 6.4, 25°C, scan rate of 10 mV·s^{-1} (a), and evolution of E_{pit} as a function $\log[Cl^-]$ for the nonpressed ($N = 0$) and ECAP'ed samples ($N = 2$ and 4) (b).

that the pitting corrosion depends on NaCl concentration. The increase of NaCl concentration leads to a decrease of the E_{pit} to more negative values accounting for a decreased resistance to pitting corrosion. This resistance increases in the same order as for the effect of scan rate: 4 passes sample < 2 passes sample < nondeformed sample.

Figure 9 displays the corrosion rate variation with the immersion time during testing of the nondeformed and ECAP'ed samples. The corrosion rates of all samples increase with increasing immersion time or with increase of ECAP passes; that is, the 4 passes ECAP'ed sample has much higher corrosion rate than the nondeformed and 2 passes ECAP'ed samples. Also, it should be noted that the corrosion rate of the 4 passes sample is about 4 times higher than the nondeformed sample after immersion for 30 days. This confirms the above

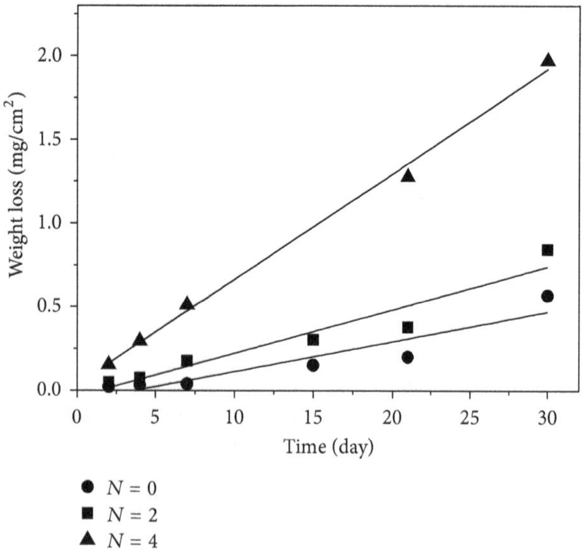

Figure 9: Corrosion rate variation with the immersion time of the nondeformed ($N = 0$) and ECAP'ed ($N = 2$ and 4), at 25°C in 3.5 wt.% NaCl solution at pH = 6.4.

results indicating that ECAP leads to the decrease of the corrosion resistance of the aluminum alloy.

4. Discussion

The corrosion behavior of a commercial aluminum alloy processed by ECAP was investigated in NaCl solution using potentiodynamic polarization and immersion tests. Polarization curves showed that the increase of ECAP passes leads, in addition to the increase of weight loss, to higher corrosion current densities and a decrease of pitting corrosion resistance. Also, electrochemical impedance spectra (EIS) tests showed that the polarization resistance (R_p) decreased with increasing ECAP passes. Therefore, corrosion resistance of the alloy remarkably diminished with increasing the ECAP number of passes. It is not easy to draw a clear conclusion about the reasons for this behavior since many contributions (grain refinement, internal stresses, second phase, etc.) should operate simultaneously. Previous works reported in the literature [36, 37] that grain refinement leads to a high density of grain boundaries which have higher energy than the bulk of the material. This enhances electron diffusion and leads to a more stable passive film and, therefore, to an enhanced corrosion resistance. Controversially, grain refinement may provide more sites to initiate corrosion which enhanced corrosion rate and consequently decreased corrosion resistance. Whether grain refinement is detrimental or not to corrosion resistance could be related to the material nature and environment [7]. It is clear from our results that the grain refinement of the ECAP'ed Al alloy to some extent was accompanied by increased susceptibility to corrosion.

It is well established that internal stresses, especially related to high dislocation accumulation generated by ECAP, may or may not result in the improvement of the corrosion resistance depending on the material [7].

ECAP leads to an intensive grain refinement accompanied by an increased dislocation density. For example, the remaining dislocation densities were equal to 0.25×10^{16} and $0.26 \times 10^{16}\ m^{-2}$ for $N = 2$ and 4, respectively. The increase of the dislocation density with the number of ECAP passes could be one of the reasons for decreasing the corrosion resistance of the deformed Al alloy. This is in agreement with the results of Schultze et al. [38].

It is known that severe plastic deformation can lead to the quick decomposition of the Al-based supersaturated solid solution [39, 40]. The start of the decomposition of supersaturated solid solution and formation of nanoprecipitates could be one of the reasons for decreasing the corrosion resistance. However, in the present work, the ECAP process leads to partial dissolution of the α-phase. In fact, since there is a large volume of insoluble eutectic constituent present, the formation or dissolution of the α-phase cannot be followed by observation of the α-phase X-ray reflections. However, a very careful measurement of the aluminum lattice parameter can potentially detect changes in iron solute contents. Thus, a reflection shift to higher 2θ angles was observed after ECAP (Figure 1(b)) indicating the decrease of the aluminum lattice parameter during ECAP. For example, the lattice parameter values of the nondeformed ($N = 0$) and ECAP'ed ($N = 4$) samples are 4.05420 and 4.04957 Å, respectively, corresponding to a relative variation of about −0.11%. The decrease of the lattice parameter is due to partial dissolution of the α-phase precipitates since the dissolution of Fe and Si in aluminum matrix decreases the lattice parameter. Indeed, the relative variations of the lattice parameter, a, with the concentration, c, of both Fe and Si, $1/a\,(\Delta a/\Delta c)$, are −13.6 [41] and −2.6% [42], respectively. The partial dissolution of the α-phase was experimentally supported by SEM investigation (Figure 4(f)). It should be noted that, besides α-phase dissolution, oriented strain introduced by the ECAP process could affect the lattice parameter value.

SEM examinations were carried out before and after immersing the nondeformed ($N = 0$) and ECAP'ed ($N = 4$) Al samples in 3.5 wt.% NaCl solution at pH 6.4 for 30 days. The SEM micrographs (Figures 10(a) and 10(b)) show obvious localized corrosion damage, particularly on 4 passes ECAP'ed sample. Figure 10(a) exhibits relatively small and hemispherical pits with different sizes, grown throughout the surface of the nondeformed sample. Severity of pitting attack is enhanced by ECAP process, which produces, in particular after 4 passes, much larger (~0.5 μm) and deeper rectangular pits. SEM micrograph of Figure 10(c) shows that the α-phase particles are the pit initiation sites. So, this attack takes place at the contact between the particles and the Al matrix. In the present work, susceptibility to corrosion results from local galvanic cells formed between the α-phase particles and the aluminum matrix. These particles act as cathodes which allowed the reduction of oxygen and drive anodic dissolution of the surrounding matrix. The particles were broken down with the ECAP passes (Figure 4(e)). Consequently, the number of the galvanic cells increases and

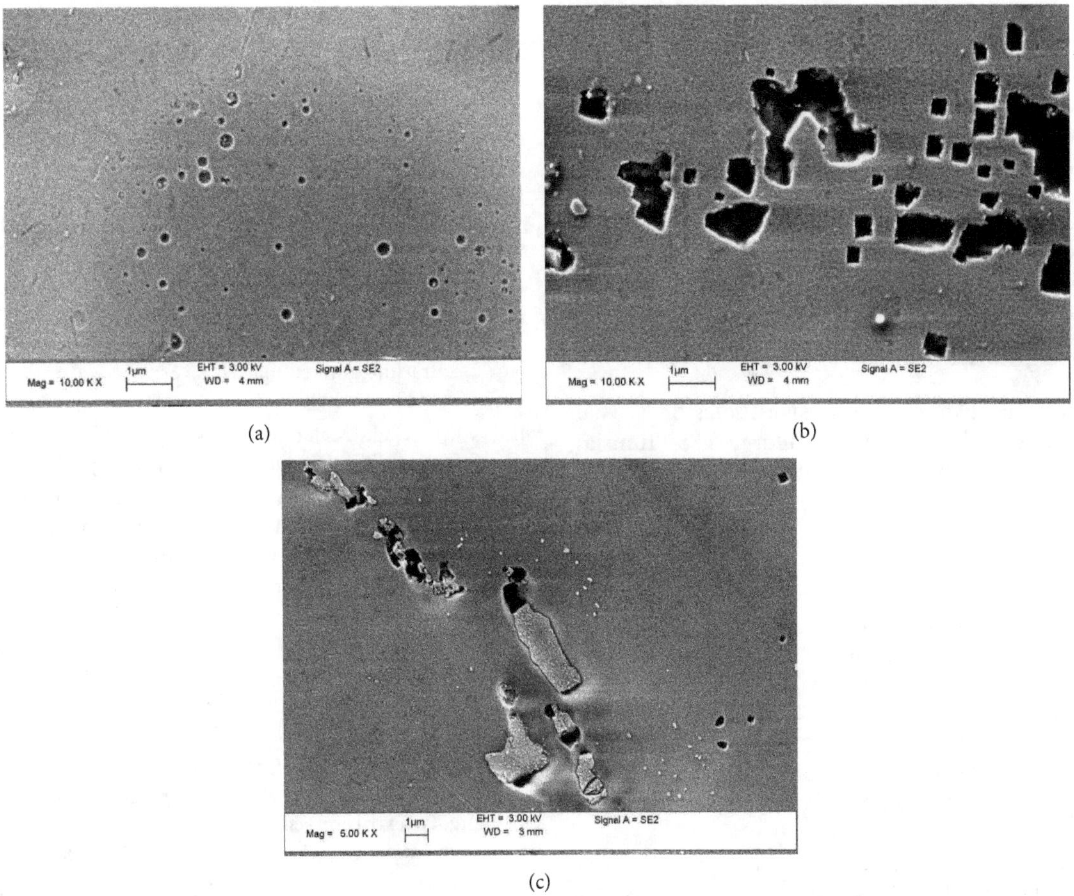

FIGURE 10: SEM micrographs of the nondeformed ($N = 0$) (a) and ECAP'ed ($N = 4$) (b) samples and the α-phase particles which are the pit initiation sites (c).

so is the contact surface between the broken particles and the aluminum matrix. This leads to enhanced dissolution of the matrix which explains the decrease of the corrosion resistance of the aluminum alloy with ECAP passes. This is in agreement with the reported results of Wei et al. [43], which showed that the fragmentation of second phase particles improved localized corrosion behavior in Al-Mn alloys processed by accumulative roll bonding (ARB). Naeini et al. [44] studied the pitting corrosion susceptibility of ultrafine grained 5052 Al alloy (AA 5052) sheets deformed by ARB. They showed that as the ARB number of passes increased, intermetallic Mg and Fe-rich particles were refined so their total surface area increased. Thus, the preferential sites for initiation of pitting corrosion increased. In contrast, Song et al. [45] found that the refinement of Si containing impurities in pure aluminum processed by ECAP reduced the susceptibility of pitting corrosion which is consistent with the 'small cathode, large anode' mechanism. Comparable results for AA 1050 aluminum alloy deformed by ECAP have been previously reported [46]. It should be noted that a critical size of particles exists in aluminum alloys below which the particles cannot act as distinct electrochemical entities [47]. For Al-Cu-Mg alloys, this critical size varied between 3 and 8 nm. For the 1100 Al alloy, the particles size was largely higher. Accordingly, it is not possible that a stable protective passive oxide film forms.

5. Conclusion

The effect of ECAP process on the corrosion behavior of a commercial aluminum alloy was investigated in NaCl solution using potentiodynamic polarization and immersion tests. The alloy contains a low volume fraction of α-AlFeSi. ECAP leads to an intensive grain refinement of the aluminum alloy to the submicrometer level accompanied by an increased dislocation density. All electrochemical tests confirm that ECAP leads to a significant decrease of the corrosion resistance of the alloy: (i) potentiodynamic curves showed that the more ECAP passes correlate with the higher corrosion current densities and the decrease of pitting corrosion resistance; (ii) electrochemical impedance spectra (EIS) tests showed that the polarization resistance (R_p) decreased with increasing ECAP passes; (iii) increase of scan rate and NaCl concentration leads to an increase and a decrease of E_{pit} to more positive and negative potentials, respectively. The increase of susceptibility to corrosion results from local galvanic cells formed between the α-AlFeSi and the aluminum matrix. The α-phase particles act as cathodes which allowed the reduction of oxygen and drive anodic dissolution of the surrounding matrix. The particles were broken down with the ECAP passes which increase the number of the galvanic cells. This leads to enhanced dissolution of the matrix which explains the decrease of the corrosion resistance of the aluminum alloy with ECAP passes.

Conflict of Interests

The authors recognized that there is no conflict of interests concerning the results obtained.

Acknowledgments

The authors would like to acknowledge Professor N. Njah from the Laboratory of Applied Metallurgy, FSS (Tunisia), and Mr. N. Bouras from the University of Sétif (Algeria) for useful helps.

References

[1] V. I. Levit, R. I. Kuznetsov, N. A. Smirnova, V. P. Chernyshev, L. S. Davydova, and M. V. Degtyarev, "Tension of single crystals of KHN77TYUR alloy under hydrostatic-pressure up to 1 GPA," *Fizika Metallov I Metaaovedene*, vol. 59, no. 1, pp. 184–191, 1985.

[2] Y. Saito, N. Tsuji, H. Utsunomiya, T. Sakai, and R. G. Hong, "Ultra-fine grained bulk aluminum produced by accumulative roll-bonding (ARB) process," *Scripta Materialia*, vol. 39, no. 9, pp. 1221–1227, 1998.

[3] J. G. Brunner, N. Birbilis, K. D. Ralston, and S. Virtanen, "Impact of ultrafine-grained microstructure on the corrosion of aluminium alloy AA2024," *Corrosion Science*, vol. 57, pp. 209–214, 2012.

[4] I. Son, H. Nakano, S. Oue, S. Kobayashi, H. Fukushima, and Z. Horita, "Effect of equal-channel angular pressing on pitting corrosion resistance of anodized aluminum-copper alloy," *Transactions of Nonferrous Metals Society of China, English Edition*, vol. 19, no. 4, pp. 904–908, 2009.

[5] M. Mabuchi and K. Higashi, "Mechanical properties of pure aluminium processed by equal channel angular extrusion," *Journal of Materials Science Letters*, vol. 17, no. 3, pp. 215–217, 1998.

[6] R. Z. Valiev and I. V. Alexandrov, "Nanostructured materials from severe plastic deformation," *Nanostructured Materials*, vol. 12, no. 1, pp. 35–40, 1999.

[7] K. D. Ralston and N. Birbilis, "Effect of grain size on corrosion: a review," *Corrosion*, vol. 66, no. 7, Article ID 075005, 13 pages, 2010.

[8] E. Kus, Z. Lee, S. Nutt, and F. Mansfeld, "A comparison of the corrosion behavior of nanocrystalline and conventional Al 5083 samples," *Corrosion*, vol. 62, no. 2, pp. 152–161, 2006.

[9] D. Song, A. Ma, J. Jiang, P. Lin, and D. Yang, "Corrosion behavior of ultra-fine grained industrial pure Al fabricated by ECAP," *Transactions of Nonferrous Metals Society of China, English Edition*, vol. 19, no. 5, pp. 1065–1070, 2009.

[10] T. S. Mahmoud, "Effect of friction stir processing on electrical conductivity and corrosion resistance of AA6063-T6 Al alloy," *Proceedings of the Institution of Mechanical Engineers C*, vol. 222, no. 7, pp. 1117–1123, 2008.

[11] A. Korchef, N. Njah, Y. Champion et al., "Material flow during equal channel angular pressing of aluminum containing Al_8Fe_2Si recipitates," *Advanced Engineering Materials*, vol. 6, no. 4, pp. 222–228, 2004.

[12] A. Korchef, Y. Champion, and N. Njah, "X-ray diffraction analysis of aluminium containing Al_8Fe_2Si processed by equal channel angular pressing," *Journal of Alloys and Compounds*, vol. 427, no. 1-2, pp. 176–182, 2007.

[13] A. Korchef, N. Njah, J. Masmoudi et al., "Evolution of the mechanical properties of aluminum containing Al_8Fe_2Si precipitates during equal channel angular pressing," *Advanced Engineering Materials*, vol. 6, no. 8, pp. 610–643, 2004.

[14] Y. Iwahashi, J. Wang, Z. Horita, M. Nemoto, and T. G. Langdon, "Principle of equal-channel angular pressing for the processing of ultra-fine grained materials," *Scripta Materialia*, vol. 35, no. 2, pp. 143–146, 1996.

[15] V. M. Segal, "Materials processing by simple shear," *Materials Science and Engineering A*, vol. 197, no. 2, pp. 157–164, 1995.

[16] B. Zaid, D. Saidi, A. Benzaid, and S. Hadji, "Effects of pH and chloride concentration on pitting corrosion of AA6061 aluminum alloy," *Corrosion Science*, vol. 50, no. 7, pp. 1841–1847, 2008.

[17] J. I. Langford, "A rapid method for analysing the breadths of diffraction and spectral lines using the Voigt function," *Journal of Applied Crystallography*, vol. 11, pp. 10–14, 1978.

[18] L. V. Azàroff and M. Buerger, *The Powder Method*, Mc Grow-Hill, New York, NY, USA, 1958.

[19] A. Rebhi, T. Makhlouf, N. Njah, Y. Champion, and J. Couzinié, "Characterization of aluminum processed by equal channel angular extrusion: effect of processing route," *Materials Characterization*, vol. 60, no. 12, pp. 1489–1495, 2009.

[20] L. Li, T. Ungár, Y. D. Wang et al., "Simultaneous reductions of dislocation and twin densities with grain growth during cold rolling in a nanocrystalline Ni–Fe alloy," *Scripta Materialia*, vol. 60, no. 5, pp. 317–320, 2009.

[21] Y. Ito and Z. Horita, "Microstructural evolution in pure aluminum processed by high-pressure torsion," *Materials Science and Engineering A*, vol. 503, no. 1-2, pp. 32–36, 2009.

[22] R. Z. Valiev, R. K. Islamgaliev, and I. V. Alexandrov, "Bulk nanostructured materials from severe plastic deformation," *Progress in Materials Science*, vol. 45, no. 2, pp. 103–189, 2000.

[23] A. Zangiabadi and M. Kazeminezhad, "Development of a novel severe plastic deformation method for tubular materials: Tube Channel Pressing (TCP)," *Materials Science and Engineering A*, vol. 528, no. 15, pp. 5066–5072, 2011.

[24] Y. Iwahashi, Z. Horita, M. Nemoto, and T. G. Langdon, "An investigation of microstructural evolution during equal-channel angular pressing," *Acta Materialia*, vol. 45, no. 11, pp. 4733–4741, 1997.

[25] Y. Iwahashi, Z. Horita, M. Nemoto, and T. G. Langdon, "The process of grain refinement in equal-channel angular pressing," *Acta Materialia*, vol. 46, no. 9, pp. 3317–3331, 1998.

[26] J. D. Gorman, A. E. Hughes, D. Jamieson, and P. K. J. Paterson, "Oxide formation on aluminum alloys in boiling deionised water and NaCl, $CeCl_3$ and $CrCl_3$ solutions," *Corrosion Science*, vol. 45, no. 6, pp. 1103–1124, 2003.

[27] Z. Szklarska-Smialowska, "Pitting corrosion of aluminum," *Corrosion Science*, vol. 49, no. 9, pp. 1743–1767, 1999.

[28] D. Song, A. Ma, J. Jiang, P. Lin, D. Yang, and J. Fan, "Corrosion behavior of equal-channel-angular-pressed pure magnesium in NaCl aqueous solution," *Corrosion Science*, vol. 52, no. 2, pp. 481–490, 2010.

[29] A. Rucinskiene, G. Bilkulcius, I. Gudaviciùté, and E. Juzeliùnas, "Magnetic field effect on stainless steel corrosion in $FeCl_3$ solution," *Electrochemistry Communications*, vol. 4, no. 1, pp. 86–91, 2002.

[30] W. S. Tait, *An Introduction to Electrochemical Corrosion Testing for Practicing Engineers and Scientists*, Pair O Docs, Racine, Wis, USA, 1994.

[31] S. S. Abd El Rehim, E. E. F. El-Sherbini, and M. A. Amin, "Pitting corrosion of zinc in alkaline medium by thiocyanate ions," *Journal of Electroanalytical Chemistry*, vol. 560, no. 2, pp. 175–182, 2003.

[32] M. A. Amin, S. S. Abd El Rehim, S. O. Moussa, and A. S. Ellithy, "Pitting corrosion of Al and Al–Cu alloys by ClO_4^- ions in neutral sulphate solutions," *Electrochimica Acta*, vol. 53, no. 18, pp. 5644–5652, 2008.

[33] M. Metikas-Hukovic and I. NMilosev, "The effect of silicate ion on the corrosion of zinc powder in alkaline solutions," *Journal of Applied Electrochemistry*, vol. 22, pp. 448–455, 1992.

[34] H. Do Duc and P. Tissot, "Rotating disc and ring disc electrode studies of tin in neutral phosphate solution," *Corrosion Science*, vol. 19, no. 3, pp. 191–197, 1979.

[35] J. Soltis, N. J. Laycock, and D. Krouse, "Temperature dependence of the pitting potential of high purity aluminium in chloride containing solutions," *Corrosion Science*, vol. 53, no. 1, pp. 7–10, 2011.

[36] Y. H. Jang, S. S. Kim, C. D. Yim, C. Lee, and S. J. Kim, "Corrosion behaviour of friction stir welded AZ31B Mg in 3·5%NaCl solution," *Corrosion Engineering Science and Technology*, vol. 42, no. 2, pp. 119–122, 2007.

[37] C. Op't Hoog, N. Birbilis, and Y. Estrin, "Corrosion of pure Mg as a function of grain size and processing route," *Advanced Engineering Materials*, vol. 10, no. 6, pp. 579–582, 2008.

[38] J. W. Schultze, B. Davepon, F. Karman, C. Rosenkranz, A. Schreiber, and O. Voigt, "Corrosion and passivation in nanoscopic and microscopic dimensions: the influence of grains and grain boundaries," *Corrosion Engineering Science and Technology*, vol. 39, no. 1, pp. 45–52, 2004.

[39] B. B. Straumal, B. Baretzky, A. A. Mazilkin et al., "Formation of nanograined structure and decomposition of supersaturated solid solution during high pressure torsion of Al–Zn and Al–Mg alloys," *Acta Materialia*, vol. 52, no. 15, pp. 4469–4478, 2004.

[40] A. A. Mazilkin, B. B. Straumal, E. Rabkin et al., "Softening of nanostructured Al–Zn and Al–Mg alloys after severe plastic deformation," *Acta Materialia*, vol. 54, no. 15, pp. 3933–3939, 2006.

[41] J. Blétry, "Effets de taille dans les solutions solides de l'aluminium avec les métaux de transition de la première série," *Journal of Physics and Chemistry of Solids*, vol. 31, no. 6, pp. 1263–1272, 1970.

[42] S. K. Pabi, "Mechanism of the dissolution of Si in Al," *Physica Status Solidi A*, vol. 50, no. 1, pp. K9–K12, 1978.

[43] W. Wei, K. X. Wei, and Q. B. Du, "Corrosion and tensile behaviors of ultra-fine grained Al–Mn alloy produced by accumulative roll bonding," *Materials Science and Engineering A*, vol. 454, pp. 536–541, 1995.

[44] M. F. Naeini, M. H. Shariat, and M. Eizadjou, "On the chloride-induced pitting of ultra fine grains 5052 aluminum alloy produced by accumulative roll bonding process," *Journal of Alloys and Compounds*, vol. 509, no. 14, pp. 4696–4700, 2011.

[45] D. Song, A. B. Ma, J. H. Jiang, P. H. Lin, and J. Shi, "Improving corrosion resistance of pure Al through ECAP," *Corrosion Engineering Science and Technology*, vol. 46, no. 4, pp. 505–512, 2011.

[46] M. Chung, Y. Choi, J. Kim, Y. Kim, and J. Lee, "Effect of the number of ECAP pass time on the electrochemical properties of 1050 Al alloys," *Materials Science and Engineering A*, vol. 366, no. 2, pp. 282–291, 2004.

[47] K. D. Ralston, N. Birbilis, M. Weyland, and C. R. Hutchinson, "The effect of precipitate size on the yield strength-pitting corrosion correlation in Al–Cu–Mg alloys," *Acta Materialia*, vol. 58, no. 18, pp. 5941–5948, 2010.

20

MCMC-Based Fatigue Crack Growth Prediction on 2024-T6 Aluminum Alloy

Xu Du,[1] Yu-ting He,[1] Chao Gao,[1,2] Kai Liu,[1] Teng Zhang,[1] and Sheng Zhang[1]

[1] *Aeronautics and Astronautics Engineering College, Air Force Engineering University, Xi'an 710038, China*
[2] *Beijing Aeronautical Technology Research Center, Beijing 100076, China*

Correspondence should be addressed to Yu-ting He; heyut666@126.com

Academic Editor: Yuqiang Wu

This work aims to make the crack growth prediction on 2024-T6 aluminum alloy by using Markov chain Monte Carlo (MCMC). The fatigue crack growth test is conducted on the 2024-T62 aluminum alloy standard specimens, and the scatter of fatigue crack growth behavior was analyzed by using experimental data based on mathematical statistics. An empirical analytical solution of Paris' crack growth model was introduced to describe the crack growth behavior of 2024-T62 aluminum alloy. The crack growth test results were set as prior information, and prior distributions of model parameters were obtained by MCMC using OpenBUGS package. In the additional crack growth test, the first test point data was regarded as experimental data and the posterior distribution of model parameters was obtained based on prior distributions combined with experimental data by using the Bayesian updating. At last, the veracity and superiority of the proposed method were verified by additional crack growth test.

1. Introduction

In aircraft structural engineering domain, the concept of damage tolerance was introduced to design key components. The prediction of fatigue crack propagation is a main work for aircraft structure service/use life management.

Accurate prediction of fatigue crack growth determined the safety of service and usage. Due to the scatter of material, the fatigue crack growth behavior of some components still creates scatter, though the components made by the same material and served under the same loading and environmental conditions.

The residual crack propagation life of the aircraft components is assessed based on the in-service observation and the mathematical model which describes the crack growth behavior of these components. The input data provided to these mathematical models are generally uncertain. There are uncertainties in the material performance, components dimensions, the measurements, and the degradation model [1]. Probability methodology is widely used to make decision by modelling these uncertainties with suitable probability distribution function. Scholars around the world have done a lot of work on the scatter of fatigue crack growth [2–4]. In addition, some stochastic fatigue crack growth models have been proposed [5, 6].

The experimental results used to characterize the model parameters can be regarded as the prior information, and it has an associated prior probability distribution function. The in-service inspection data can be used as experimental observation, which conjuncts the prior distribution of the model parameters to reduce the uncertainty in the prediction of fatigue crack growth curve for one component [7]. The posterior distribution can be used to make updated estimates of model parameter for the crack growth behavior.

The combination of prior distribution of fatigue crack growth model parameters with experimental observations can be carried out by applying the concepts of Bayesian updating. The results from updated distribution function are also termed as posterior distribution. The Markov chain Monte Carlo (MCMC) is one of the common probability simulation methods through the Bayesian updating [8]. And the Bayesian updating can be accomplished by OpenBUGS package.

Table 1: Chemical compositions of 2024-T62 (%).

Cu	Mg	Mn	Si	Fe	Zn	Ti	Cr	Other	Al
4.64	1.49	0.68	<0.5	<0.5	<0.25	<0.15	<0.1	<0.15	Balance

Table 2: Mechanical performance of 2024-T62 (%).

E/GPa	σ_b/MPa	$\sigma_{0.2}$/MPa	ζ_5/%
71	451	400	7.2

In this study, fatigue crack growth tests of 2024-T62 aluminum alloy were conducted under the constant amplitude loading. The prior distribution of model parameters was obtained based on experimental a-N data. Afterwards, a new fatigue crack growth prediction approach was proposed based on the prior distribution, in-service observation, and Bayesian updating method. Finally, the proposed approach is validated by the additional fatigue crack growth under the same test conditions.

2. Fatigue Crack Propagation Test

The 2024-T62 aluminum alloy is widely used as the main load structure in Chinese aviation industry.

2.1. Material and Specimen. Chemical compositions and mechanical performance of 2024-T62 aluminum alloy are shown in Tables 1 and 2, respectively.

With a hole in the center throughout whole specimen, it has been machined from the 2024-T62 aluminum alloy pipe material along the L-T direction. The dimensional length of specimen is shown in Figure 1, the thickness is 2.5 mm, and the diameter of center hole is 4 mm. The specimen was designed and conducted according to the ASTM standard E647 requirements [9].

2.2. Experimental Procedure. All the tests were carried out with the loading frequency of 20 Hz using a servo hydraulic universal dynamical test machine (MTS-810-500KN). All the tests were conducted under the room temperature (20°C). In the test, the measurements of the crack length use the Questar QM1 long working distance microscope (limit of resolution is 2.7 μm).

When the crack approximately propagates to 0.5 mm, start to count the number of loading cycle. At the same time, the crack length was measured by per 9000 loading cycles. It is worth noting that 0.5 mm is called initial crack length.

2.3. Experimental Results. The experimental fatigue crack propagation length from number 1 to number 4 is listed in Table 3. It is well known that the crack growth propagation region can be divided to three regions: near-threshold (ΔK_{th}) region, stable crack propagation region, and unstable (fast) crack propagation region [10]. In order to obtain more stable data, the stable crack propagation region was chosen (crack length: 0.5 mm~6 mm).

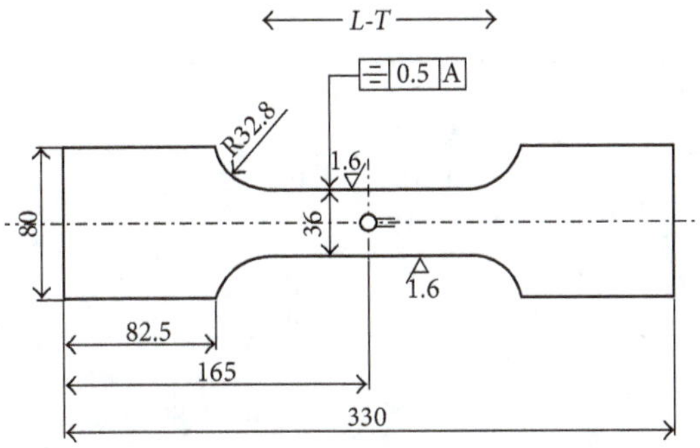

Figure 1: Geometry length of specimen.

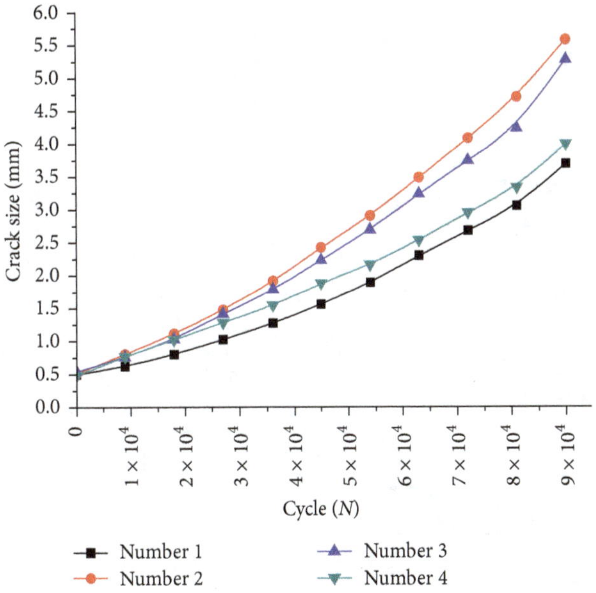

Figure 2: Experimental fatigue crack growth a-N curve.

The experimental fatigue crack propagation a-N curves for four specimens are shown in Figure 2.

As shown in Figure 2, at a given cycles, crack length has a wide range of scatter even under the same loading condition and test condition. The scatter of crack growth may be mainly attributed to the inhomogeneous material properties. Therefore, it is necessary to investigate the scatter of fatigue crack growth behavior.

3. Statistical Analyses of Crack Growth Data

The scatter of 2024-T62 aluminum alloy is analyzed by statistical analysis method. And crack growth behavior of 2024-T62 aluminum alloy is described by mathematical model.

3.1. Scatter Analysis of Crack Growth Behavior. The mean value $\overline{x}$, standard deviation σ, and coefficients of variation (COV) of specimens sample are common statistical parameters to describe the variation of statistics. Mean value is used to describe the centralized location of sample data, but σ and COV are used to describe the scatter.

TABLE 3: Crack lengths for different cycles (mm).

Specimen	Crack length										
	0	9000	18000	27000	36000	45000	54000	63000	72000	81000	90000
No. 1	0.5	0.62	0.8	1.02	1.27	1.56	1.89	2.3	2.68	3.05	3.69
No. 2	0.51	0.8	1.11	1.47	1.91	2.42	2.9	3.48	4.08	4.7	5.57
No. 3	0.54	0.76	1.03	1.41	1.78	2.23	2.69	3.23	3.74	4.23	5.27
No. 4	0.48	0.77	1.02	1.28	1.54	1.88	2.16	2.53	2.95	3.33	3.99

TABLE 4: Statistics variables of crack lengths for different cycles.

Statistics variables	Cycles									
	9000	18000	27000	36000	45000	54000	63000	72000	81000	90000
μ	0.7375	0.99	1.295	1.625	2.0225	2.41	2.885	3.3625	3.8275	4.63
σ	0.0870	0.1443	0.2168	0.3060	0.4134	0.5058	0.6080	0.7126	0.8350	1.0078
COV	0.1087	0.1343	0.1542	0.1735	0.1883	0.1934	0.1942	0.1953	0.2010	0.2005

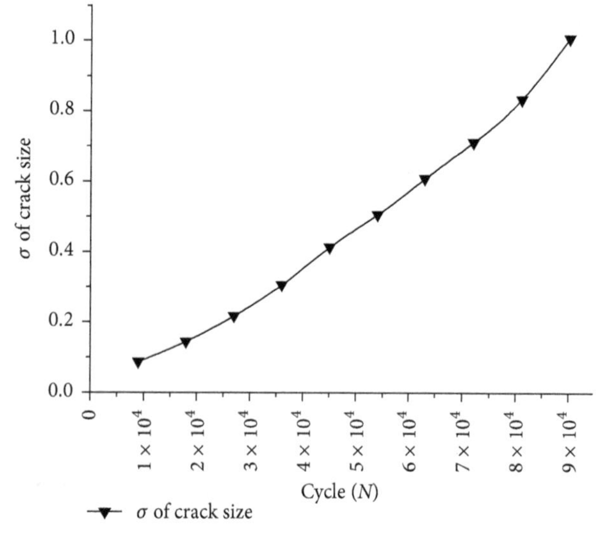

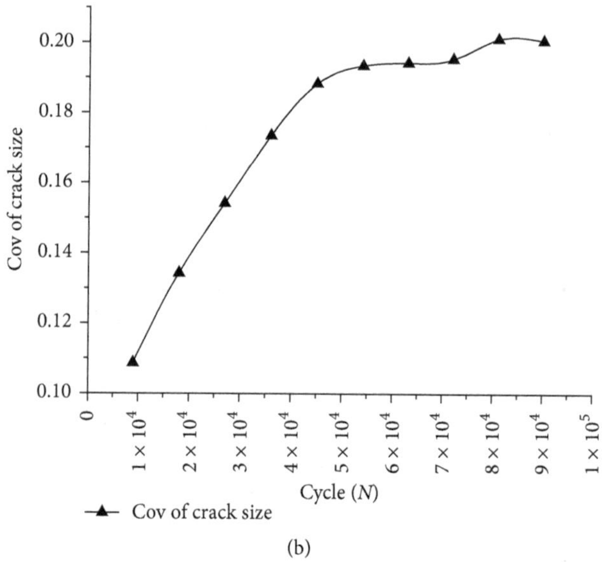

FIGURE 3: Change trend of statistical variables: (a) standard deviation σ *and* (b) coefficients of variation COV.

Mean value and standard deviation of observation $x_1, x_2, \ldots, x_n$ can be expressed as

$$\overline{x} = \frac{1}{n}(x_1 + x_2 + \cdots + x_n)$$
$$\sigma = \sqrt{\frac{\sum_{i=1}^{n} x_i^2 - n\overline{x}^2}{n-1}}. \tag{1}$$

The coefficients of variation COV (simply as C) are shown as

$$C = \frac{s}{\overline{x}}. \tag{2}$$

The statistics μ, σ, and COV were calculated based on experimental results (in Table 4). The statistical calculating formulas ((1) and (2)) are listed in Table 4.

Figure 3 gives the trend of standard deviation and COV of crack length (under same loading cycles).

As listed in Table 4 and depicted in Figure 3, standard deviation σ and coefficients of variation COV increase gradually with loading cycles increase. In a word, the scatter of crack growth process increases gradually.

3.2. Statistics of Fatigue Crack Growth Behavior. The fatigue crack growth (FCG) is modelled using Paris' law [11]. Paris' law gives the crack growth rate per stress cycle as a function of the range of stress intensity factor and material constant.

$$\frac{da}{dN} = C(\Delta K)^m, \tag{3}$$

where a is the crack length (mm), N is number of loading cycles, da/dN is the crack growth rate per loading cycle due to fatigue, ΔK is the range of stress intensity factor (in Mpa$\sqrt{\text{m}}$), and C and m are constant material properties obtained by testing standard specimens made of this material.

TABLE 5: Estimation value of model parameters.

Specimen	Parameter estimate value	
	θ_1	θ_2
Number 1	2.4312	−0.2598
Number 2	3.7303	−0.5435
Number 3	3.2953	−0.4592
Number 4	3.0991	−0.7331

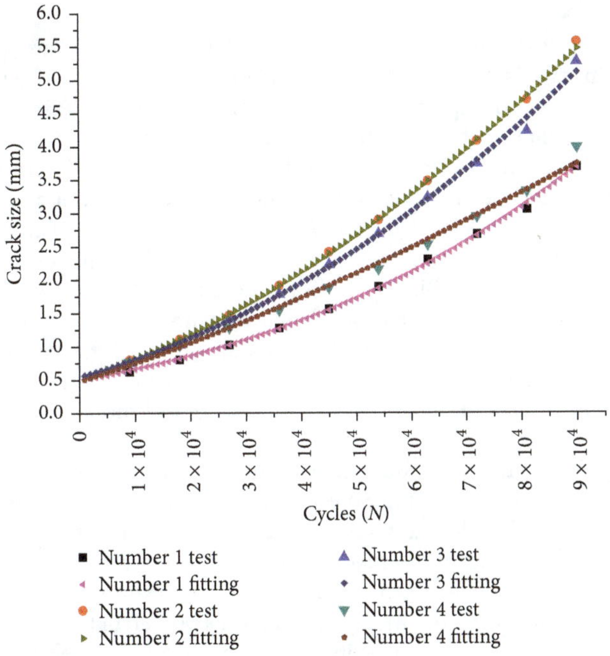

FIGURE 4: Fitting curves versus experimental results.

The crack length after N number of loading cycles is $a(N)$. The crack growth rate da/dN determines the true crack length $a(N)$ according to Paris' law (3), which has the form of $\theta_1 k(a)^{\theta_2+1}$, for some function $k(a)$. When $k(a) = a$, the solution for $a(N)$ can be expressed by [12]

$$a(N) = \frac{a_0}{\left(1 - a_0^{\theta_2}\theta_1\theta_2 N\right)^{1/\theta_2}}, \quad (4)$$

where a_0 is the initial crack length (in mm). θ_1 and θ_2 are the parameters of crack growth model for 2024-T62 aluminum alloy under constant amplitude loading. So (4) is an empirical analytical solution form for Paris' crack growth law.

Firstly, it is necessary to validate the feasibility of using (4), which can describe the crack growth behavior of 2024-T62 aluminum alloy. The estimation value of θ_1 and θ_2 can be obtained based on crack growth experimental a-N data by maximum likelihood method (MLE). The values of model parameters θ_1 and θ_2 for four specimens are listed in Table 5.

The fatigue crack growth curve can be obtained by substituting the estimation value of model parameters θ_1 and θ_2 to (4). And the obtained curve is the fitting curve of crack growth, which is shown in Figure 4.

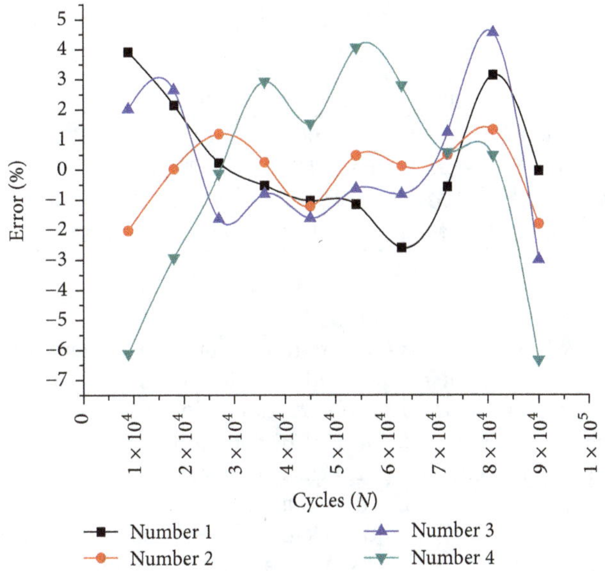

FIGURE 5: Error between experimental data and fitting curves.

It can be seen from Figure 4 that the fitting curves (based on (4)) are in good agreement with the experimental results, which can validate the feasibility of (4).

Relative error (crack growth fitting curves to experimental data) of four specimens is listed in Figure 5 and Table 6.

As listed in Table 6 and shown in Figure 5, the relative errors between the fitting curves and the experimental results were less than 7%. It means that the approach (include (4) and parameter estimating method) could be satisfactorily used in engineering. As a result, (4) is a feasible and accurate model to describe the crack growth behavior of 2024-T62 aluminum alloy under constant amplitude loads.

4. Prediction Method of Fatigue Crack Growth

In aircraft structural engineering, the common use of the media crack growth a-N experimental results represents the fatigue crack propagation behavior, which cannot consider the scatter and the distinguish of different specimens.

4.1. Prior Distribution. As we all know, crack length $a(N)$ is a nonlinear function of loading cycles numbers N. By taking natural logarithms e in (4), we can obtain the transformed real crack length $\ln[a(N)/a_0]$ as

$$\ln\left[\frac{a(N)}{a_0}\right] = -\frac{1}{\theta_2}\ln\left(1 - a_0^{\theta_2}\theta_1\theta_2 N\right). \quad (5)$$

For the experimental observed transformed crack length Y, it includes the measure error ε. And every Y_{ij} can be expressed as

$$Y_{ij} = -\frac{1}{\theta_{2i}}\ln\left(1 - a_{0i}^{\theta_{2i}}\theta_{1i}\theta_{2i}N_{ij}\right) + \varepsilon_{ij}, \quad (6)$$

where subscripts i and j represent sequence number of specimen and sequence number of experimental observations point, respectively. So Y_{ij} is the observed transformed

TABLE 6: Parameter Estimate of fatigue crack propagation curve.

Specimen	Error (%)									
	9000	18000	27000	36000	45000	54000	63000	72000	81000	90000
Number 1	3.91%	2.15%	0.22%	−0.53%	−1.04%	−1.16%	−2.60%	−0.57%	3.14%	−0.04%
Number 2	−3.95%	−1.94%	−0.79%	−1.72%	−3.16%	−1.49%	−1.84%	−1.46%	−0.65%	−3.72%
Number 3	−5.53%	−4.95%	−8.91%	−8.15%	−8.90%	−7.99%	−8.16%	−6.25%	−3.19%	−10.18%
Number 4	−2.20%	1.13%	4.04%	7.22%	5.76%	8.39%	7.08%	4.78%	4.68%	−2.43%

crack length for jth observation point of ith specimen. And a_{0i} is the initial crack length of ith specimen. ε_{ij} is the observation error for jth observation point of ith specimen, which is subjected to normal distribution ($N(0,\sigma^2)$). And error ε_{ij} for different specimens and different observations is independent identically distribution (i.i.d.).

The scatter of fatigue crack growth behavior in the model parameter is modelled by using a prior probability density function (PDF). And the assumption model parameters $\overrightarrow{\theta}_i = (\theta_{1i}, \theta_{2i})$ subjected to multinormal distribution can be expressed as

$$\overrightarrow{\theta}_i \sim \text{multi-Normal}\left(\overrightarrow{\mu}, \overrightarrow{\Sigma}\right). \quad (7)$$

Likelihood of experimental transformed crack length includes the PDF of y_{ij} and the PDF of model matrix parameter $\overrightarrow{\theta}_i = (\theta_{1i}, \theta_{2i})$.

$$y_{ij} \sim \text{Normal}\left[-\left(\frac{1}{\theta_{2i}}\right)\ln\left(1 - a_{0i}{}^{\theta_{2i}}\theta_{1i}\theta_{2i}N_{ij}\right), \sigma^2\right]. \quad (8)$$

In case of completely lack prior information, the diffuse distribution can be used to describe the prior information that can also be called noninformative prior distribution [13]. Assumption matrix parameter $\overrightarrow{\mu}$ subjected to multinormal distribution can be expressed as

$$\overrightarrow{\mu} \sim \text{multi-Noraml}\left(\overrightarrow{\mu}, \overrightarrow{\Sigma}_{\mu 0}\right),$$
$$\overrightarrow{\mu}_0 = \begin{pmatrix} 0 \\ 0 \end{pmatrix} \quad \overrightarrow{\Sigma}_{\mu 0} = \begin{pmatrix} 1000 & 0 \\ 0 & 1000 \end{pmatrix}. \quad (9)$$

Assumption parameter $\overrightarrow{\Sigma}$ subjected to inverse Wishart distribution can be expressed as

$$\overrightarrow{\Sigma} \sim \text{InverseWishart}\left(\overrightarrow{\Sigma}_0, 2\right), \quad \overrightarrow{\Sigma}_0 = \begin{pmatrix} 10 & 0 \\ 0 & 10 \end{pmatrix}. \quad (10)$$

And assumption parameter σ^2 subjected to inverse Gamma distribution is expressed as

$$\sigma^2 \sim \text{InverseGamma}(3, 0.001). \quad (11)$$

μ_1, μ_2, Σ_{11}, Σ_{12}, Σ_{21}, $\Sigma_{22,}$ and σ are the parameters of fatigue crack growth model. The posterior PDF of the model parameter is obtained using Bayesian updating, implemented using the MCMC algorithms. MCMC algorithms are a general class of computational methods used to produce samples from posterior samples. They are often easy to implement and, at least in principle, can be used to simulate from very high-dimensional posterior distributions [12]. And they have been successfully applied to literally thousands of applications. Metropolis-Hastings algorithms and Gibbs samplers are two general categories of MCMC simulation.

Briefly, OpenBUGS is a software package for Bayesian analysis of complex statistical models with the Gibbs sampler, which is a based implementation of BUGS (Bayesian Inference Using Gibbs Sample). The package contains flexible software for analyzing complex statistical models using MCMC methods. For more information about background, application, and introduction to OpenBUGS refer to Amiri and Modarres [14] and Spiegelhalter et al. [15] and Cowles. The model parameters statistics estimated from tested data (in Table 3) and above method are listed in Table 7.

One way to assess the accuracy of the posterior estimation is by calculating the Monte Carlo error (MC-error) for each parameter. This is an estimate for the difference between the mean of the sampled values (which we are using as our estimate of the posterior mean for each parameter) and the true posterior mean.

As a rule of thumb, the simulation should be run until the Monte Carlo error for each parameter of interest is less than about 5% of the sample standard deviation. As listed in Table 7, the MC-error for each parameter is all less than 5%. Therefore, we can know that the posterior distribution of model parameter is accurate.

4.2. Convergence Diagnose. Convergence diagnosis is an important work in the using of MCMC algorithm. If the convergence feature of chains is poor, the posterior distributions from chains based on MCMC are not accurate. Figure diagnosis method, checking throughout mean method, and contrasting deviation method are common methods to diagnose the convergence. And contrasting deviation method has a wide use in the convergence diagnosis analysis.

Based on the contrasting deviation method proposed by Gelman and Rubin [16], a more brief and convenient method was established by Brooks and Gelman [17]. The basic idea is to generate multiple chains starting at overdispersed initial values and assess convergence by comparing within-chain and between-chain variability over the second half of those chains. We denote the number of chains generated by L and the length of each chain by $2T$. We take as a measure of posterior variability the width of the $100(1-\alpha)\%$ credible

Table 7: Parameter Estimate of fatigue crack propagation curve.

Parameter	Mean	Standard deviation	MC-error (%)	Different percentile				
				0.025	0.05	0.5	0.95	0.975
μ_1	3.1390	0.3344	0.09	2.4860	2.638	3.1390	3.641	3.7940
μ_2	−0.4979	0.1209	0.04	−0.7369	−0.6819	−0.4977	−0.3163	−0.2605
Σ_{11}	0.4434	1.1000	0.37	0.0787	0.09277	0.2627	1.245	1.8240
Σ_{12}	−0.0881	0.3623	0.12	−0.4487	−0.301	−0.0468	0.0212	0.0507
Σ_{21}	−0.0881	0.3623	0.12	−0.4487	−0.301	−0.0468	0.0212	0.0507
Σ_{22}	0.0581	0.1555	0.06	0.0095	0.01134	0.0341	0.1653	0.2420
σ	0.0251	0.0030	0.01	0.0201	0.02073	0.0248	0.03045	0.0318

interval for the parameter of interest (in OpenBUGS, $\alpha = 0.2$). From the final T iterations we calculate the empirical credible interval for each chain. We then calculate the average width of the intervals across the L chains and denote this by L. Finally, we calculate the width l of the empirical credible interval based on all LT samples pooled together. The ratio $\widehat{R} = L/l$ of pooled to average interval widths should be greater than 1 if the starting values are suitably overdispersed; it will also tend to 1 as convergence is approached, and so we might assume convergence for practical purposes if $\widehat{R} < 1.05$.

Rather than calculating a single value of $\widehat{R}$, we can examine the behavior of $\widehat{R}$ over iteration-time by performing the above procedure repeatedly for an increasingly large fraction of the total iteration range, ending with all of the final T iterations contributing to the calculation as described above.

In the OpenBUGS software, the bgr diagram can achieve above approach. OpenBUGS automatically chooses the number of iterations between the ends of successive ranges: $\max(100, 2T/100)$. It then plots $\widehat{R}$ in red, L (pooled) in green, and l (average) in blue. As shown in Figure 6, the red line represents $\widehat{R}$, the green line represents L, and l corresponds to blue line.

In this study, three chains for each parameter were generated by OpenBUGS software. The convergence diagnosis schematic diagrams for seven model parameters were shown in Figure 6.

As shown in Figure 6, the convergence feature of model parameters is good, and the trend of model parameter with iteration increase is convergent. In other words, the posterior distributions of seven parameters are accurate and credible.

4.3. Prediction Model of Fatigue Crack Growth. The prediction of crack growth life (period) is a main thesis in the aircraft structural engineering. In this study, it is aimed to accurately predict the fatigue crack growth behavior (curve) for one target specimen using least information.

The Bayesian theorem can be written in model updating context as

$$P\left(\vec{\theta} \mid D, M\right) \propto P\left(D \mid \vec{\theta}, M\right) P\left(\vec{\theta} \mid M\right), \quad (12)$$

where $P(\vec{\theta} \mid D, M)$ corresponds to the PDF for the crack growth model M after updating with the crack growth test observations D; it is called the PDF of posterior distribution. $P(\vec{\theta} \mid M)$ is the PDF of model parameters $\vec{\theta}$ for the crack growth model M before updating; it is called the PDF of prior distribution. And $P(D \mid \vec{\theta}, M)$ is the likelihood of occurrence of the crack growth observation D given the vector of model parameter $\vec{\theta}$ and crack model M.

The fatigue crack growth experimental results can be regarded as the prior information. And the parameter prior distribution of crack growth model (4) can be obtained by above approach. According to the analysis and inference in the Section 4.1, crack growth model parameters $\vec{\theta} = (\theta_1, \theta_2)$ are subjected to multinormal $(\vec{\mu}, \vec{\Sigma})$:

$$\vec{\mu} = (3.1390, -0.4979)$$
$$\vec{\Sigma} = \begin{pmatrix} 0.4434 & -0.0881 \\ -0.0881 & 0.0581 \end{pmatrix}. \quad (13)$$

Therefore, the posterior likelihood function of model parameters for one target specimen can be calculated as

$$P\left(\vec{\theta} \mid D, M\right) \propto \prod_{z=1}^{m} \exp\left(-\frac{1}{2}\left(\frac{y_z - (1/\theta_2)\ln\left(1 - a_0^{\theta_2}\theta_1\theta_2 N_z\right)}{\sigma}\right)^2\right) \cdot H\left(\vec{\theta}\right), \quad (14)$$

where m is the number of observation point used to update the posterior distribution of model parameters. And if the first test data point was used to update the posterior distribution of parameters θ_1 and θ_2, $m = 1$.

The posterior distribution can be obtained via using Markov chain Monte Carlo (MCMC) simulation with the combination of the likelihood function (14) with prior distribution (13).

If the model parameters θ_1 and θ_2 for a specimen are determined, the fatigue crack propagation curve of this specimen can be obtained by (4). As a result, the residual growth life (period) of this specimen can be determined.

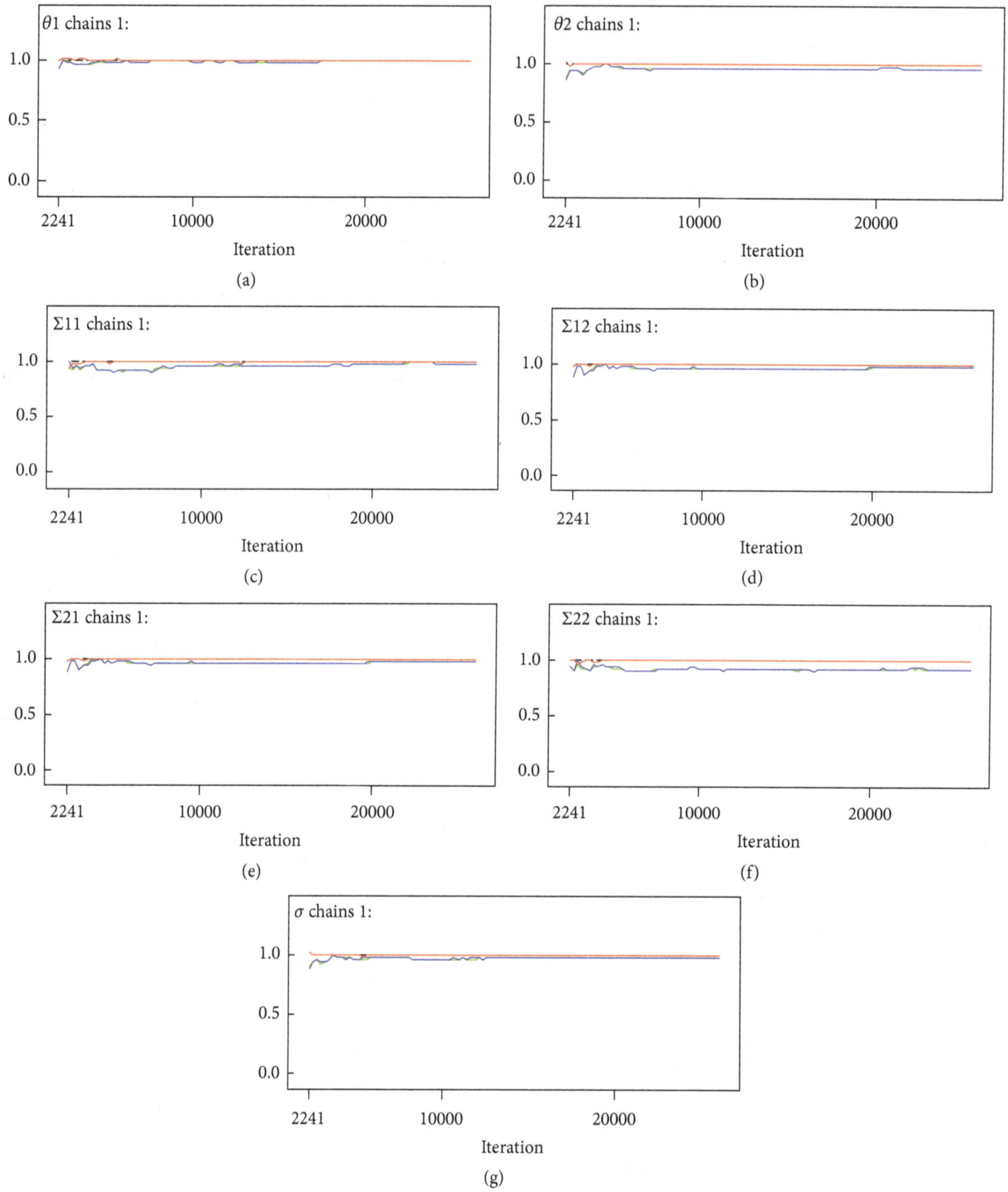

FIGURE 6: The bgr diagram of convergence diagnosis for fatigue crack growth model parameters: (a) θ_1, (b) θ_2, (c) Σ_{11}, (d) Σ_{12}, (e) Σ_{21}, (f) Σ_{22}, and (g) σ.

5. Experimental Validations

This section presents the validation of the proposed methodology via using crack propagation data from additional crack growth test.

5.1. Additional Crack Growth Test. Additional crack growth test employs the same experimental method, environmental condition, and measurement technology as the test in Section 2. Similarly, when the crack approximately propagates to 0.5 mm, start counting the number of loading cycles. In other words, the crack length is 0.5 mm when the number of loading cycles is zero.

The specific initial crack length and the crack length after 9000 cycles (number 5, number 6, and number 7 target specimen) are listed in Table 8. It is noted that the

TABLE 8: Crack lengths of initialization and first test point (mm).

Specimen	Initial crack length	Crack length after 9000 cycles
Number 5	0.5	0.66
Number 6	0.53	0.74
Number 7	0.5	0.69

experimental data (in Table 8) can represent the in-service observation during the service for aircraft structures.

5.2. Prediction of Crack Growth. The prediction of crack growth curve (number 5 specimen) combines different test data information (different data sample length and different test data). Mark different information source as Case I, Case II, and Case III. A particular introduction of Case I–Case III is presented as follows.

For Case I, the prior distribution of matrix parameter $\vec{\theta} = (\theta_1, \theta_2)$ is the only information source for the prediction of crack growth curve. So the posterior distribution mean of parameter $\vec{\theta} = (\theta_1, \theta_2)$ is (3.1390, −0.4979).

For Case II, the information source includes prior distribution of matrix parameter $\vec{\theta} = (\theta_1, \theta_2)$ and the first observation data point ($N = 9000$, $a(N) = 0.66$ mm). And the prediction model of Case II is the proposed model, which corresponds to the Section 4.3. Based on (14), the likelihood of Case II is expressed as

$$P\left(\vec{\theta} \mid D, M\right) \propto \exp\left(-\frac{1}{2}\left(\frac{y_{5,1} - \left(\ln\left(1 - a_{5,0}{}^{\theta_{5,2}}\theta_{5,1}\theta_{5,2}N_1\right)/\theta_{5,2}\right)}{\sigma}\right)^2\right) \cdot H\left(\vec{\theta}\right). \quad (15)$$

Based on the above likelihood function of posterior distribution and the OpenBUGS package, the mean value of posterior distribution of matrix parameters $\vec{\theta} = (\theta_1, \theta_2)$ can be obtained. Through the MC-error control and the convergence diagnose of the posterior distribution of the matrix parameter $\vec{\theta}$, the mean value is (2.6035, −0.3172).

For Case III, the first observation data ($N = 9000$, $a(N) = 0.66$ mm) is the only information source for the prediction of crack growth curve. The posterior distribution mean value of $\vec{\theta} = (\theta_1, \theta_2)$ is (0.8838, −2.4941), which can be estimated by using the method of Section 3.2.

For number 6 and number 7 specimen, they are also deviated to Case I, Case II, and Case III. The posterior distribution mean values of matrix parameters $\vec{\theta} = (\theta_1, \theta_2)$ for the additional three target specimens are listed in Table 10.

5.3. Additional Experimental Observation. The additional fatigue crack growth tests to 90000 loading cycles were conducted. The crack growth lengths of number 5, number 6, and number 7 specimens are listed in Table 9 (include the data of Table 8).

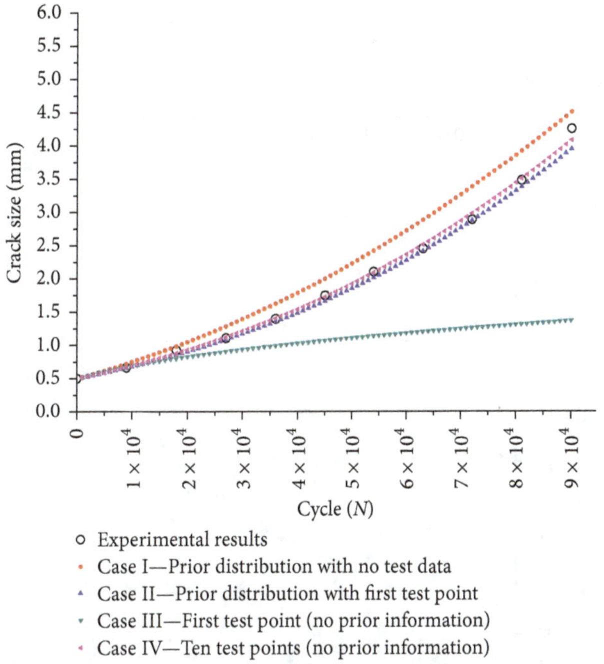

FIGURE 7: Crack propagation prediction curves of number 5 specimen.

The parameters are fitted with the same method of Section 3.2. And the information source is called Case IV. The posterior distribution of fitting parameters (θ_1, θ_2) is listed in Table 10.

The model parameter posterior distribution mean values of Case I–Case IV for number 5, number 6, and number 7 specimens are listed in Table 10.

The fatigue crack growth curve can be obtained by substituting the parameter of posterior distribution mean value to crack growth model (4). In this paper, for one target specimen, we can obtain four fatigue crack growth curves, corresponding to the above Case I, Case II, Case III, and Case IV, respectively. For example, four fatigue crack growth curves and experimental *a*-*N* observations of number 5 specimen are shown in Figure 7.

As shown in Figure 7, the fatigue crack growth curve of Case III is far away from the experimental *a*-*N* observations. And the absolute error between fatigue crack growth curve of Case II and experimental *a*-*N* observations is smaller than that of Case III. Compare Case II with Case III, we can obtain the prior information that is very important for the prediction of fatigue crack growth behavior.

The fatigue crack growth curves for number 6 and number 7 specimens are obtained by the same method and analysis steps, which are shown in Figures 8 and 9. And the experimental results of number 6 and number 7 specimen are shown in Figures 8 and 9, respectively.

As shown in Figures 7, 8, and 9, we can know that the prediction of fatigue crack growth of Case III is far away from the experimental *a*-*N* observations, so the error cannot satisfy the requirements of aircraft engineering.

TABLE 9: Crack lengths for different cycles (mm).

Specimen	Cycle										
	0	9000	18000	27000	36000	45000	54000	63000	72000	81000	90000
Number 5	0.5	0.66	0.92	1.1	1.39	1.74	2.1	2.45	2.88	3.47	4.25
Number 6	0.55	0.77	1.02	1.2	1.51	2	2.49	2.96	3.59	4.3	5.41
Number 7	0.5	0.70	0.93	1.16	1.53	1.96	2.42	2.74	3.3	4.12	4.84

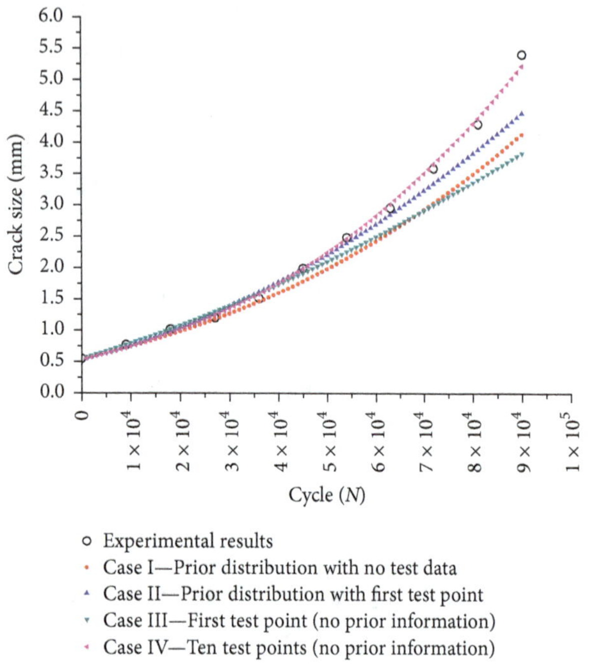

FIGURE 8: Crack propagation curves prediction of number 6 specimen.

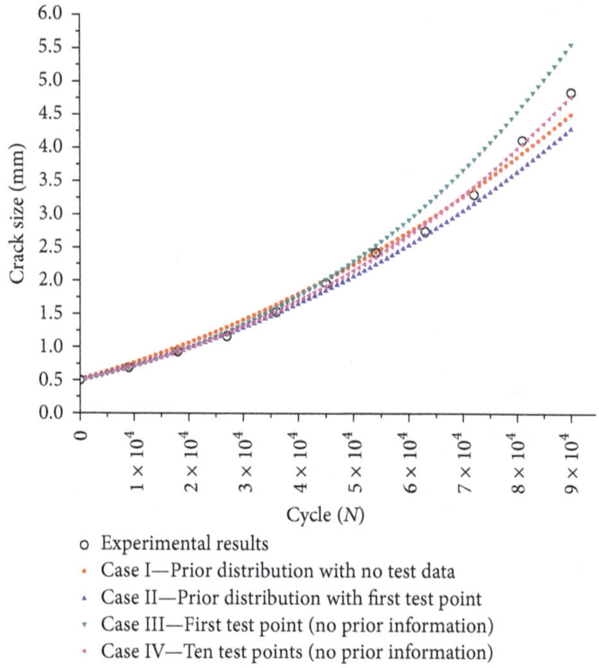

FIGURE 9: Crack propagation curves prediction of number 7 specimen.

Comparative analysis of the prediction for the fatigue crack growth of Figures 7, 8, and 9 is based on the crack growth model parameters prior distribution and observation data (Case II) and we can obtain the crack growth curve accurately. Therefore, the prediction method (Case II) proposed in this study is an advanced technology.

5.4. Error Analysis. Due to the high error of Case III, it will not be considered in the following analysis. The absolute errors (prediction value to experimental observations value) of Case I, Case II, and Case IV are analyzed in the following steps. And the absolute error is defined as

$$\text{absolute error} = \left| \frac{(\text{pridiction value}) - (\text{experiment value})}{\text{experiment value}} \right| \times 100\%. \quad (16)$$

Figure 10 shows the prediction crack life (Figure 7) versus the experimental results (Table 9) of number 5 target specimen.

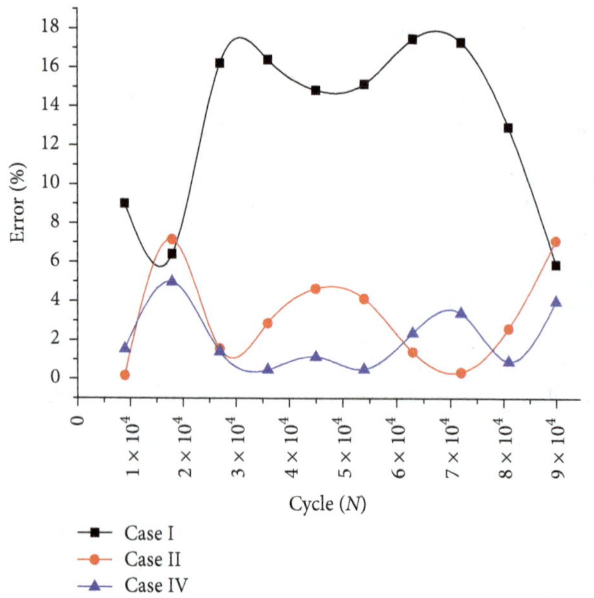

FIGURE 10: Prediction error of crack propagation curve for number 5 specimen.

TABLE 10: Parameter estimate of fatigue crack propagation curve.

Specimen	Based on prior distribution and test data				Only based on test data			
	Case I: no test data		Case II: first test point		Case III: only first test point		Case IV: ten test point	
	θ_1	θ_2	θ_1	θ_2	θ_1	θ_2	θ_1	θ_2
Number 5	3.1390	−0.4979	2.6035	−0.3172	0.8838	−2.4941	2.6938	−0.3901
Number 6	3.1390	−0.4979	3.0037	−0.4595	2.3531	−1.3182	2.9778	−0.3122
Number 7	3.1390	−0.4979	2.8857	−0.4128	3.1199	−0.2654	2.9766	−0.3517

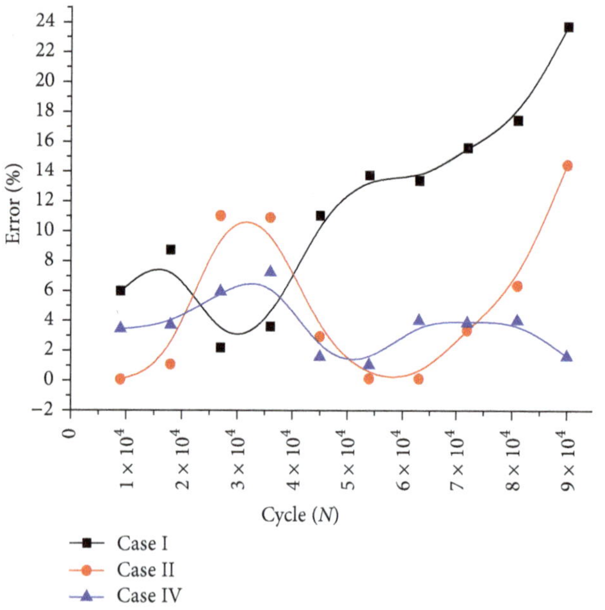

FIGURE 11: Prediction error of crack propagation curve for number 6 specimen.

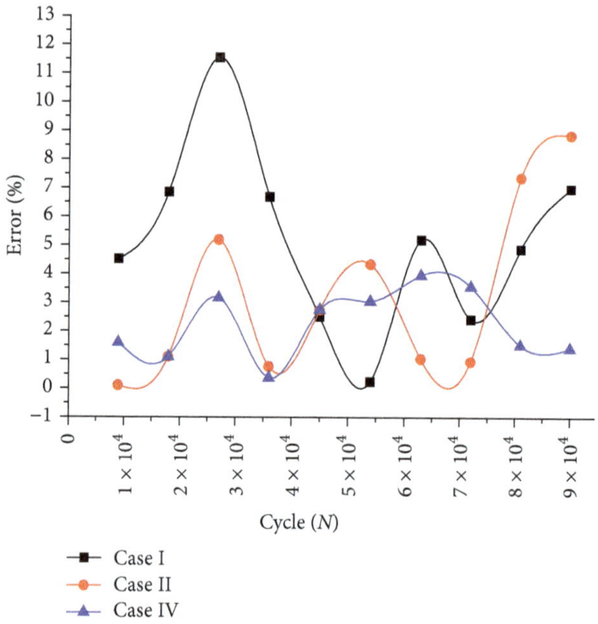

FIGURE 12: Prediction error of crack propagation curve for number 7 specimen.

As shown in Figure 10, the absolute error of Case IV is the smallest for the three cases. In other words, the prediction curve (life) of Case III is in good agreement with experimental observations. Based on the absolute error condition of Case I, it can be known that using the average *a-N* curve of number 1, number 2, number 3, and number 4 specimens to represent crack growth curve of number 5 specimen may cause bigger error.

Crack life prediction (Figures 8 and 9) versus the experimental results (Table 8) for number 6 and number 7 specimens is shown in Figures 11 and 12, respectively.

As shown in Figures 10, 11, and 12, the absolute error of Case II can satisfy the aircraft engineering accurate needs.

It is worth noting that Case I represents the mean of experimental results. In some time, the prediction of Case I may be more concise than Case II, but it will not happen all the time. In other words, it is a special circumstance, and it is determined by the stochastic and the scatter of fatigue crack growth behavior. For instance, the prediction fatigue crack growth curve of Case I (number 7 specimen, in Figure 12) is in good agreement with the experimental results.

6. Conclusion and Discussion

The paper presents a method of fatigue crack propagation by using the crack test information and in-service inspection information. The fatigue crack growth curve of 2024-T62 aluminum alloy is predicted by Bayesian theory with MCMC algorithm and OpenBUGS package. Based on the current investigation, four conclusions are drawn.

(1) The introduced model (4) can describe the fatigue crack growth behavior of 2024-T62 aluminum alloy accurately.

(2) The scatter of crack length (under same loading cycles) gradually increases with crack propagation process.

(3) Prior information of crack growth model parameter is an important information source to predict the crack growth behavior. And the MCMC is a good method to resolve the statistics estimation issue of multiparameter.

(4) The proposed approach provides an insight into fatigue crack growth curve prediction and associates the MCMC with OpenBUGS package. The predicted fatigue crack growth curve and residual life are in good agreement with the experimental observations. And the performance of proposed approach shows the superior results to the mean value of *a-N* data.

Conflicts of Interest

The authors declare no conflicts of interest regarding the publication of this paper.

Acknowledgments

This work is supported by the National Natural Science Foundation of China (Project no. 51505495). The authors gratefully appreciate the support.

References

[1] W. F. Wu and C. C. Ni, "Probabilistic models of fatigue crack propagation and their experimental verification," *Probabilistic Engineering Mechanics*, vol. 19, no. 3, pp. 247–257, 2004.

[2] X. L. Zou, "Statistical moments of fatigue crack growth under random loading," *Theoretical and Applied Fracture Mechanics*, vol. 39, no. 1, pp. 1–5, 2003.

[3] W. Li, T. Sakai, Q. Li, and P. Wang, "Statistical analysis of fatigue crack growth behavior for grade B cast steel," *Materials and Design*, vol. 32, no. 3, pp. 1262–1272, 2011.

[4] H. Y. Liou, C. C. Ni, and W. F. Wu, "Scatter and statistical analysis of fatigue crack growth data," *Journal of Chinese Society of Mechanical Engineers*, vol. 22, no. 5, pp. 399–407, 2001.

[5] J.-K. Kim and D.-S. Shim, "Variation in fatigue crack growth due to the thickness effect," *International Journal of Fatigue*, vol. 22, no. 7, pp. 611–618, 2000.

[6] W. F. Wu and C. C. Ni, "A study of stochastic fatigue crack growth modeling through experimental data," *Probabilistic Engineering Mechanics*, vol. 18, no. 2, pp. 107–118, 2003.

[7] R. Rastogi, S. Ghosh, A. K. Ghosh, K. K. Vaze, and P. K. Singh, "Fatigue crack growth prediction in nuclear piping using Markov chain Monte Carlo simulation," *Fatigue and Fracture of Engineering Materials and Structures*, vol. 40, no. 1, pp. 145–156, 2017.

[8] W. R. Gilks, S. Richardson, and D. J. Spiegelhalter, *Markov Chain Monte Carlo in Practice*, Interdisciplinary Statistics, Chapman & Hall/CRC, London, UK, 1996.

[9] Standard A. S. T. M., "E647, Standard test method for measurement of fatigue crack growth rates," *Annual Book of ASTM Standards, Section Three: Metals Test Methods and Analytical Procedures*, vol. 3, pp. 628–670, 2002.

[10] N. E. Dowling, "Mechanical behavior of materials: engineering methods for deformation," *Fracture, and Fatigue, Pearson*, 2012.

[11] N. Pugno, M. Ciavarella, P. Cornetti, and A. Carpinteri, "A generalized Paris' law for fatigue crack growth," *Journal of the Mechanics and Physics of Solids*, vol. 54, no. 7, pp. 1333–1349, 2006.

[12] M. S. Hamada, A. Wilson, C. S. Reese, and H. Martz, *Bayesian Reliability*, Springer Science & Business Media, New York, 2008.

[13] K. Mosegaard and A. Tarantola, "16 Probabilistic approach to inverse problems," *International Geophysics*, vol. 81, pp. 237–265, 2002.

[14] M. Amiri and M. Modarres, "Short fatigue crack initiation and growth modeling in aluminum 7075-T6," *Journal of Mechanical Engineering Science*, vol. 229, no. 7, pp. 1206–1214, 2015.

[15] D. Spiegelhalter, A. Thomas, N. Best et al., *OpenBUGS user manual, version 3.0.2*, MRC Biostatistics Unit, Cambridge, 2007.

[16] A. G. Gelman and D. B. Rubin, "Inference from iterative simulation using multiple sequences," *Statistical Science*, vol. 7, no. 4, pp. 457–472, 1992.

[17] S. P. Brooks and A. Gelman, "General methods for monitoring convergence of iterative simulations," *Journal of Computational and Graphical Statistics*, vol. 7, no. 4, pp. 434–455, 1998.

Permissions

All chapters in this book were first published in AMSE, by Hindawi Publishing Corporation; hereby published with permission under the Creative Commons Attribution License or equivalent. Every chapter published in this book has been scrutinized by our experts. Their significance has been extensively debated. The topics covered herein carry significant findings which will fuel the growth of the discipline. They may even be implemented as practical applications or may be referred to as a beginning point for another development.

The contributors of this book come from diverse backgrounds, making this book a truly international effort. This book will bring forth new frontiers with its revolutionizing research information and detailed analysis of the nascent developments around the world.

We would like to thank all the contributing authors for lending their expertise to make the book truly unique. They have played a crucial role in the development of this book. Without their invaluable contributions this book wouldn't have been possible. They have made vital efforts to compile up to date information on the varied aspects of this subject to make this book a valuable addition to the collection of many professionals and students.

This book was conceptualized with the vision of imparting up-to-date information and advanced data in this field. To ensure the same, a matchless editorial board was set up. Every individual on the board went through rigorous rounds of assessment to prove their worth. After which they invested a large part of their time researching and compiling the most relevant data for our readers.

The editorial board has been involved in producing this book since its inception. They have spent rigorous hours researching and exploring the diverse topics which have resulted in the successful publishing of this book. They have passed on their knowledge of decades through this book. To expedite this challenging task, the publisher supported the team at every step. A small team of assistant editors was also appointed to further simplify the editing procedure and attain best results for the readers.

Apart from the editorial board, the designing team has also invested a significant amount of their time in understanding the subject and creating the most relevant covers. They scrutinized every image to scout for the most suitable representation of the subject and create an appropriate cover for the book.

The publishing team has been an ardent support to the editorial, designing and production team. Their endless efforts to recruit the best for this project, has resulted in the accomplishment of this book. They are a veteran in the field of academics and their pool of knowledge is as vast as their experience in printing. Their expertise and guidance has proved useful at every step. Their uncompromising quality standards have made this book an exceptional effort. Their encouragement from time to time has been an inspiration for everyone.

The publisher and the editorial board hope that this book will prove to be a valuable piece of knowledge for researchers, students, practitioners and scholars across the globe.

List of Contributors

Daniela Nickel
Materials and Surface Engineering Group, Technische Universität Chemnitz, 09107 Chemnitz, Germany
Center of Engineering Materials, State Materials Testing Institute Darmstadt (MPA), Chair and Institute for Materials Science (IfW), Technische Universität Darmstadt, 64283 Darmstadt, Germany

Dagmar Dietrich, Roy Morgenstern, Ingolf Scharf and Thomas Lampke
Materials and Surface Engineering Group, Technische Universität Chemnitz, 09107 Chemnitz, Germany

Harry Podlesak
Composite Materials Group, Technische Universität Chemnitz, 09107 Chemnitz, Germany

Xiao-guo Wang
School of Materials Science and Engineering, Taiyuan University of Science and Technology, Taiyuan 030024, China
Department of Mechanical and Electrical Engineering, College of Information, Shanxi Agricultural University, Taigu 030800, China

Qiu-shu Li and Rui-rui Wu
School of Materials Science and Engineering, Taiyuan University of Science and Technology, Taiyuan 030024, China

Xiao-yuan Zhang
Department of Mechanical and Electrical Engineering, College of Information, Shanxi Agricultural University, Taigu 030800, China

Liyun Ma
Department of Mining Engineering, Luliang University, Lishi 033000, China

Zheng-bing Xiao
School of Mechanical and Electrical Engineering, Central South University, Changsha 410083, China
Collaborative Innovation Center of Advanced Nonferrous Structural Materials and Manufacturing, Central South University, Changsha 410083, China

Yuan-chun Huang
School of Mechanical and Electrical Engineering, Central South University, Changsha 410083, China
Light Alloy Research Institute, Central South University, Changsha 410012, China

Yu Liu
Light Alloy Research Institute, Central South University, Changsha 410012, China

Guangjian Peng
College of Mechanical Engineering, Zhejiang University of Technology, Hangzhou 310014, China
Department of Mechanical Engineering, Johns Hopkins University, Baltimore, MD 21218, USA

Yi Ma, Jiangjiang Hu, Weifeng Jiang and Taihua Zhang
College of Mechanical Engineering, Zhejiang University of Technology, Hangzhou 310014, China

Yong Huan
State Key Laboratory of Nonlinear Mechanics (LNM), Institute of Mechanics, Chinese Academy of Sciences, Beijing 100190, China

Zhitong Chen
Department of Mechanical and Aerospace Engineering, 3e George Washington University, Washington, DC 20052, USA

Mehrdad Vahdati and SeyedAlireza Rasouli
Department of Mechanical Engineering, K.N. Toosi University of Technology, Tehran 19395-1999, Iran

Yong Peng, Xuanzhen Chen and Guowei Liu
School of Traffic and Transportation Engineering, Central South University, Changsha 410075, China
Key Laboratory of Traffic Safety on Track, Central South University, Ministry of Education, Changsha 410075, China
National and Local Joint Engineering Research Center of Safety Technology for Rail Vehicle, Changsha 410075, China

Shan Peng
School of Traffic and Transportation Engineering, Central South University, Changsha 410075, China
Smart Transport Key Laboratory of Hunan Province, Changsha 410075, China

Chao Chen
Hunan Industry Polytechnic, Changsha 410208, China

Jiahao Li
School of Traffic and Transportation Engineering, Central South University, Changsha 410075, China
Joint International Research Laboratory of Key Technology for Rail Traffic Safety, Changsha 410075, China

Weihan Wang, Weifang Zhang, Hongxun Wang, Xiaoliang Fang and Xiaobei Liang
Science & Technology on Reliability & Environmental Engineering Laboratory, Beihang University, Beijing 100191, China

Adel Taha Abbas and Ehab Adel El-Danaf
Department of Mechanical Engineering, College of Engineering, King Saud University, Riyadh, Saudi Arabia

Mohamed Adel Taha and Mohamed Ibrahim Abd El Aal
Department of Mechanical Design and Production, College of Engineering, Zagazig University, Sharkia, Egypt

Adham Ezzat Ragab
Department of Industrial Engineering, College of Engineering, King Saud University, Riyadh, Saudi Arabia

G. Renna and P. Leo
Innovation Engineering Department, University of Salento, Via per Arnesano s,n, 73100 Lecce, Italy

G. Casalino
DMMM, Politecnico di Bari, Viale Japigia 182, 70126 Bari, Italy

E. Cerri
Department of Engineering and Architecture, University of Parma, v.le G. Usberti 181/A, 43124 Parma, Italy

J. C. Verduzco Juárez, G. M. Dominguez Almaraz and J. J. Villalón López
Facultad de Ingeniería Mećanica, Universidad Michoacana de San Nicolás de Hidalgo (UMSNH), Santiago Tapia No. 403, Col. Centro, 58000 Morelia, MICH, Mexico

R. García Hernández
Instituto de Investigaciones Metalúrgicas, UMSNH, Edificio U, Ciudad Universitaria, Gral. Francisco J. Mugica S/N, Felicitas del Rıo, 58030 Morelia, MICH, Mexico

Volker Franco Steier
Department of Automotive Engineering, University of Brasilia, 72444-240 Brasilia, DF, Brazil

Edgar Sobral Ashiuchi and José Alexander Araújo
Department of Mechanical Engineering, University of Brasilia, 70910-900 Brasilia, DF, Brazil

Lutz Reißig
Fraunhofer Institute for Mechanics of Materials IWM, Woehlerstrasse 11, 79108 Freiburg, Germany

Teng Ma
School of Materials Science and Engineering, Taiyuan University of Science and Technology, Taiyuan 030024, China
Huake Institute, Taiyuan University of Science and Technology, Taiyuan 030024, China

Junting Zhang
School of Applied Science, Taiyuan University of Science and Technology, Taiyuan 030024, China

Xiaochao Cui
School of Materials Science and Engineering, Taiyuan University of Science and Technology, Taiyuan 030024, China
School of Applied Science, Taiyuan University of Science and Technology, Taiyuan 030024, China

Xiaosi Sun
School of Materials Science and Engineering, Taiyuan University of Science and Technology, Taiyuan 030024, China

A. Muhrat, H. Puga and J. Barbosa
Centre for Micro-Electro-Mechanical Systems (CMEMS), e University of Minho, Campus of Azuŕem, 4800-058 Guimarães, Portugal

Yuanqing Wang and Zhongxing Wang
Department of Civil Engineering, Tsinghua University, Beijing 10084, China

Cui Jin-dong, Yi You-ping and Luo Guo-yun
School of Materials Science and Engineering, Central South University, Changsha 410083, China

Mária Behúlová, Eva Babalová and Miroslav Sahul
Faculty of Materials Science and Technology in Trnava, Slovak University of Technology in Bratislava, Ulica Jána Bottuč. 2781/25, 917 24 Trnava, Slovakia

Index

www.ingramcontent.com/pod-product-compliance
Lightning Source LLC
LaVergne TN
LVHW081023210926
848485LV00004B/937